# Sustainable Civil Infrastructures

**Editor-in-Chief**

Hany Farouk Shehata, SSIGE, Soil-Interaction Group in Egypt SSIGE, Cairo, Egypt

**Advisory Editors**

Khalid M. ElZahaby, Housing and Building National Research Center, Giza, Egypt

Dar Hao Chen, Austin, TX, USA

**Sustainable Civil Infrastructures (SUCI)** is a series of peer-reviewed books and proceedings based on the best studies on emerging research from all fields related to sustainable infrastructures and aiming at improving our well-being and day-to-day lives. The infrastructures we are building today will shape our lives tomorrow. The complex and diverse nature of the impacts due to weather extremes on transportation and civil infrastructures can be seen in our roadways, bridges, and buildings. Extreme summer temperatures, droughts, flash floods, and rising numbers of freeze-thaw cycles pose challenges for civil infrastructure and can endanger public safety. We constantly hear how civil infrastructures need constant attention, preservation, and upgrading. Such improvements and developments would obviously benefit from our desired book series that provide sustainable engineering materials and designs. The economic impact is huge and much research has been conducted worldwide. The future holds many opportunities, not only for researchers in a given country, but also for the worldwide field engineers who apply and implement these technologies. We believe that no approach can succeed if it does not unite the efforts of various engineering disciplines from all over the world under one umbrella to offer a beacon of modern solutions to the global infrastructure. Experts from the various engineering disciplines around the globe will participate in this series, including: Geotechnical, Geological, Geoscience, Petroleum, Structural, Transportation, Bridge, Infrastructure, Energy, Architectural, Chemical and Materials, and other related Engineering disciplines.

**SUCI series is now indexed in SCOPUS and EI Compendex.**

More information about this series at https://link.springer.com/bookseries/15140

Amin Akhnoukh · Kamil Kaloush ·
Magid Elabyad · Brendan Halleman ·
Nihal Erian · Samuel Enmon II ·
Cherylyn Henry
Editors

# Advances in Road Infrastructure and Mobility

Proceedings of the 18th International Road Federation World Meeting & Exhibition, Dubai 2021

Set 1

*Editors*
Amin Akhnoukh
Construction Management
East Carolina University
Greenville, NC, USA

Magid Elabyad
International Road Federation
Alexandria, VA, USA

Nihal Erian
KemetPro, LLC
Greenville, NC, USA

Cherylyn Henry
Charlotte, NC, USA

Kamil Kaloush
School of Engineering
Arizona State University
Tempe, AZ, USA

Brendan Halleman
International Road Federation
Brussels, Belgium

Samuel Enmon II
Street Transportation Department
City of Phoenix
Phonix, AZ, USA

ISSN 2366-3405　　　　　　　ISSN 2366-3413　(electronic)
Sustainable Civil Infrastructures
ISBN 978-3-030-79800-0　　　ISBN 978-3-030-79801-7　(eBook)
https://doi.org/10.1007/978-3-030-79801-7

© The Editor(s) (if applicable) and The Author(s), under exclusive license
to Springer Nature Switzerland AG 2022
This work is subject to copyright. All rights are solely and exclusively licensed by the Publisher, whether
the whole or part of the material is concerned, specifically the rights of translation, reprinting, reuse of
illustrations, recitation, broadcasting, reproduction on microfilms or in any other physical way, and
transmission or information storage and retrieval, electronic adaptation, computer software, or by similar
or dissimilar methodology now known or hereafter developed.
The use of general descriptive names, registered names, trademarks, service marks, etc. in this
publication does not imply, even in the absence of a specific statement, that such names are exempt from
the relevant protective laws and regulations and therefore free for general use.
The publisher, the authors and the editors are safe to assume that the advice and information in this
book are believed to be true and accurate at the date of publication. Neither the publisher nor the
authors or the editors give a warranty, expressed or implied, with respect to the material contained
herein or for any errors or omissions that may have been made. The publisher remains neutral with regard
to jurisdictional claims in published maps and institutional affiliations.

This Springer imprint is published by the registered company Springer Nature Switzerland AG
The registered company address is: Gewerbestrasse 11, 6330 Cham, Switzerland

# Contents

## Pavement Testing

**A New Low-Activity Nuclear Density Gauge for Compaction Control of Asphalt, Concrete, Soil, and Aggregate Layers in Road Construction** . . . . . . . . . . . . . . . . . . . . . . . . . . . . . . . . . . . . . . . . . . . . . 3
Linus Dep, Robert E. Troxler, and William F. Troxler Jr.

**A Study into the Benefit and Cost-Effectiveness of Using State-of-the-Art Technology for Road Network Level Condition Assessment** . . . . . . . . . . . . . . . . . . . . . . . . . . . . . . . . . . . . . . . . . . . 19
Herman Visser, Simon Tetley, Tony Lewis, Kaslyn Naidoo, Tasneem Moolla, and Justin Pillay

## Asphalt Pavement Innovations

**Paths to Successful Asphalt Road Structures Performance Based Specifications for Hot Mix Asphalt** . . . . . . . . . . . . . . . . . . . . . . . 35
Bernhard Hofko

**Study on Rejuvenators and Their Effect on Performance Characteristics of an Asphalt Concrete Containing 30% Reclaimed Asphalt** . . . . . . . . . . . . . . . . . . . . . . . . . . . . . . . . . . . . . . . . 46
Pavla Vacková, Majda Belhaj, Jan Valentin, and Liang He

**Sustainable Use of Cool Pavement and Reclaimed Asphalt in the City of Phoenix** . . . . . . . . . . . . . . . . . . . . . . . . . . . . . . . . . . . . . . . . . . . . . . 57
Ryan Stevens

**Optimizing the Adhesion Quality of Asphalt Binders Based on Energy Parameters Derived from Surface Free Energy** . . . . . . . . . . . . . . . . . . . 72
Allex E. Alvarez and Lady D. Vega

**Estimating Asphalt Film Thickness in Asphalt Mixtures Using Microscopy to Further Enhance the Performance of UAE Roadways** ..................................................... 87
Alaa Sukkari, Ghazi Al Khateeb, Waleed Ziada, and Helal Ezzat

**Effect of Crumb Rubber and Polymer Modifiers on Performance Related Properties of Asphalt Binders** ......................... 97
Joseph Dib, Nariman J. Khalil, and Edwina Saroufim

**Road Safety Leadership**

**How Do Best Performing Countries in Road Safety Save Lives on the Roads? Lessons Learned from Case Studies in Singapore** ...... 117
Alina Florentina Burlacu and Emily Tan

**Data for Road Incident Visualization, Evaluation, and Reporting (DRIVER): Case Studies of Cebu, Philippines and the Asia-Pacific Road Safety Observatory (APRSO)** ............................ 134
Florentina Alina Burlacu, Miguel Enrico III Cabag Paala, Veronica Ines Raffo, and Juan Miguel Velasquez Torres

**Road Financing Strategies**

**Is a Further Increase in Fuel Levy in Kenya Justified?** ............. 151
Evans Omondi Ochola and Jennaro Boniface Odoki

**Bridge and Structural Management**

**Aerial Robotic System for Complete Bridge Inspections** ............ 171
Antidio Viguria, Rafael Caballero, Ángel Petrus, Francisco Javier Pérez-Grau, and Miguel Ángel Trujillo

**Damage Inspection, Structural Evaluation and Rehabilitation of a Balanced Cantilever Bridge with Center Hinges** ............... 187
Chawalit Tipagornwong, Koonnamas Punthutaecha, Peerapat Phutantikul, Setthaphong Thongprapha, and Kridayuth Chompooming

**Reinforcement of Scoured Pile Group Bridge Foundations with Spun Micro Piles** ......................................... 202
Apichai Wachiraprakarnpong, Juti Kraikuan, Jirasak Watcharakornyotin, and Thanwa Wiboonsarun

**Repair of Settled Pile Bent Bridge Foundations with Spun Micro Piles** ......................................... 217
Apichai Wachiraprakarnpong, Juti Kraikuan, Wichai Yu, and Pawin Ritthiruth

## Contents

**Overview of Integral Abutment Bridge Applications in the United States** .......................................... 232
Amin Akhnoukh, Rajprabhu Thungappa, and Rudolf Seracino

### Case Studies in ITS Development

**Pacemaker Lighting Application to Prevent Traffic Congestion in a V-Shaped Tunnel and Provide Sustainable Operation: A Case Study: Eurasia Tunnel** .......................................... 243
Aşkın Kaan Kaptan and Murat Gücüyener

**Real Time Multi Object Detection & Tracking on Urban Cameras** .... 257
Rifkat Minnikhanov, Maria Dagaeva, Timur Aslyamov, Tikhon Bolshakov, and Emil Faizrakhmanov

**Effective Advanced Warning for Connected Safety Applications - Supplementing Automated Driving Systems for Improved Vehicle Reaction** .......................................... 269
Gregory M. Baumgardner, Rama Krishna Boyapati, and Amudha Varshini Kamaraj

### Traffic Planning and Forecasting - 1

**A Case Study of Rural Freight Transport – Two Regions in North Carolina** .......................................... 287
Daniel J. Findley, Steven A. Bert, George List, Peter Coclanis, and Dana Magliola

**Traffic Delay Evaluation and Simulating Sustainable Solutions for Adjacent Signalized Roundabouts** .......................................... 303
Mohamad Yaman Fares and Muamer Abuzwidah

### Safe Roads by Design - 1

**A Comparative Evaluation of the Safety Performance of Median Barriers on Rural Highways; A Case-Study** .......................................... 317
Victoria Gitelman and Etti Doveh

**Ascendi's Safety Barriers Upgrading Program** .......................................... 335
Telma Silva and João Neves

**Effectiveness of Cable Median Barriers in Preventing Cross Median Crashes and Related Casualties in the United States - A Systematic Review** .......................................... 345
Baraah Qawasmeh and Deogratias Eustace

## Integrated Transport Planning

**Transforming Infrastructure Projects Using Agile** .................. 357
Nihal Erian and Brendan Halleman

**Savings Potential in Highway Planning, Construction and
Maintenance Using BIM - German Experience with PPP** ............ 365
Veit Appelt

**Demonstrating Connectivity and Exchange of Data Between BIM
and Asset Management Systems in Road Infrastructure Asset
Management**.................................................... 379
Sukalpa Biswas, John Proust, Tadas Andriejauskas, Alex Wright,
Carl Van Geem, Darko Kokot, António Antunes, Vânia Marecos,
José Barateiro, Shubham Bhusari, and Jelena Petrović

## Greening Road Projects

**Road Construction Using Locally Available Materials**............... 395
Robert D. Friedman and Ahmed F. Abdelkader

## Practical Applications of Big Data Science

**Road Roughness Estimation Using Acceleration Data
from Smartphones**............................................. 407
Arak Montha, Attaphon Huytook, Tatree Rakmark,
and Ponlathep Lertworawanich

**Improving Traffic Safety by Using Waze User Reports** ............. 416
Raitis Steinbergs and Maris Kligis

**A Glimpse into the Near Future – Digital Twins and the Internet
of Things** .................................................... 425
Howard Shotz and James Birdsall

## Long Term Pavement and Asset Performance - 1

**Field Monitoring of Road Pavement Responses and Their
Performance in Thailand** ...................................... 441
Auckpath Sawangsuriya

**Supplementary Cementitious Materials in Concrete Industry –
A New Horizon** .............................................. 450
Amin Akhnoukh and Tejan Ekhande

**Long Term Anti Corrosion Measures for Magosaki Viaduct by Cover
Plate Method** ................................................ 458
Hiromasa Kobayashi, Yukio Usuda, Yuki Kishi, and Yukio Nagao

Contents ix

**Degradation of Friction Performance Indicator Over the Time in Highways Using Linear Mixed Models** ........................ 474
Adriana Santos, Elisabete Freitas, Susana Faria, Joel Oliveira, and Ana Maria A. C. Rocha

## Road Safety Risk Diagnosis

**Application of an Innovative Network Wide Road Safety Assessment Procedure Based on Human Factors** ......................... 493
Andrea Paliotto, Monica Meocci, and Valentina Branzi

**A Review of the Spatial Analysis Techniques for the Identification of Road Accident Black Spots and It's Application in Context to India** ............................................................. 511
Shawon Aziz and Sewa Ram

**Road Network Safety Screening of County Wide Road Network. The Case of the Province of Brescia (Northern Italy)** ............... 525
Michela Bonera, Benedetto Barabino, and Giulio Maternini

**An iRAP Based Risk Impact Analysis at National Highway-1 for a Proposed Route Connecting Coastal Areas of Bangladesh** ....... 542
Armana Sabiha Huq

## Maintenance Strategies - 1

**Improving Pavement Condition at an Accelerated Pace: The City of Phoenix Accelerated Pavement Maintenance Program** ........... 559
Ryan Stevens

**Smart Infrastructure Asset Management System on Metropolitan Expressway in Japan** ................................. 575
Hirotaka Nakashima, Taishi Nakamura, Yusuke Hosoi, Koji Konno, and Hinari Kawamura

## Decarbonizing Road Transport

**Evaluation of the $CO_2$ Reduction Effect of Low Rolling Resistance Asphalt Pavement Using the Fuel Consumption Simulation Method** ... 589
Yu Shirai, Atsushi Kawakami, Masaru Terada, and Kenji Himeno

**Green Energy Sources Based on Thermo-Electrochemical Cells for Electricity Generating from Transport, Engineering Buildings and Environment Waste Heat** ............................. 602
Igor N. Burmistrov, Nikolay V. Kiselev, Elena A. Boychenko, Nikolay V. Gorshkov, Evgeny A. Kolesnikov, and Stanislav L. Mamulat

## Long Term Pavement and Asset Performance - 2

**An Approach to Estimate Pavement's Friction Correlation Between PCI, IRI & Skid Number** .................................... 611
Carlos J. Obando, Jose R. Medina, and Kamil E. Kaloush

**Effect of Different Aggregate Gradations on Rutting Performance of Asphalt Mixtures for UAE Roadways** ...................... 622
Khalil Almbaidheen, Ghazi Al-Khateeb, Waleed Ziada, and Myasar Abulkhair

**Evaluation of AASHTO Mechanistic Empirical Design Guide Inputs to the Performance of Tennessee Pavements** .................... 634
Onyango Mbakisya, Msechu Kelvin, Udeh Sampson, and Owino Joseph

**Influence of Dynamic Analysis on Estimation of Rutting Performance Using the Fixed Vehicle Approach** .......................... 647
Gauri R. Mahajan, Radhika Bayya, and Krishna Prapoorna Biligiri

## Street and Highway Designs for CAVs - 1

**A Review on Benefits and Security Concerns for Self Driving Vehicles** .............................................. 667
Gozde Bakioglu and Ali Osman Atahan

**Contrast Ratio of Road Markings in Poland - Evaluation for Machine Vision Applications Based on Naturalistic Driving Study** ................................................ 676
Tomasz E. Burghardt and Anton Pashkevich

**San Diego Bus Rapid Transit Using Connected and Autonomous Vehicle Technology** ............................................ 691
Dan Lukasik and Dmitri Khijniak

## Maintenance Strategies - 2

**Introduction to the RoadMark System Low Cost, Fast and Flexible Data Collection for Rural Roads** ............................ 707
Mark J. Thriscutt

**Similarity Between an Optimal Budget in Pavement Management & an Equilibrium Quantity of Demand-Supply Analysis in Economics** ............................................ 719
Ponlathep Lertworawanich

**Integrating Flexible Pavement Surface Macrotexture to Pavement Management System to Optimize Pavement Preservation Treatment Recommendation Strategy** .................................... 735
Seng Hkawn N-Sang, Jose Medina, and Kamil Kaloush

# Contents

## Safe Roads by Design - 2

**MASH TL-3 Development and Evaluation of the Thrie-Beam Bullnose Attenuator** ................................................. 755
Robert Bielenberg, Ron Faller, and Cody Stolle

**Safety Performance Evaluation of Modified Thrie-Beam Guardrail** .... 772
Robert Bielenberg, Ron Faller, and Karla Lechtenberg

**In-Service Performance Evaluation of Iowa's Sloped End Treatments** ....................................................... 789
Cody Stolle, Jessica Lingenfelter, Khyle Clute, and Robert Bielenberg

**A Synthesis of 787-mm Tall, Non-proprietary, Strong-Post, W-beam Guardrail Systems** ........................................... 805
Scott Rosenbaugh, Robert Bielenberg, and Ronald Faller

**A Synthesis of MASH Crashworthy, Non-proprietary, Weak-Post, W-beam Guardrail Systems** ........................................... 821
Scott Rosenbaugh, Robert Bielenberg, and Ronald Faller

**The Safety Highway Geometry Based on Unbalanced Centripetal Acceleration** ............................................... 837
Creso de Franco Peixoto and Maria Teresa Françoso

**Stiffening Guidance for Temporary Concrete Barrier Systems in Work Zone and Construction Situations** ......................... 850
Karla Lechtenberg, Chen Fang, and Ronald Faller

**Safety Performance Evaluation of a Non-proprietary Type III Barricade for Use in Work Zones** ................................. 868
Karla Lechtenberg, Ronald Faller, Jennifer Rasmussen, and Mojdeh Asadollahi Pajouh

## ITS Design and Implementation Strategies

**Unified ITS Environment in the Republic of Tatarstan** ............. 881
Rifkat Minnikhanov, Maria Dagaeva, Sofya Kildeeva, and Alisa Makhmutova

## Road Pricing and Tolling

**Assessment of the Potential Implementation of High-Occupancy Toll Lanes on the Major Freeways in the United Arab Emirates** ......... 897
Ahmed Shabib, Mahmoud Khalil, and Muamer Abuzwidah

## Driver Behavior Strategies

**Posted Road Speed Limits in Abu Dhabi: Are They Too High? Should They Have Been Raised? Evidence Based Answers** ... 915
Francisco Daniel B. Albuquerque

## Transport Responses to the Pandemic

**The Impact of the COVID-19 Pandemic on Mobility Behavior in Istanbul After One Year of Pandemic** ... 933
Ali Atahan and Lina Alhelo

**The Evaluation of the Impacts on Traffic of the Countermeasures on Pandemic in Istanbul** ... 950
Mahmut Esad Ergin, Halit Ozen, and Mustafa Ilıcalı

## New Approaches to Performance Delivery

**From Reactive to Proactive Maintenance in Road Asset Management** ... 963
Timo Saarenketo and Vesa Männistö

**Observing How Influence of Nature Phenomena Against Inside Tunnel by Air Pressure Information** ... 975
Kensaku Kawauchi, Yumi Watanabe, Yuichi Mizushima, and Takeo Hosokai

**Road Asset Management: Innovative Approaches** ... 986
Soughah Salem Al-Samahi and Fernando Varela Soto

## Traffic Planning and Forecasting – 2

**Evaluating the Efficiency of Constructability Review Meetings for Highway Department Projects** ... 997
Amin K. Akhnoukh, Minerva Bonilla, Nicolas Norboge, Daniel Findley, William Rasdorf, and Clare Fullerton

**A Novel Method for Aggregate Tour-Based Modeling with Empirical Evidence** ... 1007
Yanling Xiang, Shiying She, Meng Zheng, Heng Liu, and Huanyu Lei

**Multilayer Perceptron Modelling of Travelers Towards Park-and-Ride Service in Karachi** ... 1026
Irfan Ahmed Memon, Ubedullah Soomro, Sabeen Qureshi, Imtiaz Ahmed Chandio, Mir Aftab Hussain Talpur, and Madzlan Napiah

**Congestion on Canada's Busiest Highway, 401 Problems, Causes, and Mitigation Strategies** ... 1039
Abdul Basith Siddiqui

## Multi-Stakeholder Transportation Strategies

**A Holistic Approach for the Road Sector in Sub-Saharan Africa** ..... 1053
Tim Lukas Kornprobst, Ulrich Thüer, and Yana Tumakova

## Innovations in Road Materials

**The Introduction of Micro - & Nanodispersed Fillers into the Bitumen
Binders for the Effective Microwave Absorption (for the Road,
Airfield & Bridge Pavements)** .............................. 1071
Stanislav Mamulat, Igor Burmistrov, Yuriy Mamulat, Dmitry Metlenkin,
and Svetlana Shekhovtsova

**Rheological Properties of Rubber Modified Asphalt Binder
in the UAE** ................................................ 1083
Mohammed Ismail, Waleed A. Zeiada, Ghazi Al-Khateeb, and Helal Ezzat

**Recycling Waste Rubber Tires in Pervious Concrete Evaluation
of Hydrological and Strength Characteristics** .................... 1098
Sahil Surehali, Avishreshth Singh, and Krishna Prapoorna Biligiri

**Incorporation of CFRP and GFRP Composite Wastes in Pervious
Concrete Pavements** ........................................ 1112
Akhil Charak, Avishreshth Singh, Krishna Prapoorna Biligiri,
and Venkataraman Pandurangan

**Asphalt Modified with Recycled Waste Plastic in South Africa
Encouraging Results of Trial Section Performance** ............... 1125
Simon Tetley, Tony Lewis, Waynand Nortje, Deane Koekemoer,
and Herman Visser

## Climate Resilient Road Design – 1

**GIS Aided Vulnerability Assessment for Roads** .................. 1139
Berna Çalışkan, Ali Osman Atahan, and Ali Sercan Kesten

**Investigation of Historical and Future Air Temperature Changes
in the UAE** ............................................... 1148
Reem N. Hassan, Waleed A. Zeiada, Muamer Abuzwidah,
Sham M. Mirou, and Ayat G. Ashour

**Climate Teleconnections Contribution to Seasonal Precipitation
Forecasts Using Hybrid Intelligent Model** ...................... 1167
Rim Ouachani, Zoubeida Bargaoui, and Taha Ouarda

**Development of Pavement Temperature Prediction Models
for Tropical Regions Incorporation into Flexible Pavement
Design Framework** ......................................... 1181
Chaitanya Gubbala, Krishna Prapoorna Biligiri,
and Amarendra Kumar Sandra

## Impacts of Transport Investments

**Experiences of High Capacity Transport in Finland** . . . . . . . . . . . . . . . . 1197
Vesa Männistö

**Impacts of Transportation Infrastructure Investments and Options
for Sustainable Funding** . . . . . . . . . . . . . . . . . . . . . . . . . . . . . . . . . . . . . . . 1207
Daniel J. Findley, Steven A. Bert, Weston Head, Nicolas Norboge,
and Kelly Fuller

## Climate Resilient Road Design – 2

**Climate Resilient Urban Mobility by Non-motorized Transport** . . . . . . 1225
Kigozi Joseph

**Author Index** . . . . . . . . . . . . . . . . . . . . . . . . . . . . . . . . . . . . . . . . . . . . . . . . 1237

# Pavement Testing

# A New Low-Activity Nuclear Density Gauge for Compaction Control of Asphalt, Concrete, Soil, and Aggregate Layers in Road Construction

Linus Dep[(✉)], Robert E. Troxler, and William F. Troxler Jr.

Troxler Electronic Laboratories, Research Triangle Park, Durham, NC, USA
ldep@troxlerlabs.com

**Abstract.** Density is used as a measure for quality control of compaction of asphalt, concrete, soil, and aggregate layers in road construction. For in-place density measurements, nuclear gauges have been successfully used for decades because of the good measurement properties, such as precision and sensitivity of the nuclear technique. The technique is nondestructive and is a direct method for measuring density. The widespread use of nuclear gauges has been limited because of the nuclear regulatory requirements associated with the usage of radiation sources.

To overcome this limitation, in 2016, a nuclear gauge using a low-activity radiation source and a high-efficiency detector was developed. This gauge used mostly for compaction testing of soil and aggregate layers and is exempt from the nuclear regulatory requirements in the United States. Using similar technology, a new nuclear gauge was developed recently to measure density of asphalt and concrete layers in addition to soil and aggregate layers. This gauge is also exempt from the nuclear regulatory requirements in the United States. The new gauge has the option of using a separate probe based on an electromagnetic method for measuring the moisture content of soils and aggregates. This paper presents the design features, measurement properties, and safety aspects of the new nuclear gauge and its associated moisture probe. Comparison of the new low-activity nuclear gauge and a conventional nuclear gauge for measurement of density of asphalt layers, and density and moisture of soil and aggregate layers of thickness up to 30-cm (12-in.) are also presented.

**Keywords:** Compaction · Soil · Asphalt · Nuclear · Gauge · Gamma

## 1 Introduction

Compaction to a desired density of the soil, aggregate, asphalt, and concrete layers is a key factor in determining the durability of road foundation and pavements. Density, which is a direct measure of compaction is monitored during the entire compaction phase for quality control and quality assurance (QC/QA).

There are several methods of measuring in-place density of foundation layers (ASTM D1556/1556M, ASTM D2167, and ASTM D2937), in-place moisture of foundation layers (ASTM D2216, ASTM D4959, ASTM D4643, and ASTM D4944),

© The Author(s), under exclusive license to Springer Nature Switzerland AG 2022
A. Akhnoukh et al. (Eds.): IRF 2021, SUCI, pp. 3–18, 2022.
https://doi.org/10.1007/978-3-030-79801-7_1

and in-place density of asphalt pavement layers (ASTM D2726, ASTM D1188, and ASTM D6857). These methods require extraction of a sample from the test layers and by concept provide accurate measurements since mass and volume are directly measured. These methods are not effective for QC/QA purposes because of the long execution times, the destructive nature, and the errors associated with the handling of granular materials.

The nuclear method, which is the gamma-ray scattering method for density and the neutron scattering method for moisture determination, is a nondestructive in-place measurement method suitable for statistical process control. The nuclear methods were originally developed to measure density and moisture in compacted soil layers in roads (Krueger 1950; Belcher et al. 1950; Pieper 1949). Later, the gamma-ray scattering method for measuring the density of asphalt layers was evaluated (Hughes 1962; Sloane 1962). With the improvements to design and instrumentation, rugged and dependable nuclear density gauges (NDG) have been used for over six decades for nondestructive in-place density and moisture measurements of foundation and pavement layers of roads (ASTM D6938 and ASTM D2950).

The nuclear method of measuring the density of foundation and pavement layers has many advantages: (1) it is a nondestructive measurement, (2) it is an in-place measurement, (3) the measurement is independent of material composition and hence is used on soils, aggregates, concrete, and asphalt, (4) it has good sensitivity to density, (5) the measurement has the desired accuracy and precision for the application, (6) the measurements can be made within a short duration, (7) the density determination is based on gamma-rays (photons) scanning a large volume and is suitable for heterogeneous materials, (8) measurements can be made on hot or cold conditions of the layer, (9) the moisture levels encountered during compaction do not affect the measurement accuracy, and (10) it is a portable instrument.

The NDG is a controlled device because of the amount or activity of the radioactive material it uses and is thus subjected to nuclear regulatory requirements. Some of the requirements in the United States include a license to own and operate, radiation safety training and radiation exposure monitoring for gauge operators, and permission for multi-state usage. These regulations have become a burden for some gauge users and prevent others from using the gauge. For decades, a license-exempt nuclear density gauge was pursued by the industry.

The first license-exempt nuclear density gauge, Troxler Model 4590, was introduced in the United States in 2015 (Dep et al. 2016). This gauge, a low-activity nuclear density gauge (LNDG/S) was specially designed for soil and aggregate density measurements. This gauge has only the direct transmission modes of measuring density and uses a non-nuclear probe for moisture measurements. The measurement performance of this gauge for density measurement of soil and aggregate layers are similar to that of a conventional NDG (Dep et al. 2016; Baker and Meehan 2018; Berney and Mejias 2019; Dep et al. 2021a). The measurement method and measurement precision are given in ASTM D8167.

Recently, a license-exempt low-activity nuclear density gauge (LNDG), Troxler Model 4540, for in-place density measurement of foundation and pavement layers was introduced in the United States (Fig. 1). As with conventional NDGs, the new LNDG has both backscatter and direct transmission modes to measure density of soil, aggregate, concrete and asphalt layers. The measurement process and measurement properties of the direct transmission modes to measure density of soil and aggregate layers are similar to those of the LNDG/S. The measurement properties of the backscatter mode to measure density of asphalt layers are also similar to those of the conventional NDG (Dep et al. 2021b).

This paper presents an overview of the principle of gamma-ray density measurement, the design features, measurement properties, and safety aspects of the new LNDG, and the associated non-nuclear moisture probe. Comparison of the new LNDG with a conventional NDG for fast and slow measurements of density of asphalt layers, and density and moisture measurement of soil layers of thickness up to 30 cm (12 in.) are also presented. A new method of using the non-nuclear probe for soil and aggregate layers is also included.

**Fig. 1.** Low-activity nuclear density gauge (Troxler Model 4540) (left) and the non-nuclear moisture probe (right). The moisture probe wirelessly communicates with the gauge.

## 2 Low-Activity Nuclear Density Gauge for Asphalt and Soil Density Measurements

### 2.1 General Description

The LNDG uses the gamma-ray scattering method for density measurements. In this method, a gamma-ray source is placed on or in the material and a gamma-ray detector is used to measure the scattered gamma-ray rates or intensities. At energies in which Compton scattering is the predominant gamma-ray interaction with matter, the gamma-ray scattering rate depends on the material density and is independent of the chemical composition of the material under test. With a suitable calibration, the gamma-ray scattering rate can be used for the density determination.

Figure 2 illustrates the basic components of the LNDG. The LNDG uses a low-activity gamma-ray source located in the lower end of an extendable rod. The activity of the Cs-137 radioisotope gamma-ray source is nearly two orders of magnitude lower than that used in a conventional NDG. The LNDG uses a high-efficiency energy-selective gamma-ray detector. Hence, a spectrometric method is used for selectively measuring the scattered gamma-rays. The source, detector, and instrumentation are similar to those used in the LNDG/S (Dep et al. 2016).

The LNDG can be operated in both backscatter and direct transmission modes. In normal use, the backscatter method is used for measuring density of asphalt layers, and a direct transmission mode is used for measuring the density of soil and aggregate layers. In the backscatter mode, a truly nondestructive method, the source is placed just above the surface of the asphalt layer (Fig. 2-right). In the direct transmission mode, a 19-mm (0.75-in.) diameter access hole is prepared, and the source is placed at the desired depth approximately equal to the thickness of the test layer (Fig. 2-left).

To account for the radioactive decay of the gamma-ray source, which causes the emission rate to decrease over time, the detected gamma-ray rates are corrected using a measurement taken in a standard set condition. This measurement, scaled per minute, is called the "standard count" ($C_{std}$) and is acquired at the beginning of the test day. Furthermore, because of the low-activity gamma-ray source, the gamma-rays from the natural radioactivity in the test material and the surrounding environment are corrected for by using a separate measurement scaled per minute called the "background count" ($C_{bgd}$).

## 2.2 Measurement Process

The density measurement process begins with the preparation of the test location, which is the same procedure as that for a conventional NDG gauge (Manual of Operation and Instruction for Model 4540 2020). The test surface on which the gauge is placed should be flat and free of debris. For asphalt, the gauge is placed directly on the test surface. For soil and aggregate layers, an access hole is made by hammering a 19-mm (0.75-in.) diameter drill rod to about 50-mm (2-in.) deeper than the desired measurement depth. Two measurements are made: 1) source rod is lowered to the desired depth $i$, for example, backscatter, and the gamma-rays are counted for a duration. This measurement, which is scaled per minute, is the measurement count ($C_{meas,i}$), and 2) the source rod is moved to the background measurement configuration, and the gamma-rays are counted for a duration and scaled per minute ($C_{bgd}$). The scattered gamma-ray rate ($C_{scat,i}$) is given by

$$C_{scat,i} = C_{meas,i} - C_{bgd} \tag{1}$$

The density of the test material is then determined from using the following relationship:

$$\frac{C_{scat,i}}{C_{std}} = A_i exp^{-B_i \rho} + C_i \tag{2}$$

where $A_i$, $B_i$, and $C_i$ are the calibration constants for the depth mode $i$, which is determined during the factory calibration of the gauge. The density $\rho$ is then determined using the measured counts and the pre-determined constants.

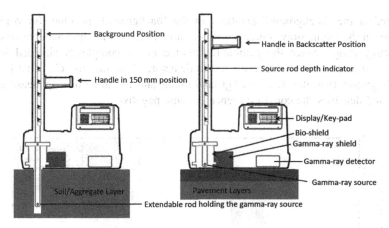

**Fig. 2.** Schematics of the low-activity nuclear density gauge (Troxler Model 4540): direct transmission mode in soils or aggregate layers (left) and backscatter mode on asphalt pavement (right).

## 2.3 Safety and Radiation Exposure to Gauge Operator

Humans are exposed to radiation from terrestrial, cosmogenic, and anthropogenic sources (Fig. 3). The average person in the United States receives 3.2 mSv (320 mrem) from ubiquitous natural background originated from terrestrial and cosmogenic sources (NCRP Report 160 2009).

The annual radiation dose received by a nuclear gauge operator can be estimated by knowing the radiation dose profile for a gauge, which is provided by the gauge manufacturer, and the activities involved in the normal usage of a gauge such as transporting, operating, storing, and servicing a gauge. A gauge operator acquiring 5,000 density measurements per year using an LNDG, receives a dose of about 0.15 mSv (15 mrem). This dose is about 15% of the allowable exposure to a member of public in the United States from sources other than the natural background.

## 2.4 Density Calibration

The density calibration is the process of finding a mathematical relationship between the gauge response and the density of a material. Similar to the calibration of the LNDG/S, LNDG uses the same three density calibration standards used for the conventional NDG. The calibration standards are cuboid in shape and made of magnesium, a mix of magnesium and aluminum, and aluminum, covering a density rage from 1,700 to 2,700 kg/m$^3$ (110 to 160 lb/ft$^3$). The density of the calibration standards has previously been directly determined by measuring the mass and volume of the cuboids. The density values assigned to the standards are traceable to the standards in the US National Institute of Standards and Technology.

Figure 4-a shows the calibration curves for an LNDG and an NDG. Note that the gauge response of an LNDG is the actual scattered gamma-ray count rate (Eq. 1): the difference between the count rate measured at the desired depth and the count rate measured at the 'background' position. In the 'background' position, for complete shielding of the gamma-ray source requires a large mass of shielding material resulting in a heavy gauge. Since the gamma-ray source is not completely shielded in this position, some gamma-rays traverse to the detector. Therefore, the $C_{bgd}$ that is measured is greater than the 'true' background count and for some deeper measurement depths and densities the count difference become negative.

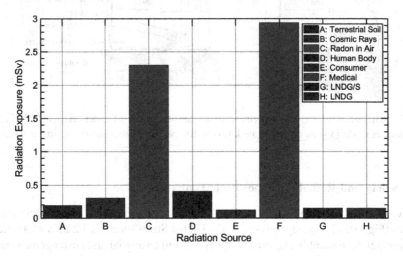

**Fig. 3.** Annual radiation exposure to an average person in the United States from various activities.

## 3 Gauge Measurement Properties

### 3.1 Sensitivity

The sensitivity, the instrument response to a unit change in the measurement property, can be determined from the slope of the calibration curve (Fig. 4-a). For example, for the backscatter mode of measuring density, the sensitivity of the LNDG is about 20 counts per min per kg/m$^3$. In general, for all depth modes, it was found that the LNDG shows a slightly higher sensitivity than that of an NDG.

### 3.2 Precision

The precision of the density measurement depends on the depth mode and the density of the test material. The main factor that affects the precision is the variability of the measure of scattered gamma-ray intensity or "counts". The variability of the counts follows the Poisson distribution as they are originated from the spontaneous decay of a radioactive source. The precision can be estimated using the gamma-ray counts taken during calibration incorporating Monte Carlo methods. Figures 4-b, 4-c, and 4-d indicate the measurement precision of an LNDG for backscatter, 150-mm (6-in.) and 300-mm (12-in.) depth modes for various combinations of measurement time ($T_m$) and background time ($T_b$). Also shown are the precision results for a conventional NDG (Troxler Model 3440). The precision of the LNDG density measurements can be improved by measuring for a longer duration.

The precision (repeatability and reproducibility [R&R]) of density measurements of asphalt layers using the backscatter mode of an LNDG is similar to that of an NDG (ASTM D2950). The R&R of an LNDG was determined from a pilot study (Dep et al. 2021b) where measurements were made by four operators, each operating a separate LNDG with $T_m$ = 1 min and $T_b$ = 1 min. Multiple density measurements were made on four asphalt slabs with densities ranging from 1,920 to 2,560 kg/m$^3$ (120 to 160 lb/ft$^3$). The R&R of the LNDG density measurement was about 22 kg/m$^3$ (1.4 lb/ft$^3$) and is similar to that for an NDG.

The R&R of the density measurements of soils and aggregate layers of an LNDG is similar to that of an NDG (ASTM D8167; ASTM D6938). The precision was determined for the 150-mm (6-in.) direct transmission mode using measurements made by ten operators, each operating a separate LNDG/S. Multiple density measurements were made on three soil types with densities ranging from 1,930 to 2,230 kg/m$^3$ (121 to 139 lb/ft$^3$). The R&R of the density measurement of an LNDG/S range from 5 to 13 kg/m$^3$ (0.3 to 0.8 lb/ft$^3$) and is similar to that for an NDG.

### 3.3 Accuracy

The accuracy is the closeness of a measurement to that of a measurement using an acceptable method. The NDG density measurement is an industry-accepted method. Therefore, LNDG measurements were compared with that of NDGs for evaluating the measurement accuracy.

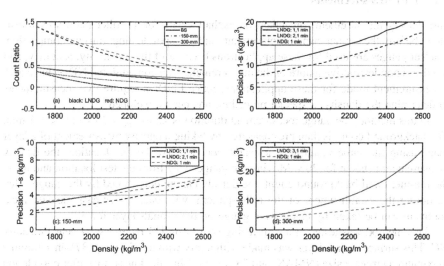

**Fig. 4.** Calibration curves and density measurement precision for an LNDG.

10 L. Dep et al.

For asphalt density, the accuracy was determined from density measurements taken by an operator using an LNDG and an NDG. One operator, operating the gauges in backscatter mode, measured the density of eight slabs. The density measurement of the two types of gauges showed good correlation (coefficient of determination, $R^2 \approx 1$) and agreement (root mean square error, RMSE $\approx 13$ kg/m$^3$ (0.8 lb/ft$^3$)) (Dep et al. 2021b).

For soils and aggregate layer densities, the accuracy of an LNDG operating in the direct transmission modes is similar to that of an LNDG/S. The measurement accuracy of the LNDG/S direct transmission mode meets industry requirements (Dep et al. 2021a).

### 3.4 Depth of Measurement for Backscatter Mode

When measuring the density of layers made of asphalt, soils, and aggregates of varying size and geometry, it is important to use a large enough sample volume. For nondestructive testing instruments such as nuclear gauges, the depth of measurement (DOM) characterizes the scanning depth of probing fields (gamma-rays or photons). For nuclear density gauges, DOM can be determined from measuring the density of two-layer structures such as a glass layer on top of an aluminum layer.

The DOM for LNDG operating in backscatter mode was determined by Dep et al. (2021b). The DOM of an LNDG is about 90 mm (3.5 in.) and is similar to that of a conventional NDG.

## 4 Density Measurement of Asphalt Concrete

### 4.1 Fast Measurements

Fast density measurements are taken on asphalt layers for 1) roller pattern establishment before the actual asphalt placement and 2) spot testing during compaction. Nuclear gauges are used for these applications because of the sensitivity, accuracy, and rapid and nondestructive nature of the measurement. The conventional NDGs have a 15 s measurement mode for these applications. Since LNDGs require two measurements to determine the density, a special procedure has been developed.

In this method, the gauge measurement duration is set to $T_m = 30$ s and $T_b = 1$ min. The background count is then pre-measured by taking a density measurement of a layer made using the same material at a nearby location. At the test location, the density measurement is taken for 30 s. Moving the gauge away from the test location and using the pre-measured background count, the density can be determined by the gauge. That is, during the establishment of the roller-pattern, with each roller pass, density measurement can be taken by placing the gauge on the asphalt layer for 30 s.

This method was tested at a resurfacing project on US Highway 64 near Pittsboro, NC. This surface course layer was made with a mix with a 9.5-mm (0.375-in.) nominal maximum aggregate size (NMAS) and 38-mm (1.5-in.) thickness. Just after completion of rolling, 16 locations were selected. First, a density measurement was taken at a nearby location with $T_m = 30$ s and $T_b = 1$ min duration to capture the background count. Then, at each test location, two 30 s measurements were made using an LNDG and two 15 s measurements were made using an NDG. For both gauges, the density readings were obtained using the factory calibrations- no layer specific offsets were incorporated. Figure 5-a shows the results of the first set of fast measurements, and Fig. 5-b shows the average of the two readings for the LNDG and NDG.

The measurement precision is determined using the duplicate measurements at each location. The precision for both the LNDG and NDG is about 16 kg/m$^3$ (1 lb/ft$^3$) at 1-standard deviation. Using the averages at each location, the average difference between the two methods was $16 \pm 18$ kg/m$^3$ ($1.0 \pm 1.1$ lb/ft$^3$). The LNDG density measurements showed good correlation and agreement to that by the NDG ($R^2 = 0.8$; RMSE = 24 kg/m$^3$ [1.5 lb/ft$^3$]).

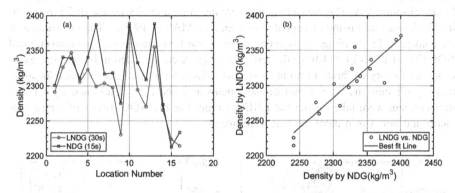

**Fig. 5.** Fast density measurements on asphalt.

## 4.2 Slow Measurements

Slow measurements, which are measurements with longer duration and hence better precision, are used for QC/QA purposes. To achieve the best accuracy of the density measurement, the LNDG requires taking a specific background measurement at each test location. The total measurement time for an LNDG can be from 2 to 4 min.

### 4.2.1 Surface Course (Fine Graded Mix)

The density testing was conducted on a section of a new road construction in Greensboro, NC. The measurements were obtained on a surface course of 38-mm (1.5-in.) thickness made using an asphalt mix with an NMAS of 9.5 mm (0.375 in.). The measurements were taken at 12 locations in a ramp section. Three operators used two LNDGs and one conventional NDG (Troxler Model 3440) for the test. All gauges were operated in backscatter mode using the factory calibration- no layer specific offsets were used. At each test location, two $T_m = 1$ min and $T_b = 1$ min measurements and one $T_m = 2$ min and $T_b = 1$ min measurement were taken using the two LNDG gauges. For the NDG, two 1 min measurements and one 4 min measurement were taken.

Figure 6-a shows the average densities determined by each gauge for the 12 locations. Using the replicate measurements, the density measurement precision (repeatability) was determined. For the LNDG, repeatability at 1-standard deviation is about 18 kg/m$^3$ (1.1 lb/ft$^3$) for the $T_m$ = 1 min and $T_b$ = 1 min measurement and about 13 kg/m$^3$ (0.8 lb/ft$^3$) for the $T_m$ = 2 min and $T_b$ = 1 min measurement. For the NDG, repeatability at 1-standard deviation is about 11 kg/m$^3$ (0.7 lb/ft$^3$) for the 1 min measurement. The reproducibility at 1-standard deviation of an LNDG $T_m$ = 1 min and $T_b$ = 1 min density measurement is about 21 kg/m$^3$ (1.2 lb/ft$^3$). The density measurement sensitivity of an LNDG is similar to that of an NDG (slope $\approx$ 0.9). The LNDG measurements correlated and agreed with those taken using an NDG ($R^2$ = 0.8; RMSE $\approx$ 16 kg/m$^3$ [1 lb/ft$^3$]).

### 4.2.2 Binder Course (Coarse Graded Mix)

The density testing was conducted on a section of a new road construction project in Fredericksburg, VA. The measurements were taken on a binder course of 76-mm (3.0-in.) thickness made using an asphalt mix with an NMAS of 19-mm. The measurements were taken at 18 locations. Two operators used one LND gauge and one conventional NDG (Troxler Model 3440) for the test. All gauges were operated in backscatter mode using the factory calibration: no layer specific offsets were used. At each test location, two $T_m$ = 1 min and $T_b$ = 1 min measurements and one $T_m$ = 2 min and $T_b$ = 1 min measurement were taken using the LNDG gauge. For the NDG, two 1 min measurements and one 4-min measurement were taken.

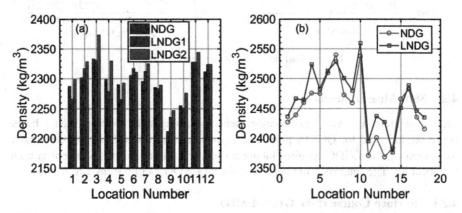

**Fig. 6.** Slow density measurements on asphalt: (a) 9.5-mm mix and (b): 19-mm mix.

Figure 6-b shows the average densities determined by each gauge for the 18 locations. Using the replicate measurements, the density measurement precision (repeatability) for the LNDG at 1-standard deviation is about 22 kg/m$^3$ (1.4 lb/ft$^3$) for the $T_m$ = 1 min and $T_b$ = 1 min measurement and about 18 kg/m$^3$ (1.1 lb/ft$^3$) the $T_m$ = 2 min and $T_b$ = 1 min measurement. For the NDG, repeatability at 1-standard deviation is about 13 kg/m$^3$ (0.8 lb/ft$^3$) for the 1 min measurement. The density measurement sensitivity of an LNDG is similar to that of an NDG (slope $\approx$ 1). The measurements taken using the LNDG correlated and agreed well with those taken using the NDG ($R^2$ = 0.9; RMSE $\approx$ 26 kg/m$^3$ [1.6 lb/ft$^3$]).

# 5 Density and Moisture Measurement of Soil and Aggregate Layers

A review of previous studies of using an LNDG/S with the moisture probe (Troxler Model 6760) is described in Dep et al. (2021a). The wet density (WD) measurements using LNDG/S showed a sensitivity similar to that of an NDG. The moisture measurements using the probe showed an acceptable sensitivity compared with that of other acceptable moisture measurement methods. The WD and moisture measurements using LNDG/S and the moisture probe showed good correlation and agreement to those by other acceptable methods (Dep et al. 2016; Baker and Meehan 2018; Berney and Mejias 2019; Dep et al. 2021a).

## 5.1 Density Measurements of Soils

In November 2020, testing was conducted at a compacted site for a new shopping complex near Pittsboro, NC. The testing was conducted a few days after compaction.

The measurements were taken at 6 locations. Two operators used one LNDG gauge, one LNDG/S gauge and one conventional NDG (Troxler Model 3440) for the test. All gauges were operated in 150-mm (6-in.) and 300-mm (12-in.) direct transmission modes using the factory calibration- no layer specific offsets were used. At each test location: for 150-mm (6-in.) mode, two $T_m = 2$ min and $T_b = 1$ min measurements were taken using the LNDG and LNDG/S gauges; for the 300-mm (12-in.) mode, two $T_m = 3$ min and $T_b = 1$ min measurements were taken using the LNDG and LNDG/S gauges; and for both depth modes, one 4-min measurement was taken using the NDG gauge.

Figures 7-a and 7-b show the results of the test. The density measurement sensitivity of an LNDG is similar to that of an LNDG/S (slope $\approx 0.9$). The measurements taken using an LNDG correlated and agreed well with those taken using an LNDG/S ($R^2 \approx 1$; for 150-mm (6-in.) mode RMSE $\approx 4$ kg/m$^3$ (0.2 lb/ft$^3$); for 300 mm (12-in.) mode RMSE $\approx 25$ kg/m$^3$ [1.5 lb/ft$^3$]). The density measurement sensitivity of an LNDG is similar to that of an NDG (slope $\approx 0.9$). The measurements taken using and LNDG correlated and agreed well with those taken using an NDG ($R^2 \approx 0.9$; for 150-mm mode (6-in.) RMSE $\approx 27$ kg/m$^3$ (1.7 lb/ft$^3$); for 300-mm (12-in.) mode RMSE $\approx 36$ kg/m$^3$ (2.3 lb/ft$^3$)).

## 5.2 Moisture Measurements of Soils Using the Non-Nuclear Moisture Probe

### 5.2.1 General Description

The moisture probe (Troxler Model 6760) is described in Dep et al. (2021a). The probe responds to the dielectric properties of the material. As the dielectric constant of a soil increases with the increase in the moisture content, the response of the probe, with a suitable calibration, can be used for determining the moisture density (M), which is the mass of water in the soil per unit volume. For construction grade soils and aggregates of reasonable range of moisture content, the response depends linearly on M.

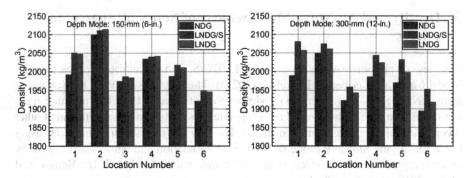

**Fig. 7.** Wet density measurements of soils using three types of nuclear gauges in Pittsboro, NC: (a) 150 mm mode and (b) 300 mm mode.

The probe is inserted into an access hole in soil to determine M. The probe body is designed such that it can be inserted to an access hole of the same geometry as that used for LNDG density measurement. The M of the material can be read within a few seconds after inserting the probe into the access hole.

In order to determine the gravimetric moisture content %M, the LNDG is used for determining the WD and then the probe is used to determine M. The gravimetric moisture content (%M) is further computed using the following relation

$$\%M = \frac{100 * M}{(WD - M)} \qquad (3)$$

The probe has built-in calibration curves for two soil types: 1) soils with high clay content (clay curve) and 2) other soils and aggregates (general curve). Initially, once the soil type is identified in the field, the correct calibration curve should be selected. This calibration is linear in M and only provides a baseline moisture content. An adjustment to the intercept or a field specific offset is required to determine the moisture content. This offset is determined by taking measurements at three or more sites of known %M, by another acceptable method such as oven-drying, and WD. Once this offset is determined, it is saved in the gauge and can be incorporated for the duration of that project.

### 5.2.2 Moisture Measurement Precision

The precision (R&R) of the moisture measurement is similar to that of an NDG (Dep et al. 2021a; ASTM D 6938). The precision was determined using measurements taken by eight operators, each operating a separate moisture probe and an LNDG/S. Multiple moisture measurements were made on three soil types with moisture ranging from 3.4% to 15.2%. For a single measurement, the R&R of %M ranges from 0.4% to 0.8%.

### 5.2.3 Moisture Measurement Accuracy

When using the probe with the LNDG, the access hole is used first for LNDG WD measurement. The source rod insertion and removal can deform the shape of the access hole. Hence, it is possible to develop an air-gap between the sensing area of the probe and the wall of the access hole affecting the probe response. Also, at a location of a compacted soil or aggregate layer, it is possible to have a localized moisture distribution. In a previous study, when conducting the tests of a clayey-sand structure, some test locations showed this effect (Dep et al. 2021a). Therefore, by taking two or more moisture readings at locations close to the base of the LNDG can minimize such errors.

In this study, a new sampling method for the probe and the typical method is evaluated. In this new method, at each location, moisture testing is performed at three points close to the gauge base. The first point uses the access hole after a gauge density measurement. The other two points are about 150-mm (6-in.) from the first access hole and in a pattern such that the holes locate the vertices of an equilateral triangle. Then, soil samples were collected around each of these access holes using a 76-mm (3-in.) diameter auger for moisture determination by oven-drying method.

### 5.2.4 Moisture Measurements of Soils

In May 2021, WD and moisture (M) testing was conducted at a compacted site for a ramp near Garner, NC, connecting I-40 West to a section of I-540 under construction. The soil type that was classified according to AASHTO was of silty clay type with some gravel (A-2, A-5, A-7-5). The testing was conducted about four weeks after compaction.

An LNDG and a moisture probe were used for the testing at eight different locations. The LNDG was operated in 150-mm (6-in.) direct transmission mode for WD measurement. The probe used the clay calibration curve for M measurement. At each test location, an access hole was made for LNDG WD measurement and the moisture probe measurement. Using this access hole, three WD measurements were made with measurement times of $T_m = 2$ min and $T_b = 1$ min to minimize the testing uncertainty. The gauge was set aside, and the moisture measurement was made using the same access hole. Then, moisture measurements were made at the two other nearby points-150-mm (6-in.) from the first hole at the vertices of an equilateral triangle. Samples were then collected centered at each of the access holes using a 76-mm (3-in.) diameter auger. The moisture content of these samples was determined by the oven-drying method (ASTM D2216).

Figure 8-a shows the moisture distribution near each test location as determined from oven-drying method. As expected, some locations showed moisture variances as high as 2%.

The moisture measurements were analyzed in two ways: first as a measurement at each access hole (1-Point) and second as average values for the three access holes for each location (3-Points). Figure 8-b shows the results for M measurement by using the probe with the soil-specific offset determined on site and the oven-drying method with WD from the LNDG. The average difference between the two methods was $0 \pm 7$ kg/m$^3$ ($0.0 \pm 0.4$ lb/ft$^3$). The 3-Points averaging method showed an improved correlation ($R^2 = 0.9$) between the probe and oven-drying with WD from LNDG. Also, the 3-Points averaging method showed an improvement in the agreement between the two methods as seen from the RMSE values (1-Point RMSE = 10 kg/m$^3$ [0.6 lb/ft$^3$]; 3-Points RMSE = 6 kg/m$^3$ [0.4 lb/ft$^3$]). Using the 3-Points averaging, M can be determined with an accuracy less than 16 kg/m$^3$ (1 lb/ft$^3$).

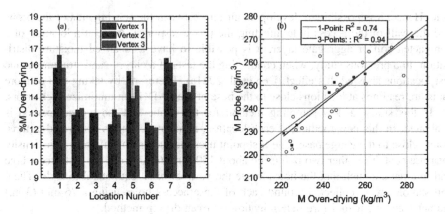

**Fig. 8.** Non-nuclear moisture probe measurements: (a) Local variance of moisture near each test location and (b) Moisture density measurements: moisture probe vs oven-drying/LNDG.

## 6 Conclusions

This work provides an analysis of the measurement properties and field performance of a low-activity nuclear density gauge with both backscatter and direct transmission measurement modes and its associated non-nuclear moisture probe.

The following findings and conclusions were drawn from this work:

- The density measurement sensitivity of the LNDG is acceptable for testing compaction of asphalt and soil foundation and pavement layers of roads. The sensitivity to density is similar to that of an NDG.
- The density measurement precision of the LNDG, when operating in backscatter and direct transmission modes, meets the industry requirement. For the same measurement precision, measurement time of an LNDG is longer than that of an NDG.
- The total measurement times, including time for background measurement, for the LNDG are 2 to 3 min for backscatter to 200-mm (8-in.) depth modes and 4 min for 230-mm to 300-mm (9-in. to 12-in.) depth modes. When operating an LNDG in backscatter mode during compaction, a fast 0.5 min measurement can be taken for spot testing.
- The depth of measurement of an LNDG when operating in backscatter mode is about 90-mm (3.5-in.) and is similar to that for an NDG.
- When operating on hot asphalt layers, LNDG showed stable density measurements.
- The repeatability and reproducibility of the LNDG density measurement is similar to that of an NDG.
- Density measurements that were taken with the LNDG and an NDG showed a strong correlation and agreement. This is expected because both types of nuclear gauges use the same measurement method: gamma-ray scattering.
- The associated non-nuclear moisture probe response depends on the moisture content and composition of the soil and aggregate. The response is linear in the moisture range of compaction-ready soil or aggregate layers.

A New Low-Activity Nuclear Density Gauge for Compaction Control of Asphalt

- A soil or aggregate specific calibration is required for the probe to accurately measure moisture in foundation layers. Most cases, a field determined offset to the factory calibration is sufficient.
- The probe showed adequate sensitivity in measuring moisture in foundation layers.
- Based on the preliminary data, the probe multi-points moisture density measurement method is promising and showed within 16 kg/m$^3$ (1 lb/ft$^3$) agreement to the accepted oven-drying method. Further studies are needed to assess the effectiveness of this method.

The LNDG is a nuclear gauge that is exempt from licensing and other nuclear requirements in the United States. The annual radiation exposure to a gauge operator from using an LNDG is a small fraction of the ubiquitous exposure from natural background radiation. Hence, with fewer or less restrictive regulations, it will be easier for geotechnical testing agencies and others in the field around the world to obtain and use the LNDG. The increase of density monitoring during compaction by using LNDGs will assist in producing durable foundations and pavements of roads.

**Acknowledgements.** The authors gratefully acknowledge Superior Paving Corp., VA, Thompson-Arthur Paving Company, NC, North Carolina Department of Transportation, and Geo Technologies, Inc., NC, for their support throughout this research by providing test sites and quality control data.

# References

ASTM D1188-07: Standard test method for bulk specific gravity and density of compacted bituminous mixtures using coated samples. ASTM International, West Conshohocken, Pennsylvania (2015)

ASTM D1556/1556M-15e1: Standard test method for density and unit weight of soil in place by sand-cone method. ASTM International, West Conshohocken, Pennsylvania (2015)

ASTM D2167-15: Standard test method for density and unit weight of soil in-place by the rubber balloon method. ASTM International, West Conshohocken, Pennsylvania (2015)

ASTM D2216-10: Standard test method for laboratory determination of water (moisture) content of soil and rock by mass. ASTM International, West Conshohocken, Pennsylvania (2010)

ASTM D2487-17e1: Standard practice for classification of soils for engineering purposes (Unified Soil Classification System). ASTM International, West Conshohocken, Pennsylvania (2017)

ASTM D2726: Standard test method for bulk specific gravity and density of non-absorptive compacted bituminous mixtures. ASTM International, West Conshohocken, Pennsylvania (2019)

ASTM D2937-10: Standard test method for density of soil in-place by the drive-cylinder method. ASTM International, West Conshohocken, Pennsylvania (2010)

ASTM D2950/2950M-14: Standard test method for density of bituminous concrete in place by nuclear methods. ASTM International, West Conshohocken, Pennsylvania (2014)

ASTM D4643-17: Standard test method for determination of water content of soil and rock by microwave oven heating. ASTM International, West Conshohocken, Pennsylvania (2017)

ASTM D4944-18: Standard test method for field determination of water (moisture) content of soil by the calcium carbide gas pressure tester. ASTM International, West Conshohocken, Pennsylvania (2018)

ASTM D4959-16: Standard test method for determination of water content of soil by direct heating. ASTM International, West Conshohocken, Pennsylvania (2016)

ASTM D6857/6857M-18: Standard test method for maximum specific gravity and density of asphalt mixtures using automatic vacuum sealing method. ASTM International, West Conshohocken, Pennsylvania (2018)

ASTM D6938-10: Standard test method for in-place density and water content of soil and soil-aggregate by nuclear methods (shallow depth). ASTM International, West Conshohocken, Pennsylvania (2010)

ASTM D8167/8167M-18a: Standard test method for in-place bulk density of soil and soil-aggregate by a low-activity nuclear method (shallow depth). ASTM International, West Conshohocken, Pennsylvania (2018)

Baker, W.J., Meehan, C.L.: A comparison of in-place unit weight and moisture content measurements made using nuclear based methods and the drive cylinder method. In: IFCEE 2018: Advances in Geomaterial Modeling and Site Characterization, pp. 1–11 (2018)

Belcher, D.J., Cuykendall, T.R., Sack, H.S.: The measurements of soil moisture and density by neutron and gamma-ray scattering. Technical Development Report Number 127, Civil Aeronautics Administration Technical Development and Evaluation Center, Indianapolis, Indiana (1950)

Berney, E., Mejias, M.: Comparative testing between electrical impedance and hybrid nuclear devices for measuring moisture and density in soils. Transp. Res. Rec. J. Transp. Res. Board **2657**, 29–36 (2019)

Dep, L., Troxler, R.E., Troxler, W.: New low-activity nuclear gauge for soil wet density measurement with low regulatory burden. In: Proceedings of 2nd International Road Federation Asia Regional Congress & Exhibition, Kuala Lumpur, Malaysia, pp. 445–456 (2016)

Dep, L., Troxler, R., Sawyer, J., Mwimba, S.: A new low-activity nuclear density gauge for soil density measurements. J. Test. Eval. **50** (2021a). Published ahead of print, 12 May 2021

Dep, L., Troxler, R., Mwimba, S., Croom, C., Langston, W.: Quality control of hot mix asphalt pavement compaction using in-place density measurements from a low-activity nuclear gauge. In: Proceedings of International Airfield and Highway Pavements Conference 2021 (2021b). Published online: June 04, 2021

Hughes, C.S.: Investigation of a nuclear device for determining the density of bituminous concrete. J. Assoc. Asphalt Paving Technol. **31**, 400–417 (1962)

Krueger, P.G.: Soil density by gamma-ray scattering, MS Dissertation: Cornell University, New York (1950)

Manual of operation and instruction for Model 4540 (2020): Troxler Electronic Labs, Durham, North Carolina

NCRP Report 160: Ionizing radiation exposure of the population of the United States. National Council on Radiation Protection and Measurements, Bethesda, Maryland (2009)

Pieper, G.F. Jr.: The measurement of moisture content of soil by the slowing of neutrons. MS Dissertation: Cornell University, New York (1949)

Sloane, R.L.: Development of a nuclear surface density gage for asphaltic pavements. Highway Res. Board Bull. **360** (1962)

Troxler, R.E., Dep, W.L.: Methods, systems, and computer program products for measuring the density of material (2009). U.S. Patent 7,569810 B1, filed June 29, 2007, and issued August 4, 2009

# A Study into the Benefit and Cost-Effectiveness of Using State-of-the-Art Technology for Road Network Level Condition Assessment

Herman Visser[(✉)], Simon Tetley, Tony Lewis, Kaslyn Naidoo, Tasneem Moolla, and Justin Pillay

ARRB Systems, Pinetown, South Africa
herman.visser@arrbsystems.com

**Abstract.** Pavement condition assessment data is the cornerstone of any pavement management system (PMS). It enables expenditure to be optimally utilized to maximize benefit and reduce costs. The accuracy and frequency of the data plays a crucial role in the selection of appropriate repair measures to not only improve the road network condition but prolongs the effects of the improvement.

Over the past decade, the manner in which this data is gathered has transformed from being based purely on visual condition assessments into the use of semi and full automated measuring devices using lasers and digital imaging which drastically reduces subjectivity of data and increases collection efficiency.

This study investigates and quantifies the economic lifecycle benefits of using devices such as the Network Survey Vehicle (NSV), Falling Weight Deflectometer (FWD) and Traffic Speed Deflectometer Device (TSDD) in comparison to that using manual visual condition assessment methods. These devices measure the structural and functional condition of the pavement and include parameters such as roughness, texture depth, rut depth, deflection, digital imaging and automated crack detection by the TSDD.

HDM-4 has proved to be the ideal tool to predict life cycle cost and benefit of a road network and has been utilized in this study. The outcome of the study shows that the cost state-of-the-art road assessment technology used for network level condition assessment is justified by the benefit or savings in expenditure by road agencies.

Whilst this study is based in South Africa, the results and findings will be applicable globally.

**Keywords:** Comprehensive · Pavement · Condition · Assessment · Cost · Benefit · HDM-4

## 1 Introduction

Road asset management relies to a large extent on the accuracy and frequency of road condition data gathered from the road network. This data is analysed in a PMS (Pavement Management System) to rank the condition of the roads within the network,

© The Author(s), under exclusive license to Springer Nature Switzerland AG 2022
A. Akhnoukh et al. (Eds.): IRF 2021, SUCI, pp. 19–31, 2022.
https://doi.org/10.1007/978-3-030-79801-7_2

20 H. Visser et al.

identify appropriate remedial measures and to apportion available funding to roads in the network in such a way that the funds are most effectively utilised.

This paper focuses on the most cost-effective ways to obtain this essential data; it compares the techniques currently used in South Africa, which are mainly based on visual condition assessments, with other more recently developed methods, which enable a much more comprehensive, and largely automated, data collection.

Three scenarios are compared in this paper, firstly by collection of data using a visual assessment only, secondly, by including Falling Weight Deflectometer (FWD) measured deflection together with, automated roughness, rut depth, texture measurements and high resolution digital imaging from a Network Survey Vehicle (NSV), and, thirdly, using a Traffic Speed Deflectometer Device (TSDD), which captures integrated condition data, including continuous deflection measurements, roughness, rut depth, automatic crack detection, imaging and texture in a single pass at traffic speeds.

The ability of these three scenarios to provide the most economical cost benefit solution for the management of a road network are compared in this study by utilising the Highway Development and Maintenance Management System (HDM-4), a decision making tool developed by the World Bank for verifying the engineering and economic viability of investments in road projects.

## 2 Details of Selected Road Network

A simulated "road network" consisting of several roads, with a total length of approximately 66 kms, located in the Kwa-Zulu Natal and Eastern Cape provinces, was created for this study. The data required for these roads in order to analyse the economic life cycle in HDM-4 is traffic volume, pavement structure, various condition parameters and geometric properties. Details of the road network are presented in Table 1.

**Table 1.** Details of road network analyzed in HDM-4

| Road Description | Surfacing Type | Length (km) |
|---|---|---|
| P 21 Primary lane | Asphalt | 35.6 |
| R72 Eastbound Slow lane | Asphalt | 7.65 |
| R72 Eastbound Fast lane | Asphalt | 7.65 |
| R72 Westbound Slow lane | Asphalt | 7.65 |
| R72 Westbound Fast lane | Asphalt | 7.65 |
| | **Total** | **66.2** |

The pavement structure and condition of these roads is obviously variable whereas HDM-4 requires uniformity in order to accurately define and analyse economic life cycle effects. The process of dividing the road network into uniform sections is discussed in detail later in the paper.

## 3 Comparison of Condition Data Capturing Methods

In this study three different methods of assessing the condition of the candidate road sections in the network were used, firstly by means of a manual visual assessment, secondly using a combination of NSV and FWD, and finally utilising an iPAVe (intelligent pavement assessment vehicle) TSDD that, is capable of measuring simultaneous and integrated structural and functional condition data.

The first method included in this study was undertaken by means of a detailed visual inspection based on South African standards in accordance with the COTO TMH9 2016 Part B. This manual provides a standard method for the visual condition assessment of flexible paved roads at network level for use in Pavement Management Systems. Different distress types are classified and detailed descriptions of degree and extent of distress are given. The assessments are based on surface and structural distress descriptions as well as functional items such as assessment of riding quality and drainage etc.

Modes of distress are defined as deformation, cracking and disintegration of the surfacing. Each of these modes of distress may occur in one (or more) typical manifestations, these are called the various "types" of distress, for example crocodile (alligator) cracks, surface cracks, block cracks, transverse cracks and longitudinal cracks are included in the cracking mode whilst rutting and undulation would be typical deformation modes. During the visual assessment the various types of distress are further defined into "degree" and "extent". Degree 1 indicates that no attention is required, Degree 3 indicates that maintenance is required in the near future, while Degree 5 indicates that immediate attention is required. Similarly, "extent" is rated Extent 1, indicating an isolated occurrence of distress, Extent 3 as intermittent occurrence, and Extent 5 as extensive occurrence.

Segment lengths into which the visual assessments are divided may differ but are usually regarded as lengths of 5 kms for rural roads and 2 kms for peri-urban roads and freeways with urban roads typically being segmented by intersection.

The second method of assessing the condition of the candidate roads was by analysing data captured by a NSV, as well as by the analysis of deflection measurements carried out using a FWD. The NSV, as shown in Fig. 1, has the capability of collecting rut depth, texture depth and roughness data together with digital imaging at traffic speeds. The data is linked to accurate linear measurement systems and GPS coordinates which enables the collection of geometric properties Fig. 2.

**Fig. 1.** Network survey vehicle

First introduced in Europe, FWD has been in use in the United States since the 1980s (Wang and Birken 2014). It has since become the most widely used deflection measurement device in the world, including South Africa, where the FWD has been used since the '90s for project and network level assessments. The device attempts to simulate a moving wheel load by dropping a short impulse load onto the pavement. The pavement response to the load creates a deflection bowl, as shown in Fig. 3, which is measured by geophones that are positioned at various distances from the load centre.

**Fig. 2.** Falling weight deflectometer (FGSV 2004)

Historically, the FWD has generally been considered to be the most suitable device for deflection surveys. The stationary testing at each test point does however raise traffic accommodation and safety concerns whilst the stop-start nature of the FWD test means that deflections are usually performed at a low frequency – typically at 200 m intervals for network level assessments.

The third method of assessing the condition of the candidate roads was by means of a TSDD which enables continuous deflection measurements and other condition parameters to be taken at traffic speeds of up to 80 km/h. The iPAVe version of the TSDD that was used in this study is shown in Fig. 3. It uses a patented Doppler laser beam technology, also shown in Fig. 3, to measure the vertical displacement velocity at various offsets from the loaded wheel. The "area under the curve" method by Muller and Roberts (2013) is used to convert deflection slopes to a deflection bowl that represents the pavement's response to the wheel load of the TSDD.

The particular iPAVe device used in this study features a 7 laser Doppler beam in front of the left rear wheel. The iPAVe includes spatially geo-referenced integrated laser crack detection, digital imaging and a Class 1 laser profilometer in addition to the traffic speed deflectometer used for structural measurements. This allows for a comprehensive pavement assessment that simultaneously provides integrated functional and structural condition data. Compared to the FWD, the TSDD has the ability to measure a much higher deflection measurement frequency (every 25 mm) at traffic speed allowing for significantly improved production and safety to the road user and operator. The iPAVe TSDD with typical condition assessment output is shown in Fig. 3.

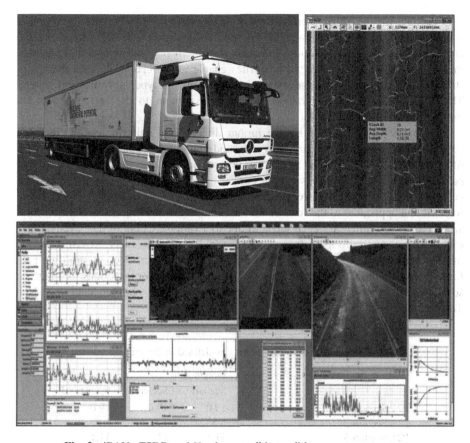

**Fig. 3.** iPAVe TSDD and Hawkeye toolkit condition assessment output

## 4 Assessment Methodology

In the case of this study, each of roads shown in Table 1 was regarded as a portion in a road network. In the **first scenario** the data was limited to visual assessments carried out by trained and experience assessors as per the stipulations of TMH9 whilst driving the roads at 20 kms per hour.

In the **second scenario** the visual assessments were undertaken by post-rating high definition photographs taken by the NSV at 20 m intervals along the selected roads. The visual assessment data was supplemented by FWD deflection measurements every 200 m as well as by continuously measured roughness, texture depth and rutting obtained from the NSV.

The data for the **third scenario** relies on the output from the iPAVe which is similar to the second scenario, with the exception of being collected by a single vehicle and being capable of continuous deflection measurements and high definition automated crack detection. These two properties enable identification of exact areas with structural distress and accurate estimation of cracking type, severity and extent. Thereby providing objective rather than subjective input to the sensitive decision making algorithms of HDM-4.

The roads listed in Table 1 were further divided into uniform sections for each scenario based on pavement structure and condition data. A uniform section is identified as a section of road that is uniform in roughness, deflection measurement and pavement structure. This allows for accuracy in defining each segment of the road network to ensure that the proposed remedial works, as identified in HDM-4, are relevant and applicable to the road condition. The length and number of each uniform section identified differs for each scenario due to a variable extent of data available for each scenario, i.e. uniform sections for Scenario 1 are based solely on visual assessment whereas those for Scenarios 2 and 3 are also based on deflection, profiling and, for scenario 3, automated crack detection. The identified uniform sections for the three scenarios are shown in Table 2 and an inventory of the condition data shown in Table 3.

**Table 2.** Details of uniform sections as defined in the HDM-4 model

| Uniform Section | Length (km) | | |
|---|---|---|---|
| | Scenario 1 | Scenario 2 | Scenario 3 |
| P21 Section 1 | 19.0 | 19.0 | 19.0 |
| P21 Section 2 | 5.5 | 5.5 | 5.5 |
| P21 Section 3 | 11.1 | 11.1 | 11.1 |
| R72 Eastbound Fast Lane Section 1 | 1.3 | 1.5 | 1.0 |
| R72 Eastbound Fast Lane Section 2 | 1.6 | 2.4 | 2.4 |
| R72 Eastbound Fast Lane Section 3 | 0.6 | 1.6 | 1.3 |
| R72 Eastbound Fast Lane Section 4 | 4.2 | 2.2 | 1.1 |
| R72 Eastbound Fast Lane Section 5 | N/A | N/A | 1.8 |
| R72 Eastbound Slow Lane Section 1 | 1.2 | 1.4 | 1.4 |
| R72 Eastbound Slow Lane Section 2 | 1.6 | 1.9 | 1.6 |
| R72 Eastbound Slow Lane Section 3 | 0.6 | 1.7 | 3.3 |
| R72 Eastbound Slow Lane Section 4 | 4.3 | 1.3 | 1.5 |
| R72 Eastbound Slow Lane Section 5 | N/A | 1.4 | N/A |
| R72 Westbound Fast Lane Section 1 | 6.0 | 6.0 | 6.0 |
| R72 Westbound Fast Lane Section 2 | 1.6 | 1.6 | 1.6 |
| R72 Westbound Slow Lane Section 1 | 5.7 | 5.7 | 5.7 |
| R72 Westbound Slow Lane Section 2 | 1.9 | 1.9 | 1.9 |
| Total | 66.2 | 66.2 | 66.2 |

**Table 3.** Inventory of available data for each scenario

| Uniform Section | Length (km) | | |
|---|---|---|---|
| | Scenario 1 | Scenario 2 | Scenario 3 |
| Visual Condition Assessment | Yes | Yes | Yes |
| Roughness | No | Yes | Yes |
| Rutting Measurements | No | Yes | Yes |
| Cracking*[2] | No | No | Yes |
| Deflection Measurements | No | Yes | Yes |
| Texture Depth | No | Yes | Yes |
| Structural Number* | Yes* | Yes | Yes |

Note: *[1] The structural number (SN) is derived from layer thickness and coefficients for Scenario 1 whilst this is calculated from actual data for scenario 2 and 3. *[2] Cracking from post rated visual assessment in Scenario 2 and from automated crack detection in Scenario 3

Scenario 1 has a degree and extent rating of various distresses such as cracking (crocodile, longitudinal and transverse), ravelling, rutting, potholes, patching, edge break, undulation, binder condition, bleeding etc. The types of distress required for HDM-4 input include cracking, ravelling, rutting, potholes, texture and edge break area. The limitations of using visual condition assessment data, as done for Scenario 1, is that the extent and degree rating (1 to 5) is subjective and does not provide an accurate determination of these distresses as required by HDM-4. A basic estimation method of converting extent and degree into surface area was used to populate and define the HDM-4 road network for Scenario 1. These parameters are more accurately estimated for Scenario 2 and 3 using the additional laser equipment and imaging of the NSV and iPAVe.

Scenario 2 and 3 does not make use of the manual visual assessment data for input into HDM-4, but instead makes use of superior accuracy instrumented survey data collected by the NSV and iPAVe for all the required condition input mentioned above. Scenario 2 makes use of the cracking distress rating obtained from the post rated visual condition assessment data to estimate cracking input. In Scenario 3, the automated crack detection functionality of the iPAVe, in combination with the ARRB Systems Hawkeye Toolkit, is used to accurately calculate the cracked percentage of the road from the laser survey. The NSV and iPAVe data used for Scenario 2 and 3 respectively, uses the same technology and equipment that results in similar IRI, texture depth and rutting measurements.

The deflections of the FWD are recorded as a single measurement every 200 m (5 per km) which provides a mere 2.5% coverage if 5 m coverage per deflection point is assumed whereas the iPAVe provides continuous deflection measurements processed to 5 m intervals (200 per km). This allows for 100% coverage and precise identification of locations where changes in structural conditions occur. It also enables localised areas of failure or poor structural condition to be identified. Maximum deflections from the FWD and iPAVe are used to calculate the SN using the built-in HDM-4 functionality.

**Table 4.** Summary of road network condition as defined in HDM-4 model

| Uniform Section | Scenario 1 | | | | Scenario 2 | | | | Scenario 3 | | | |
|---|---|---|---|---|---|---|---|---|---|---|---|---|
| | Rut (mm) | IRI (mm/m) | $Y_{max}$ (mm) | SN | Rut (mm) | IRI (mm/m) | $Y_{max}$ (mm) | SN | Rut (mm) | IRI (mm/m) | $Y_{max}$ (mm) | SN |
| P21 Section 1 | 5 | 1.5 | N/A | 4.6 | 2.6 | 1.9 | 0.21 | 8.5 | 2.6 | 1.9 | 0.21 | 8.4 |
| P21 Section 2 | 10 | 3 | N/A | 4.2 | 8.4 | 3.4 | 0.72 | 3.9 | 8.4 | 3.4 | 0.65 | 4.2 |
| P21 Section 3 | 5 | 2 | N/A | 4.6 | 3.8 | 2.4 | 0.23 | 8.1 | 3.8 | 2.4 | 0.21 | 8.5 |
| R72 EB FL S1 | 10 | 2.5 | N/A | 3.9 | 4.5 | 2.3 | 0.56 | 4.6 | 4.4 | 1.8 | 0.64 | 4.2 |
| R72 EB FL S2 | 15 | 3 | N/A | 3.9 | 6.7 | 3.4 | 0.34 | 6.4 | 6.3 | 3.3 | 0.39 | 5.8 |
| R72 EB FL S3 | 10 | 3.5 | N/A | 5.5 | 5.3 | 3.3 | 0.32 | 6.6 | 5.5 | 3.0 | 0.33 | 6.5 |
| R72 EB FL S4 | 15 | 3.5 | N/A | 4.7 | 12.1 | 3.8 | 0.31 | 6.7 | 12.0 | 3.3 | 0.43 | 5.5 |
| R72 EB FL S5 | N/A | N/A | N/A | N/A | N/A | N/A | N/A | N/A | 13.0 | 4.2 | 0.19 | 9.0 |
| R72 EB SL S1 | 15 | 2.5 | N/A | 4.4 | 5.3 | 2.0 | 0.77 | 3.8 | 5.3 | 2.0 | 0.77 | 3.8 |
| R72 EB SL S2 | 20 | 3 | N/A | 2.8 | 12.0 | 4.2 | 0.59 | 4.7 | 11.8 | 3.7 | 0.62 | 4.4 |
| R72 EB SL S3 | 15 | 3.5 | N/A | 4.9 | 7.9 | 3.5 | 0.61 | 4.4 | 9.0 | 3.7 | 0.53 | 4.8 |
| R72 EB SL S4 | 15 | 3.5 | N/A | 6.5 | 9.4 | 3.8 | 0.42 | 5.5 | 9.5 | 3.8 | 0.35 | 6.2 |
| R72 EB SL S5 | N/A | N/A | N/A | N/A | 19.8 | 4.6 | 0.35 | 6.2 | N/A | N/A | N/A | N/A |
| R72 WB FL S1 | 5 | 2.5 | N/A | 4.5 | 4.5 | 2.7 | 0.33 | 6.5 | 4.5 | 2.7 | 0.31 | 6.7 |
| R72 WB FL S2 | 10 | 4.5 | N/A | 4.8 | 7.1 | 4.4 | 0.32 | 6.6 | 7.1 | 4.4 | 0.31 | 6.7 |
| R72 WB SL S1 | 10 | 3 | N/A | 4.5 | 5.5 | 3.3 | 0.58 | 4.5 | 5.8 | 3.3 | 0.60 | 4.4 |
| R72 WB SL S2 | 5 | 5 | N/A | 4.8 | 5.3 | 5.4 | 0.38 | 5.9 | 5.6 | 5.4 | 0.44 | 5.4 |

As seen in Table 4, there is a significant discrepancy between the visually assessed rutting and IRI when compared to the measured values. The structural number calculations for Scenario 1 are based on assumption rather than measurement as per Scenario 2 and 3 with the difference being significant to the HDM-4 output.

## 5 Economic Analysis

As discussed previously, the aim of this study is to determine the differences in cost to the roads agency when making use of different road condition assessment methods i.e. basic manual visual assessment, semi-automated data collection and fully automated evaluation. Better quality and higher accuracy/frequency data used as input into PMS allows for better informed expenditure decisions to optimise the life cycle of a road network (Zhang et al. 2010). An economic analysis and comparison of the three scenarios has been used to quantify the capital and recurring cost over a typical life cycle of the simulated road network. In this study an analysis period of 20 years and a discount rate of 8%, as per South African standard practice, is assumed for all three scenarios.

HDM-4 is used in this study to define the road network, work standards and strategic analysis for each scenario. The program applies each work standard over the analysis period of each uniform section in order to optimise and determine which work standard is most economical. The work standard alternatives and their maximum intervals used for this study is shown in Table 5.

**Table 5.** Work standard alternatives available for each scenario

| Work Standards | Interval (years) |
| --- | --- |
| Do Nothing (Minimum maintenance) | Annually |
| Single Seal Surfacing | 5 |
| Double Seal Surfacing | 5 |
| Asphalt Overlay | 7 |
| Light Rehabilitation | 10 |
| Rehabilitation | 15 |
| Reconstruction | 20 |

The unit cost of each alternative was estimated using current industry related construction costs. The unit cost of the economically preferred alternative is applied to each uniform section over the analysis period and discounted to a present value. This total discounted cost for each scenario consists of a recurring cost and capital cost. The recurring cost represents the maintenance cost over the analysis period with the capital cost representing the initial cost of the selected remedial intervention that are implemented on the road sections.

The "Maximise Net Present Value" (NPV) optimisation method function in the HDM-4 strategic analysis was used to perform the economic life cycle analysis on the

28    H. Visser et al.

road network for each of the three scenarios. This optimisation function allows the selection of the most cost-effective set of maintenance and improvement standards over the analysis period.

The economic indices that are of importance to this study are the discounted costs of each scenario. These enables the determination of possible savings in agency expenditure on the road network, which can be compared to the operational cost of data collection for each alternative. The discounted costs are shown in Table 6, and is expressed in South African Rand (ZAR) per km, in order to compare with the data collection cost.

**Table 6.** Discounted recurring and capital cost for each scenario

| Scenario | Overall cost for 66.2 km's | Discounted Cost (ZAR) per km | | | Savings per km relative to Scenario 1 |
|---|---|---|---|---|---|
| | | Recurring | Capital | Total Cost | |
| 1. Visual Assessment | R 94,076,118 | R 665,284 | R 756,020 | R1,421,304 | – |
| 2. FWD & NSV | R 87,111,265 | R 576,033 | R 740,046 | R1,316,079 | R 105,225 |
| 3. iPAVe TSDD | R 86,108,129 | R 567,904 | R 733,020 | R1,300,924 | R 120,381 |

The savings per kilometre relative to the total discounted cost for Scenario 1 are R 105,225 and R 120,381 per kilometre for Scenarios 2 and 3 respectively.

The higher cost for Scenario 1 is directly related to the input data collected during the visual condition assessment. These input parameters required by HDM-4 are estimated from degree and extent ratings and are often less accurate (as illustrated in Table 4) due to being based on subjective human observation rather than actual measurement fact. A conservative approach is often followed to counter the risk of inaccurate measurement data leading to either over-expenditure on the road network or, more likely, reduced maintenance capability if budgets are constrained.

Scenario 2 provides a significant discounted cost savings of R 105,225 per kilometre compared to Scenario 1 due to the accurate rutting, IRI, texture and deflection measurements obtained from the FWD and NSV devices. The largest contribution to the cost savings is from recurring cost; this highlights the importance of obtaining accurate instrument data for future maintenance decisions at network level.

The discounted cost savings of R 120,381 per kilometre for Scenario 3 is approximately 15% higher than for Scenario 2 with a cost savings of R 105,225 per kilometre. The high accuracy of automated crack detection laser technology allows for a precise measurement of cracking of the road as well as enabling the percentage of wide cracks to be identified. The latter being a highly sensitive and critical input for HDM-4.

Based on the findings of this study, there is a clear and significant cost benefit or savings using the latest road condition assessment technology to obtain data for input to PMS. There is however a cost implication for using the NSV, FWD and iPAVe devices.

The cost of data collection for the three scenarios have been estimated using market related prices. Scenario 1 involves two certified visual assessors in a passenger vehicle driving at 20 km/h and occasionally stopping to inspect the road condition and fill in visual assessment forms or electronic device. An all-inclusive estimated operational cost of R 163 per km has been calculated for Scenario 1.

The cost for Scenario 2 includes the operation of a FWD and NSV vehicle and involves operators for each device. A safety vehicle behind the FWD due to the stop-start nature of the test has also been factored into the cost. It also includes a post-rating visual assessment of the road. The total estimated operational cost of Scenario 2 is R 2,250 per km. The production rate of the FWD, which is the limiting factor for this scenario, is estimated at approximately 200 km per day. The FWD test is performed every 200 m or 5 times per km.

Scenario 3 only involves the iPAVe TSDD and operators with no safety vehicle required due to the 60-80 km/h testing speed. The total estimated operational cost of Scenario 3 is R 3,500 per km. The production rate of the iPAVe is estimated at approximately 400 km per day.

## 6 Summary of Findings

A comparison between discounted cost savings per km and operational cost for each scenario compared to Scenario 1 is listed in Table 7.

**Table 7.** Comparison of savings vs cost for each scenario

| Scenario | Cost incl. Data Collection for 66.5 km's | Discounted Cost per km (ZAR) | Operation cost per km (ZAR) | Total Cost per km (ZAR) | Total Savings per km (ZAR) |
|---|---|---|---|---|---|
| 1. Visual Assessment | R 94,086,907 | R 1,421,304 | R 163 | R 1,421,467 | – |
| 2. FWD & NSV | R 87,223,788 | R 1,316,079 | R 1,700 | R 1,317,779 | R 103,688 |
| 3. iPAVe | R 86,339,794 | R 1,300,924 | R 3,500 | R 1,304,424 | R 117,043 |

It is clear that there is a significant benefit or saving in road agency expenditure even with the comparatively high cost of the road condition assessment equipment compared to conventional methods.

The benefit or savings for using the iPAVe TSDD is R 13,355 per km more than using the FWD and NSV combination even though the operation cost of the iPAVe is double that of the NSV and FWD combination. If this benefit were to be applied to the entire South African paved road network, a saving of over R 2 Billion would be realised. When compared to the manual visual assessment, the use of iPAVe TSDD would return a R 19 Billion saving.

30 H. Visser et al.

In addition to the savings in cost, the iPAVe TSDD provides high frequency deflection measurements that can, in combination with the digital imaging, identify localised failures, locate shallow underground services and roads that are structurally deficient but showing no visual distress.

## 7 Conclusions and Recommendations

The use of state-of-the-art road assessment technology to collect road condition data, instead of conventional methods, not only has economic benefits but ensures a PMS with highly accurate condition data enabling better informed decisions on the most suitable repair strategy – in terms of both cost and technical appropriateness.

The aim of any roads agency is to optimise expenditure on their road network by minimising total transport cost and optimising expenditure to ensure the lowest possible road user cost. A decrease in expenditure without compromising the road network condition or increasing the road user cost can thus be seen as a double benefit. This study has shown that increasing the quality and accuracy of road condition data does result in a significant cost saving to the road authority, the road user and most importantly to the National fiscus.

The main conclusion from this study is that accurate road condition assessment data is crucial in order to correctly define the road network and make appropriate expenditure decisions to optimise the life cycle management of the roads within the network. The benefit of using the FWD and NSV combination or iPAVe devices is clearly shown with a cost savings of approximately R 103,688 (+− US $ 7,400) and R 117,043 (+− US $ 8,400) per kilometre, respectively. The 100% coverage full spectrum iPAVe TSDD provides superior road condition assessment data compared to the FWD and NSV combination enabling more accurate input to the HDM-4 analysis with resultant cost saving in the order of R 13,355 (+− US $ 1,000) per km. In addition, the high deflection frequency of the iPAVe eliminates the need for additional routine deflection testing at project level, except for localised investigations.

It is recommended that this limited study be extended to a larger road network which includes a wider variety of road conditions. The use of fixed length segments should be investigated in addition to uniform sections as is used in this study.

**Acknowledgements.** The following people have assisted with sourcing and analysing the data required for this study and I would like to thank the following:
- Francois van Aswegen from GIBB for providing comprehensive data for the R72 Settlers Way road.
- Yeshveer Balaram from ARRB Systems for providing comprehensive data for the P21 road.

# References

Muller, W.B., Roberts, J.: Revised approach to assessing traffic speed deflectometre data and field validation of deflection bowl predictions. Int. J. Pavement Eng. **14**(4), 388–402 (2013). https://doi.org/10.1080/10298436.2012.715646

Zhang, H., Keoleian, G., Lepech, M., Kendall, A.: Life cycle optimization of pavement overlay systems. J. Infrastruct. Syst. **16**, 310–322 (2010). https://doi.org/10.1061/(ASCE)IS.1943-555X.0000042

Committee of Transport Officials Manual for Visual Assessments of Road Pavements. The South African National Roads Agency SOC Limited PO Box 415, Pretoria, 0001 (2016)

Wang, M.L., Birken, R.: Sensing solutions for assessing and monitoring roads. In: Sensor Technologies for Civil Infrastructures, vol. 2 (2014). Applications in Structural Health Monitoring

Research Society for Roads and Traffic. FGSV Working Paper "Load Capacity", Part C2 "FWD, Evaluation and Evaluation", AK 4.8.2, draft as of February 2004 (2004)

# Asphalt Pavement Innovations

# Paths to Successful Asphalt Road Structures Performance BasedSpecificationsforHotMixAsphalt

Bernhard Hofko[✉]

Vienna University of Technology (TU Wien), Vienna, Austria
bernhard.hofko@tuwien.ac.at

**Abstract.** Current and future challenges in road engineering, including but not limited to increasing heavy goods traffic, scarce natural resources, climate crisis and shrinking public budgets, bring the urgent need for economically and ecologically efficient road pavements. To ensure long-lasting roads, high-quality mix design and quality assurance/control are crucial preconditions. This includes realistic simulation of the stress situation in the field by state-of-the-art laboratory testing of materials.

Therefore, this paper focuses on performance-based test methods that address the effective mechanical characteristics of bituminous materials, which may be introduced into national requirements within the framework of European HMA specifications. These test methods comprise low-temperature tests, i.e. the tensile stress restrained specimen test (TSRST) or the uniaxial tensile strength test (UTST), stiffness and fatigue tests, i.e. the four-point bending beam test (4PB) or the uniaxial tension-compression test (DTC), as well as methods to determine permanent deformation behaviour employing dynamic triaxial cyclic compression tests (TCCT).

These tests are used for the performance-based mix design and subsequently implemented in numerical pavement models to predict in-service performance. Thus, load-induced stresses and mechanogenic effects on the road structure can be simulated, and improved predictions of the in-service performance of flexible pavements over their entire service lives are possible.

**Keywords:** Asphalt · Performance-based contracts · Mix-design optimization

## 1 Introduction

For the optimisation of flexible road pavements, research efforts in the past decades have been focused on the setup and implementation of performance-based test methods for hot mix asphalt (HMA) and their implementation into performance prediction models. While the recipe-based approach and empirical test methods count for material characteristics that have been found to correlate with engineering properties indirectly, performance-based test methods describe fundamental engineering properties, which are directly correlated to field performance.

By 2007, harmonised European Standards (EN) for testing asphalt materials were introduced in all CEN member countries within the European Union. These EN

© The Author(s), under exclusive license to Springer Nature Switzerland AG 2022
A. Akhnoukh et al. (Eds.): IRF 2021, SUCI, pp. 35–45, 2022.
https://doi.org/10.1007/978-3-030-79801-7_3

36    B. Hofko

standards distinguish between the recipe-based (empirical) mix design approach and the fundamental, performance-based approach. Although both approaches aim to realise well-performing, structurally optimised pavements, an essential advantage of the performance-based approach is based on the laboratory assessment of physically sound material parameters.

These key performance parameters of HMA include (i) viscoelastic material behaviour (complex modulus and phase angle), (ii) fatigue resistance under repeated load cycles, (iii) resistance to cracking at low temperatures and (iv) resistance to rutting at high temperatures. These material parameters can be used to specify the mix properties within an advanced type testing procedure required to meet customised quality standards for materials defined in tender documents and mix design (Blab and Eberhardsteiner 2009).

In the HMA test standard series EN 12697-xx, key performance HMA properties are addressed by different performance tests, summarised in Table 1.

To identify the rutting behaviour at elevated temperatures, cyclic axial load tests with or without confining pressures (TCCT Triaxial Cyclic Compression Test or UCCT uniaxial Cyclic Compression Test) are specified. The low-temperature behaviour is tested by the Tensile Stress Restrained Specimen Test (TSRST) and the Uniaxial Tensile Strength Test (UTST), respectively. For characterising the stiffness and fatigue performance of asphalt mixtures, different tests are described in the standards, including bending tests (e.g. two-point 2PB or four-point 4PB) and direct and indirect tensile tests. Furthermore, the European HMA specification EN 13108-x offers different categories for these performance-based HMA properties, which may be introduced as fundamental or performance-based HMA requirements into the national specifications.

Such performance-based HMA specifications require more complex mix design and type testing procedures. However, in combination with these performance-based HMA specifications, mechanistic models allow a more reliable prediction of in-service performance of HMA pavement structures. The objectives of these advanced pavement design models are to enable the simulation of thermo- and load-induced stresses and mechanogenic effects and thus improved forecasts of the in-service performance of flexible and semi-rigid pavements.

**Table 1.** Key performance indicators for HMA according to EN 12697-xx

| Layer | Stiffness | Fatigue | Low-temperature cracking | Permanent deformation |
|---|---|---|---|---|
| Surface | x | (x) | x | x |
| Binder | x | (x) | x | x |
| Base | x | x | (x) | (x) |
| Test procedure | 2-Point-Bending test with trapezoidal specimen (2PB-TR) 2-Point-Bending test with prismatic specimen (2PB-PR) 3-Point-Bending test (3PB) 4-Point-Bending test (4PB) Cyclic indirect tensile test (CIT-CY) Direct tension-compression test (DCT) | Cyclic indirect tensile test (CIT-CY) 4-Point-Bending test (4PB) | Temperature Stress Restrained Specimen Test (TSRST) Uniaxial tension stress test (UTST) Uniaxial Cyclic tension stress test (UCTST) | Triaxial cyclic compression test (TCCT) Uniaxial cyclic compression test (UCCT) |
| EN standard | EN 12697-26 | EN 12697-24 | EN 12697-46 | EN 12697-25 |

x...performance characteristic mandatory, (x)...additional performance characteristic.

## 2 Low-Temperature Performance

Traditional studies and guides on pavement design have primarily concentrated on classical fatigue cracking that considers failure to initiate at the bottom of the base layers induced by a large number of small repeated traffic-loadings (Blab and Eberhardsteiner 2009, Wistuba et al. 2006). However, this simple approach does not always fit the reality since two different major types of crack damage occur in flexible pavements: cracks that start at the bottom of the base layer and grow upwards generally named fatigue cracks, and surface cracks that are initiated on top of the pavement by thermal loading. Thermally-induced stresses combined with traffic loading may exceed the critical tensile strength and lead to surface-initiated top-down cracking along the wheel paths.

In the field of low-temperature behaviour, the research activities have been focused on developing appropriate test methods to better understand and identify the fracture mechanisms utilising laboratory experimentation and assess further the risk of temperature cracking for different bituminous materials which are exposed to stress and temperature.

Low-temperature cracking of flexible pavements results from thermal shrinkage during cooling, inducing tensile stress in the asphalt. To simulate the situation in flexible pavement layers, the following test methods on asphalt mix specimens according to the EN 12697-46 are employed:

- Tensile Stress Restrained Specimen Test (TSRST): while the deformation of the specimen is restrained, the temperature is reduced by a pre-specified cooling rate;
- Uniaxial Tensile Strength Test (UTST): to assess the risk of low-temperature cracking, the stress induced by thermal shrinkage is compared with the respective tensile strength;

The target parameters found by TSRST are the fracture temperature ($T_{crack}$) and the corresponding fracture stress ($\sigma_{crack}$). An illustration of the test procedure of the TSRST is given in Fig. 1.

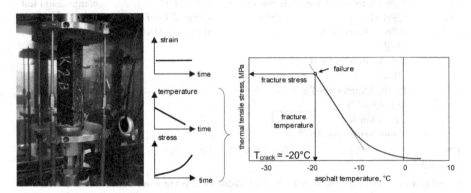

**Fig. 1.** TSRST: experimental setup and illustration of result (Blab and Eberhardsteiner 2009)

The UTST is an isothermal process at specified temperatures (e.g. +10, +5, −5, −15 and −25 °C). After stress-free cooling of the asphalt to the testing temperature, the UTST is performed by applying a constant deformation rate (1 mm/min) until the specimen fails.

Combining the results of TSRST and UTST, the tensile strength reserve (Δβ) is found, a traditional target parameter for low-temperature cracking (Fig. 2). The tensile strength reserve provides information on the amount of stress due to traffic loading that can be introduced into a pavement in addition to thermal stress before the tensile strength is reached.

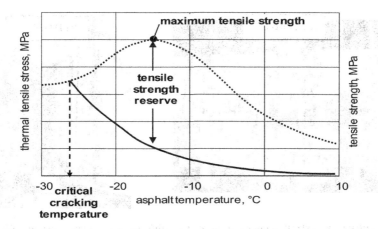

**Fig. 2.** Superposition of TSRST and UTST results to derive tensile strength reserve (Blab and Eberhardsteiner 2009)

## 3 Stiffness and Fatigue Behavior

Stiffness and fatigue testing, where repeated stress is applied on a test specimen, has been an essential topic in pavement engineering for decades. Comprehensive research was carried out, e.g. by RILEM and the US Strategic Highway Research Program (SHRP), where a sophisticated layout of test methods for asphalt mix testing and design has been developed. These initiatives have started a broad discussion on new ways to optimise fatigue testing procedures further and interpret test results. Presently two standards, EN 12697-24 (fatigue) and EN 12697-26 (stiffness) specify the methods for characterising the stiffness and fatigue performance of asphalt mixtures by different test methods, including bending tests and direct and indirect tensile tests (Di Benedetto et al. 2001; Hofko et al. 2012).

These test methods aim to derive two material characteristics: the material's stiffness, expressed by the norm of the complex asphalt modulus ($|E^*|$), and the fatigue performance, expressed by the number of permissible load repetitions ($N_{f/50}$).

The traditional fatigue criterion of asphalt mixtures is linked to the number of load cycles until half the initial stiffness is reached. The evolution of stiffness modulus vs. the number of load repetitions is plotted to derive the fatigue function. This function provides essential information for the derivation of fundamental relationships between mix composition and stiffness properties and serves as input for material and pavement structure optimisation.

From the EN test methods following two methods were selected to perform stiffness and fatigue tests on asphalt mixtures:

- the four-point bending-beam-test (4PBBT) (Fig. 3a) and
- the direct tension-compression test (DTCT). (Fig. 3b).

**Fig. 3.** 4PBBT & DTCT equipment used for stiffness and fatigue testing (Blab and Eberhardsteiner 2009)

Figure 4 shows typical stiffness tests on a stone mastic asphalt SMA 11 used for surface layers performed at different temperatures and loading frequencies. Results are the master curve of the complex modulus E* at a reference temperature of, e.g. +15 °C (Fig. 4a), and the frequency-independent representation of the loss modulus E" and the storage modulus E' in a so-called Cole-Cole plot (Fig. 4b). Consequently, these test results describing the temperature and frequency-dependent material response of asphalt can be used to compute thermal and load induce stresses and strains in the asphalt layers using a numerical pavement model.

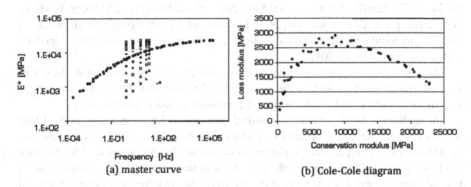

**Fig. 4.** Stiffness master curve of an SMA 11 derived from a 4PB

For predicting the fatigue damage, tests with a high number of repeated cyclic loading are performed. Such tests can be carried out under stress or strain-controlled conditions providing typical fatigue curves shown in Fig. 5 for an asphalt concrete with a 22 mm maximum nominal aggregate size (AC 22) at 30 Hz and +20 °C. From such curves, the permissible load repetitions ($N_{perm}$) are obtained to describe the theoretical lifetime within an analytically based pavement design method based on fatigue laws.

Fatigue tests according to EN standards are carried out at three different strain levels at 20 °C and 30 Hz with a sixfold repetition. Consequently, the strain $\varepsilon_6$ at $10^6$ permissible load repetitions is calculated from the semi-logarithmic regression curve as a characteristic fatigue parameter of the tested material according to Fig. 6. In the given example, the parameter for the tested AC 22 is $\varepsilon_6 = 145$ µm/m.

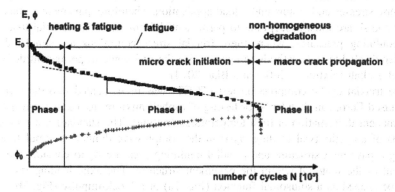

**Fig. 5.** Phases of a fatigue curve from a strain-controlled 4PB (acc. to Di Benedetto et al. 2001).

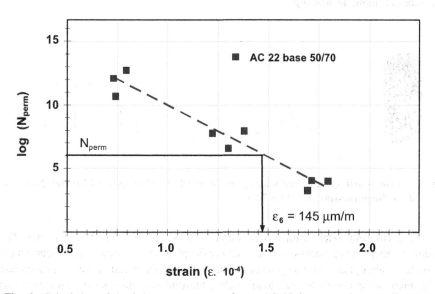

**Fig. 6.** Calculation of the fatigue parameter $\varepsilon_6$ for an AC 22 from a strain-controlled 4PB

## 4 Permanent Deformation Behavior

Another challenging topic in flexible pavement research is the fundamental description of the performance of asphalt mixtures at elevated temperatures. To better understand the permanent deformation behaviour, tests that realistically simulate in-situ stress conditions and traffic loads are necessary. Permanent deformation can be related to the material characteristics of asphalt mixtures at high temperatures in combination with deviatoric stresses and strains under load application. Therefore, pavement surface and binder course are most susceptible to permanent deformation. Cyclic axial load tests with confining pressures, where these triaxial stress conditions are simulated, are considered reliable test methods to characterise the resistance to permanent deformations of asphalt mixtures (Hofko and Blab 2014).

The triaxial cyclic compression test (TCCT) was implemented into the series of harmonised European Standards for testing of asphalt mixtures to assess the resistance to permanent deformation at high temperatures (rutting). The standard test procedure consists of a cyclic axial loading $\sigma_A(t)$ in the compressive domain to simulate a tire passing a pavement structure and a radial confining pressure $\sigma_c$ to consider the confinement of the material within the pavement structure. The axial loading $\sigma_A(t)$ can either be shaped as a sinusoidal function (Fig. 7a) or a block-impulse (Fig. 7b).

The standard states that the confining pressure $\sigma_C$ can either be held constant or oscillate dynamically without providing more specific information. However, the TCCT recommended for performance testing is loaded by a sinusoidal axial at a constant confinement loading.

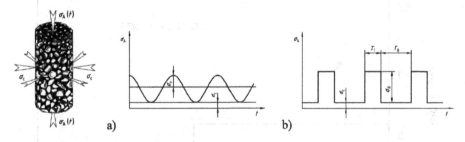

**Fig. 7.** a) sinusoidal shaped axial loading and b) block-impulses as axial loading, both with constant confining pressure (EN 12697-25)

The permanent axial strain $\varepsilon_{ax}(n)$ is analysed vs. the number of load cycles. The resulting creep curve shows two characteristic phases: a primary nonlinear and a secondary creep phase with a quasi-constant incline of the creep curve. The creep rate $f_c$ in micrometre per meter per load cycle (μm/m/n) can now be determined as an incline of the linear approximation function fitted to the quasi-linear part of the creep curve. Figure 8 gives an example for the creep rate $f_c$ calculated for a surface layers material AC 11 with two different binders, an unmodified and a polymer-modified binder.

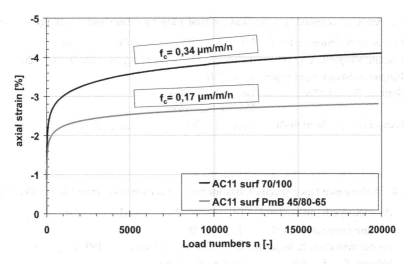

**Fig. 8.** Creep curves for an AC 11 with two different binders

## 5 Asphalt Mix Requirements

Based on these standardised performance-based tests, performance-based specifications have been implemented in some EU member states as so-called fundamental requirements. These specifications are no longer recipe-based, addressing only volumetric properties in an asphalt mix but include performance parameters. Depending on the layer, performance parameters are specified with respective classes to take project-specific conditions like climate and traffic volume into account. One example for implementation into national specification is shown in Table 2, Table 3 and Table 4 for surface, binder and base layer material, respectively.

**Table 2.** Performance-based specifications for surface layer asphalt concrete (AC) (FSV, 2018)

| Parameter/performance level | R1 | R2 | R3 | R4 | R5 |
|---|---|---|---|---|---|
| Fracture temperature in °C | $T_c \leq -30$ | $T_c \leq -25$ | $T_c \leq -30$ | $T_c \leq -25$ | $T_c \leq -20$ |
| Fatigue resistance $\varepsilon_6$ in µm/m | $\varepsilon_{6\text{-NR}}$ (no requirements) | | | | |
| Stiffness $S_{mix}$ in MPa | Declare $S_{min}$-value Declare $S_{max}$-value | | | | |
| Creep rate $f_{cmax}$ in µm/m/n | $f_{cmax} \leq 0.2$ | | $f_{cmax} \leq 0.4$ | | |

44 B. Hofko

**Table 3.** Performance-based specifications for binder layer asphalt concrete (AC) (FSV, 2018)

| Parameter/performance level | V1 | V2 | V3 | V4 |
|---|---|---|---|---|
| Fracture temperature in °C | $T_c \leq -25$ | $T_c \leq -20$ | $T_c \leq -25$ | $T_c \leq -20$ |
| Fatigue resistance $\varepsilon_6$ in μm/m | $\varepsilon_6 \geq 130$ | | | |
| Stiffness $S_{mix}$ in MPa | Declare $S_{min}$-value Declare $S_{max}$-value | | | |
| Creep rate $f_{cmax}$ in μm/m/n | $f_{cmax} \leq 0.2$ | | $f_{cmax} \leq 0.4$ | |

**Table 4.** Performance-based specifications for base layer asphalt concrete (AC) (FSV, 2013)

| Parameter/performance level | E1 | E2 | E3 | E4 |
|---|---|---|---|---|
| Fracture temperature in °C | $T_c \leq -20$ | | | |
| Fatigue resistance $\varepsilon_6$ in μm/m | $\varepsilon_6 \geq 190$ | $\varepsilon_6 \geq 130$ | $\varepsilon_6 \geq 190$ | $\varepsilon_6 \geq 130$ |
| Stiffness $S_{mix}$ in MPa | Declare $S_{min}$-value Declare $S_{max}$-value | | | |
| Creep rate $f_{cmax}$ in μm/m/n | $f_{cmax} \leq 0.4$ | | $f_{cmax} \leq 0.6$ | |

## 6 Conclusions

Today, road pavements have to last longer and endure high traffic loads under challenging climatic conditions to be ecologically and economically sustainable. Moreover, traffic densities, axle loads and tire pressures will increase during the coming decades. To guarantee a long lifetime of flexible and semi-rigid pavement structures, optimising pavement materials and asphalt mixtures is getting more important to avoid damages and subsequently minimise costs for road maintenance.

Therefore, the prediction of in-service performance of road pavements during their lifetime is one of the main challenges of pavement research. For flexible pavements, the key performance characteristics are fatigue and low temperature, as well as permanent deformation behaviour at elevated temperatures. Enhanced test methods, so-called performance-based tests, to address these fundamental characteristics are implemented in the latest European standards. Fundamental or performance-based specifications for HMA may be implemented by the road authorities. These performance tests are used on the one hand to improve the mix design process of asphalt mixtures significantly. On the other hand, they provide material input parameters for numerical models employed to predict the in-service performance of flexible pavement structures more reliably.

In combination with enhanced binder tests, the implementation of performance-based asphalt mix specifications are the future way to create a sustainable road infrastructure for the decades to come.

# References

Blab, R., Eberhardsteiner, J.: Methoden der Strukturoptimierung flexibler Straßenbefestigungen (Performance-based optimization of flexible road structures). Progress Report of the CD Laboratory at the Institute of Road Construction, Vienna University of Technology, Vienna (2009). (in German)

Di Benedetto, H., Partl, M.N., Francken, L., De La Roche, C.: Stiffness testing for bituminous mixtures. J. Mater. Struct. **34**, 66–70 (2001).https://doi.org/10.1007/BF02481553

Findley, W.N., Kasif, O., Lai, J.: Creep and Relaxation of Nonlinear Viscoelastic Materials. Dover Publications Inc., Mineola (1989)

Hofko, B., Blab, R., Mader, M.: Impact of air void content on the viscoelastic behavior of hot mix Asphalt. In: Proceedings of the 2nd International Workshop on 4PB, University of California, CA (2012)

Hofko, B., Blab, R.: Enhancing triaxial cyclic compression testing (TCCT) of hot mix asphalt by introducing cyclic confining pressure. Road Mater. Pavem. Design **15**(1) 16–34 (2014)

Wistuba, M., Lackner, R., Blab, R., Spiegl, M.: Low-temperature performance prediction of asphalt mixtures used for long-life pavements – new approach based on fundamental test methods and numerical modeling. Int. J. Pavem. Eng. **7**(2), 121–132 (2006)

Austrian Transportation Research Society FSV: Performance Based Specifications for Hot Mix Asphalt Layers, RVS 08.16.06. Vienna, Austria (2018)

# Study on Rejuvenators and Their Effect on Performance Characteristics of an Asphalt Concrete Containing 30% Reclaimed Asphalt

Pavla Vacková[1]([⊠]), Majda Belhaj[1], Jan Valentin[1], and Liang He[2]

[1] Faculty of Civil Engineering, Czech Technical University in Prague University, Prague, Czech Republic
Pavla.vackova@fsv.cvut.cz
[2] School of Civil Engineering, Chongqing Jiaotong University, Chongqing, China

**Abstract.** The use of reclaimed asphalt has become a common practice and necessity in many developed countries whereas there are several options how to deal with elevated content of such material. Well-developed are solutions where reclaimed asphalt is cold fed by up to 20%. Higher portions of reclaimed asphalt are still challenging and need additional focus to become a regularly used standard. Quite common in last few years has been the use of rejuvenators, which are special chemical compounds helping to partly restore the degraded bitumen and secure sufficient performance behavior of asphalt mix containing higher portions of reclaimed asphalt. Recent study focused on use of several market-established rejuvenators and some newly developed variants, which are predominantly based on renewable materials. A typical asphalt concrete AC16 has been selected and 30% virgin aggregate was replaced by reclaimed asphalt of 0/8 mm grading. A scenario with 50% reclaimed asphalt was assessed too but is not part of this paper. Rejuvenators were added in a content between 5–7% of the degraded bitumen content present in the reclaimed asphalt. Functional tests have been performed focusing on water susceptibility, deformation behavior (stiffness) and resistance to crack propagation. At the same time bitumen extracted from the reclaimed asphalt was mixed with rejuvenator and selected bitumen tests were performed to find or confirm potential relations between binder and asphalt mix behavior.

**Keywords:** Asphalt mix · Reclaimed asphalt · Rejuvenator · Performance · Stiffness · Complex modulus · Moisture susceptibility

## 1 Introduction

Use of reclaimed asphalt (RA) in asphalt mixtures is a common trend in most countries, which can be characterised as technologically advanced with respect to road infrastructure development. Recycling of construction materials saves direct primary costs like new material purchasing and at the same time facilitates savings of other costs like energy, material disposal and transport as well. Those factors are connected to environment-friendly approaches, primarily in the case of emission, greenhouse gas production reduction, mitigation of waste disposal etc. (Zaumanis 2019). Material disposal can subsequently be reduced solely to waste or by-product materials, which

---

© The Author(s), under exclusive license to Springer Nature Switzerland AG 2022
A. Akhnoukh et al. (Eds.): IRF 2021, SUCI, pp. 46–56, 2022.
https://doi.org/10.1007/978-3-030-79801-7_4

have no way of further utilisation – e.g. the recycled material from asphalt pavements containing tar can be hardly reused in new asphalt mixture due to very high PAH emissions harming the environment and health (Devulapalli 2019).

Use of elevated RA content can be challenging from point of view of asphalt mixture characteristics. If the RA content is too high and the design is not properly tested, the final mixture can show worsen durability and lower resistance to water immersion, higher risk of thermal induced or fatigue-related cracks, increased brittleness etc. All these negative factors can be reduced or even eliminated by proper mix design and sufficient testing. The use of RA leads to rheological changes in asphalt mixture. The higher the added amount of recycled asphalt is, the stiffer and more brittle mixture behaviour can be expected. One option how to prevent this state is the utilization of suitable rejuvenators or softer bituminous binders (Mogawer 2012).

Generally, rejuvenators are plant or chemical bases additives used to rejuvenate the oxidized bituminous binder by increasing the maltenes constitutions of the RA bituminous binder. Hence, its viscosity and stiffness are reduced, and its ductility is increased. All the above mentioned actions and factors have allowed a significant increase in the amount of RA used in the asphalt mixture (Ali 2016).

## 2 Asphalt Mixture Variants

In this paper the results for asphalt concrete regularly used for binder layers with maximum particle size of 16 mm – $AC_{bin}$ 16 are presented. The mix variants had all to fulfill national product requirements which are specified by the technical standard CSN 73 6121 and follow categories given in EN 13108-1. Six different types of rejuvenators were applied in the mixture with 30% reclaimed asphalt to determine the individual influence of each additive. Some of them are commercially available, first three representatives mentioned in Table 1 are new experimental developments, which are firstly assessed by this study.

**Table 1.** Description of dosed rejuvenators

| Additive | Dosage of binder in RA | Base of the additive |
|---|---|---|
| A | 5, 7, 14% | Additive based on waste fats collected in waste water treatment plants and/or in slaughterhouses *(experimental product)* |
| B | 7, 10% | Additive based on refined vegetable rapeseed oil and waste fats *(experimental product)* |
| C | 7% | Additive based on refined vegetable rapeseed oil and used engine oil *(experimental product)* |
| RF | 6% | Oil-based additive *(commercial product)* |
| SR | 6, 12% | Pine wood extract (lignin), obtained from a natural renewable source - crude tall oil, which is a by-product of paper industry *(commercial product)* |
| AF | 7% | Cationic type surfactant based on phosphoric esters dissolved in fatty acid esters and paraffinic oils *(commercial product)* |

48  P. Vacková et al.

The rejuvenators were dosed to asphalt mix variants according to (i) nomograms from manufactures or (ii) nomograms based on binder testing (the rejuvenators were dosed in different concentration to bituminous binder recovered (distilled) from RA and tested for softening point and penetration whereas penetration was selected as the preferred characteristic). Additionally, the concentration of some of the rejuvenators varied to evaluate the amount of additive on asphalt mixtures properties.

The rejuvenators were used with bituminous binder 50/70 (softening point R&B = 48,8 °C; penetration @25 °C = 56 dmm). The same binder without any rejuvenator was used as a reference asphalt mixture.

All asphalt mixtures were produced with the same composition using the same input materials (aggregate, RA, filler and binder) in order to minimize the influence of these materials on variability of the resulting characteristics.

## 3 Test Results

The following characteristics were determined on each asphalt mix and are presented further in this paper:

- **Volumetric characteristics** (maximum bulk densities according to EN 12697-5, bulk densities according to EN 12697-6, voids contents of asphalt mixtures according to EN 12697-8);
- **Dynamic modulus** determined according to EN 12697-26 by non-destructive test method (4PB-TR) for selected test temperatures (0, 10, 20 and 30 °C) and frequencies (range of 0.1 Hz–50 Hz);
- **Stiffness modulus** determined according to EN 12697-26 by non-destructive test method (IT-CY) for selected test temperatures;
- **Resistance to thermal induced crack propagation** according to the modified test procedure based on EN 12697-44 (three-point bending test performed on semicircular test specimens).

### 3.1  Air Void Content

Air voids content is one of the essential parameters of an asphalt mixture, which undoubtedly influences all other performance characteristics. Air voids content should to be sufficiently high to avoid rutting, but at the same time sufficiently low to avoid the penetration of water and small particles into the road surface structure. The test specimen used for this test were prepared according to EN 12697-30 with impact compactor (2 × 50 blows from each side) at the temperature of 150 °C.

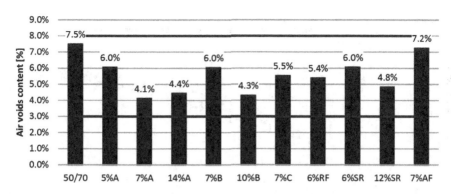

**Fig. 1.** Air void content results of $AC_{bin}$ 16 mix variants

Air voids content for $AC_{bin}$ 16 should be in the Czech Republic between 3.0 and 8.0% for control testing (CSN 73 6121 2019). Air voids content for the reference mixture with 50/70 paving grade and no rejuvenator was 7.5% (Fig. 1), while in case of rejuvenated mixtures, the parameter decreased for all tested variants. Thus, the results showed an improvement in workability of the variants with rejuvenators.

This decrease in voids content is variable depending on the type of the rejuvenator as well as its content within the mixture. For instance, the higher amount of the rejuvenator is, the less air voids have been found. On the other hand, for the same content (7% rejuvenator related to the bitumen content in RA), statistical results indicate a significant difference between the compared types of rejuvenators. Unpredictably, the addition of 7% of "A" showed a slightly lower air void than the use of 14% for the same rejuvenator.

When similar amount of rejuvenator was used (~7%), the mix variants reached similar voids content with two exceptions – rejuvenator A, which improved the workability if compared to others, and rejuvenator AF, which on the other hand evinced worsen workability. The results for the variant with 7% rejuvenator A are unusual, because they are out of expected line in order to two other concentrations.

### 3.2 Dynamic Test Results

The experimental results of the dynamic modulus $|E^*|$ vs. the reduced frequency $\delta$ for virgin (unaged) asphalt mixtures show that all the mixtures containing different rejuvenators, except the one with SR, are shifted to the right downward (decreased stiffness) across all reduced frequencies comparing the reference master curve. It is interesting, that all the added rejuvenators seem to have the same or very similar effect on the viscoelastic nature of the mixture at higher reduced frequencies/low temperatures since they all do have the same and superposed to each other curves, while at lower reduced frequencies/high temperatures the curves seem to differ slightly following this order: SR > RF > AF > C > A > B (Fig. 2).

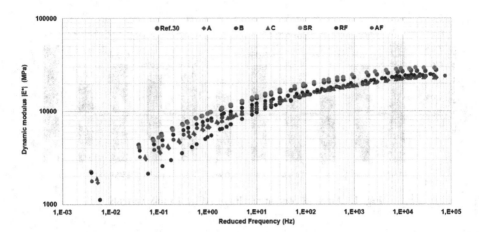

**Fig. 2.** Dynamic modulus master curves

The master curve of the mixture with SR had shown the highest stiffness for virgin mixture across all the reduced frequencies comparing all rejuvenators and higher stiffness comparing the referential mixture across the higher reduced frequencies.

Most of the rejuvenators seem to work, but some are more effective at reducing stiffness than others, such as rejuvenator B. In fact, the asphalt mixture with rejuvenator B has shown to be better at reducing stiffness over the whole course of the master curve, but also reduce the stiffness too greatly at high temperatures/low reduced frequencies, thus lowering the continuous grade range of the rejuvenated binder too much – this might be necessarily not a weakness. According to the dynamic modulus this rejuvenator acts effectively against the stiffening effect of aged binder in the reclaimed asphalt.

Stiffness is nevertheless only one aspect how rejuvenators should affect the behavior of aged binders, the other being the phase angle (to examine changes in viscoelasticity). Therefore, it is essential to review the phase angle master curves as well, see Fig. 3.

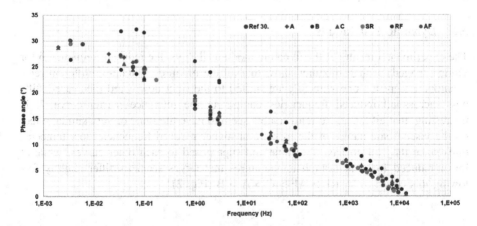

**Fig. 3.** Phase angle master curves

From the master curves of the phase angles, it is evident that the rejuvenator B has the highest phase angle compering to all the other rejuvenators as well as the referential mixture along all the frequencies. It systematically has a greater viscous component than the other mixes. On the other hand, results show that all the rejuvenators have similar results as the referential mixture at the higher reduced frequencies/low temperatures, while at the lowest reduced frequencies/highest temperatures, results show that only mixtures with rejuvenators RF and C have lower phase angles than the referential mixture. It can be seen as well that at the medium reduced frequencies, none of the mixtures with rejuvenators have lower phase angle than the referential mixture following this order: A > C > AF > SR > RF.

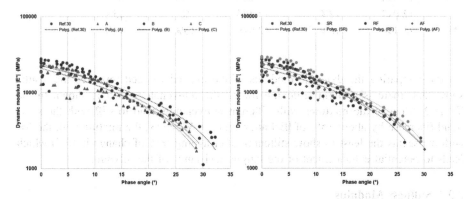

**Fig. 4.** Comparison of dynamic modulus and phase angle

Black space diagrams are very powerful since they do not require any shifting to create the curves from measured data at different temperature, like the case of the master curves, and still plotting the phase angle against the dynamic modulus |E*| regardless the test temperature. They can be expressly used to show if the phase angle is changing positively, or negatively as well if the stiffness is decreasing or increasing.

From the Black space diagram shown in Fig. 4, it is noticed that at high temperatures, the data from phase angle 25° to 30° and from |E*| 5000 MPa to 1000 MPa, the mixture with rejuvenator RF has shown a reduction in the phase angle at low |E*| values, which illustrates the softness of this additive in the mixture. While, at the low temperatures, phase angle 0° to 5° and |E*| 25 000 MPa to 20 000 MPa, the trend-lines show softening for all the results of the rejuvenators, yet the change in phase angles are different. The results are shared into two groups: rejuvenators such as SR tend to soft yet increase the phase angle, thus increasing the viscous nature of the binder at low temperature and rejuvenators from the second group such as A, B, C, RF and AF tend to soft the mixture and decrease the phase angle and thus makes the binder within the mixture more elastic in nature (Table 2).

52    P. Vacková et al.

**Table 2.** |E*| % Difference between rejuvenators and referential for unaged and aged mixture

| Name | Dynamic modulus at 20 °C, 10 Hz | | | | Change in dynamic modulus after ageing (%) |
|---|---|---|---|---|---|
| | Unaged | Unaged | Aged | Aged | |
| | Dynamic modulus (MPa) | Difference from ref (%) | Dynamic modulus (MPa) | Difference from ref (%) | |
| 50/70 | 13 710 | 0% | 15 166 | 0% | 10% |
| A | 10 541 | 23% | 12 371 | 18% | 15% |
| B | 9643 | 30% | 12124 | 20% | 20% |
| C | 10 579 | 23% | 12 589 | 17% | 16% |
| SR | 14 394 | −5% | 15 609 | −3% | 8% |
| RF | 12 047 | 12% | 14 143 | 7% | 15% |
| AF | 11 198 | 18% | 12 949 | 15% | 14% |

By comparing the dynamic modulus of the unaged to aged mixtures the results show that, there is a slight stiffening for all the rejuvenators as it is shown in the Table 2. Most of the mixtures with different rejuvenators tend to stiff with the same trend (increase by about 15% of |E*| after ageing). Whereas, the mixture with the SR additive seems the least to show stiffening after ageing (8% of change in |E*|) which leads to potential enhancement of the fatigue resistance of the mixture.

### 3.3   Stiffness Modulus

Stiffness modulus is a factor that can prevent early degradation of the asphalt mixture and the asphalt multilayer pavement system if reaching reasonable ranges. It is significantly important especially when softening or rejuvenating agents for the bituminous binder are used. The stiffness results for different temperatures are presented in the Fig. 5.

Variants containing rejuvenators had lower stiffness comparing to the reference mix regardless of the test temperature, which confirms the purpose and expectations of their use. This reduction of stiffness modulus due to rejuvenators grants better workability during mix production and shall have a positive effect on fatigue life. Generally, higher stiffness does not essentially mean that the asphalt mix will have a better overall performance and life-time. When the stiffness is too high (especially elevated RA content with degraded bitumen and without sufficient softening), the mixture can be more susceptible to cracking caused by low temperature of sudden climatic changes.

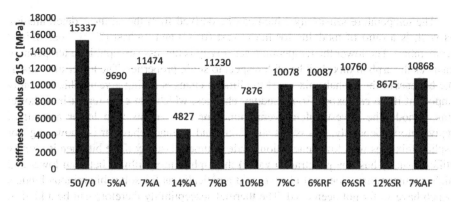

**Fig. 5.** Stiffness modulus results of AC$_{bin}$ 16 mix variants

As it was said the use of rejuvenators decreased the stiffness modulus. As same as when the higher amount of rejuvenator was used, the more significant was the decrease. Only the stiffness of variant with 7% of rejuvenator A caused the increase of stiffness. This mixture had even the lowest voids content, which means, that there probably were or might have been some problems during preparation of the mix variant/test specimens. For the other rejuvenators (B and SR), it applies, that increase in the rejuvenator content in the asphalt mix led to a less stiff mixture.

One of the primary objectives of the addition of a rejuvenator is to restore the maltenic part of the asphalt binder and deal with the brittleness of high-RAP mixtures at low temperatures. Results showed that the rejuvenator influenced positively the behavior of the mixtures at a low temperature (i.e. 0 °C), enlisting significant decrease in stiffness for higher rejuvenator content (Table 3).

**Table 3.** Stiffness modulus results of AC$_{bin}$ 16 mix variants

| Mix variant | Stiffness modulus [MPa] | | | Temp. sensitivity |
|---|---|---|---|---|
| | 0 °C | 15 °C | 27 °C | |
| 50/70 | 25338 | 15337 | 6574 | 3.9 |
| 5%A | 17995 | 9690 | 3593 | 5.0 |
| 7%A | 22965 | 11474 | 4470 | 5.1 |
| 14%A | 13178 | 4827 | 1624 | 8.1 |
| 7%B | 21363 | 11230 | 4098 | 5.2 |
| 10%B | 17078 | 7876 | 2563 | 6.7 |
| 7%C | 21558 | 10078 | 4236 | 5.1 |
| 6%RF | 20360 | 10087 | 3666 | 5.6 |
| 6%SR | 20938 | 10760 | 4522 | 4.6 |
| 12%SR | 18909 | 8675 | 3490 | 5.4 |
| 7%AF | 18502 | 10868 | 4599 | 4.0 |

The temperature sensitivity which can be derived from the stiffness determination as well, is a ratio of modulus for the lowest (0 °C) and highest (27 °C) determined temperature. The lower the temperature sensitivity is the less susceptible to temperature changes the asphalt mix shall be with respect to strain properties. The lower the value is the better will the asphalt mix resist to climatic changes during a year. This is mainly important if we can expect some sudden changes, which can even repeat in a shorter time-period. The lowest values were found for the variants with lower amount of rejuvenators. Generally, this is not surprising and similar behaviour is known e.g. for asphalt mixtures with low-penetration grade bitumen. Usually if the bitumen is already stiffer (usually has low penetration value) the asphalt mix stiffness between lower and higher temperatures does not change as dramatically as for softer bituminous binders which have so far not been aged. The thermal susceptibility therefore can be a kind of another indicator how well a rejuvenator is changing the properties of the degraded bitumen/asphalt mix (Fig. 6).

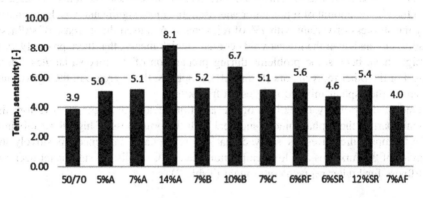

**Fig. 6.** Temperature sensitivity results of stiffness modulus of $AC_{bin}$ 16 mix variants

### 3.4 Resistance to Thermal Induced Crack Propagation

Resistance to thermal induced (frost) cracking was determined according to modified test method based on the standard EN 12697-44. The method was adjusted to meet the options of Czech technical environment. The modified test procedure is in detail described e.g. in (Vacková 2020), but the most important differences are: (i) test specimens are compacted according to EN 12697-30 with a diameter of 100 mm, (ii) loading rate is decreased to 2.5 mm/min and (iii) besides the calculation of fracture toughness determination of fracture energy till the maximum force and till the full crack propagation (including the unloading part) is provided as well.

The fracture toughness is not that easily expected as other strength characteristics – the stiffness in almost all cases decreases if rejuvenators are used, but the fracture toughness is influenced by brittleness more than stiffness of the specimen. For some applications of the selected rejuvenators the decrease is visible, but most of the variants reached very similar fracture toughness as the reference mix (Fig. 7).

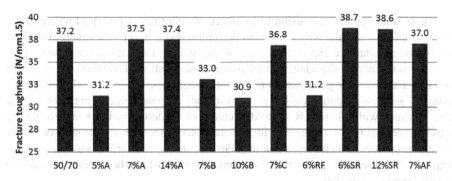

**Fig. 7.** Fracture toughness results of $AC_{bin}$ 16 mix variants

The differences in fracture energy values are more visible. This is one of the reasons and a clear justification, why the fracture energy was added to the test method, even though the EN standard does not include it. The fracture energy (mostly) increases with the increasing content of rejuvenator in an asphalt mixture. The highest values were reached by the variants with higher amount of rejuvenator, but the noticeable increase which was reached for the mixtures with A was not repeated for the other rejuvenators – the increase for twofold content of SR rejuvenator was only ∼16% and for the rejuvenator B even drop in the energy characteristics was determined (−20%) (Fig. 8).

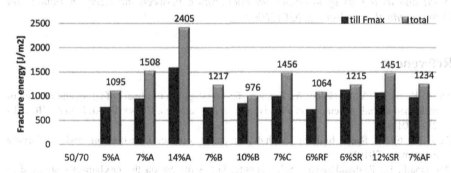

**Fig. 8.** Fracture energy results of $AC_{bin}$ 16 mix variants

## 4 Conclusion

The paper presents part of the research held on asphalt mixtures with different based rejuvenators (bio and oil). The required or expected properties of rejuvenators are not strictly defined (at least not in the Czech Republic), so the proper testing of final asphalt mixtures is crucial to understand the differences between the individual rejuvenators and to qualify their effect on asphalt mix performance. The study also covered experimentally prepared rejuvenators.

There is no general conclusion of influence of either bio or oil based rejuvenators. Every additive acts differently regardless the origin and it is very important to settle at the beginning what we expect from certain rejuvenator. For example, the best workability was reached by the additive A (if similar concentration of additives is compared) and also reached very good results of fracture toughness, but on the other the "softening" of stiffness modulus was lower than for other variants with concentration around 7%. Mixtures with additive B showed decrease of modulus (stiffness and dynamic), which is profitable for mixtures with elevated RA content, but also showed lower fracture properties, which means that the added value of lower brittleness has not been acquired. This this also applies to RF rejuvenators, with which the mixture evinced the lowest fracture properties.

If higher or lower concentration of rejuvenators is used the obvious trend is followed – the workability is improved (the air voids content decreases), the stiffness modulus decreases, but on the other hand the temperature sensitivity of stiffness modulus increases and the fracture properties is improved. If higher amount of rejuvenators is used the risk of rutting is increased, so it is necessary to pay attention to his property, but the rutting test was not part of this research study. Also higher amount of rejuvenator has significant financial impact on the final price of asphalt mixtures. So not only properties of asphalt mixture have to be in balance, but also financial part of the mixture is very important.

*This paper was elaborated within the activities of project no. SGS21/046/OHK1/1T/11 supported by The Czech Student Grant Competition (SGS) and as a part of international cooperation between the Czech Republic and China within the project no. 8JCH1002.*

# References

Ali, A.W., Mehta, Y.A., Nolan, A., Bennert, T.: Investigation of impact of aging and amount of reclaimed asphalt pavement on effectiveness of rejuvenators. Constr. Build. Mater. **110**, 211–217 (2016)

CSN 73 6121: Road building–Asphalt Pavement courses–Construction and conformity assessment (2019)

Devulapalli, L., Kothandaraman, S.K., Sarang, G.: A review on the mechanisms involved in reclaimed asphalt pavement. Int. J. Pavement Res. Technol. **12**(2), 185–196 (2019). https://doi.org/10.1007/s42947-019-0024-1

Mogawer, W., Bennert, T., Sias Daniel, J., Bonaquist, R., Austerman, A., Booshehrian, A.: Performance characteristics of plant produced high RAP mixtures. Road Mater. Pavement Des. **13**(1), 183–208 (2012). https://doi.org/10.1080/14680629.2012.657070

Vacková, P., Valentin, J., Belhaj, M.: What information can be provided by the asphalt crack propagation test done on semicylindric specimens? IOP Conf. Ser. Mater. Sci. Eng. **960**, 4 (2020)

Zaumanis, M., Boesiger, L., Kunz, B., Cavalli, M.CH., Poulikakos, L.: Determining optimum rejuvenator addition location in asphalt production plant. Constr. Build. Mater. **198**, 368–378 (2019). https://doi.org/10.1016/j.conbuildmat.2018.11.239

# Sustainable Use of Cool Pavement and Reclaimed Asphalt in the City of Phoenix

Ryan Stevens[(✉)]

The City of Phoenix, Phoenix, AZ, USA
ryan.stevens@phoenix.gov

**Abstract.** In 2020, The City of Phoenix began work on two projects to improve sustainability and livability: The Cool Pavement Pilot Program (CPPP) and Reclaimed Asphalt Pavement (RAP) trials.

Faced with long-term projections of rising urban temperatures and an increased frequency of dangerous heat waves, jurisdictions are seeking ways to reduce pavement temperatures to improve sustainability and resilience. The City initiated a pilot program to evaluate a colored pavement coating, building off previous work in Los Angeles. The City enlisted help from Arizona State University researchers and scientists for a comprehensive evaluation of the CPPP. The project includes thirty-six miles of pavement across the City. After extensive public outreach, installation occurred between June and October 2020. The CPPP resulted in unprecedented media interest, positioning Phoenix in the spotlight for technology and innovation both locally and nationally. Cool Pavement may measurably improve comfort and livability in the City.

In 2019, approximately twenty-one percent of all asphalt pavement in the United States included Reclaimed Asphalt Pavement (RAP). The City's stance on RAP recently changed because of improved methodologies for ascertaining the quality and performance of RAP and is exploring its use. The City has partnered with Arizona State University to study and test the feasibility of using RAP. Trial projects have been constructed with RAP in slurry seals, micro-surfacing and overlays. The City also developed two sites to create RAP from asphalt millings generated in day-to-day maintenance operations.

**Keywords:** Sustainability · Asphalt · Pavement · Maintenance · Urban heat island · Innovation · Community

## 1 Introduction

The City of Phoenix, Arizona is the fifth largest city in the United States. Approximately 1.7 million people reside in Phoenix, occupying an area of 1,340 $km^2$. As the capital of the State of Arizona, Phoenix is the centre of the Phoenix-Mesa-Scottsdale Metropolitan Statistical Area, which has a population of approximately 4 million residents. As such, Phoenix's street network is heavily utilized, with residents, commuters, and visitors traveling within, to, and through Phoenix as part of their daily activities.

As a large city, Phoenix's 7,800 km street network is critical infrastructure facilitating the safe transport of people, goods, and services using many modes of

© The Author(s), under exclusive license to Springer Nature Switzerland AG 2022
A. Akhnoukh et al. (Eds.): IRF 2021, SUCI, pp. 57–71, 2022.
https://doi.org/10.1007/978-3-030-79801-7_5

58     R. Stevens

transportation including vehicles, bicycles, public transit, and micro-mobility solutions. To maintain pavement infrastructure, Phoenix utilizes a Pavement Management System to prioritize preservation and rehabilitation treatments using available funding.

Always seeking to improve how things are done, Phoenix is actively investigating new processes and technologies, particularly in sustainability and addressing heat. As one of the hottest cities in the United States, Phoenix must actively seek solutions to address the issues of urban heat, including investigation how to cool the existing infrastructure. Phoenix has identified cool pavement coatings as one technology that can help address heating due to the urban heat island effect. Phoenix has worked with manufacturers, contractors, and researchers to undertake the Cool Pavement Pilot Program (CPPP) to explore cool pavement coatings. This pilot program is believed to be, currently, one of the largest in the world for this technology. In the CPPP, a light-coloured, polymer -modified asphalt sealcoat has been identified as a product which could allow the Phoenix to address urban heat. Sustainability is also important for Phoenix. By working with industry, Phoenix has identified that the use of Reclaimed Asphalt Pavement (RAP) in the surface layers of asphalt pavement can perform as well as asphalt mixes using virgin aggregate. Because RAP is a recycled material, its use will decrease the energy consumption and $CO_2$ emissions involved in the manufacturing process, making asphalt pavement more sustainable and environmentally friendly. This paper will discuss the steps used to develop these two important pilot programs, the challenges and opportunities presented during implementation, and report current progress.

## 2   Phoenix Street Network

Phoenix's 7,800 km street network includes arterial/collector streets and local streets. Table 1 details the average number of vehicular travel lanes and the Average Annual Daily Traffic (AADT) for the various street classifications throughout Phoenix.

**Table 1.**  City of Phoenix street classification parameters

| Street classification | Number of vehicular lanes | Width (m) | AADT |
|---|---|---|---|
| Arterial | 5–7 | 22.5–31.7 | 20,000–60,000 |
| Major Collector | 3–4 | 15.2–19.5 | 8,000–20,000 |
| Minor Collector | 2–3 | 12.2–15.2 | 1,500–8,000 |
| Local | 2 | 9.8–8.5 | 300–1,500 |

In the city's pavement management system, the street network is split into two sub-networks:

- Arterial Street Network: includes Arterial and Major Collector streets
- Local Street Network: includes Minor Collector and Local streets

Figure 1 shows how the overall street network is split into its sub-networks.

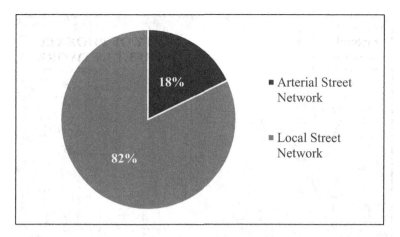

**Fig. 1.** Phoenix sub-network split

By using sub-networks, the pavement management system analysis can use separate cost/benefit models. Resurfacing specifications and funding resources can also vary based on the street classification.

Phoenix's street layout mainly follows a consistent grid layout, with the Arterial Street Network primarily aligned South to North and West to East at 1.6 km (one mile) increments. However, Phoenix's location in the Salt River Valley creates natural disruptions to the grid pattern in some areas. Each member of the City Council represents a specific region of Phoenix called a district. Thus, Phoenix is made up of eight Council Districts of approximately equal population, each with their own priorities, concerns, demographics, and character. Figure 2 shows the City of Phoenix street network and the eight Council Districts.

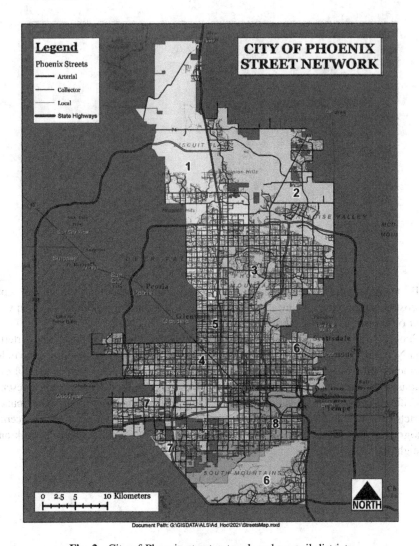

**Fig. 2.** City of Phoenix street network and council districts

## 3 Pavement Condition

Phoenix uses a pavement management system methodology to assess pavement conditions and perform a network-level analysis using various budget scenarios. Phoenix has used automated systems to measure pavement distresses since 2008 and in 2018, upgraded to an Automated Road Analyzer 9000 (ARAN). Pavement distress data are collected on a continual basis, taking approximately two years to complete data collection of both sub-networks. As pavement condition data are collected, the data are

Sustainable Use of Cool Pavement and Reclaimed Asphalt in the City of Phoenix    61

processed to rate, classify, and quantify the pavement distresses used for further analysis. Data are then exported to the pavement management system for further processing. An overall Pavement Condition Index (PCI) is also calculated to report general road conditions for comparison throughout Phoenix. The PCI is used as a composite index to easily communicate information about pavement quality to the public and decision makers. Using 2018 data as a baseline, the PCI scale and percent of the network in each condition state is shown in Table 2.

**Table 2.** Pavement condition distribution

| PCI Value | Condition | Amount of Street Network |
|---|---|---|
| 90–100 | Excellent | 0.58% |
| 70–89 | Good | 39.78% |
| 45–69 | Fair | 54.75% |
| 20–44 | Poor | 4.88% |
| 0–19 | Very Poor | 0.00% |

## 4 Preservation and Rehabilitation Treatments

Phoenix utilizes many preservation and rehabilitation treatments to maintain the pavement. The treatments 'toolbox' includes treatments to address streets from good to poor condition, to keep good streets in good condition, and restore streets in poor condition. The street classifications for which the various treatments can be used are demonstrated in Table 3.

**Table 3.** Pavement preservation and rehabilitation treatments used in Phoenix

| Treatment Name | Rehabilitation or Preservation | Applicable street classifications | | | |
|---|---|---|---|---|---|
| | | Arterial | Major Collector | Minor Collector | Local |
| Mill & Overlay | Rehabilitation | ✓ | ✓ | ✓ | ✓ |
| Microsurfacing | Preservation | ✓ | ✓ | | |
| Slurry Seal | Preservation | | | ✓ | ✓ |
| Chip Seal | Preservation | | | ✓ | ✓ |
| Sealcoat | Preservation | | | ✓ | ✓ |
| Fog Seal | Preservation | ✓ | ✓ | ✓ | ✓ |

## Mill & Overlay

Mill & Overlay is a street rehabilitation technique that requires the removal of the top layer of asphalt pavement using the grinding action of a large milling machine. After the top layer is removed, a new layer of pavement, approximately 2.5 cm thick, is placed on top of the milled pavement. Phoenix typically uses SBS polymer-modified and tire-rubber modified asphalt binders in the mill & overlay, with an effective asphalt binder grade of PG76–22. As a part of mill & overlay projects, curb ramps are evaluated and retrofit to meet American with Disabilities Act (ADA) standards where required.

## Microsurfacing

Microsurfacing is a mix of a polymer-modified asphalt emulsion, water, other chemicals, and small aggregate to replace fine aggregates lost and provide a sealed wearing surface to the existing street. Two mixes are primarily used: Type II with 4.75 mm maximum nominal aggregate size and Type III with 9.5 mm maximum nominal aggregate size. Microsurfacing is used for the Arterial Street Network. As a part of microsurfacing projects, curb ramps are evaluated and retrofit to meet current ADA standards where required.

## Slurry Seal

Like microsurfacing, slurry seal is a mixture of asphalt emulsion, water, chemicals and aggregate to provide a new wearing surface. Unlike microsurfacing, the asphalt emulsion breaks, or hardens, primarily by evaporation, not chemical processes. Two mixes are primarily used: Type II with 4.75 mm maximum nominal aggregate size and Type III with 9.5 mm maximum nominal aggregate size. Slurry seal is used on the Local Street Network.

## Chip Seal

Chip seal is a treatment where nearly uniform-sized aggregate is applied on top of a polymer-modified asphalt membrane. Phoenix uses a chip seal where the aggregate is pre-coated with asphalt binder. This is done to promote better aggregate retention. Chip seal is used in a high-volume and low-volume traffic formulation depending on pavement distresses and is used on the Local Street Network in industrial and residential settings. Two mixes are primarily used; Low Volume with 12.5 mm maximum aggregate size and High Volume with 19.0 mm maximum aggregate size.

## Sealcoat

Various asphalt sealcoat products are used to treat streets in good condition. The products address multiple distresses ranging from hairline cracking, early loss of fines, and initial oxidation. Sealcoat products are used on the Local Street Network, as the friction characteristics are not appropriate for the higher speed Arterial Street Network.

## Fog Seal

Fog seal is an asphalt emulsion used to prevent early oxidation. Traditional and polymer-modified fog seals with or without rejuvenator are used depending on the circumstance. As a liquid product, without fillers or aggregate, fog seal can be used on all street classifications.

## 5 Cool Pavement Pilot Program

Phoenix is one of the hottest large cities in the Southwest region of the United States, consistently reaching temperatures greater than 37 °C during the summer months. In 2020, Phoenix experienced 34 days where the high temperature was greater than 43 °C. Extreme hot temperatures can have a negative impact on people's health and causes an increased energy usage to power air conditioners for most of the day. Though this issue of heat is complex, Phoenix is examining how the built environment can be improved to help combat increasing heat, particularly due to the Urban Heat Island Effect (UHI), as part of a larger climate action plan.

Cool pavement technologies can be effective tools to make the built environment more heat resilient. Since Phoenix's roadway and building infrastructure is mostly built out, implementing cool pavement materials (such as pervious pavement) is cost prohibitive. Rather, Phoenix is exploring the use of cool pavement coatings on existing pavement as a tool to combat UHI. In Fall 2019, Phoenix participated in a workshop hosted by the Federal Highway Administration and the City of Los Angeles (LA) to learn about cool pavement coatings, and LA's experience with such technologies. Upon further discussions with industry and the material supplier, Phoenix decided to implement a series of trial projects using cool pavement coating technology. This series of projects is called the Cool Pavement Pilot Program (CPPP).

To develop the CPPP, it was important to understand both the technical and human parameters that would be important to select candidate streets. Since Phoenix occupies an area of 1,340 km$^2$, it was decided that projects in the CPPP would be spread throughout the city. In this way, the trials would be on streets that get different traffic levels, varying degrees of precipitation, varying degrees of natural or artificial shade and varying demographics. Since the coating is classified as a sealcoating treatment, it was also determined that candidate streets would be in good condition, having received a mill & overlay or slurry seal treatment within the previous two years. It was also decided that local streets or minor collector streets, which are closer to homes where nighttime human activities occur and cooling is desired, would be considered for the CPPP. Phoenix used the pavement management system and project records to identify the streets in good condition as candidates and made field visits to verify the roadway surfaces would be appropriate for sealcoating. From the candidate streets, the City Council offices and the Mayor identified the neighborhoods where there may be need for potential cooling and where residents would be receptive to living in CPPP areas for the trial. Additionally, a city-owned parking lot in a city park was selected to be included in the CPPP as a testbed for product application and direct comparison to a traditional, black-colored sealcoat product. The selected CPPP locations are shown in Fig. 3.

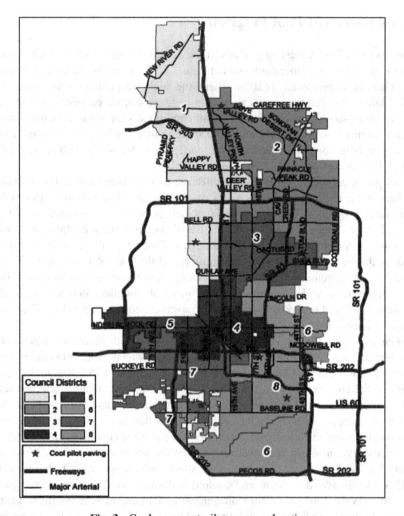

Fig. 3. Cool pavement pilot program locations

Phoenix also engaged with Arizona State University (ASU) to be research partners in the CPPP. ASU is responsible to taking the various temperature measurements and analysis that will allow Phoenix to assess if this technology is beneficial to the city and its residents. Across all nine pilot locations, approximately 528,000 m$^2$ of asphalt pavement was sealed with the cool pavement coating.

*Material and Application*
The sealcoat material used in the CPPP is a polymer-modified, asphalt-based sealcoat emulsion that is light in color. This is achieved by using polymers and different pigments to create a product that is gray instead of black. Once the product sets, it takes on a color like cured portland cement concrete. The high albedo of this material is the primary mechanism which reflects solar energy away from the pavement more than

dark-colored coatings. The material is asphalt-based, which makes it compatible with Phoenix's existing asphalt surfaces and adheres strongly to existing asphalt pavement surfaces and can be reused as RAP when milled for future application. The material has a Solar Reflectance Index (SRI) greater than 0.3; new asphalt pavement has an SRI generally less than 0.01; where the SRI is defined as a number from 0 to 1.0 that measure a material ability to reflect solar energy back into the atmosphere, being 0 total absorption and 1.0 total reflectance.

Prior to the Phoenix's CPPP, the material supplier recommended only a squeegee method of application. This had been the only method used in LA. This application method however limits productivity of the process and is labor intensive. The squeegee application also made it difficult to control and calibrate the application rate of the product which could result in uneven shades of the finished material on the pavement surface. The squeegee application enabled LA to apply approximately 96,150 $m^2$ of this material in 5 years. Considering the size of the CPPP, Phoenix determined that it was necessary to have the contractor work with the manufacturer and supplier to evaluate the material properties and construction practices and explore the possibility of utilizing a spray application method. The project team worked closely together to determine the best adjustment to the installation process and calibrated the material application rate. The material was applied to the pavement surface in two coats with a total application rate of 0.95 L per square meter. This allowed Phoenix to apply more material in a shorter amount of time, resulting in less impacts to residents due to construction restrictions. Additionally, spray application facilitated a verifiable, repeatable and reliable quality control and quality assurance for the project.

*Resident Notifications and Public Outreach*
Phoenix planned and undertook a thorough public engagement process to inform residents about the inclusion of their neighborhoods in the CPPP, educate them about the construction process, and the project outcomes, particularly that their streets would look grey rather than the darker appearance associated with traditional seal coats. A unique challenge arose, however, due to the global SARS-CoV-2 pandemic (COVID 19). Due to public health concerns, traditional in-person public meetings with residents about the CPPP would not be possible. Phoenix did not know how long COVID-19 would impact this type of public outreach, but knew it was important to continue with planned work and to engage with the public on the project. The decision was made to virtually inform and engage the public utilizing a WebEx platform and in addition to creating a project webpage. This was the first time the Street Transportation Department utilized a virtual platform for public engagement, which proved be effective. Lessons learned from this virtual public meeting were used to hold many additional meetings for other transportation-related project through the remainder of the pandemic and will remain an option beyond the pandemic. During construction traditional outreach methods were used including project signage at each site and physical door hangers for construction notices. In addition to these traditional public outreach tools, digital notices were issued on social media about the CPPP. Phoenix also utilized its social media platforms to steer the public to project-specific websites and virtual meeting events. The magnitude of the project made news in different television and digital press media.

## Construction

The CPPP began construction in June 2020; the first location was the parking lot of city-owned Esteban Park. Installation of the coating for the neighborhood streets occurred between August 2020 to November 2020. The treated pavement area for all CPPP locations is 528,627 $m^2$ and is shown in Table 4.

**Table 4.** Cool pavement pilot program location information

| Location Designation | Council District | Pavement Area Treated ($m^2$) |
| --- | --- | --- |
| Esteban Park | 8 | 12,709 |
| 2–19 | 7 | 55,162 |
| 11–22 | 4 | 80,409 |
| 11–30 | 8 | 45,803 |
| 18–16 | 5 | 39,939 |
| 26–28 | 3 | 61,993 |
| 29–37 | 3 | 84,934 |
| 33–18 | 1 | 50,946 |
| 59–23 | 2 | 96,732 |

The Esteban Park installation was unique from the other eight locations in the following ways:

- The location was a parking lot, not local or minor collector streets.
- The location was also treated with a traditional asphalt sealcoat as a comparison.
- The cracks in the asphalt were sealed due to the pavement condition compared to the neighborhood streets which didn't need crack sealing.
- The coating was applied using the buggy and manual squeegee method, not the spay application method used for the neighborhood streets.

## Data Collection and Analysis

To evaluate the effectiveness of the CPPP as a tool to mitigate the UHI, Phoenix has partnered with ASU to measure and analyze various data. The various data sets for treated and untreated streets include:

- Pavement surface temperatures
- Temperature gradient within the pavement
- Near-surface air and radiant temperatures
- Solar Reflection Index over time
- Resident experience and perception survey

A report of the findings from the ASU study is set to be released later in 2021, however, preliminary results indicate positive outcomes with significantly reduced surface temperatures, higher reflectance and small reduction in air temperatures. The formal results from the ASU are not yet available at the time of writing this paper.

City staff measured SRI and pavement surface temperature for five locations, outside of the analysis performed by ASU. The SRI measurements presented in Fig. 4

Sustainable Use of Cool Pavement and Reclaimed Asphalt in the City of Phoenix    67

were taken immediately after the coating was applied and compared to adjacent, uncoated streets. Surface temperature readings taken in the afternoons between August and November 2020, are presented in Fig. 5.

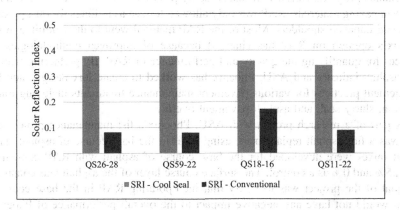

**Fig. 4.** Solar reflection index measurement at installation

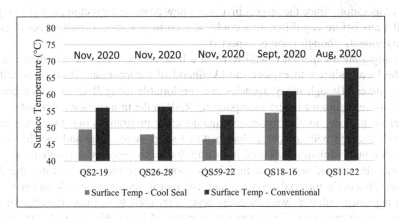

**Fig. 5.** Treated vs. untreated pavement surface temp

For the subset of five locations, roads treated with cool seal were cooler than the untreated roads by 7 °C on average. This temperature difference was also observed to be greater at higher ambient temperatures. Phoenix theorizes, that by cooling the asphalt pavement, the pavement will age less and need maintenance less frequently. As part of the ASU research, temperature readings will be taken at the top and bottom of the asphalt pavement to measure temperature gradients on streets with and without the cool pavement coating.

## 6 Reclaimed Asphalt Pavement in Phoenix

Phoenix is exploring the use of Reclaimed Asphalt Pavement (RAP) in its pavement maintenance program. Because of discouraging prior experience, RAP was prohibited for use in paving materials. RAP was only allowed for use as millings for dust proofing alleys and unpaved shoulders. Most of the RAP material went to the landfill as waste. Phoenix's opinion on RAP has changed because of improved methodologies and practices for quantifying the quality and performance of RAP. By partnering with the local asphalt industry and ASU, Phoenix has worked to create mix designs and best management practices for various pavement maintenance treatments including micro-surfacing, slurry seal, and asphalt pavement overlay.

As part of a research project with ASU, Phoenix's fist maintenance project with RAP was a full asphalt replacement, using RAP in the base course of asphalt. Three asphalt mixes were developed for the base course of asphalt with RAP contents of 15%, 25% and 0% as a control. The surface course layer of the asphalt was consistent. The goal of the project was to verify that incorporating RAP in the base course of asphalt would not have any negative impact to the overall performance of Phoenix's mix design for pavements. The surface course of this project, however, utilized virgin material with 0% RAP. The project was constructed in November 2018 and is still under observation. Since the project involves a new pavement structure, it will likely be several years before any differences could be seen in pavement condition between the three pavements, though none is expected.

Having gained experience with RAP in a base-course mix, Phoenix then worked with the local industry to explore the likelihood of incorporating RAP in its mill & overlay mixes. Though many agencies are comfortable using RAP as a base course, Phoenix is being innovative by incorporating RAP in the thin surface course of its mill & overlay projects in the pavement maintenance program. Prior to construction, Phoenix partnered with a local asphalt material supplier to create and test the performance of typical asphalt mixes used in its mill & overlay projects. An asphalt mix with RAP was also created and tested. Initial laboratory test results indicated that the RAP mix met Phoenix's material property and volumetric requirements. A balanced mix design approach utilizing two laboratory tests were used to evaluate the relative performance of all the mixes. This approach consists of Illinoi Flexibility Index Test (I-FIT) AASHTO TP124, which measure resistance to cracking, and The Hamburg Wheel-Track Test AASHTO T324 which measures the susceptibility to rutting. Generally, though higher asphalt grades make a mix more flexible and resistant to cracking, the tradeoff is that the mixes become more susceptible to rutting or permanent deformation. Lower binder grades make the mixes stiffer, hence more susceptible to cracking and resistant to rutting. The balanced mix design approach seeks to find the optimum binder content and properties that will make the mix best resist rutting and cracking. Four mixes were created to the same aggregate gradation, with different binder grades and types. One of the mixes was the 20% by weight of the virgin replaced with RAP material and 2.3% of the binder weight replaced with a rejuvenating agent. The interaction of the cracking index and rut depth is plotted to establish performance criteria ranges from 'acceptable' to 'high performance'. This allowed

Phoenix to compare the overall and relative performance and quality of the mixes. These criteria are defined in Table 5. Results from the testing are shown in Fig. 6.

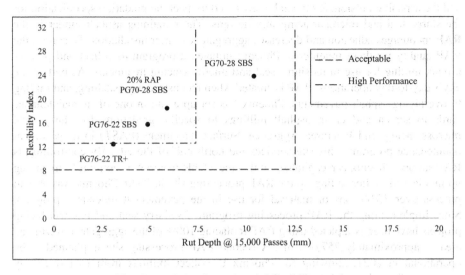

**Fig. 6.** Asphalt mix rutting and flexibility results

**Table 5.** Performance criteria thresholds for rutting and flexibility testing

| Level | Flexibility Index | Rut Depth (mm) 15,000 passes |
|---|---|---|
| High Performance | > = 15 | < = 7.5 |
| Acceptable | > = 8 | < = 12.5 |

Based on test results, all of Phoenix's available asphalt mixes fall in the 'acceptable' range. The proposed mix using 20% RAP fell in line with mixes that did not include RAP. Prior to conducting these tests, Phoenix's used the PG76–22 TR+ base asphalt in its overlay mixes which fell just short of the high-performance range because of its stiffness. Phoenix adopted the PG76–22 SBS binder which indicated high performance as its primary base binder in its overlay program. Though still acceptable, the PG70–28 binder had a greater rut depth, which is expected from a softer binder. Since the binder in RAP is aged, it is typically very stiff. So combining RAP and a rejuvenator in a mix with the softer binder decreases the rutting potential and moves the mix into the 'high performance' range, along with the newly adopted mix with PG76–22 SBS. Based on the encouraging test results, a 1.6 km project on a heavy industrial corridor was the first to use the RAP mill & overlay mix as a new wearing surface. Construction took place in October 2020; as of this writing the overlay is performing as well, matching or exceeding the performance of non-RAP mixes placed around the same time of year. Four additional projects of similar size using the 20% RAP mix will be constructed in 2021 to make sure Phoenix's material suppliers and contractors can produce and place the material consistently.

In addition to using RAP in asphalt pavement mixes, Phoenix has explored its use in microsurfacing and slurry seal treatments. RAP can be used to replace 100% of the aggregate in these treatments, since there is no interaction between the asphalt in RAP and the asphalt emulsion. RAP can be processed to meet the gradation specification for the slurry seal and microsurfacing mix designs. The remaining asphalt binder in the RAP encourages adhesion and decreases aggregate loss after installation. To control the RAP quality for these treatments, Phoenix initiated a program to collect and process asphalt milling for use in its slurry seal and microsurfacing treatments. As part of the city's day-to-day operations, RAP is created when reconstructing, patching, and milling to overlaying asphalt pavements. Phoenix has set up a site in one of its maintenance yards to screen and crush asphalt millings to specific gradations for slurry seal, microsurfacing, and Fractured Aggregate Surface Treatment (FAST) in its pavement maintenance program. This site services the north half of the city. By controlling its RAP supply, Phoenix can capitalize on this recyclable material, keeping it from ending up in a landfill. After setting up the RAP processing site in 2020, Phoenix was able to process over 7,700 tons of material for use in the pavement maintenance program. Since implementing the RAP processing program, 23 slurry seal and microsurfacing projects have been completed using RAP replacing 100% of the aggregate, covering an area of approximately 757,000 $m^2$. A second RAP processing site is planned to be operational in 2021, allowing for Phoenix to collect millings from its day to day operations from the south half of the city.

Using RAP has sustainable benefits because its use preserves natural resources as less new aggregates or asphalt need to be mined, it saves landfill capacity used for dumping the valuable RAP material, reduces the cost associated with producing new material, and reduces Green House Gas emission of asphalt pavement production. Additionally, Phoenix is projected to realize a 10% reduction in material cost when RAP is fully implemented enabling it to maintain more streets.

# 7 Conclusions and Future Work

While working to maintain its roadway infrastructure, Phoenix is also focused on making the city better by investigating cooling and sustainable materials to use in its pavement maintenance program. Initial results from the CPPP shows that cool pavement coatings can reduce the pavement surface temperature. This could be beneficial to the pavement itself, by decreasing thermal gradients in the pavement structure. Cool pavement coatings can also help reduce the urban heat island effect, improving safety and livability in Phoenix. Future research is being planned to quantify the effect on air temperature. Even a small decrease in air temperatures can cause decreases in energy savings and improve the health of residents. Phoenix will also be examining similar asphalt-based products, and different colors while still maintaining a high solar reflectance. Although cool pavement coatings are not the primary solution to increasing urban heat, when used in conjunction with other heat-mitigation efforts and policies cool pavement can be an effective tool to help cool down neighborhoods while keeping pavement in good condition. Phoenix is actively working with ASU to identify additional research topics for the CPPP Phase II, while ASU plans to public their research results from Phase I later in 2021.

To be more sustainable, Phoenix is being innovative by taking its asphalt millings, processing them to create RAP, and using the material in the asphalt surface treatments and rehabilitation treatments. Asphalt pavement is one of the most recyclable materials in the world. By accepting and expanding its use from to include surface layers and resurfacing treatments, Phoenix can make its paving program more sustainable. Phoenix is continuing to pilot RAP asphalt mixes in the mill & overlay program, with the goal of eventually using RAP asphalt mixes as a standard mix for pavement construction and rehabilitation. This will allow Phoenix to reduce the amount of virgin aggregate that needs to be mined, replacing up to 20% of the aggregate with RAP. For slurry seal and microsurfacing treatments, Phoenix will be finalizing the specifications for the use of RAP and seeks to use 100% RAP in all slurry seal and microsurfacing treatments within the next three years. By the end of 2021, Phoenix hopes to have a second RAP processing site operational so that all asphalt millings throughout the city can be collected and processed to create RAP.

# Optimizing the Adhesion Quality of Asphalt Binders Based on Energy Parameters Derived from Surface Free Energy

Allex E. Alvarez[⊠] and Lady D. Vega

Universidad Industrial de Santander, Bucaramanga, Santander, Colombia
allex.alvarez@uis.edu.co

**Abstract.** In the search to improve the performance and response of asphalt binders for the fabrication of asphalt mixtures, different additives are currently added either to the asphalt binder or the asphalt mixture, whose effect is mainly evaluated by macroscopic tests. However, less frequently these analyzes focus on the evaluation of the microscopic behavior of the mixture, its components, or their interaction. In this context, the objective of this study is to evaluate the feasibility of optimizing the adhesion quality of asphalt binder-aggregate interfaces for both modified- and residual-asphalt binders, based on the surface free energy (SFE) of the constituent materials. For this purpose, this study evaluated asphalt binders—here called as modified-asphalts—obtained from the mixture of neat asphalt binder and: (*i*) natural mineral filler, (*ii*) manufactured filler, and (*iii*) warm mix asphalt additives; in addition, residual-asphalts obtained from asphalt emulsions were analyzed. The SFE components of all asphalt binders were determined by means of the Wilhelmy plate method, whereas the Universal Sorption Device was used to determine the SFE of the five aggregates characterized. Analysis of four energy parameters—derived from the SFE components—suggests that optimization of the adhesion quality of asphalt binder-aggregate interfaces formed with modified-asphalt binders is feasible by using a specific optimal modifier dose. Similar results were reported for some aggregate-residual-asphalt combinations. Future research is recommended to verify at the asphalt mixture level the possible improvement in its quality of adhesion, based on the optimal doses here reported.

**Keywords:** Surface free energy · Adhesion quality · Modified asphalts ·
Asphalt residue · Pavements

## 1 Introduction

The roads constitute an important infrastructure element for the development and economic growth of each nation, since they not only connect the traffic of the different main cities but help the producers of each region with the delivery of their products to consumers all around the country. Worldwide, a large part of these roads are built employing flexible pavement structures, whose improvement suggests the need to continuously enhance parameters such as the performance of asphalt mixtures. In this search, it is reasonable to focus on the physical and chemical properties of both the

© The Author(s), under exclusive license to Springer Nature Switzerland AG 2022
A. Akhnoukh et al. (Eds.): IRF 2021, SUCI, pp. 72–86, 2022.
https://doi.org/10.1007/978-3-030-79801-7_6

asphalt binder (or asphalt) and aggregates, and assess their overall response, owing that the performance of the asphalt mixture highly depends on the adhesion quality developed between these materials.

In addition, it is relevant to note that to date, there are multiple materials added to asphalt binder which causes the modification of the neat asphalt binder properties (Kim 2009). These materials comprise the numerous types of mineral-filler (i.e., commonly defined as passing 200 sieve material; 0.075 μm), used in the fabrication of asphalt mixtures, which are present with different purposes like increasing stability and stiffness, along with the asphalt volume to reduce the optimal asphalt content, augment the rutting resistance, and modify the asphalt-aggregate adhesion quality in asphalt mixtures (Li et al. 2017; Little et al. 2017). These mineral fillers include those of natural (i.e., rock crushing dust) and manufactured origin (e.g., lime, Portland cement, ashes, slag, and sulfur), whose effect has been mainly analyzed through macroscopic approaches. In particular, sulfur has been studied again in recent years, after its use in paving mixtures a few decades ago. When combining sulfur with asphalt, an improvement in penetration, softening point, viscosity, and adhesion has been evinced, as well as an increase in the mechanical response of the corresponding asphalt mixtures in comparison to those that do not contain sulfur (Gawel 2000). On the other hand, the production of warm mix asphalt (WMA) involves the use of technologies that can be arranged into two main groups: additives and foaming. These technologies aim to decrease the viscosity of the asphalt, allowing the reduction of the fabrication temperature of the asphalt mixture compared to the temperatures required for production of conventional hot mix asphalt. Lastly, asphalt emulsions are an efficient alternative to produce cold asphalt mixtures as well as build both prime coats and tack coats, among other applications.

The aforementioned materials are analyzed in the present study within a common framework corresponding to the current need of identifying and applying solid technical criteria that allow determining their most appropriate dose for the fabrication of asphalt mixtures. In this context, the objective of this work focuses on evaluating the feasibility of optimizing the adhesion quality of asphalt-aggregate interfaces formed with modified- and residual-asphalts, by means of the materials surface free energy (SFE). In the work, modified-asphalts are termed as those asphalts added with natural mineral filler, manufactured filler (i.e., lime and sulfur), and WMA additives. In this fashion, the goal is to strengthen the quality of the modified asphalt binders, while making the best use of the employed materials, and potentially provide additional criteria to develop asphalt mixtures with improvements in the asphalt-aggregate adhesion quality.

This study is based on previous research, conducted in the exploratory scheme of the dose optimization of different materials additioned to asphalt binders. These previous efforts include the optimization of the addition rate of sulfur (Alvarez et al. 2017), natural mineral fillers (Alvarez et al. 2019b), and emulsifiers for the manufacture of asphalt emulsions (Alvarez et al. 2019a). On the basis of the mentioned approaches, in this continuation study the analysis related to optimizing the addition of both natural and manufactured fillers (i.e., sulfur) as well as residual-asphalts was unified.

In addition, for the first time, the possible optimization of the inclusion of lime as a manufactured filler is explored. For this purpose, the calculations of the energy parameters—derived from SFE—were merged through a common set of five aggregates aiming to bestow comparisons of the different materials across the analyzes.

## 2 Methods and Materials

This section presents some basic theoretical elements about SFE and describes the determination of the SFE components, computation of energy parameters—derived from the SFE components of asphalt and aggregates—, and the materials used in the study. The SFE is a thermodynamic property defined as the minimum amount of energy that a material requires to be fractured and create a new surface of unit area under vacuum conditions; this property endorses the evaluation of adhesion quality between different materials (Bhasin and Little 2007). According to the Good-Van Oss-Chaudhury theory, the total SFE of a material, $\Gamma$, is described by two components: (i) a non-polar ($\Gamma^{LW}$), and (ii) a polar component ($\Gamma^{AB}$), where the latter is made up of an acid ($\Gamma^+$) and a basic ($\Gamma^-$) monopolar component. In regards to this theory, the total SFE of a material can be calculated using the Eq. (1) (Van Oss 1994).

$$\Gamma = \Gamma^{LW} + 2\sqrt{\Gamma^+ \Gamma^-} = \Gamma^{LW} + \Gamma^{AB}(J/m^2) \tag{1}$$

In this work, the Wilhelmy plate method was used to determine the asphalt SFE components. This method encompasses the determination of the contact angle between the asphalt (thin film coating a glass plate) with different probe liquids (i.e., water, ethylene glycol, formamide, and glycerin) whose SFE components are known (Hefer et al. 2006). On the other hand, the SFE components of the aggregates were determined by means of the Universal Sorption Device (USD) through the implementation of the method suggested by Bhasin and Little (2007).

Once the SFE components of both asphalt binders and aggregates have been obtained, it is possible to calculate the four energy parameters analyzed in this work. The first parameter is the work of adhesion in dry condition $\left( W_{AB}^{dry} \right)$ (Eq. 2); this parameter quantifies the amount of energy that must be administered when propagating an existing crack at the interface of two materials (e.g., asphalt binder and aggregate) and creating two surfaces of unit area under an ideal moisture free condition. In Eq. (2), the sub-indices A and B refer to the asphalt and aggregate, respectively (Bhasin and Little 2007).

$$W_{AB}^{dry} = 2\sqrt{\Gamma_A^{LW} \Gamma_B^{LW}} + 2\sqrt{\Gamma_A^+ \Gamma_B^-} + 2\sqrt{\Gamma_A^- \Gamma_B^+}(J/m^2) \tag{2}$$

Secondly, with the work of adhesion in wet condition $\left( W_{ABW}^{wet} \right)$, parameters closer to reality are analyzed since asphalt mixtures get in contact with water due to climatic variables. Through Eq. (3), the susceptibility of the asphalt-aggregate interfaces to the loss of adhesion can be calculated due to the water disruptive effect at the interface—water is labeled in this equation with subindex W—. In addition, to calculate the work of adhesion in wet condition, the interfacial energies $\gamma_{AW}, \gamma_{BW}$, and $\gamma_{AB}$ were determined by application of the Eq. (4). For the work of adhesion in wet condition, negative results are usually attained, indicating that in the presence of water the interface is prone to be disrupted without application of external energy (Bhasin et al. 2006).

$$W_{ABW}^{wet} = \gamma_{AW} + \gamma_{BW} - \gamma_{AB}(\text{J}/\text{m}^2) \tag{3}$$

$$\gamma_{ij} = \Gamma_i + \Gamma_j - 2\sqrt{\Gamma_i^{LW}\Gamma_j^{LW}} - 2\sqrt{\Gamma_i^+\Gamma_j^-} - 2\sqrt{\Gamma_i^-\Gamma_j^+}\,(\text{J}/\text{m}^2) \tag{4}$$

The $ER_1$ index (Eq. (5)), relates the work of adhesion in dry condition and the work of adhesion in wet condition. This parameter allows assessing the proneness to deterioration of the asphalt-aggregate interface in the presence of moisture. In a previous work (Bhasin et al. 2007), it was elucidated that high values of this parameter are associated with a better interaction between the materials and a greater resistance to moisture damage. Finally, the spreading coefficient ($SC$), which is calculated through the Eq. (6), measures the wettability or coating capacity of the asphalt over the aggregate (Salazar et al. 2014). Corresponding computation includes the asphalt work of cohesion, $W_{AA}$,—computed through Eq. (2)—, which indicates the molecular attraction existing between the particles of the material.

$$ER_1 \text{index} = \left| \frac{W_{AB}^{dry}}{W_{ABW}^{wet}} \right| \tag{5}$$

$$SC = W_{AB}^{dry} - W_{AA}(\text{J}/\text{m}^2) \tag{6}$$

Regarding the materials used in the study, Table 1 presents the five aggregates employed and a brief geological description of these materials. Moreover, five asphalt cements from Ecopetrol S.A. refinery in Barrancabermeja (Colombia) were used, which classified as penetration asphalts (60–70 $^1/_{10}$ mm) based on the specifications from the Instituto Nacional de Vías (INVIAS 2013). These asphalts are included as different materials, based on representative differences exhibited by their SFE components (Table 1) and their variation in energy parameters derived from their combination with the aggregates used. The SFE of the aggregates and the neat asphalt binders have been earlier quantified (Alvarez and Caro 2009; Alvarez et al. 2019b). Table 2 shows the asphalt cement modifiers (i.e., natural filler, manufactured filler, and WMA additives), and the materials handled to produce the asphalt emulsions evaluated in this investigation. These emulsions were prepared from BAR 5 asphalt. Upon fabrication, the residual-asphalt was recovered following the protocol described in the ASTM D7497–07 standard (furnace evaporative technique) (ASTM International 2016).

**Table 1.** Total SFE and SFE components of the aggregates and asphalts used

| Convention | | $\Gamma$ (J/m$^2$) | SFE components (J/m$^2$) | | | Aggregate description/asphalt production |
|---|---|---|---|---|---|---|
| | | | $\Gamma^{LW}$ | $\Gamma^+$ | $\Gamma^-$ | |
| Aggregates | ALB | 0.0554 | 0.0359 | 0.0003 | 0.3635 | Gravel, with predominance of sandstone and siltstone |
| | BCM | 0.1554 | 0.0416 | 0.0042 | 0.7652 | Gravel, with predominance of sandstone and mudstone |
| | GUA | 0.1222 | 0.0428 | 0.0021 | 0.7372 | Gravel, with predominance of sandstone and shale |
| | RIO | 0.1125 | 0.0371 | 0.0025 | 0.5630 | Gravel, with predominance of amphibolite |
| | RIS | 0.2121 | 0.0521 | 0.0035 | 1.8548 | Gravel, with predominance of basalt |
| Asphalts | BAR 1 | 0.0134 | 0.0127 | 0.0000 | 0.0090 | Asphalt produced in 2013 |
| | BAR 2 | 0.0176 | 0.0166 | 0.0001 | 0.0019 | Asphalt produced in 2015 |
| | BAR 3 | 0.0115 | 0.0000 | 0.0000 | 0.0000 | Asphalt produced in 2013 |
| | BAR 4 | 0.0166 | 0.0176 | 0.0001 | 0.0019 | Asphalt produced in 2015 |
| | BAR 5 | 0.0131 | 0.0083 | 0.0034 | 0.0017 | Asphalt produced in 2016 |

**Table 2.** Description of both modifiers and materials used for the manufacture of asphalt emulsions

| Modifier | Convention | Description |
|---|---|---|
| Natural mineral filler | Filler 1 | Gravel, with predominance of sandstone and lutite |
| | Filler 2 | Gravel, with predominance of marl and limestone |
| | Filler 3 | Quartz-latite, latite, and tuff |
| Manufactured filler | Lime | Hydrated lime, 85% purity |
| | Sulfur | Conventional commercial product, solid phase |
| Additives for WMA production | Carnauba | Wax of natural origin obtained from the leaves of the Brazilian Copernicia cerifera palm tree |
| | Sasobit$^®$ | Chemical additive: synthetic paraffin wax produced through the Fischer-Tropsch process |
| | Evotherm$^®$ | Chemical additive designed to enhance coating, adhesion, and workability |
| Materials employed in asphalt emulsions | AP | Adhesion promoter additive |
| | PP | Cationic latex polymer-SBR |
| | E1 | Emulsifier type 1: N-fatty alkyl polypropylene polyamines |
| | E2 | Emulsifier type 2: amines, N-tallow alkyl trimethyl endy amines, N-alkyl-tallow trimethyl endy, ethoxylated 2-ethylhexaonic acid amines, tallow alkyl |

Table 3 presents the volumetric concentrations of the mineral filler used to fabricate the mastics (asphalt-mineral filler combination); these mastics were generated using BAR 1- and BAR 2-asphalts. The volumetric concentration ($VC$) is defined, in accordance with Eq. (7), from the filler volume ($V_f$) and the asphalt volume ($V_a$). Likewise, Table 4 displays the modifier to asphalt mass ratio ($m_m/m_a$) of WMA additives that were combined with BAR 3- and BAR 4-asphalts, along with different combinations and doses of materials used to produce the asphalt emulsions.

$$VC = \frac{V_f}{V_f + V_a} \tag{7}$$

**Table 3.** Volumetric concentrations of mineral-filler evaluated

| Modifier | Convention | Volumetric concentration | |
|---|---|---|---|
| | | BAR 1 Asphalt | BAR 2 Asphalt |
| Natural mineral filler | Filler 1 | 0.12, 0.18, 0.24, 0.38 | 0.15, 0.18, 0.28, 0.38 |
| | Filler 2 | 0.12, 0.18, 0.24, 0.38 | 0.19, 0.28, 0.38 |
| | Filler 3 | 0.13, 0.19, 0.25, 0.39 | 0.17, 0.21, 0.29, 0.39 |
| Manufactured filler | Lime | 0.03, 0.07, 0.10, 0.17, 0.24 | 0.05, 0.10, 0.17, 0.24 |
| | Sulfur | 0.00, 0.05, 0.11, 0.25, 0.34 | – |

**Table 4.** Doses of WMA additives and emulsifiers

| Modifier | Dose (%) | |
|---|---|---|
| Carnauba wax | 1, 3, 5 | $m_m/m_a$ |
| Sasobit® | 1, 3, 5 | |
| Evotherm® | 0.1, 0.3, 0.5 | |
| E1 | 0.2, 0.3, 0.4 | Emulsifier dose per total mass of asphalt emulsion |
| E1 + AP | 0.2, 0.3, 0.4 | |
| E1 + PP | 0.2, 0.3, 0.4 | |
| E2 | 0.2, 0.3, 0.4 | |
| E2 + AP | 0.2, 0.3, 0.4 | |
| E2 + PP | 0.2, 0.3, 0.4 | |

# 3 Results and Analyzes

The results of the study are presented below based on the four energy parameters analyzed: (*i*) work of adhesion in dry condition, (*ii*) work of adhesion in wet condition, (*iii*) $ER_1$ index, and (*iv*) spreading coefficient.

## 3.1 Work of Adhesion in Dry Contidion

Figure 1 presents the work of adhesion in dry condition values for some of the modified- and residual-asphalt combinations (see Table 2). These findings suggests that for each modified- and for most of the residual-asphalts, exists an optimal dose of additive and emulsifier, respectively, for which a maximum value of work of adhesion in dry condition is accomplished (i.e., optimum adhesion quality). This maximum value is critical for the lifespan of asphalt mixtures, since high values of this parameter are directly associated with asphalt mixtures with higher resistance to fracture (i.e., adhesive failure) at the asphalt-aggregate interface (Cong et al. 2017). Similar results were registered for the remaining material combinations evaluated—information not shown in the figure owing to space constraints—.

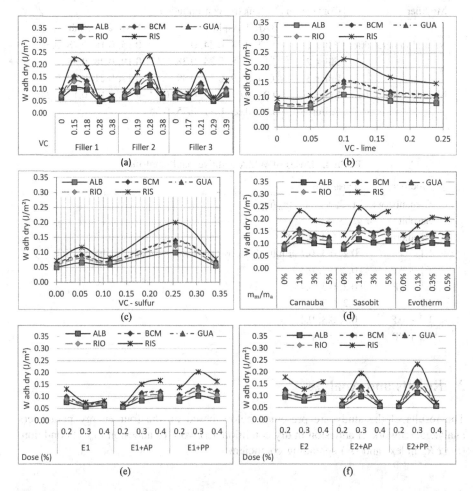

**Fig. 1.** Work of adhesion in dry condition for the interfaces formed with the modified-asphalts: a) BAR 2-natural filler, b) BAR 2-lime, c) BAR 1-sulfur, d) BAR 3-WMA additive and the residual asphalt of the emulsions fabricated with the emulsifier: e) E1, and f) E2.

For the materials combinations that include modified asphalts, at the optimal doses of modifier the work of addition in dry condition was significantly improved, doubling, or even tripling their values compared to the neat asphalt binder; such is the case, for example, of combinations that include sulfur (Fig. 1c). Notably, the BAR 1-sulfur-RIS combination without modifier (i.e., volumetric concentration equals to 0), exhibited a work value of 0.073 $J/m^2$, whereas with an optimal volumetric concentration of 0.25, a response of 0.2 $J/m^2$ was computed, achieving an increase of 274%. In addition, combinations including modified asphalts were rendered where the work of adhesion in dry condition assessed for some doses other than the optimal, diminished as compared to the material without the modifier.

On the other hand, in the residual-asphalts recovered from the emulsions that include the adhesion promoter and polymers, optimal values of work of adhesion in dry condition and their respective optimal doses were successfully identified. For instance, for combination BAR 5-E2 + PP-RIS, the work of adhesion in dry condition increment was 335% when changing the emulsifier dose from 0.2 to 0.3%. Nevertheless, for the combination of materials that solely employed the emulsifier for the manufacture of emulsions, an optimal value was not obtained and the work of adhesion in dry condition dropped. Namely, for the BAR 5-E1-RIS combination, the energy parameter value was 0.131 $J/m^2$ with an emulsifier dose of 0.2% and plummeted to 0.075 $J/m^2$ when rising to a dose of 0.3%—reduction of 42.7%.

On the other hand, the data presented in Fig. 1 suggest that the dose of a particular additive leading to the optimal value of work of adhesion in dry condition remains the same for all the aggregates; despite that, these optimum values show differences. As an example, while the RIS aggregate led to the best adhesion quality at the interface, the ALB had the poorest results.

## 3.2 Work of Adhesion in Wet Condition

Figure 2 introduces the work of adhesion in wet condition results for some materials combinations including both modified- and residual-asphalt binders. These data also allows to identify an optimal dose of mineral-filler, lime, sulfur, and WMA additive that leads to an optimum adhesion quality. This adhesion quality is reflected by a minimum absolute value of work of adhesion in wet condition that is related to a minimal moisture damage susceptibility at the asphalt-aggregate interface. Similar trends were determined for the other materials combinations evaluated; nonetheless, owing to space restrictions, these are not shown.

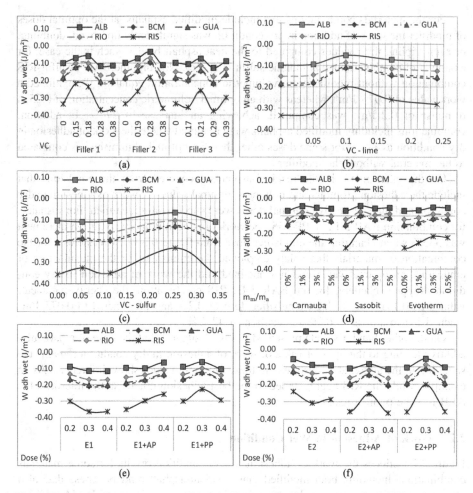

**Fig. 2.** Work of adhesion in wet condition for the interfaces formed with the modified asphalts: a) BAR 2-natural filler, b) BAR 2-lime, c) BAR 1-sulfur, d) BAR 3-WMA additive and the residual asphalt of the emulsions fabricated with the emulsifier: e) E1, and f) E2.

For residual asphalts, the data presented in Figs. 2(e) and 2(f) suggests that the optimal dose was given for combinations E1 + PP, E2 + AP, and E2 + PP, with an emulsifier content of 0.3%. This dose is consistent with that obtained for the work of adhesion in dry condition. On the other hand, for residual asphalts corresponding to emulsions produced merely with emulsifiers E1 and E2, an optimal value was not reported, and the absolute value of the work of adhesion in wet condition increased, representing a greater susceptibility to moisture damage. At the same time, according to the data presented in Fig. 2(e), for combination E1 + AP, although an optimum could not be distinguished with the dose used, an improvement was noted in the response quantified by this energy parameter.

Similarly, when evaluating the work of adhesion in wet condition for certain doses else than the optimal, the response of some materials combinations including modified asphalts lessen compared to the combination including the neat asphalt binder. These results concur with previously reported data (Prowell et al. 2005), suggesting that the addition of some mineral-fillers might boost the susceptibility to moisture damage of hot mix asphalt mixtures, while endorsing the need for analysis of the doses and combinations of materials for the fabrication of asphalt mixtures. In addition, the ALB aggregate provided the best material responses, resulting in the lowest absolute values of work of adhesion in wet condition throughout the studied materials combinations. Inversely, the RIS aggregate revealed responses associated with the lowest resistance to moisture damage.

### 3.3 $Er_1$ Index

Figure 3 shows the $ER_1$ index results for some of the modified- and residual-asphalt binders studied. The $ER_1$ index data suggests that for each modified asphalt, there is an optimal additive dose (e.g., mineral-filler), reflected in a maximum index value. These $ER_1$ index values are indicative of modified-asphalt-aggregate interfaces with high resistance to both fracture and moisture damage. Previous work (Bhasin et al. 2006) suggested that $ER_1$ index values below 0.5 are related to hot mix asphalt mixtures with a high susceptibility to moisture damage, whilst values above 1.5 are associated with a low proneness to the same distress. It is worth to highlight that although not all the figures of the analyzed materials are herein disclosed due to space limitations, these also validate the current results.

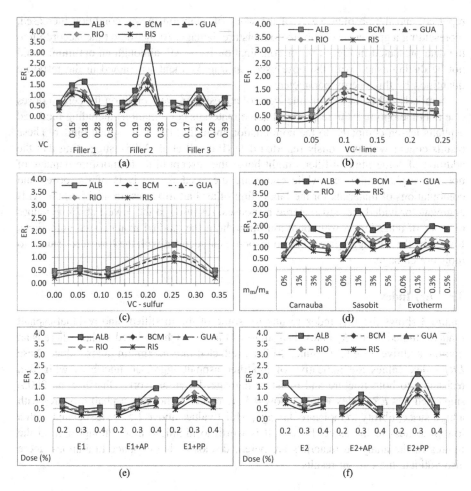

**Fig. 3.** $ER_1$ index for the interfaces formed with the modified asphalts: a) BAR 2-natural filler, b) BAR 2-lime, c) BAR 1-sulfur, d) BAR 3-WMA additive, and the residual asphalt of the emulsions fabricated with the emulsifier: e) E1, and f) E2.

Regarding material combinations including the residual asphalts, they performed in a similar way to that discussed in the former energy parameter, with a decrease in response for the combinations that included the emulsifiers, E1 and E2. This premise can be observed, for instance, in reviewing the results of these combinations with aggregate ALB, for which a reduction of 56% and 48%, respectively, was detected when the emulsifier dose was incremented from 0.2 to 0.3%. Optimal values of emulsifier dose for material combinations including E1 + PP, E2 + AP, and E2 + PP, were also identified at the 0.3% of emulsifier dose.

## 4 Spreading Coefficient

Figure 4 presents the spreading coefficient values determined for some of the modified- and residual-asphalt binders analyzed. Analogous to the previous energy parameters, for each additive there is an optimal dose, conducive to a maximum value of the spreading coefficient, which denotes an optimal capacity of the modified asphalt to coat the aggregate surface. The responses of materials combinations not included, due to space limitations, are like those here presented and confirms the analyzes.

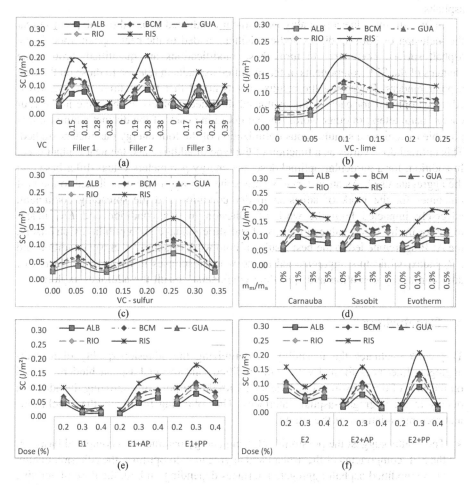

**Fig. 4.** Spreading coefficient for the interfaces formed with the modified asphalts: a) BAR 2-natural filler, b) BAR 2-lime, c) BAR 1-sulfur, d) BAR 3-WMA additive and the residual asphalt of the emulsions fabricated with the emulsifier: e) E1, and f) E2.

Nevertheless, for the residual asphalts (Figs. 4(e) and 4(f)) obtained from emulsions that contain materials E1, E1 + AP and E2, an optimal spreading coefficient value was not attained for the analyzed doses. In addition, a diminishment in the spreading coefficient was observed for combinations including emulsions fabricated with emulsifiers E1 and E2. Based on these results, additional research is further recommended to explore emulsifier contents lower than those here studied. In addition, for the energy parameters previously analyzed, its maximum—optimum—value changed when modifying the aggregate. This response was also obtained for the spreading coefficient. For this parameter, the RIS aggregate led to the best outcome along all the modified asphalts.

In overall, the adhesion quality assessment conducted is based on the premise that the inclusion of both modifiers and emulsifiers, added to the neat asphalt, modify the asphalt surface chemistry. In turn, these modifications are quantified through the changes of the SFE components of the modified asphalts in relation to those on the neat asphalt binders. Regarding the results, it was noted that the additive dose acknowledged as optimal for each of the evaluated modified asphalts, is the same in the four energy parameters; in other words, by incorporating each of the distinct additives, it is possible to achieve a simultaneous optimization of the four energy parameters studied. This optimal dose varies as a function of the type of additive and the neat asphalt used. In addition, for a particular optimal dose, the energy parameters values differ based on the aggregate used. As an instance, whereas the RIS aggregate contributed to the best outcomes for the work of adhesion in dry condition and the spreading coefficient, adversely impacted both the work of adhesion in wet condition and $ER_1$ index, resulting in the worst values in terms of moisture damage susceptibility. Conversely, the ALB aggregate led to the best results of both work of adhesion in wet condition and $ER_1$ index, while this aggregate led to the worst responses in terms of both work of adhesion in dry condition and spreading coefficients for all the additives explored.

## 5 Conclusions

This work evaluates the feasibility of optimizing the adhesion quality of asphalt-aggregate interfaces formed with both modified- and residual-asphalt binders—based on the SFE of the constituent materials—, with the purpose of improving the adhesion quality at the aggregate-asphalt interfaces. The following highlights the conclusions and recommendations of the study:

- The results suggest the possibility of using energy parameters—computed based on the SFE components of the asphalt and aggregate—to optimize the adhesion quality at the modified asphalt-aggregate interfaces depending on both the dose of additive and type of additive. In addition, it was proved that for each of the evaluated additives, the optimal dose determined from the four energy parameters analyzed is the same, indicating the potential of a simultaneous improvement in the different adhesion quality parameters.
- The additives and asphalt binders studied presented optimal doses that contributed to enhancements in the adhesion quality of the aggregate-modified asphalt

interfaces. These improvements lead to greater resistance to adhesive failure, as well as less susceptibility to moisture damage, and better aggregate coating capacity.

- Despite the aggregate type did not modify the optimal additive dose, when changing the aggregate used there was a clear variation in the values achieved by the energy parameters (i.e., there is a relative change in the adhesion quality according to the aggregate used).
- Regarding the residual asphalts recovered from the asphalt emulsions, it was possible to differentiate emulsifier contents that permit optimizing the adhesion quality developed with the aggregate. However, some aggregate-residual asphalt combinations were identified, for which an increase in the proportion of emulsifier yielded to a decrease in adhesion quality. These combinations should be hallmark of further investigations in the light of understanding these specific responses.
- The results encourage the need of future work to assess whether the optimal doses identified from energy parameters—analysis of asphalt-aggregate interfaces—contribute to optimal responses in terms of performance at the level of asphalt mixtures.

**Acknowledgements.** The authors express their gratitude to Iván J. Paba for his contribution in the fabrication of the asphalt emulsions that were analyzed in the present study. In addition, the authors would like to acknowledge Katy L. Gómez, Diana C. Gómez, Andrea C. Rodríguez, Edgardo J. Díaz, Oriana P. Daza, Sharick M. Vides, Leidy V. Espinosa, Ricardo A. Mejia, and Asmirian M. Perea for their contribution in the laboratory work required toward the SFE measurements of the asphalt materials.

# References

Alvarez, A.E., Caro, S.: Determinación de la energía superficial libre de cementos asfálticos colombianos. Ingeniería e Investigación **29**(2), 20–24 (2009)

Alvarez, A. E., Díaz, E. J., Daza, O. P., Vides, S.M., Espinosa, L.V. : Análisis de asfaltos empleando parámetros de energía derivados de mediciones de energía superficial libre. Peruvias, Perúvias, Lima, 36–42 (2017)

Alvarez, A.E., Espinosa, L.V., Perea, A.M., Reyes, O.J., Paba, I.J.: Adhesion quality of chip seals: comparing and correlating the plate-stripping test, boiling-water test, and energy parameters from surface free energy. J. Mater. Civ. Eng. **31**(3), 1–11 (2019)

Alvarez, A.E., Gomez, K.L., Gomez, D.C., Reyes-Ortiz, O.J.: Optimising the effect of natural filler on asphalt-aggregate interfaces based on surface free energy measurements. Road Mater. Pavement Des. **20**(7), 1548–1570 (2019)

ASTM International: ASTM D-7497-09: Standard practice for recovering residue from emulsified asphalt using low temperature evaporative technique. ASTM International, West Conshohocken, PA, 1–2 (2016)

Bhasin, A., Howson, J., Masad, E., Little, D.N., Lytton, R.L.: Effect of modification processes on bond energy of asphalt binders. In: Transportation Research Board 86th Annual Meeting, Washington, D.C., pp. 1–14

Bhasin, A., Little, D.N.: Characterization of aggregate surface energy using the universal sorption device. J. Mater. Civ. Eng. **19**(8), 634–641 (2007)

Bhasin, A., Masad, E., Little, D., Lytton, R.: Limits on adhesive bond energy for improved resistance of hot mix asphalt to moisture damage. Transp. Res. Rec. **1970**, 3–13 (2006)

Cong, L., Peng, J., Guo, Z., Wang, Q.: Evaluation of fatigue cracking in asphalt mixtures based on surface energy. J. Mater. Civ. Eng. **29**(3), 1–6 (2017)

Gawel, I.: Sulphur-modified asphalt. In: Yen, T.F., Chilingarian, G.V. (eds.) Asphaltenes and asphalts. Elsevier, Los Angeles, pp. 515–535 (2000)

Hefer, A.W., Bhasin, A., Little, D.N.: Bitumen surface energy characterization using a contact angle approach. J. Mater. Civ. Eng. **18**(6), 759–767 (2006)

INVIAS: Especificaciones generales de construcción de carreteras y normas de ensayo para carreteras. Bogotá D.C (2013)

Kim, R.Y.: Modeling of Asphalt Concrete. McGraw-Hill, New York (2009)

Li, C., Chen, Z., Wu, S., Li, B., Xie, J., Xiao, Y.: Effects of steel slag fillers on the rheological properties of asphalt mastic. Constr. Build. Mater. **145**, 383–391 (2017)

Little, D.N., Allen, D.H., Bhasin, A.: Modeling and Design of Flexible Pavements and Materials. Springer, Cham (2017)

Prowell, B.D., Zhang, J., Brown, E.R.: Aggregate properties and the performance of superpave-designed hot mix asphalt, NCHRP report 539 National Cooperative Highway Research Program-Transportation Research Board, Washington, D.C (2005)

Salazar, J., Pacheco, J.F., Jiménez, M.J.: Determinación de la resistencia a la tracción del asfalto (BBS) y trabajo de adhesión (WaL,S) de los ligantes asfalticos, mediante determinaciones de ángulo de contacto. Métodos y Materiales **4**(1), 17–23 (2014)

Van Oss, C.J.: Interfacial Forces in Aqueous Media. Marcel Dekker Inc., New York (1994)

# Estimating Asphalt Film Thickness in Asphalt Mixtures Using Microscopy to Further Enhance the Performance of UAE Roadways

Alaa Sukkari, Ghazi Al Khateeb$^{(\boxtimes)}$, Waleed Ziada, and Helal Ezzat

University of Sharjah, Sharjah, United Arab Emirates
galkhateeb@sharjah.ac.ae

**Abstract.** The asphalt binder coating aggregate particles in hot-mix asphalt (HMA) is considered the key factor that is responsible for the durability of asphalt mixtures and hence the service life of Asphalt Concrete (AC) layers of asphalt pavements. This portion of the asphalt binder is what so-called the asphalt film in the mixture. Estimating the thickness of this film experimentally has been a challenging process for asphalt researchers as it requires high accuracy and needs very careful procedures in the analysis part. In the current Superpave specifications, the asphalt film thickness is not considered as one of the mix design requirements for asphalt mixtures but instead, the Voids in Mineral Aggregate (VMA) is considered in the mix design.

This research offers a new methodology to image and analyze the microstructure of the asphalt mix for quality and performance purposes. Using a stereo microscope is preferred to SEM, mainly because the electrons in the SEM device detect the fillers with a resolution like the asphalt, providing a challenge when thresholding the image. The binary image produced is then analyzed to obtain the aggregate shape descriptors and voids area. Measurements of the asphalt thickness can only be done manually, a single and two step measurement process is explained in the paper.

**Keywords:** Film thickness · Microscopy · Image analysis · Durability

## 1 Introduction

In Superpave, the minimum VMA requirement which has been used since the Superpave was developed back in 1993 is the one that is supposed to control and ensure an adequate level of asphalt binder content to provide acceptable durability in the asphalt mixture. However, there are several defects in the VMA criteria as reported by several researchers such as Al-Khateeb (2018), Al-Khateeb and Shenoy (2019), and Kandhal et al. (1998). For example, asphalt mixtures with coarse aggregate gradations have difficulty meeting the Superpave minimum VMA criteria although these mixtures have thick asphalt films. In other words, the minimum VMA requirement is Superpave does not ensure a minimum asphalt film thickness around the aggregate particles for acceptable durability.

The idea of asphalt binder thickness was proposed by Francis Hveem to estimate the asphalt content in the mixture. Hveem defined it as the ratio of effective asphalt

© The Author(s), under exclusive license to Springer Nature Switzerland AG 2022
A. Akhnoukh et al. (Eds.): IRF 2021, SUCI, pp. 87–96, 2022.
https://doi.org/10.1007/978-3-030-79801-7_7

volume to the aggregate surface area. Hveem assumed that the asphalt binder is uniform around the aggregate particles regardless of their sizes. This assumption was proven to be wrong by several researchers. They found that asphalt binder is thicker around finer aggregate particles, and fillers are embedded in it. Hveem used surface area factors, and effective asphalt binder content to estimate the asphalt binder coating around the aggregates. Recent research work has shown that aggregate shape, gradation, and surface area are affecting the thickness greatly.

The direct measurement of the thickness of the asphalt binder films coating the aggregate particles in the asphalt mixture is therefore the accurate way of ensuring acceptable durability levels for asphalt mixtures. Although this process requires extensive effort and time and accuracy in taking the scanning images and in the analysis part, the results and findings are worth the effort.

Although in previous research studies, the importance of replacing the VMA requirements with asphalt binder film thickness has been stressed, much of the research was theoretical, and the experimental attempts have been very limited and when done, they were limited to several scanning images on few asphalt mixture specimens. In addition, no experimental attempt has been done to study the asphalt mastic itself where very fine particles (filler material) are coated by asphalt films. Previous studies assumed that the asphalt mastic coating large aggregate particles are part of the asphalt films coating these particles. In this study, the accurate methodology will be outlined to measure the thickness of the asphalt films even those around the very fine particles in the mixture to accurately estimate the film thicknesses in the mixture and obtain the distribution over the different aggregate particles in the entire gradation of the mixture. This will be accomplished by microscopic scanning techniques and using accurate image analysis procedures to finally come up with accurate results and findings.

## 2 Background

The concept of thin-film thickness gained traction when the asphalt mixtures designed to meet the VMA requirements did not meet the performance requirements expected. This led the researchers to criticize the Superpave requirements.

Kandhal et al. (1998) reviewed the requirements of VMA in Superpave criteria. They stated that the voids in mineral aggregates (VMA) requirements for coarse-graded mixes could not be attained, and thus the mixes are not accepted under the specified criteria, while dense-graded mixes had acceptable voids in mineral aggregates (VMA). The flawed criteria required revision and researchers started proposing asphalt film criteria. Asphalt binder thickness is directly affected by aggregate size. Coarser aggregates have a thinner asphalt film compared to finer aggregates. Kandhal et al. concluded in their studies that (i) The Superpave criteria requires revision, (ii) An average film thickness of 8 microns is recommended at the time of research.

Elseifi et al. (2008) studied the concept of asphalt film thickness by microscopic imaging and scanning electron microscopy (SEM). The images were analyzed using image processing software to show the asphalt film. The results showed that smaller aggregates had a thicker asphalt film than the coarser aggregates. The research proved previous research findings that asphalt binder is not uniformly distributed among the aggregates and that fillers are embedded in the asphalt binder, forming an asphalt mastic. The research also showed that air voids tend to be between the aggregate and the asphalt binder, creating a weakness in the mixture.

As for performance researchers attempted to link binder thickness to the performance of the asphalt mixtures. Researchers linked the binder thickness to fatigue, rutting, and moisture damage. Kandhal and Chakraborty (1996) studied the influence of the asphalt film on the short- and long-term aging of asphalt mixes. They used the polynomial regression analysis to find the relationship between binder film thickness and resilient modulus after aging. The relationship showed that stiffness increased with the decrease in the binder film due to aging, (ii) short- and long-term aging is directly linked to binder-film thickness, viscosity increases as the binder film decreases, the complex modulus ($G^*$) decreased with the increase of the binder film.

Li et al. (2009) studied the trend of the effect of the binder film on the performance of pavements. Samples were collected from Mn Road for the research. They compared rutting, transverse cracking, and the international roughness index (IRI) to binder-film thickness, respectively. The results showed that transverse cracking decreases with an increase in the asphalt binder film, in stiffer binders. Rutting depth in softer binders, ex. PG 58-22, increased as the binder film thickness increased. International roughness index (IRI) and top-down cracking increased as the binder film increased in stiffer binders, ex. PG 60-70. This is due to stiffer binders having a rougher surface and are prone to cracking. The researchers concluded that the binder film thickness plays an important role in asphalt performance and requires more testing before proposing a criterion.

Attia et al. (2009) studied the binder film as a design parameter. They reported that while coarser mixes have low VMA, these mixes have a high asphalt binder film thickness. The objective of their study is to review the multiple methods developed to attempt to calculate the asphalt binder film thickness. Samples were collected from Nebraska highways, with visible distresses. The researchers compared six different models to calculate the asphalt binder film. They concluded that the equations did not show any differences in the theoretical calculation of the asphalt binder film. The samples with a higher binder film did not oxidize and resisted aging. The researchers also cautioned against the increase of the binder film as it weakens the interlock between the aggregates.

Sengoz and Topal (2007) studied the effect of binder films on the aging of asphalt mixes. The researchers prepared samples according to the Superpave™ criteria. The researchers varied the binder film thickness from 4.9 to 11 microns and used regression methods to study the effect. The results showed that stiffness and aging increased as the binder film thickness fell below 9–10 microns. The recommendations based on this study state that a 4% air void content is optimum, and 9–10 microns had the best results in durability.

Wu et al. (2012) studied the effect of the asphalt binder on short- and long-term aging. They took field samples from HMB15 roads in Britain. The testing was done according to Superpave™. The researchers applied quadratic regression models to represent the findings. The result showed that specimens with higher binder films recovered faster from short-term aging and resisted long-term aging better.

Oliver (2011) studied the effect of the binder film and the effect of mineral fillers in the asphalt mastic on fatigue using samples obtained from Australian (AustRoads) and Victorian roads (VicRoads). The researcher noted that a slight increase (1%) in bitumen content resulted in a significant (150%) increase in fatigue resistance and a decrease in rut resistance. The researcher noted that altering the film thickness by adding mineral fillers did not result in any performance improvement. The tests showed that failure tendency increased when the film thickness was decreased.

Sengoz and Agar (2007) studied the effect of binder films on asphalt susceptibility to moisture damage. The researchers used a modified Lottman test to measure the moisture damage susceptibility of various compacted specimens. The results indicated as the binder film thickness decreases; the moisture damage tendency increases.

Jiang et al. (2020) studied the effect of binder-film thickness on aging and fatigue susceptibility in unmodified and modified asphalt mixes. The researchers measured the film thickness using 2D energy-dispersive X-ray spectroscopy. They proposed an unconventional new technique to measure the film thickness, as well as a new index, binder-film thickness/standard deviation of binder-film ($Tm/SD_t$), to evaluate the effect of binder films on the performance. The readings have shown that modification of the mixes altered the binder film distribution greatly; modified asphalt mixes have a higher resistance to fatigue due to higher average binder film; increasing the binder film increased durability; altering the asphalt content, without altering the asphalt gradation, resulted in significant differences binder-film distribution.

Multiple models were developed to estimate the film thickness in the asphalt mixtures. These models do not consider the presence the fines in the asphalt mixtures, which affect the accuracy of the estimation. The models rely on the assumptions that the aggregates are spherical and have a uniform asphalt film surrounding them. Both assumptions were proven to be wrong by Elseifi et al. (2008).

Al-Khateeb (2018) investigated the Superpave criteria using the VMA requirements. The researcher used SGC test data to study the relationships and proposed a model for calculating the asphalt binder film thickness. The results showed that there is a poor relationship between asphalt binder film thickness and VMA. This implies that coarser mixes might have adequate binder film thickness, but inadequate VMA, and vice versa for finer gradations. The relationship between the VFA parameter and binder film thickness was higher. The researcher also reported that coarser gradations have a higher binder film thickness compared to uniform gradations. The results supported the need to revise the Superpave criteria.

Panda et al. (2016) proposed two empirical models to estimate the aggregate surface area. The researchers used imaging techniques to develop the second model. The researchers argued that such techniques only scanned particles of size 4.75 mm or higher. The surface area factor was added to extrapolate the finer aggregate sizes. The results of the models were consistent with the previous models developed.

Heitzman (2007; 2005) proposed developed models for calculating film thickness. He argued that the standard models developed to calculate the asphalt film thickness did not consider the aggregate specific gravity. He proposed two models to calculate asphalt film thickness in the two-dimensional and three-dimensional planes. The two models were an improvement over the standard model. The models took into consideration the aggregate shapes, specific gravity, and binder content. Heitzman proposed an 8 to14-micron specification range based on his calculation. He also concluded that the mixture durability is different between fine and coarse mixes.

Al-Khateeb and Shenoy (2019) proposed a model to calculate the asphalt binder film thickness using regression analysis and experimental data. The model is dependent on aggregate gradation, and asphalt binder content. The researchers also concluded that the model can be used before the mix to estimate the asphalt binder thickness and can be used for quality assurance purposes.

## 3 Methodology

### 3.1 Specimen Preparation

A Cardi coring machine will be used to prepare 4″ cores. IPC Global Auto Saw II will be used to cut the specimens into 2 cm thick slices. A manual saw was fitted with an industrial diamond blade and was used to cut the samples along their length to produce an 11.6 × 7.6 cm (±0.5 cm) rectangle with 2 cm thickness. This cutting procedure, although unconventional, has facilitated capturing images with the stereomicroscope. Figure 1 shows the specimen after cutting.

**Fig. 1.** Asphalt specimens after cutting.

### 3.2 SEM Imaging

SEM imaging is extensively used in research to understand the behavior of the asphalt on the microstructure level. SEM was used to capture images like stereo microscopy., due to the non-conductivity of the asphalt mixture, the specimen was gold-coated a day prior to SEM imaging. The images were analysed using the method outlined in the section below. Figures 2 shows a SEM image. The figures show that using SEM to estimate the film thickness is unfavoured. The electrons cannot distinguish between the filler imbedded in the asphalt and the asphalt itself.

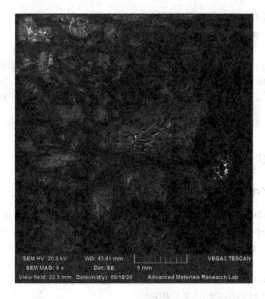

**Fig. 2.** SEM image of UAE's asphalt mix.

## 3.3 Microscopy Imaging

Samples are placed under a reflective stereomicroscope with an LED attachment and a digital camera, as shown in Fig. 3. The imaging procedure requires calibration of the microscope before capturing the images. The calibration is required later to set the scale with the software. An image with the scale drawn must be saved and used to set the scale on the imaging software. All the other images must have similar magnification and calibration, otherwise, new calibration and scale must be prepared.

**Fig. 3.** Stereo microscope with LED attachment.

The Procedure for scanning the images is divided into multiple steps as will outlined below.

The first step was to experiment with different magnification magnitudes to determine the best image in terms of clarity, contrast, and brightness. The best magnification for our material is 14×–16×, with a calibrated resolution of 0.294 pixels/μm. AmScope is used as digital camera software.

The second step is to capture images from the surface of the sample. Aggregates sizes 9.5–25 mm require more than 1 image. Aggregates less than 9.5 mm need between 2 and 20 images to cover the aggregate.

A total of 20–40 images of the aggregate and its surrounding area are taken and then stitched using the microscope's software. Image stitching is combining 2 or more images into 1 image. The software scans the images and combines the pixels in the images to "stitch" them, or group them, into 1 image. Image sizes range between 5 × 7 mm$^2$ to 6 × 8 mm$^2$. A Sample might require more than 300 images to be fully captured. 50–60% of the surface is captured as a representative sample.

### 3.4 Image Analysis

The software used in this research is ImageJ/FIJI. ImageJ is a Java-based image processing program developed at the National Institutes of Health and the Laboratory for Optical and Computational Instrumentation.

The third step is to switch the image into an 8-bit binary image using the analysis software. This is the most crucial step, as the software cannot detect all the filler particles in the image, because the fillers have a slightly similar resolution to asphalt, thus the process must be done manually. Certain software offers multiple algorithms to produce binary images.

Figure 4 shows binary images produced automatically and manually. The size of the aggregates in the images will differ based on the area covered by the software scanning the pixels of the aggregates. Figure 5 shows a simplified step to produce binary images.

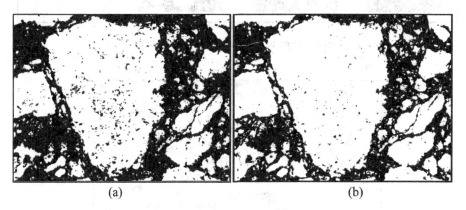

**Fig. 4.** (a) Automatic thresholding, and (b) Manual thresholding.

The images are then scanned to find the edge of the aggregates, and finally, the size of the aggregates is analyzed and saved in an excel sheet. ImageJ/FIJI particle analysis plugin gives the area and shape descriptors of the aggregates and voids.

Since the study aims to measure the filler-free asphalt binder thickness, there two methods to measure the thickness. The first method is shifting the measurement line next to the filler particle. This method is timesaving and offers accurate readings.

The second method is measuring the distance through the filler particle, then subtracting the length of the filler from the total length. This method is accurate but extremely time-consuming, especially when 400 readings are needed to stabilize the average reading. The average difference between the two methods is <1 micron or 1.0%.

The film thickness measurement should be from the edge of the aggregate to the adjacent aggregate, in the same way as Elseifi et al. (2008) and Jiang et al. (2020). If the edges are scanned and the ImageJ draws an outline of the aggregate, the outer outline should be considered in the thickness measurement, as shown in Fig. 6.

For aggregates of sizes $\geq 9.5$ mm, 20 to 30 readings of all the sides are representative. For aggregates of sizes between 4.75–1.18 mm, 4–8 readings are representative. Aggregates of sizes 0.6–0.15 mm 2–3 readings are representative. A specimen should have between 400 readings to stabilize the average, minimum, and maximum film thickness.

In certain mixes, like the gap-graded mix, air voids present in the image need further analysis. The length of the air voids must be measured first, then an average must be calculated and deducted from the effect binder thickness. This requires using the raw un-analyzed image to get the air voids length measurements, as ImageJ/FIJI tends to fill the voids with black/grey (Table 1).

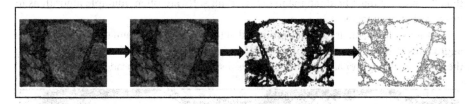

**Fig. 5.** Steps to produce an analysed image.

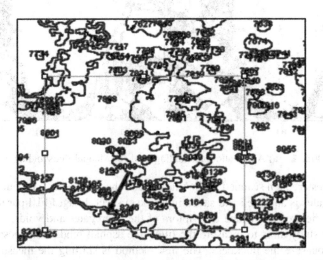

**Fig. 6.** Measuring asphalt thickness around an aggregate.

**Table 1.** Readings required. Per aggregate size.

| Size of aggregates | Readings |
| --- | --- |
| $\geq 9.5$ mm | 20 to 30 |
| 4.75–1.18 mm | 4 to 8 |
| 0.6–0.15 mm | 2 to 3 |

In certain AC mixtures, like the gap-graded mixture, air voids present in the image need further analysis. The length of the air voids must be measured first, then an average must be calculated and deducted from the effect binder thickness. This requires using the raw un-analyzed image to get the air voids length measurements, as ImageJ/FIJI tends to fill the voids with black/grey. The second method involves following the same image analysis procedure outlined above, but the thresholding color should be changed to red to scan the air void surface area. Using the circle surface area equation, the average radii of the air voids are calculated and subtracted from the average value of the air voids.

## 4 Results

The analysis shows a non-uniform distribution of the asphalt binder around the larger aggregates, contradicting the assumption that the asphalt binder is uniformly distributed around the aggregates in the mix.

The images show that the asphalt binder is thicker around the larger aggregates with an average of 80 microns while smaller and filler aggregates had an average of 35 microns. The filler particles and finer aggregates tend to be close to each other which due to the mixing and compaction procedure. The skewness is due to the non-uniformity of the asphalt film thickness around the aggregates. Larger aggregates have thicker film thickness compared to the film thickness around the fillers, due to the proximity of the fillers around the larger aggregates.

## 5 Conclusions

- This research presents a proposed methodology to scan the asphalt under the microscope and analyze the images to obtain the effective asphalt binder thickness around the aggregate particles in the asphalt mix.
- This method can also be applied to study the microstructure of the asphalt for quality purposes to ensure the aggregates are undamaged during the compaction, the asphalt binder is properly mixed, and the air voids are up to a specific area.
- The results show that the 8-to-14-micron range do not agree with the results in this research. Microscopic imaging showed film thickness to be more than 100 microns.
- Air voids appeared near the boundary between the aggregates and asphalt as well as in the asphalt mastic itself.

96 A. Sukkari et al.

**Acknowlegement.** The authors acknowledge the support of the Research Institute of Science and technology (RISE) for the support in this research.

# References

Al-Khateeb, G.G.: Conceptualizing the asphalt film thickness to investigate the Superpave VMA criteria. Int. J. Pavement Eng. **19**(11), 957–965 (2018). https://doi.org/10.1080/10298436. 2016.1224414

Al-Khateeb, G.G., Shenoy, A.: Mixture-property-independent asphalt film thickness model. Mater. Today Commun. **19**, 482–486 (2019). https://doi.org/10.1016/j.mtcomm.2017.11.007

Attia, M., Abdelrahman, M.A., Molakatalla, U., Salem, H.M.: Field evaluation of asphalt film thickness as a design parameter in Superpave mix design. Int. J. Pave. Res. Technol. **2**, 205–210 (2009)

Elseifi, M.A., Al-Qadi, I.L., Yang, S.H., Carpenter, S.H.: Validity of asphalt binder film thickness concept in hot-mix asphalt. Transp. Res. Rec. **2057**, 37–45 (2008)

Heitzman, M.A., Schaefer, V., Thompson, R.B.: Development of new film thickness models for hot mix asphalt (2005)

Heitzman, M.: New film thickness models for iowa hot mix asphalt (2007)

Jiang, J., Li, Y., Zhang, Y., Bahia, H.U.: Distribution of mortar film thickness and its relationship to mixture cracking resistance. Int. J. Pave. Eng. 1–10 (2020)

Kandhal, P.S., Chakraborty, S.: Effect of asphalt film thickness on short- and long-term aging of asphalt paving mixtures. NCAT Report 96-01 (1996)

Kandhal, P., Foo, K., Mallick, R.: A critical review of VMA requirements in superpave (1998)

Li, X., Williams, R.C., Marasteanu, M.O., Clyne, T.R., Johnson, E.: Investigation of in-place asphalt film thickness and performance of hot-mix asphalt mixtures. J. Mater. Civ. Eng. **21**(6), 262–270 (2009)

Oliver, J.: The effect of binder film thickness on asphalt cracking and ravelling. Road and Transport Research. **20**, 3–13 (2011)

Panda, R.P., Das, S.S., Sahoo, P.K.: An empirical method for estimating surface area of aggregates in hot mix asphalt. J. Traffic Transp. Eng. (Engl. Edn.) **3**(2), 127–136 (2016). https://doi.org/10.1016/j.jtte.2015.10.007

Sengoz, B., Topal, A.: Minimum voids in mineral aggregate in hot-mix asphalt based on asphalt film thickness. Build. Environ. **42**(10), 3629–3635 (2007). https://doi.org/10.1016/j.buildenv. 2006.10.005

Sengoz, B., Agar, E.: Effect of asphalt film thickness on the moisture sensitivity characteristics of hot-mix asphalt. Build. Environ. **42**(10), 3621–3628 (2007). https://doi.org/10.1016/j. buildenv.2006.10.006

Wu, J., Nur, N.I., Jun, C., Yun, L., Hainin, M.R., Airey, G.D.: Aging properties of HMB15 with varying binder film thicknesses. Sains Malays. **41**(12), 1595–1603 (2012)

# Effect of Crumb Rubber and Polymer Modifiers on Performance Related Properties of Asphalt Binders

Joseph Dib[1(⊠)], Nariman J. Khalil[2], and Edwina Saroufim[2]

[1] Department of Civil and Environmental Engineering, University of Nevada
at Reno, Reno, USA
joseph.dib@nevada.unr.edu
[2] Department of Civil and Environmental Engineering, University of Balamand,
Tripoli, Lebanon

**Abstract.** Crumb rubber modified bitumen (CRMB) has been generally used to improve the high temperature performance of asphalt mixtures in the past decades. Besides, the addition of a certain percentage of polymers into asphalt binders has been known to enhance their engineering characteristics. This study aims to investigate the rutting and fatigue cracking performance of CRMB, and polymer modified asphalt binders by means of Dynamic Shear Rheometer (DSR), Multiple Stress Creep Recovery (MSCR), and Bending Beam Rheometer (BBR). Different percentages of Crumb rubber (5, 10, 15, and 25%) obtained using the ambient grinding process were blended with a PG70-16 asphalt binder using the wet process. Small percentages ranging from 0.5 to 2% of polymers (LSBS, PPA, Elvaloy, and CBE) were added to a PG64-16 asphalt binder. It is demonstrated that mixing a specified percentage of crumb rubber from scrap tires, with the proposed optimized laboratory protocol, will result in an improved high-temperature performance reflecting higher rutting resistance complying with certain needs of regions with warm climates, especially the Mediterranean area. However, crumb rubber content should be restricted to 10% to limit the effect of the CR on the low-temperature performance grade. In the case of polymer modification, significant benefits were observed for rutting resistance of the resulting modified asphalt binders without jeopardizing fatigue cracking resistance criteria.

**Keywords:** Crumb rubber · Polymers · Performance grading · MSCR · Penetration · Softening point

## 1 Introduction

Asphalt binder grade determination in Lebanon depends on the "conventional" test strategies: the penetration, softening point, and viscosity testing. These tests do not reflect the climate, nor traffic conditions and aging. The Performance Grading (PG) system is a comprehensive reaching system that permits the detection of such factors in choosing the right asphalt binder grade for the right place. This method has proved its superiority over other systems in North America (Liu et al. 2018). In

---

© The Author(s), under exclusive license to Springer Nature Switzerland AG 2022
A. Akhnoukh et al. (Eds.): IRF 2021, SUCI, pp. 97–113, 2022.
https://doi.org/10.1007/978-3-030-79801-7_8

addition, it is possible to apply such a system in Lebanon and to perform the transition from old criteria to advanced standards. This transition is deemed necessary for a better practice on the paved sections over utilizing the regular techniques (Morea et al. 2014). However, adopting these standards in Lebanon results in the need for a modification of the current asphalt binder grade utilized because it does not satisfy the climatic conditions all over the country (Khalil et al. 2009). Two of the widely used modifiers are rubber and polymers. Lebanon is not an asphalt binder producer country, and it does not contain any active refineries. The country relies mainly on importing different oil products to satisfy the local market needs. Different sources were identified, according to the Lebanese Ministry of Energy and Water - the General Directory of Petroleum. A previous work by (Khalil et al. 2009) resulted in the pavement design temperatures for Lebanon. The purpose of the current study is to explore the modification of the available asphalt binder to meet PG grade requirements for different climatic zone in Lebanon and the Mediterranean area.

The auto industry produces a large number of scrap tires every year, which ends up stockpiled in landfills. This represents a serious environmental concern on several levels (Torretta et al. 2015; Tsai et al. 2017). In Lebanon, there is no restrictions on what exactly should be done with tires after their service life. However, it has been proven in several countries that the utilization of crumb rubber (CR) for asphalt binder modification is an economic and effective practice to alleviate the problem (Mturi et al. 2014). On the other hand, polymer modified asphalt binders have proven to be effective in providing enhanced quality and improved properties (Xu et al. 2016).

In spite of several research conducted in the field of crumb rubber modified bitumen (CRMB) and polymer modified asphalt binders, there is a lack of enough knowledge in Lebanon. The main objective of this research is to identify the performance enhancement that is possible to obtain when adding CR and polymers. In this regard, the following objectives were set:

1. Evaluate the effect of CR and polymers on the basic rheological characteristics of the studied asphalt binders such as penetration, softening point, and viscosity.
2. Investigate the temperature susceptibility of the modified asphalt binders.
3. Find modified bitumen blends that satisfy Superior Performing Asphalt Pavements requirements for performance grading.
4. Determine and compare the studied binder to SHRP criteria for rutting, fatigue, and thermal cracking resistance.
5. Assess the effect of CR on the high temperature performance using the MSCR (multiple stress creep recovery) test.

## 2   Literature Review

CRMB and polymer modified asphalt binders present better performance than neat asphalt binder and produce pavements with better mechanical behavior. The conventional and rheological test results show that the addition of modifiers such as CR and polymers could lead to a significant improvement on the high temperature performance of the resulting mixture (Labaki and Jeguirim 2017; Lesueur 2009; Liu et al. 2018).

The variability in improvement relies on many factors such as modifier type, content, surface characteristics of the rubber particles or polymer (Khan et al. 2016; Presti 2013), and blending conditions including stress level and temperature (Magar 2014). For instance, the polymer network plays a considerable role in the stress-dependent behavior of the polymer modified asphalt binders, more specifically at high stress levels. Whereas at low stress levels, normally associated with conventional PG, the polymer plays the role of a filler that stiffens the asphalt binder (Anderson et al. 2010). It is also described that the behavior of asphalt binders at low stress levels is linear, when in fact this behavior becomes more non-linear as the stress level increases (Angelo 2009; Jafari et al. 2015). It was noted that the same behavior was observed with CRMB (Divya et al. 2012).

The United States Federal Highway Administration (FHWA) proposed the Multiple Stress Creep Recovery (MSCR) test to better evaluate the high temperature behavior of modified asphalt binders. The non-recoverable compliance was proposed as a replacement of the rutting factor parameter since it cannot properly assess the rutting resistance capacity of modified asphalt binders. Several researchers carried out MSCR tests on neat binders, polymer modified binders, and CRMB (Nejad et al. 2012; Angelo 2009). The results validate a better correlation to mixture rutting using MSCR test parameters compared with the existing Superpave binder criteria. In addition, the relationship of the non-recoverable creep compliance with laboratory wheel tracking test results was better established with the high stress levels of the MSCR test (Wasage et al. 2011), generally used to relate field rutting performance to laboratory outcomes.

# 3 Materials and Testing

## 3.1 Materials

### 3.1.1 Neat Asphalt

Two different asphalt binders are used in this study: PG70-16 and PG64-16. They will be referred to as source 1 and source 2, respectively. These grades are currently utilized in Lebanon based on availability. The binder obtained from source 1 was modified with CR, whereas the binder from source 2 was modified with polymers. The complete scope of this research will include the modification of both binder types with all modifiers in future studies. For the purpose of this paper, the relative performance variation will be used to compare the overall impact of modifiers.

### 3.1.2 Crumb Rubber

The crumb rubber is obtained using the ambient grinding process with particles passing sieve #10. It was attained from 3R recycling factory located in South Lebanon. According to the results of literature review (Neto et al. 2006), the ambient grinding method was chosen due to the benefits of the obtained surface and its feasible application in Lebanon. In addition, 10, 15, and 20% CR content (by weight of asphalt binder) are selected. ASTM (American Society for Testing Materials) defined asphalt rubber as a blend of asphalt cement which contains at least 15% of CR by the weight of the total blend. The research team found this threshold a good starting point to characterize the behavior of CRMB with a variation of ±5%, before moving to higher or lower percentages in future studies.

Three different kinds of CRMB were prepared by wet process at 180 ± 5 °C. CR was added within the first 15 min, and the total mixing time is 45 min ± 1 min with shearing rate at 4000 r/min.

### 3.1.3 Polymers

Polymers usually utilized for asphalt binder modification are two types: elastomers and plastomers. Polymers used in this research are: LSBS (styrene–butadienestyrene), PPA (polyphosphoric acid), Elvaloy, and CBE with relatively small percentages ranging from 0.5% to 2% to evaluate their effect on the overall performance of asphalt binders taking into consideration their economic feasibility in Lebanon. The percentages used are optimum quantities found in literature review recommended for each type (Bulatović et al. 2012).

## 3.2 Testing

### 3.2.1 Aging Procedure

Short-term aging of asphalt binder is due to oxidation and volatilization that occurs in batch plant and drum mix at about 150 °C and during construction. It is simulated using the Rolling Thin-Film Oven (RTFO) according to ASTM D4402. The asphalt binder obtained from RTFO was aged in the Pressure Aging Vessel (PAV) following AASHTO R28 procedure to simulate long-term aging. More specifically, the asphalt binder was exposed to air at high pressure of around 304 psi (2.1 MPa) and temperature of 194 to 230 °F (90 to 110 °C) for 20 h in the PAV. A mass of 50 g of the RTFO-aged asphalt binder was poured into each of the PAV sample pans. After 20 ± 1 h, the sample pans were heated before placing them in a vacuum oven at 170 for 30 min at a pressure of 2.1 psi (15 kPa) to remove air bubbles.

### 3.2.2 Penetration, Softening Point, Viscosity

Conventional parameters including penetration, softening point, and viscosity are used to classify the asphalt binders under study (Angelo 2009). The procedure is performed on the un-aged sample tested for penetration at 25 °C, softening point, and viscosity following ASTM D5, ASTM D36, and ASTM D4402, respectively. The penetration test is an indication of asphalt binder's consistency with simple testing tools and certain testing conditions consisting of a 100 g load and measuring the penetration value after 5 s at 25 °C. The ring and ball test is used to determine the softening point, or the temperature at which the asphalt binder shifts from a solid phase to a fluid phase. This parameter was normally utilized by highway agencies to determine the maximum temperature of usage on roadways. A higher softening point would normally illustrate lower temperature susceptibility. The rotational viscosity of asphalt binders is determined to ensure that at normal plant temperatures the binder will have a low viscosity enough to be readily pumped or piped. It is not expected for the binder, while pumping and before mixing, to be subjected to aging, therefore the rotational viscosity is tested for the original binder. It is also used as part of the viscosity grading system.

### 3.2.3 Rheological Properties of Asphalt Binder

To fully characterize the impact of rubber particles and polymers on the rheological behaviour of asphalt binders, many parameters are required, more specifically rutting parameter (G*/sinδ), fatigue parameter (G*.sinδ), and low-temperature testing. A sample is aged in the RTFO, followed by shear testing according to ASTM D6373 and ASTM D7175 using a dynamic shear rheometer (DSR), and a 25 mm parallel plate with 1 mm gap to determine the complex modulus (G*) and the phase angle (δ). The test temperature ranged from 58 °C to 82 °C. Another part of the aged binder is placed in the PAV for long term aging, followed by DSR testing to determine its rheological properties using an 8 mm parallel plate and 2 mm gap, and thereafter determine the grade of the asphalt binder used. The oscillation frequency is 10.5 rad/s and the oscillation strain amplitude is 12% for un-aged binder, 10% for short term aging asphalt, and 1% for long term aged binders.

Asphalt binders at low temperatures are too stiff and difficult to adhere on the parallel plate while tested using the DSR. Therefore, asphalt binder properties at low temperatures were assessed using the Bending Beam Remoter (BBR) test according to AASHTO T313. This procedure was used on the PAV residue to determine how much a binder creeps or deflects under a constant load at a constant temperature. Asphalt binder at low temperatures acts more like an elastic solid than a viscous fluid. This test describes how stresses are expected to increase in an asphalt binder as the temperature decreases. The BBR data is used to generate the thermal stress curve. The samples tested were beam shaped with dimensions of $0.04 \times 0.5 \times 4.9$ in. ($6.25 \times 12.5 \times 125$ mm). The beam theory was used in the BBR procedure to calculate the stiffness of the asphalt beam sample under a static load (100 g) applied at the middle of the beam. The sample was placed in a controlled temperature bath. The resistance of the asphalt binder to creep loading was designated by the creep stiffness, and the change in asphalt stiffness (relaxation rate) was designated by m-value. The creep stiffness and m-value were determined after 60 s of loading that simulates 2 h of loading at 50 °F (10 °C) cooler temperature.

### 3.2.4 Multiple Stress Creep Recovery Test

MSCR test is used to evaluate the rutting performance of asphalt binders at high temperature. According to AASHTO TP70, the non-recoverable creep compliance (Jnr) is calculated as a measure of the asphalt binder's contribution to mixture permanent deformation behaviour and percent recovery (R%) reflects the elasticity of asphalt binder. This test will be performed at the determined high temperature for each asphalt binder obtained from the DSR testing. The test is conducted by the repeated loading for the duration of 1 s followed by 9 s of recovery period using the DSR system. Two stress levels of 0.1 and 3.2 kPa are applied and 10 cycles are conducted for each stress level. The method to calculate these parameters is performed by obtaining for each of the ten cycles the adjusted strain value at the end of the creep portion (denoted $\varepsilon_1$) and the adjusted strain value at the end of the recovery portion of each cycle (denoted $\varepsilon_{10}$) using Eqs. (1) and (2), respectively.

$$\varepsilon_1 = \varepsilon_c - \varepsilon_0 \tag{1}$$

$$\varepsilon_{10} = \varepsilon_r - \varepsilon_0 \tag{2}$$

Jnr and R% are then calculated using Eqs. (3) to (6).

$$Jnr(Stress\ Level, N) = \frac{\varepsilon_{10}}{Stress\ Level} \tag{3}$$

$$Jnr_{Stress\ Level} = \frac{\sum_1^{10}(Jnr(Stress\ Level, N))}{10} \tag{4}$$

$$\varepsilon_r = \frac{(\varepsilon_1 - \varepsilon_{10}) \times 100}{\varepsilon_1} \tag{5}$$

$$R_{Stress\ Level} = \frac{\sum_1^{10}(\varepsilon_r(Stress\ Level, N))}{10} \tag{6}$$

Where:

$\varepsilon_0$: Initial strain in the creep stage.

$\varepsilon_r$: Final strain in the recovery stage.

The calculation method for 0.1 kPa and 3.2 kPa stress levels are similar. The test temperature was based on the high PG temperature of each asphalt binder. The test was performed on the RTFO-aged samples using two replicates for each binder.

## 4 Results and Discussion

### 4.1 Penetration, Softening Point, and Viscosity

Results of Penetration, Softening point and Viscosity have been presented in Table 1. Based on obtained results, it could be concluded that CR had considerable effect on the stiffness of the studied asphalt binders, whereas polymers had little effects on those properties. This behavior could be attributed to the low percentage of polymers used in this study. According to Table 1, the penetration has been decreasing by increasing the CR content, while using different types of polymers did not affect the penetration value. This means that temperature susceptibility of CRMB has been reduced by increasing the CR content, where asphalt binders with less temperature susceptibility have more resistance to rutting and low temperature cracking.

Softening point values were higher when increasing the CR content compared to the Neat Binder (NB). This trend was not observed when adding various types of polymers to the neat asphalt binder labeled as "A". Similar trend is observed with the viscosity of CRMB which exceeds the maximum allowable criteria set by the Strategic Highway Research Program (SHRP) of 3.0 Pa.s (Magar 2014). This outcome raise attention to monitoring to the handling of CRMB in refineries, especially when utilizing more than 15% CR content. Literature showed that rubber crumbs can swell up to 3 to 5 times its original size due to the absorption of maltenes components of the bitumen which leave a higher proportion of asphaltenes increasing the resulting asphalt binder viscosity (Eberhardsteiner et al. 2015).

**Table 1.** Penetration, softening point, and viscosity of the modified asphalt binders.

| Sample ID | NB | CRMB 10% | CRMB 15% | CRMB 20% | A | B | C | D | E |
|---|---|---|---|---|---|---|---|---|---|
| Composition | Neat-S2 | CR | CR | CR | Neat-S1 | LSBS | Elvaloy | PPA | 7686 |
| Percent (%) | – | 10 | 15 | 20 | – | 2 | 0.5 | 1 | 2 |
| Penetration (100 g, 5 s, 0.1 mm) | 61 | 55 | 40 | 35 | 31 | 29 | 30 | 30 | 29 |
| Softening point (°C) | 49 | 53 | 56 | 60 | 44 | 51 | 50 | 51 | 51 |
| Viscosity (Pa.s) | 0.537 | 1.505 | 2.201 | 3.125 | – | 0.900 | 0.875 | 0.775 | 0.450 |

## 4.2 PG Results

The accumulation of non-recoverable deformation in asphalt mixtures in response to repeated load applications at high temperature leads to pavement rutting. To address this issue, PG specifications are set on the high-temperature stiffness of asphalt binder using $G^*/\sin \delta$ parameter for un-aged and RTFO-aged asphalt binders. The contribution of asphalt binder to rutting is minimized when a minimum of $G^*/\sin \delta$ of 1.00 kPa for un-aged and 2.00 kPa for RTFO-aged are at the proper grade temperature. To resist rutting, an asphalt binder should be stiff enough not to deform too much and elastic enough to be able to return to its original shape after loading. Therefore, the complex shear modulus divided by sinus of the phase angle should be large; in other words, to guarantee a better resistance to rutting, a higher value of $G^*$ (stiffer binder) and a lower value of $\delta$ (more elastic) are preferable.

Figures 1 and 2 show the ratio of the complex modulus $G^*$ and the phase angle $\delta$ of the original and RTFO-aged asphalt binders with different percentages of CR. It can be noticed that the resistance to rutting increases with the increase of CR percentage. In addition, the susceptibility to temperature variation decreased with the addition of 10% CR compared to the neat binder, whereas no further improvement was observed with further addition of CR. The short-term aging showed a similar behavior; however, it is noted that the presence of CR accelerates the stiffening rate of asphalt binders. However, it is difficult to assess high temperature performance accurately by PG temperature alone. This type of testing cannot simulate real loading condition well, and $G^*/\sin \delta$ cannot characterize the delayed elastic deformation and recovery capability of asphalt binders, which makes it harder to fully characterize the viscoelastic rheological properties of CRMB.

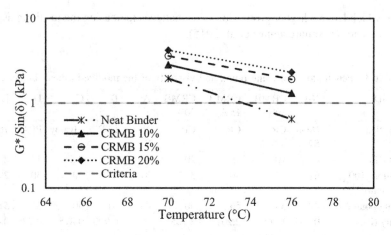

**Fig. 1.** G*/sin(δ) versus temperature for un-aged sample.

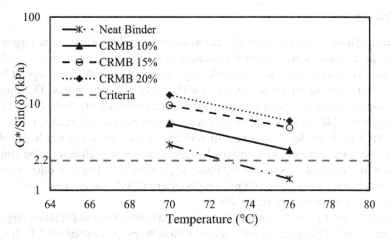

**Fig. 2.** G*/sin(δ) versus temperature for short-term aged sample.

Due to the accumulation of loads after the pavement has been in-service for some time, cracking can occur at lower to moderate pavement temperatures. The DSR test method allows calculating G*.sin δ to characterize asphalt binder stiffness after long-term aging. The maximum specification limit of G*.sin δ to minimize the contribution to fatigue cracking is 5,000 kPa. A maximum value was considered in PG specification because low values of G*.sin δ indicate low energy dissipation. The DSR testing for the long-term aged sample was similar to DSR testing on original and RTFO-aged binder. However, this test should be performed using 8 mm plate diameter and 2 mm gap. To resist fatigue cracking, known as alligator cracking, the asphalt binder should be flexible and not brittle (a lower G* value is required). It should be able to dissipate energy by rebounding and not cracking, and be sufficiently flexible at lower temperatures (i.e., intermediate temperatures), where fatigue cracking is most likely to occur (a lower phase angle δ is required). For that purpose, the complex shear modulus viscous portion G*.sin δ should be a minimum. Fatigue cracking is mostly common at the late age of the pavement service life, and therefore it is performed on long-term aged asphalt binders. Figure 3 shows the critical fatigue temperatures for the CRMB versus the neat binder. The addition of CR improved fatigue cracking resistance by lowering the intermediate temperature of the modified asphalt binders; however, the addition of CR by more than 10% made the resulting binders more susceptible to temperature variation.

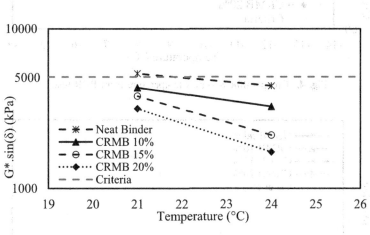

**Fig. 3.** G*sin(δ) versus temperature for long-term aged sample.

A general rule of thumbs indicates that the Creep Stiffness (S) almost doubles for every 6° decrease in temperature. This behavior was observed for all asphalt binders. The flatter the curve of temperature versus S(t), the fewer thermal stresses are developed with decrease in temperature. Almost all asphalt binders modified or non-modified have similar slopes as shown in Fig. 4.

The slopes of the stiffness versus temperature curve or "m-value" are plotted in Fig. 5. A minimum threshold value of 0.3 for m-value and a maximum of 300 MPa for the Creep Stiffness are specified by Superpave specifications. As the slope of the asphalt binder stiffness curve flattens, the ability of the asphalt pavement to relieve thermal stresses by flow decreases. The observation in Fig. 4 is validated in Fig. 5, where all asphalt binders have almost a similar slope. Nonetheless, the stiffness of CRMB with 15% and 20% increased and the m-value decreased significantly. This behavior means that with the addition of CR asphalt binders becomes more susceptible to low-temperature cracking with lower stress relief rate.

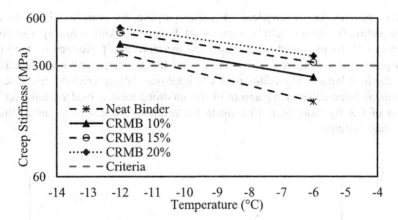

**Fig. 4.** Creep stiffness versus temperature from BBR test.

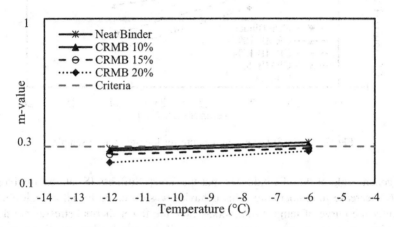

**Fig. 5.** m-value versus temperature from BBR test.

Determining the Performance Grade (PG) of asphalt binders can better characterize and predict the impact of rubber and polymer modifications. All modified and non-modified asphalt binders were graded using the Superpave PG method following Table 1 in AASHTO M320. Results are summarized in Figs. 6 and 7 for CRMB, and polymer modified asphalt binders, respectively. The neat binder from source 1 was found to have a PG70-16, and the addition of 10% CR bumped it to PG76-16. CRMB with 15% and 20% resulted in a PG82-10 bumping the high temperature two grades up. Hence, the low temperature was affected beyond 10% CR. The intermediate temperature decreased with the increasing amount of CR, which signifies an improved fatigue cracking resistance of asphalt binders containing crumb rubber. All testing results were checked against the precision and bias determined in the appropriate AASHTO standard for each test method.

Regarding polymer modification, the neat binder used had a PG64-16 obtained originally from source 2 refinery and transported to Lebanon. The addition of polymers, regardless of their type and percentages, bumped the high temperature one grade without affecting the intermediate and low temperatures. This would improve the rutting resistance without risking fatigue or thermal cracking.

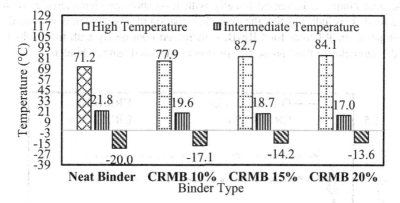

Fig. 6. True PG of CRMB.

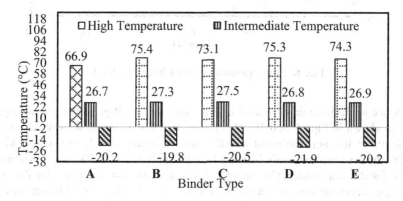

Fig. 7. True PG of polymer modified asphalt binder.

## 4.3 MSCR Results

The standard PG grading does not allow proper evaluation and comparison of the results of the high temperature performance and rutting resistance of CRMB. Therefore, the MSCR is applied to provide the necessary tools for this evaluation after the addition of different percentages of crumb rubber. This test was found not feasible for the polymer modified asphalt binders at this stage of the research, since there was no significant variation in the resulting PG with the percentages of polymers used.

An expanded segment of the strain versus time plot (i.e., confined to first loading-unloading cycle) is shown for the CRMB and neat binder at 0.1 kPa stress level in Fig. 8. It should be noted that the average of test results for two replicates is shown.

The recovery of the asphalt binder was enhanced with the addition of CR, where for the neat binder almost no recovery was observed. The recovery was evidently observed with the addition of 20% CR, almost immediately after unloading. The strain build-up in the loading stage started to turn into a non-linear behavior with 15% and 20% CR. For the neat binder, it increased linearly with time showing characteristics of linear viscous fluid. This indicates that the main deformation of neat asphalt is viscous flow at high temperature. Whereas for CRMB, it increased nonlinearly with time, which signifies the viscoelastic fluid properties of rubber modified asphalt binders.

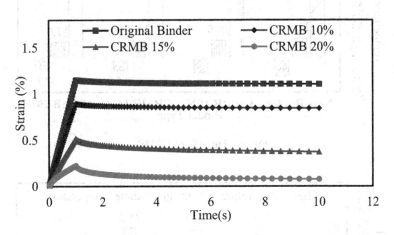

**Fig. 8.** Strain variation with 0.1 kPa stress-level.

The accumulated strain decreased with the increase of CR percentage in the asphalt binder as shown in Figs. 9 and 10. This proves that CR increased the elasticity of the asphalt binder. It is noticeable that with the increase of the stress level from 0.1 kPa to 3.2 kPa, the accumulated strain increased significantly from 12.5% to a maximum of 479.6% for the neat binder. On the other hand, a significant difference for the strain percentage accumulated is shown when the CR is added to the asphalt binder. In other words, the stress level has a great effect on accumulated strain, and the strain growth rate increases with stress level increasing. This behaviour validates that the main type of creep deformation of neat asphalt binder is viscous flow at the loading stage. Only a small amount of deformation is recoverable, leading to a significant permanent deformation. For rubber modified asphalt binders, viscoelastic flow is the main type of creep deformation, and the asphalt binder has the ability to recover after unloading, resulting in lower permanent deformation with the increasing presence of CR.

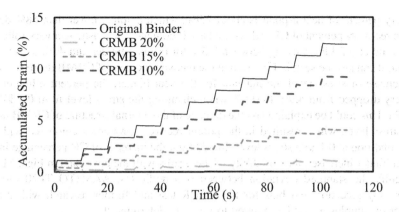

**Fig. 9.** Accumulated Strain for 0.1 kPa stress-level.

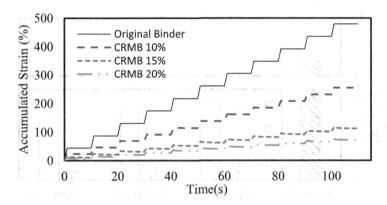

**Fig. 10.** Accumulated Strain for the rubber modified Asphalt binders at 3.2 kPa stress-level.

As the above results showed, the accumulated strain can characterize the permanent deformation/rutting properties of asphalt under repeated loading. Moreover, the recovery portion of the unloaded segment can reflect the delayed elastic characteristics of CRMB. The recoverable creep compliance can be used to assess deformation resistance properties of asphalt binders. The higher Jnr, the larger permanent deformation under repeated loading. Hence, the deformation recovery percent can be utilized to characterize the elastic properties of asphalt binders, where the higher R% reflect better elastic properties which is important in rutting resistance. The calculation method for Jnr and R% is detailed in Sect. 3.2.4. Figures 11 and 12 summarize the results obtained.

Figure 11 indicates a negative trend between Jnr and CR content. The non-recoverable creep compliance of neat asphalt is much larger than that of CRMB at different stress levels which signifies better resistance to deformation of CRMB. At 0.1 kPa, Jnr decreased from 11 kPa for original binder to 0.1 kPa with 20% CR content. As well, Jnr varied between 14 kPa for original binder and 1 kPa with 20% CR content at 3.2 kPa. Jnr values are inversely proportional to the percentage of CR with the values of Jnr decreasing with the addition of crumb rubber. Figure 12 shows the deformation recovery percent of neat asphalt is about 4.9%, which is much lower than 68% deformation recovery percent of CRMB with 20% CR content. The elastic recovery followed same trend at 3.2 kPa, varying between 0.5% for original binder and 46% at 20% CR content. It can also be seen in Fig. 12 that the recovery percent of CRMB decreases with the increase of stress level, noting that for the neat binder, the percent difference in recovery dropped from 4.9% to 0.5% when changing the stress level from 0.1 kPa to 3.2 kPa. This could be explained by the damage of the internal structure of CRMB under high stress level, which resulted in the poorer recovery properties compared with the same condition at 0.1 kPa stress level. In addition, the increase of CR percentage in the asphalt binder increased the variability of the results with the error bars in Figs. 11 and 12 signify the standard deviation between two replicates. AASHTO TP70 did not contain any precision and bias for the MSCR test and further research with higher number of samples would be required to properly determine it.

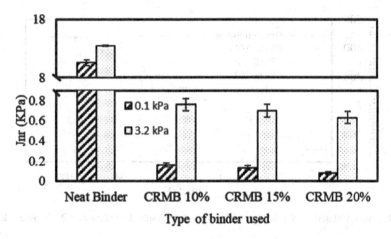

**Fig. 11.** Non-recoverable creep compliance of CRMB.

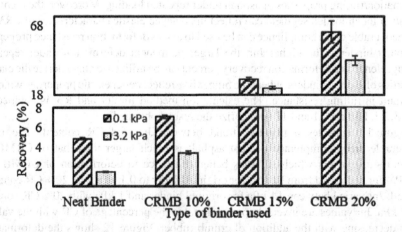

**Fig. 12.** Deformation recovery percent of CRMB.

# 5 Conclusions and Recommendations

In this study, different percentages of crumb rubber modifier were added to a PG70-16 neat binder and several types and content of polymers were added to a PG64-16 neat binder, both typically used in Lebanon. Penetration, softening point, viscosity, and conventional PG grading were investigated for the resulting asphalt binders. In addition, the high temperature performance of different CRMB were inspected by using DSR and MSCR tests. Based on the results of this limited laboratory investigation, the following findings and conclusions are summarized:

- The experimental results of the penetration test and rutting parameter of the conventional PG did not correlate in the case of polymer modification, whereas the results lined up with CR modifier with the decrease of penetration value and increase in the high temperature grade when adding more CR.
- The addition of CR should be limited to 10% in order not to risk thermal cracking resistance. In addition, the viscosity value reaches the limiting criteria of 3.0 Pa.s at 15% and exceeds it with 20% CR.
- The viscosity and low temperature were not impacted by the addition of polymers; hence the high temperature performance was enhanced regardless of the type of polymer used, validating the optimum percentage used of each type to ensure good rutting resistance.
- CR modifier improved the rutting resistance of the neat binder as indicated by the increased percent recovery values. The degree of improvement was proportional to CR content. Also, the rutting parameter from conventional PG testing reflected lower permanent deformation, thus improve the high temperature behavior of asphalt binder to a higher working temperature range.
- Based on the non-recoverable creep compliance and deformation recovery percent obtained by the MSCR test, it can be found that there is a considerable improvement of rutting resistance for the CRMB compared to the neat binder. MSCR test on polymer modified asphalt binders should be performed in the next stage of this research to better distinguish their high temperature performance since the conventional PG methods cannot evaluate it accurately.
- More stress levels testing (i.e., 10 and 25 kPa) should be performed with different CR mesh size (from different suppliers in Lebanon) in order to properly investigate the linearity or non-linearity and the viscoelastic behavior of the modified asphalt binders by plotting creep compliance and percent recovery against stress level, respectively.
- Achieving the required PG for different climatic zones in Lebanon and the surrounding area is feasible. Further efforts should be made to rely on the PG system to characterize imported asphalt binders and then modify it based on its terminal use. The rehabilitation of existing non-active refineries, coupled with the addition of the appropriate logistics to enable asphalt binder modification, should be the next step in implementing a complete system for better performing roadways in Lebanon. This methodology has been successfully established by many highway agencies around the world.

# References

Anderson, M., D'Angelo, J., Walker, D.: MSCR: a better tool for characterizing high temperature performance properties. Asphalt Inst. **25**(2), 15–23 (2010)

Angelo, J.D.: The relationship of the MSCR test to rutting. Road Mater. Pavement Des. **10**, 61–80 (2009)

ASTM D2872: Standard test method for effect of heat and air on a moving film of asphalt (rolling thin-film oven test). American Society for Testing Materials (ASTM) International, West Conshohocken, PA (2012)

ASTM D4402/D4402M: Standard test method for viscosity determination of asphalt at elevated temperatures using a rotational viscometer. American Society for Testing Materials (ASTM) International, West Conshohocken, PA (2015)

ASTM D5/D5M: Standard test method for penetration of bituminous materials. American Society for Testing Materials (ASTM) International, West Conshohocken, PA (2015)

ASTM D6648-08: Standard test method for determining the flexural creep stiffness of asphalt binder using the bending beam rheometer (BBR). American Society for Testing Materials (ASTM) International, West Conshohocken, PA (2016)

AASHTO R 28: Standard Practice for Accelerated Aging of Asphalt Binder Using a Pressurized Aging Vessel (PAV). American Association of State Highway and Transportation Officials (AASHTO), Washington, D.C. (2016)

AASHTO T 313: Standard Method of Test for Determining the Flexural Creep Stiffness of Asphalt Binder Using the Bending Beam Rheometer (BBR). American Association of State Highway and Transportation Officials (AASHTO), Washington, D.C. (2019)

AASHTO TP 70: Standard Method of Test for Multiple Stress Creep Recovery (MSCR) Test of Asphalt Binder Using a Dynamic Shear Rheometer (DSR). American Association of State Highway and Transportation Officials (AASHTO), Washington, D.C. (2013)

AASHTO M 320: Standard Specification for Performance-Graded Asphalt Binder. American Association of State Highway and Transportation Officials (AASHTO), Washington, D.C. (2017)

Bulatović, V.O., Rek, V., Marković, K.J.: Polymer modified bitumen. Mater. Res. Innov. **16**(1), 1–6 (2012)

Divya, P.S., Gideon, C.S., Murali Krishnan, J.: Influence of the type of binder and crumb rubber on the creep and recovery of crumb rubber modified bitumen. J. Mater. Civ. Eng. **25**(4), 438–449 (2012)

Eberhardsteiner, L., et al.: Influence of asphaltene content on mechanical bitumen behavior: experimental investigation and micromechanical modeling. Mater. Struct. **48**(10), 3099–3112 (2015)

Jafari, M., Babazadeh, A., Aflaki, S.: Effects of stress levels on creep and recovery behavior of modified asphalt binders with the same continuous performance grades. Transp. Res. Rec. **2505**(1), 15–23 (2015)

Khalil, N., Bahia, H., Clopotel, C.: Developing a performance grading system for asphalt binders in Lebanon. In: Sixth International Conference on Maintenance and Rehabilitation of Pavements and Technological Control (MAIREPAV6). Turin, Italy (2009)

Khan, I.M., Kabir, S., Alhussain, M.A., Almansoor, F.F.: Asphalt design using recycled plastic and crumb-rubber waste for sustainable pavement construction. Proc. Eng. **145**, 1557–1564 (2016)

Labaki, M., Jeguirim, M.: Thermochemical conversion of waste tyres-a review. Environ. Sci. Pollut. Res. **24**(11), 9962–9992 (2017)

Lesueur, D.: The colloidal structure of bitumen: consequences on the rheology and on the mechanisms of bitumen modification. Adv. Colloid Interface Sci. **145**(1–2), 42–82 (2009)

Liu, G., Li, L., Wu, S., van de Ven, M.: Studying the effect of soft bitumen on the rheological properties of reclaimed PMB binder by using the DSR. Int. J. Pavement Eng. **19**(8), 706–712 (2018)

Magar, N.R.: A study on the performance of crumb rubber modified bitumen by varying the sizes of crumb rubber. Int. J. Eng. Trends Technol. **14**(2), 50–56 (2014)

Morea, F., Zerbino, R., Agnusdei, J.: Wheel tracking rutting performance estimation based on bitumen Low Shear Viscosity (LSV), loading and temperature conditions. Mater. Struct. **47** (4), 683–692 (2014)

Mturi, G.A., O'Connell, J., Zoorob, S.E., De Beer, M.: A study of crumb rubber modified bitumen used in South Africa. Road Mater. Pavement Des. **15**(4), 774–790 (2014)

Nejad, F.M., Aghajani, P., Modarres, A., Firoozifar, H.: Investigating the properties of crumb rubber modified bitumen using classic and SHRP testing methods. Constr. Build. Mater. **26** (1), 481–489 (2012)

Neto, S.A.D., Farias, M.M.D., Pais, J.C., Pereira, P.A.: Influence of crumb rubber gradation on asphalt-rubber properties. In: Asphalt Rubber Conference, pp. 679–692. Palm Springs, California, USA (2006)

Presti, D.L.: Recycled tyre rubber modified bitumens for road asphalt mixtures: a literature review. Constr. Build. Mater. **49**, 863–881 (2013)

Subhy Torretta, V., Rada, E.C., Ragazzi, M., Trulli, E., Istrate, I.A., Cioca, L.I.: Treatment and disposal of tyres: two EU approaches. A review. Waste Manag. N.Y. **45**, 152–160 (2015)

Tsai, W.-T., Chen, C.-C., Lin, Y.-Q., Hsiao, C.-F., Tsai, C.-H., Hsieh, M.-H.: Status of waste tires' recycling for material and energy resources in Taiwan. J. Mater. Cycles Waste Manage. **19**(3), 1288–1294 (2017)

Wasage, T., Stastna, J., Zanzotto, L.: Rheological analysis of multi-stress creep recovery (MSCR) TEST. Int. J. Pavement Eng. **12**, 561–568 (2011)

Xu, O., Xiao, F., Han, S., Amirkhanian, S.N., Wang, Z.: High temperature rheological properties of crumb rubber modified asphalt binders with various modifiers. Constr. Build. Mater. **112**, 49–58 (2016)

# Road Safety Leadership

# How Do Best Performing Countries in Road Safety Save Lives on the Roads? Lessons Learned from Case Studies in Singapore

Alina Florentina Burlacu[1(✉)] and Emily Tan[2]

[1] Global Road Safety Facility, The World Bank, Washington, DC, USA
`fburlacu@worldbank.org`
[2] TSM Consultancy Pte Ltd, Singapore, Singapore

**Abstract.** We are still in an era where road engineers are focused only on design standards and creating roads for motor vehicles. It is necessary to change this mentality in order to have safer streets, and Singapore is a good example of success story for building a safer road infrastructure for safer communities, especially through the coordination of different institutions. Singapore is considered one of the best performing countries globally and regionally in terms of road safety. Road safety management rules and regulations implemented in the country have resulted in significant strides in managing the effects of collision factors related to roadway design, human behavior, and vehicle attributes. As a result, road safety statistics have shown that fatalities on the Singapore road network have been steadily declining over the past decade. This is leading to a desire on the part of neighboring countries to follow Singapore's example and learn from its experience. This paper will show examples and case studies on why Singapore roads are not only considered the safest in the region, but they also rank among the safest globally, through a strong road safety management overseeing safe road infrastructure, enforcement, vehicle safety.

**Keywords:** Road safety · Road safety management · Road safety engineering · Road safety audits (RSA) · Red light cameras · Singapore

## 1 Introduction

With the rapid development of urban mobility, urban roads have become composite systems used by multiple users, including vehicle drivers and passengers, motorcyclists, pedestrians and cyclists. The increasing number of vehicles and road users, together with increased traffic speeds, led to defects and problems of road design and traffic management allow more and more road crashes and injuries. Unequivocally, the safety of road users' life and property should be of the highest priority of the design and management of road agencies, and actions have been taken by the authorities and organizations among many countries in the world. With its steadily declining numbers of road fatalities and serious injuries over the past decade, Singapore's road system is not only considered the safest in the Southeast Asia region, but also ranks high in the world with respect to road safety.

© The Author(s), under exclusive license to Springer Nature Switzerland AG 2022
A. Akhnoukh et al. (Eds.): IRF 2021, SUCI, pp. 117–133, 2022.
https://doi.org/10.1007/978-3-030-79801-7_9

## 1.1 Road Safety Data of Singapore

In Asia and the Pacific, it is estimated that vulnerable road users (VRUs) such as pedestrians, cyclists and motorcyclists account for approximately 50% to 75% of all road traffic fatalities. Globally, the South-East Asia region accounts for the second highest road fatality rate, surpassed only by Africa. In 2016, there were approximately 20.7 road traffic fatalities per 100,000 people in the South-East Asia region (WHO 2018), and the road traffic fatality rate in the South-East Asia region was approximately 13% greater than the overall global road traffic fatality rate in 2016.

As a comparison, the safety of Singapore roads stands out from the rest of Southeast Asian countries. Road safety statistics provided by the Singapore Traffic Police (2019) show that there is a downward trend in the number of road traffic fatalities (as presented in Fig. 1): over a 10-year period (2009–18), Singapore was able to reduce the number of fatalities on its roads by approximately 32%, to 124 fatalities in 2018. In addition, the overall number of collisions resulting in injury over a 10-year period has experienced decreases and increases, which is fluctuating between 80 to 95 collisions per 10,000 vehicles. By combining these data, we can deduce that overall there is a decrease in the number of fatalities and the number of collisions resulting in injury in Singapore roads.

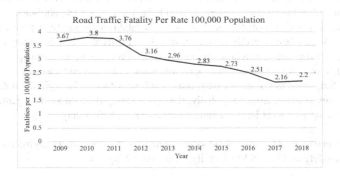

**Fig. 1.** Fatalities per 100,000 population – Singapore

The good performance of Singapore roads in terms of road safety (with only 2.8 fatalities per 100,000 population, according to WHO, 2018) can be attributed to a number of factors such as lower speeds, increased driver awareness and enforcement, safer vehicles, safer roads and better post-collision care, and these factors are considered as some of the key elements of the road system: road users, roads and roadsides, vehicles and travel speeds in the Safe System Approach. The main highlight of the Safe System Approach is that while it recognizes the need for responsible road user behavior, it also accepts that human error is inevitable. It therefore aims to create a road transport system that makes allowance for errors and minimizes the consequences - in particular, the risk of death or serious injury.

## 1.2 Agencies and Organizations Contributing to Road Safety in Singapore

With the idea of Safe System Approach, Land Transport Authority in Singapore (LTA) are responsible for the planning, design, construction, operation, maintenance and regulation of roads in Singapore. Roads are designed in accordance to design standards set out by LTA. This entails a number of design specifications documents which take road safety into consideration when LTA developed the design standards, and helped create a safer road environment for all road users. In addition, the enforcement of traffic regulations falls under the responsibility of the Singapore Traffic Police (TP) who helps to guarantee road safety from the perspective of legislation.

The Road Safety division of the Singapore Traffic Police holds the responsibility of delivering road safety education to the public. They work hand in hand with various community groups to promote road safety for all road users. The traffic police reaches the masses using the following communication channels: Television Commercials, Radio Commercials, Newspaper Advertisement, Publicity materials, Internet and Info-Communication Technology, etc.

A number of other non-governmental organizations play a role in delivering road safety education to the public. These include the Singapore Road Safety Council (SRSC) and the Automobile Association of Singapore (AAS). The SRCS was established in 2009 to focus on road safety issues. The SRCS is funded mainly through private donations and small government grants.

In this paper, three essential managements conducted to make Singapore a best-performing countries in road safety are explained, namely Road Safety Management, Speed Management and Road Infrastructure Management. In succession, several case studies in Singapore regarding road safety are introduced to depict a holistic image of how Singapore constantly improves road safety from the perspective of creating safer road design as well as treating existing dangerous road design.

# 2 Road Safety Management

The first and fundamental pillar of the first Decade of Action's Global Plan is road safety management (UNRSC 2011). The World Report on Road Traffic Injury Prevention (Peden 2004) and the Global Plan highlight that a systematic and planned approach is required to improved road safety performance. The most effective way countries and organizations can improve road safety performance is by establishing an effective road safety management system.

In Singapore, the LTA are responsible for the provision and maintenance of road facilities, and vehicle safety, whereas the Traffic Police are entrusted with the responsibility of enforcing traffic regulations, publicity and education.

## 2.1 Road Safety Audits (RSA)

The first important component of Singapore's road safety management is RSAs. The Austroads guidelines define an RSA as a "formal examination of a future road or traffic project, an existing road, or any project which interacts with road users, in which an independent, qualified team assesses the crash potential and safety performance". (Austroads 2019a, b) RSAs can lead to safer roads by removing or treating safety hazards and promoting the incorporation of safety features or collision-reduction features.

RSAs were introduced to Singapore in 1998 by the LTA. RSAs are also known as a Project Safety Reviews (PSR) in Singapore. The PSR is not a design check, it is a review of the safety and adequacy of the design. The PSR is an independent review and assessment of the project team's assertion that the proposed road system is safe to use (LTA 2019a, b).

The PSR process in Singapore consists of four stages to which an audit of a road scheme is required:

- Planning - Preliminary Design Safety Submission
- Design – Detailed Design Safety Submission
- Construction – Temporary Traffic Control Safety Submission
- Completion – Post Construction Safety Submission

An independent safety review report is required for each of the above-mentioned stages, highlighting safety deficiencies and proposing remedial countermeasures. Table 1 provides a summary of the PSR process and responsibilities held by various stakeholders.

**Table 1.** PSR stages

| Stage | Role | Responsibility |
|---|---|---|
| Preparation of Design Details | Contractor/Traffic Consultant | Prepare design details and Project Brief |
| | LTA project team | In-principle approval of design details |
| Conduct of Safety Review | Independent safety review team | Prepare draft safety review report |
| | LTA project team Contractor/Traffic Consultant | Review and accept draft safety review report |
| | Independent safety review team | Prepare final safety review report |
| Preparation of Response | Contractor/Traffic Consultant | Prepare response to recommendations in safety review report |
| | LTA project team | Approve Safety Submission |
| Audit | LTA Safety Division | Audit Safety Submission |
| Endorsement | PSR Committee (Roads) | Endorse Safety Submission |
| Implementation | Contractor/Traffic Consultant | Implement design details on site |

## 2.2 Red Light Camera System

Red Light Camera System is a powerful approach of Singapore's traffic safety management. Red light camera studies (Retting et al. 2010) have observed changes in red light violations at non-camera sites, i.e. at similar signalized intersections in the same communities that were not equipped with RLC. Large reductions in violation rates, of an order resembling those at camera sites, were observed at these sites, indicating a spill over or 'halo' effect. This shows that in some cases the implementation of RLC can have a positive effect on nearby areas.

As of 2019, there are a total of 240 red light cameras located across Singapore. Figure 2 provides a visual representation of the location of red-light cameras in Singapore.

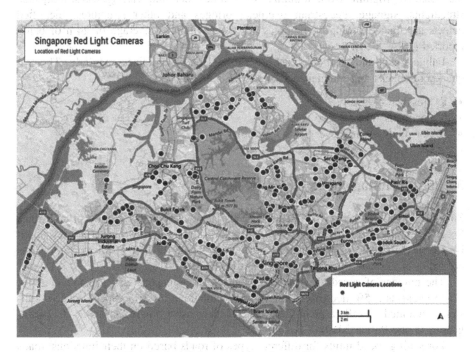

Fig. 2. Red light camera locations - Singapore

The existing network of RLC's was introduced to the Singapore road network in 1986. Since their installation, the number of vehicles on the road has increased significantly from under half a million vehicles in 1987 to over 900,000 vehicles in 2018. Furthermore, the number of driving license holders has increased considerably from 735,480 holders in 1987 to over two million in 2018.

The number of Red-Light violation cases recorded in 2018 was 53,910, representing a 15.7% increase from 46,599 cases in 2017. The number of red-light running collisions increased slightly by 2.6% to 120 collisions in 2018, from 117 collisions in 2017. Although an increase in the number of violations has been recorded between 2018 and 2019, over a long-term period there is a decline.

## 2.3 LTA Road Maintenance Program

Road maintenance is vital for the preservation of a road's safety and quality. The frequent maintenance of a road can lead to a reduction in the probability that a collision will occur due to hazards located on the road. Furthermore, regular maintenance ensures that the level of safety is adequate for all road users.

In Singapore, the LTA operates a comprehensive road maintenance program, which encompasses the maintenance of roads, roadside features and pedestrian facilities. The maintenance program covers a number of facilities including carriageways, footpaths, streetlights, signage, pedestrian overhead bridges and other structures. Identified defects are treated promptly, potholes are fixed within an average of 24 h of being reported (LTA 2019).

Public roads and road facilities are checked and maintained regularly. The frequency of checks and maintenance of a road is as follows:

- Expressways – Daily
- Major roads – Every two weeks
- Minor roads – Once every two months

## 3 Speed Management

Speed is considered a key risk factor in collision severity and causation of collisions. Speed accounts for approximately 30% of road deaths in high-income countries, and it is estimated that speed is the main contributory factor for half of all road collisions occurring in low and middle-income countries (WHO 2018).

The number of speeding violations detected in Singapore was 156,157 in 2018. This represents a 5% decrease from 164,319 violations in 2017. Speed related collisions amounted to 719 cases in 2018, a 5.6% decrease from 762 cases in 2017 (Singapore Traffic Police 2019).

For setting speed limits for different types of roads based on their functions, roads in Singapore are classified into five categories (LTA 2019): Expressways, Major Arterial Roads, Minor Arterial Roads, Primary Access Roads and Local Access Road.

### 3.1 Setting Speed Limits

Speed limits should be evidence-led and self-explaining, seeking to reinforce the driver's assessment of what is a safe speed to travel. Speed limits should encourage drivers to be self-compliant. Drivers should view the speed limit as the maximum speed and not a target speed.

There is a strong correlation between the probability of a collision and speed as well as the severity of injuries sustained due to a collision. Research has shown that the probability of a collision can be reduced by 5% with every 1 mph/h (1.6 km/h) reduction in average speeds (Taylor, Lynam and Baruya 2000).

When setting an appropriate speed limit for a road, a number of essential factors are taken into consideration:

- History of collisions
- Road geometry and engineering
- Road function
- Composition of road users
- Existing traffic speeds
- Road environment

In addition, there are a number of approaches used when setting appropriate speed limits, which include the following:

- Engineering Approach – the base speed limit of a road is set according to a number of factors including the 85th percentile speed, the design speed of the road and other conditions.
- Expert System Approach – speed limits are determined by the use of computer programs which take into consideration a number of factors relating to road conditions.
- Safety Systems Approach – speed limits are based on the type of collision which can occur, the likely severity of the collisions, and the human body's tolerances to the forces of the collision.

## 3.2 Speed Cameras

Speed cameras have long been used as an effective method of reducing traffic speeds and thus reducing casualties and collisions. A number of speed camera studies have shown that speed cameras can lead to a reduction in the number of collisions. In some cases, collisions can be reduced by as much as 17% (Department for Transport 2002).

Speed cameras are one of the speed enforcement measures utilized in Singapore. There are four different types of speed cameras used in Singapore, Fixed Speed Camera (FSC), Police Speed Laser Camera (PSLC), Mobile Speed Camera (MSC) and Average Speed Camera (ASC). In total, there are a total of 87 speed camera locations across Singapore.

There is a perception amongst certain members of the public that speed cameras are only a revenue generating apparatus used by authorities. Although the use of speed cameras can generate significant income for authorities, their effectiveness in reducing collisions is undeniable.

Figure 3 shows the number of speeding violations recorded by the Singapore Traffic Police between 2010 and 2018. A noticeable decline in the number of violations is observed between 2014 and 2015. Speeding violations declined from 278,545 in 2014 to 186,838 in 2015 representing a 33% decrease. This decline can be directly attributed to measures implemented by the Singapore Traffic Police, such as the installation 20 speed cameras at 11 locations on 1st March 2015. But it is difficult to tell whether this decline is entirely due to the introduction of new speed cameras or other confounding factors.

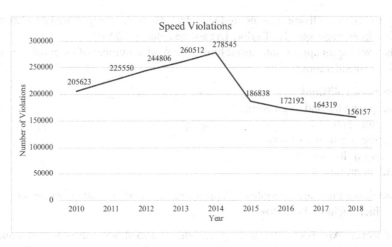

**Fig. 3.** Speeding violations 2010–2018 – Singapore

## 3.3 Low Speed Zones

Lower speed limits are enforced in designated zones. These include School Zones and Silver Zones where speed limits are set at 40 km/h. School Zones are roads in close proximity to a school or roads located between School Zone signs. Silver Zones are located in residential neighborhoods. In addition to lower speed limits (when possible), a number of safety features are implemented in order to enhance road safety for children or elderly pedestrians. For example, Silver Zone features are the following:

- Rest points on the road divider, to assist elder pedestrians crossing in two stages.
- Upon entering the Silver Zone, road signs are displayed in addition to three rumble strips on the road; this is to slow down drivers.
- Other traffic features, including chicanes, reduced lane widths and gentle curves along sections of the road.

## 3.4 Traffic Calming Measures

Traffic calming is a system which uses varying design and management strategies to achieve a balance of traffic on roads and streets, particularly those with VRUs. Traffic calming measures not only act to slow down vehicles, but also improve safety and convenience for pedestrians. This can be for example achieved by implementing measures used to shorten crossing distances for pedestrians. Traffic calming measures can include the following:

- Vertical and horizontal measures such as road humps, chicanes, road narrowing.
- Optical measures which include the use of rumble strips, shortened sightlines, road surface changes including color and texture.
- Changes to road environment including the abundant use of vegetation, placement of street furniture.
- Lower speed limit zones, which includes school zones or silver zones.

In Singapore, Traffic calming measures can be found on a number of streets. For example, center refuge islands, center dividers and chevron markings can be easily spotted as measures of slowing down vehicles and guaranteeing pedestrians' safety. Besides, physical measures (i.e., road humps) are also widely used by Singapore to calm traffic. It is worth pointing out that hatching is also used as a traffic calming measure to provide better driver visibility and protect jaywalkers at the turn of a road.

In short, various traffic calming measures are adopted in Singapore roads as a cost-effective way to improve the safety of roads.

# 4 Road Infrastructure Management

"Safer roads and safer mobility" is the second pillar of the UN Global Plan for the Decade of Action for Road Safety 2011–2020. The necessity to increase the safety and quality of road networks for all roads users especially the most vulnerable is highly emphasized in this pillar. In order to achieve this, measures such as improved safety-conscious planning, design, construction and operation of roads need to be adopted.

## 4.1 Collision Factors

Road infrastructure, which includes roadway and roadside design elements, can have a significant role in determining the risk of a traffic collision occurring. A traffic collision can be caused directly due to a defect on the road or in some cases misleading elements of the road environment can lead to human error. Road safety design concepts such "Forgiving Roads" and "Self-explaining" are significant and should be adopted in the design of roads in order to reduce and minimize the risk of traffic collisions.

There is a strong correlation between poor road designs and a higher risk of a collision. Roads should be self-explaining, providing guidance to roads users on what they should be doing. Collisions can be triggered by negative road engineering factors or misleading road environments which lead to collisions through human error.

## 4.2 Safer Roadsides Through Forgiving Roads

Roadsides play an important role in the safety of roads. A well-design and safety-oriented road, also called forgiving road, aims not only to prevent collisions from occurring, but also to reduce the damage inflicted upon a driver and other passengers if a collision does happen due to human errors. A forgiving road is largely based on how the roadside is designed and equipped, as a number of inadequately design roadside elements pose a risk to drivers. In short, forgiving roads reduce the consequences of driver error. The common measures of providing safer roadsides and forgiving roads are as follows:

Vehicle Impact Guardrail (VIG)

Vehicle Impact Guardrails (VIG) or safety barriers are a form of forgiving roadside treatment, which are designed to prevent vehicles from running off the carriageway. However, VIGs can also act as a hazard in circumstances where installation is incorrect or the wrong type of VIG is used. The VIG ends are considered hazardous when the end is not properly anchored to the ground, or when it does not flare away from the carriageway (La Torre 2012). Collisions with an 'unforgiving' VIG end can results in fatal consequences.

VIGs can be classified as energy-absorbing and non-energy-absorbing, this is dependent on whether it is a tangent terminal or a flared terminal. Tangent terminals are aligned parallel to the carriageway edge and are energy absorbent, they are designed to stop a vehicle. Flared terminals deviate from the alignment of the carriageway edge, typically they are not designed to dissipate significant amounts of energy when a head on collision occurs.

Rumble Strips

Rumble strips are a road safety feature designed to alert drivers to potential hazards. When a vehicle comes into contact with the rumble strips, a "rumbling" sound is transmitted in addition to a vibrating effect. In Singapore, rumble strips are typically used in pedestrian priority areas such as Silver Zones. Three yellow rumble strips are installed in order to alert vehicle drivers of pedestrians.

Raised Profile Marking

Raised profile markings function similarly to rumble strips. They can be installed along a carriageway in order to prevent vehicles deviating from the carriageway. This road safety feature is particularly useful in preventing drowsy or distracted drivers from causing a potential collision, as drivers are alerted by the vibration and sound emitted from the strips.

The LTA provides guidance on where raised profile markings should be provided. Raised profile markings are to be provided for the following:

- "From the start of the shoulder marking at the exit road to 10 m behind the gore area;
- From the start of chevron to 10 m after the gore area;
- From the start of the deceleration lane along the expressway next to pave shoulder;
- Continuously along the expressway shoulder lane next to slow lane." (LTA 2017)

Design specification for raised profile markings provided by within the LTA is presented in Fig. 4.

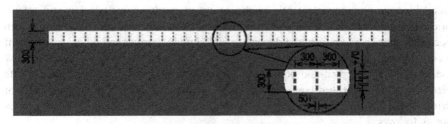

**Fig. 4.** Raised profile marking

Crash Cushions

Crash cushions are considered highly effective in reducing the consequences of a crash. The crash cushion is designed to absorb the impact of the collision. Crash cushions are usually positioned in front of diverging roads, and along expressways and major arterial roads.

The use of crash cushions has been very effective in some cases. A 40% reduction in injury crashes was observed at study sites in Birmingham, England. The treated site also experienced a reduction in the number of fatal and serious crashes from 67% to 14% (TMS Consultancy 1994).

### 4.3 Self-explaining Roads

Self-explaining roads are designed so that the appropriate speed and driving behavior required on the road is self-evident to drivers. Self-explaining roads can be implemented through a number of measures such as appropriate road markings and roadside features.

Self-explaining roads have been successfully implemented in a number of urban areas around the world. According to a study undertaken by Charlton et al. (2010), the implantation of self-explaining roads resulted in a significant reduction in vehicle speeds. The study area was segregated into two sections, one receiving treatment measures such as increased landscaping and limiting forward visibility, the other area received no treatment. The results of the study showed a significant reduction in vehicles speeds for the section which had received the 'Self-explaining road treatments'.

## 5 Case Studies

To have a clear understanding of the reasons why Singapore roads system is the safest in the world, we may learn from the microscopic cases of the identified dangerous elements which are hazardous to road users and cases of sustainable safer roads in Singapore.

### 5.1 Identifying Dangerous Existing Streets

A number of methods can be utilized to assess the safety of a street or road. This includes RSAs and collision hotspot schemes. An RSA can be undertaken for an existing road or route on the road network. The aim of an RSA of an existing road is to identify hazards which may cause a collision in the future. Remedial measures can be applied to hazards identified during the safety audit.

Collision data is an important factor in identifying and treating collision hotspots. Collision hotspots can be identified by looking at a baseline period, which is usually 3–5 years. Clusters are identified by defining a minimum threshold in a defined radius. A collision hotspot can then be defined as an area which the number of collisions exceed the pre-determined threshold during the baseline period.

Since 2005, the LTA have implemented a road safety initiative known as the Black Spot Programme. The key objective of the Black Spot Programme is to identity, monitor and treat locations with a high number of traffic collisions. The Black Spot Programme has proven to be a success with an average of five to 10 location being removed per year due to the collision rate falling below the defined threshold level. The programme has resulted in a 75% drop in collisions over a three-year period in some cases (LTA 2014).

## 5.2 Dangerous Road Designs

Road alignment has a significant influence on the safety of a road, which includes the dimension of radii, ratio of consecutive curves, dimensions of vertical curves and sight distance conditions. (Mohammed 2013) Badly designed roads can be a major contributory factor leading to a collision, which includes road defects and misleading road environments.

A number of different road elements such as markings, signs, road geometry, lighting, road surface, and traffic and speed management are essential to maintain the safety of a road. The guidance and standards provided do promote safe design, but without comprehensive audits significant safety hazards can be neglected.

Hazards such as the one presented in Fig. 5 pose a risk to drivers. Without RSAs, hazard like this can go unnoticed until it's too late. (LTA 2019) There is inadequate visibility along the vertical curve of the road displayed in Fig. 5. Driver's visibility and sight stopping distance is greatly affected, as drivers are unable to see and react to potential hazards which lay ahead.

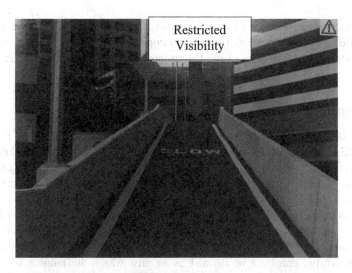

**Fig. 5.** Inadequate visibility along vertical curve - Singapore

## 5.3 Dangerous Pedestrian and Cyclist Infrastructure

Pedestrians and cyclists are considered the most VRUs. Surprisingly, it is very common in many countries to focus on road design to accommodate vehicles and neglect the need for safe pedestrian and cyclist infrastructure. VRUs which include pedestrians, cyclists and motorized two- and three-wheelers account for over half of all road deaths globally. In South-East Asia motorized two- and three-wheelers account for the highest proportion of road deaths - approximately 43% (WHO 2018).

Although Singapore has been making strides in promoting walking and cycling, pedestrians and cyclists are still very vulnerable on roads. The number of collisions on Singapore roads involving pedestrians was 1,036 in 2018, where elderly pedestrians accounted for 25% of pedestrian collisions. Pedestrian fatalities in 2018 were 40, where 62.5% were elderly pedestrians. Jaywalking is attributed to 40% of collisions involving elderly pedestrians (Singapore Traffic Police 2019). Although the quality and provision of pedestrian and cyclist infrastructure is considered better than that of neighboring countries, inferior pedestrian and cycling facilities can still be found around the country.

Figure 6 shows two pedestrian crossings with tactile paving. The crossing directed towards the main carriageway on the left, formed part of a crossing across the main carriageway which is now removed, but the dropped kerb and tactile paving remain in place. The unremoved dropped kerb and tactile paving can be confusing and pose a risk to pedestrians, especially visibility impaired pedestrians.

**Fig. 6.** Unremoved crossing – Singapore

Inadequate design of pedestrian infrastructure is an unfortunate but recurring theme on some streets in Singapore. The common design errors observed include: no dropped kerbs and tactile paving at pedestrian crossing areas, incorrect positioning of tactile paving, inadequate widths of pedestrian footpaths and dangerous level changes along edge of pedestrian footpaths.

## 5.4 Safer Design Principles: Sustainable Cycling and Pedestrian Infrastructure

Singapore has implemented a range of measures such as Certificate of Entitlement (COE) and road pricing to maintain adequate capacity on roads. In recent times, initiatives such as Car-Lite have been introduced in order to promote sustainable modes of transport and reduce dependency of motor vehicles. A number of roads have been redeveloped and equipped with infrastructure to support sustainable modes of transport such as walking and cycling. Bencoolen Street in Singapore is one example, having recently undergone a transformation to become more pedestrian and cycle friendly.

Bencoolen Street was formerly a four-lane street with no cycle infrastructure and narrow pedestrian footpaths. The past street configuration is presented in Fig. 7.

**Fig. 7.** Previous street configuration - Bencoolen Street, Singapore

The new revitalized Bencoolen Street provides pedestrians with wider footpaths and cyclists have segregated cycle paths. The carriageway width has been reduced to two lanes in order to accommodate the provision of a cycling path and wider pedestrian footpaths. Figure 8 and Fig. 9 show Bencoolen Street after improvements have been implemented. Figure 8 shows the reduced carriageway width of Bencoolen Street. A cycle lane can be seen in red adjacent to the pedestrian footpath. The provision of segregated cycle lanes, cycle parking and wide pedestrian footpaths has made the area more sustainable and inclusive to pedestrians and cyclists.

How Do Best Performing Countries in Road Safety Save Lives on the Roads? 131

**Fig. 8.** New street configuration – Bencoolen Street, Singapore

**Fig. 9.** Improved pedestrian facilities - Bencoolen Street, Singapore

## 6 Conclusions

This paper was elaborated from the perspective of various key management measures regarding traffic safety (i.e. Road Safety Management, Speed Management, Road Infrastructure Management), and shows how Singapore enjoys a high level of safety on its road network due a number of complementing factors and not just one factor, and how it has also made significant progress in addressing the road safety issue faced by the nation, making the Singapore approach to road safety a representative example of the well-known Safe System Approach. One of the highlights of this paper is the effective road safety management system in Singapore, based especially on the tight coordination between the LTA and Traffic Police, which has as a core focus the extensive use of road safety audits and frequent maintenance for road infrastructure, together with a strong and reliable red light camera system spread across the entire country. Speed management is also given top priority, through evidence-led and self-explaining speed limits enforced by 87 speed camera locations across Singapore, together with clearly marked low speed zones and evidence-based traffic calming measures.

The lessons from Singapore may inspire road engineers and make them realize the imperious need to change the conventional mentality of only focusing on design standards and creating roads for motor vehicles, and not for all types of road users, especially those most vulnerable. Singapore is a good example of success story for building a safer road infrastructure, for safer communities. At the same time, the cases of Singapore roads also arouse the fact that there is no single solution to tackle road safety, but collective ones. These include safer roads through improved design and maintenance of the road network, better road management systems, high vehicle standards, together with improved communication, education and awareness. Measures implemented in Singapore can be easily replicated in other cities and countries in the region, but also globally.

**Acknowledgements.** This analysis was undertaken thanks to support from Bloomberg Philanthropies and the World Bank through the Global Road Safety Facility (GRSF), under the Bloomberg Philanthropies Initiative for Global Road Safety 2015–2019.

# References

Austroads: Guide to Road Safety - Part 6: Managing Road Safety Audits (2019a)
Austroads: Guide to Road Safety – Part 6A: Implementing Road Safety Audits (2019b)
Charlton, S.G., Mackie, H.W., Baas, P.H., Hay, K., Menezes, M., Dixon, C.: Using endemic road features to create self-explaining roads and reduce vehicle speeds. Accid. Anal. Prev. **42**(6), 1989–1998 (2010)
Department for Transport: A Cost Recovery System for Traffic Safety Cameras-First Year Report. Department for Transport, London (2002)
La Torre, F.: Forgiving roadsides design guide (No. 2013/09) (2012)
LTA (2014). Factsheet: Enhancing Safety on Our Roads for All Road Users. Press Room. https://www.lta.gov.sg/apps/news/page.aspx?c=2&id=91de65eb-fea5-48f6-b101-5573dc68face
LTA: Procedures on Importation and Registration of a Car in Singapore (2017)
LTA: Certificate of Entitlement (COE)|Vehicle Quota System|Owning a Vehicle|Roads & Motoring (2019). https://www.lta.gov.sg/content/ltaweb/en/roads-and-motoring/owning-a-vehicle/vehicle-quota-system/certificate-of-entitlement-coe.html
LTA: Code of Practice - Street Work Proposals Relating to Development Works (2019)
LTA: Road Maintenance Programme|Maintaining Our Roads & Facilities|Road Safety & Regulations|Roads & Motoring (2019). https://www.lta.gov.sg/content/ltaweb/en/roads-and-motoring/road-safety-and-regulations/maintaining-our-roads-and-facilities/road-maintenance-programme.html
LTA: PSR (ROADS) PROCESS (2019)
LTA: Road Safety Guide Book - Temporary Road Works for LTA Projects (2019)
Mohammed, H.: The influence of road geometric design elements on highway safety. Int. J. Civil Eng. Technol. **4**, 146–162 (2013)
Peden, M.: World report on road traffic injury prevention (2004)

Retting, R.A., Ferguson, S.A., Hakkert, A.S.: Effects of red light cameras on violations and crashes: a review of the international literature. Traffic Inj. Prev. 4(1), 17–23 (2003)

Singapore Traffic Police. Statistics (2019). https://www.police.gov.sg/news-and-publications/statistics?category=RoadTrafficSituation#content

Taylor, M.C., Lynam, D.A. Baruya, A.: TRL Report 421 – The Effects of Drivers' Speed on the Frequency of Road Accidents. TRL, Crowthorne (2000)

TMS Consultancy: Research on loss of control accidents on Warwickshire motorways and dual carriageways. Coventry (1994)

WHO: Global status report on road safety 2018. World Health Organization (2018)

WHO: The top 10 causes of death (2018). https://www.who.int/news-room/fact-sheets/detail/the-top-10-causes-of-death

WHO: Road Safety - Speed (2018)

# Data for Road Incident Visualization, Evaluation, and Reporting (DRIVER): Case Studies of Cebu, Philippines and the Asia-Pacific Road Safety Observatory (APRSO)

Florentina Alina Burlacu[1]([⊠]), Miguel Enrico III Cabag Paala[2],
Veronica Ines Raffo[3], and Juan Miguel Velasquez Torres[1]

[1] The World Bank Global Road Safety Facility, Washington, DC, USA
fburlacu@worldbank.org
[2] The World Bank Global Road Safety Facility, Manila, Philippines
[3] The World Bank Global Road Safety Facility, Buenos Aires, Argentina

**Abstract.** Robust crash data systems are key to understand the nature and magnitude of the road safety problem and to propose appropriate solutions. However, resource constrained governments oftentimes lack adequate data systems or tools for collecting, storing, managing, analyzing, and reporting historic crash data. As a result of previous World Bank efforts, and with the aim of tackling this problem, a free open-source system for recording and visualizing geo-referenced crash data was developed and piloted. This system, Data for Road Incident Visualization, Evaluation, and Reporting (DRIVER), does not have licensing costs for clients and can be deployed by agencies interested in improving their crash data management systems. Originally in response to the absence of sound crash data in the Philippines, the World Bank, working with the Government of the Philippines (GoP), developed and piloted DRIVER starting 2014. After successfully completing the pilot phase in Cebu, and later in Manila, all relevant government agencies signed a memorandum of understanding (MOU) on data input, sharing, and use in 2016, marking the beginning of DRIVER as the national level road crash data management platform. The GoP is working to scale-up DRIVER at national level with World Bank support, with pilots starting in other countries around the world. This tool complements the work recently started under the Asia-Pacific Road Safety Observatory (APRSO) which aims to support countries in generating reliable crash data. DRIVER's successful implementation reinforces the growing demand for data and emphasizes the need for localization, capacity-building, and developing robust road safety institutional arrangements.

**Keywords:** Road safety · Road safety management · Data · Database systems · Observatory · APRSO · DRIVER Crash Database

## 1 Introduction

Road crash fatalities and injuries lead to major human, social, and economic losses, especially in low- and middle-income countries, which suffer 93 percent of the road deaths (World Health Organization 2018). These losses contribute to poverty at the

© The Author(s), under exclusive license to Springer Nature Switzerland AG 2022
A. Akhnoukh et al. (Eds.): IRF 2021, SUCI, pp. 134–147, 2022.
https://doi.org/10.1007/978-3-030-79801-7_10

national level by limiting economic growth, and at the individual level by driving families into poverty through the death or disability of the breadwinner. This makes road safety, especially of vulnerable road users, a crucial component of inclusive and sustainable mobility.

In the Philippines, at least 35 people are killed on the road every day due to road crashes (Philippine Statistics Authority 2021). From 2009 to 2019, road crash fatalities have been increasing at an average rate of 5%, instead of decreasing according to the Decade of Action for Road Safety 2011–2020 target (Fig. 1).

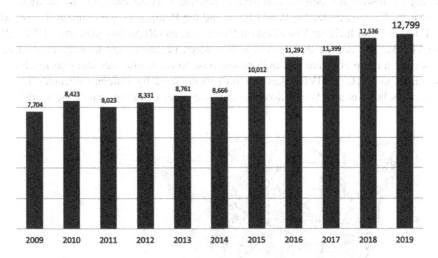

**Fig. 1.** Road crash fatalities in the Philippines from 2009 to 2019

While road crashes affect all ages, most road crash victims belong to the most productive age segments of the economy (Philippine Statistics Authority 2015) (Fig. 2).

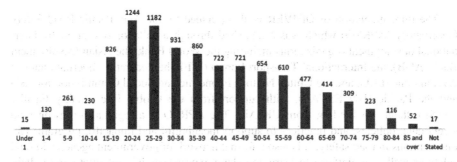

**Fig. 2.** Crash fatalities by age group in the Philippines, 2015

Crashes have profound effects economically costing the equivalent of 2.6% of the Philippines' Gross Domestic Product (GDP), accounting for loss of life, medical expenses, and grief and suffering, as well as socially by affecting the most vulnerable (pedestrians, cyclists and motorcyclists) who are in many cases the poorest also.

These warrant the urgent and effective implementation of road safety measures. However, there remains substantial underreporting of road crashes which prevents the government from understanding the gravity of the road safety issue. The lack of disaggregated and georeferenced data also prevents the government to understand where and when crashes happen, who are involved in crashes, and how and why crashes happen. This ultimately translates to a lack of evidence-based measures to making the roads safer and the continuing increase of crash fatalities and injuries.

It is in this regard that the World Bank and the Philippines government developed the Data for Road Incident Visualization Evaluation and Reporting platform - DRIVER (see Fig. 3). The aim of this system is to collect, organize, and manage crash data, supporting a more effective and holistic response to crash fatalities and injuries. Since its launch in 2016, DRIVER has not only been used by different agencies in the Philippines but has also been piloted in multiple other countries across the world.

**Fig. 3.** Metro Manila crashes as shown in DRIVER in 2015–2019

The implementation of DRIVER is also pursued by the Asia Pacific Road Safety Observatory (APRSO) which was established through a collaborative effort by international development organizations including the World Bank, the Asian Development Bank (ADB), the International Transport Forum (ITF), the Fédération Internationale de l'Automobile (FIA), and the United Nations Economic and Social Commission for Asia and the Pacific (UN ESCAP), with support from the Global Road Safety Facility (GRSF), thanks to funding from UK Aid. The APRSO is a regional forum on road safety data, policies, and practices to support meaningful and impactful programs and interventions in road safety. It is also a formal network of government agencies in road safety as well as a platform to learn and share experiences in road safety and collaborate with experts in various aspects of safety. The Philippines government is currently an active member of the APRSO.

This paper aims to demonstrate how DRIVER has addressed challenges in collecting and managing road safety data, particularly on how it has been implemented in the Philippines and led to changes in the previous road safety data management process, and in the past year has expanded to other countries as well as the APRSO.

## 2 Road Safety Data in the Philippines

In the Philippines, multiple government agencies collect road crash data, resulting to a wide range of systems that are neither integrated nor standardized. This translates to a lack of coordinated and evidence-based programs in road safety.

One of the main sources of data in the country is vital statistics data from the Philippine Statistics Authority (PSA). The Philippines has a functioning civil registration system with established and standardized procedures in reporting deaths and as part of the national coordinating body for civil registration, the PSA collects data from local civil registries and then publishes this in an annual report. Specifically, they collect data on road crash fatalities from death certificates. While the PSA can be a reliable source of fatality data, their data is not detailed enough to develop meaningful programs in road safety. A number of gaps have been observed:

1) Road crash fatality data is not updated regularly and takes at least a year to collate. When crash fatality data are released and updated, road and traffic conditions have already changed.
2) The data is not geo-referenced which means that it is difficult to locate exactly where fatal crashes are happening.
3) The level of data aggregation makes it difficult to develop targeted and evidence-based road safety measures.
4) Data sharing among agencies is not mandated or flexible although PSA is able to share their data through their reports or on a request basis by the concerned agency.

The latest literature on death registration in the Philippines even shows that in 2010 only 66% of deaths are registered while this is expected to improve by 80% on 2019. (Canadian Department of Foreign Affairs, Trade and Development et al. 2014).

Aside from the PSA, hospitals in the Philippines record crash injuries and fatalities through the Online National Electronic Injury Surveillance System (ONEISS) of the Department of Health (DOH). A sample ONEISS form is shown in Fig. 4. The data then follows a three-step process of collection, validation and cleaning, and publication.

For hospitals with internet connection, they are able to directly record their data in ONEISS. For those without internet, the data is sent to the DOH for uploading to ONEISS. The data collected by DOH through ONEISS include detailed injury data such as the nature and cause of injury. They also use the International Classification of Diseases-10 (ICD-10) code in classifying the severity of injuries as well as the 30-day threshold for classifying a crash fatality, as recommended by the WHO.

While ONEISS is able to collect detailed injury data, it is still in the process of scaling-up in the Philippines. Only a percentage of hospitals are reporting to ONEISS. At the same time, the data in the hospitals are not integrated with the data from the police. There are numerous instances when crashes are recorded by hospitals but not by police and vice versa. There are also fields that are solely reported by the police

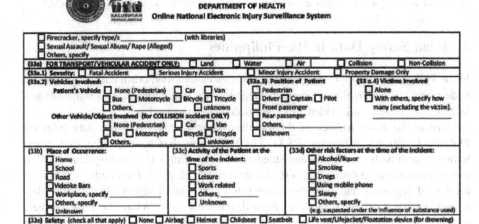

**Fig. 4.** Department of Health's Online National Electronic Injury Surveillance System (ONEISS)

(e.g. circumstances leading to a crash) and hospitals (e.g. type of injury). These realities point to a need to integrate the two to address gaps in data and make meaningful decisions in road safety.

The primary source of crash data in the country comes from the Philippine National Police (PNP). The PNP is the national police force of the Philippines and has multiple levels of offices from local precincts to regional headquarters and the national headquarters. The PNP is the sole agency mandated to conduct investigation on crashes. The traffic investigator records crash data onto the Incident Report Form (IRF) (see Fig. 5). The form is primarily for prosecution purposes and is designed to identify victims, suspects, and narrative details behind a crash. These forms are then encoded by a Crime Registrar into the Crime Incident Reporting System (CIRS) or E-Blotter which is an electronic database system for recording all crime incidents.

The CIRS is a Microsoft Access database and is installed at every police station. Records are then sent to the PNP headquarters which are collated in a central database. CIRS can only be accessed by police personnel and crash data are not opened to the public. Furthermore, crash statistics are not published regularly.

Under the PNP, a road safety lead department was created, which is the PNP-Highway Patrol Group (HPG). The HPG primarily reports road crash data on national roads and expressways. There is an HPG national headquarters at the PNP and under them are regional headquarters. Each regional office compiles road crash data through spot reports for the HPG which is then submitted to the national office for collating and monthly reporting. These spot reports are basically narratives of each crash which is different from the IRF. The collated data is then published as a nationwide summary of road crashes.

Crash fatality data in HPG is severely underreported and only makes up approximately 10% of crash fatalities recorded in the civil registration system. Like the rest of

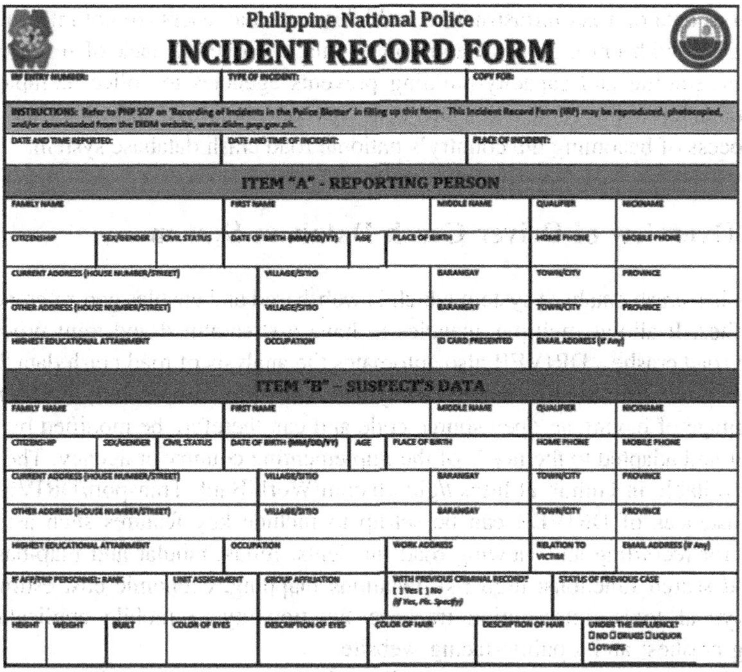

**Fig. 5.** Philippine National Police

the PNP, HPG is only able to report fatalities within 24 h of a crash. The data from HPG is not integrated with the rest of the data from PNP and from the health sector.

Furthermore, the Department of Public Works and Highways (DPWH), which is the national government body in charge of road infrastructure, had also maintained a road crash database called the Traffic Accident Recording and Analysis System (TARAS). In 2004, the DPWH developed and implemented TARAS which was meant to aid the department in identifying and analyzing blackspots. The data from TARAS come from the police, being collected individually by regional DPWH personnel. DPWH then encodes the forms completed by the police into the system. In 2013 however, TARAS was discontinued because of sustainability issues.

Finally, in the absence of a reliable national database, various city-level government agencies have implemented their own database systems for recording crashes. Local governments are able to capture data that are amicably settled and will not be reported to the police or to the hospitals.

Since there are several independent database systems from each government agency, with varying purposes, reporting standards and methods, the delivery of evidence-based road safety programs has become increasingly difficult. Furthermore, the lack of a central and integrated repository for data prevents government from systematically understand road safety challenges and conduct meaningful and impactful analysis. This lack of coordination in collecting crash data also translates to the lack of sharing of data among agencies, researchers, advocates, and the academe. This also means that other types of data such as data on road infrastructure, vehicles, and road users are not integrated and used together with crash data. In addition to this, the over-all lack of resources and equipment, staffing and capacity-building prevents agencies to collect complete and uniform data. It is in this context that DRIVER was conceptualized and piloted and is now in the process of becoming the country's national road crash database system.

## 3 An Overview of Driver Crash Database System

DRIVER is a crash database system which is web-based and enables geo-referencing of road crashes. It allows multiple agencies to have a streamlined and joint process for reporting road crashes. DRIVER also automates the analysis of road crash data through its suite of powerful visualization and analytical tools. Most importantly, DRIVER has the advantage of having an open source code and can therefore be modified by a local developer and adapted to the needs of the implementing country or agency. The source code is available in Github at https://github.com/WorldBank-Transport/DRIVER.

All instances of DRIVER can be set-up to include key features such as: a web interface for recording and viewing road incidents; robust tabular and map-based filtering and search functions; high risk locations mapping, economic cost estimations, and analytical tools; intervention tracking functionality; a mobile application for recording crashes; and a public-facing website.

DRIVER's primary aims are to facilitate integration of crash data reporting of multiple agencies, standardize terms, definitions, institutional arrangement for reporting, to provide analytical tools to support evidence-based investments and policies in road safety, and to monitor and evaluate impact of interventions. In the Philippines, DRIVER is publicly available at roadsafety.gov.ph. Materials and resources regarding DRIVER can be accessed at the Global Road Safety Facility (GRSF) website or www.roadsafetyfacility.org.

DRIVER was first developed in 2012. A prototype of DRIVER was used by the Cebu City Traffic Office (CCTO formerly CITOM) to aid the planning of the Cebu Bus Rapid Transit System in the Philippines. The prototype replaced the paper and logbook-based recording of crashes and enabled the government to visualize and analyze crashes on a map.

During this pilot, the government has witnessed the power of merely recording crashes onto a map and how this can inform evidence-based decisions in road safety. By simply mapping where crashes happen, the government was able to identify high-risk locations and conduct further investigations. Some of the analyses that were conducted include comparing the actual risk with the perceived high-risk locations and a comparison between high-risk locations and concentration of traffic enforcement assignments.

Given the pilot platform's success in Cebu and the discontinuation of TARAS, the DOTr sought the assistance of the World Bank to support the scaling up of the pilot program to cover the whole of the Philippines and to expand the analytical features of the platform. DRIVER is now being conceptualized as the country's national road crash database systems.

Substantial inter-agency discussions were held regarding mobile and web-based platform data entry fields and platform specifications. The discussions drew upon existing data collection forms developed by DPWH, PNP, MMDA, and DOH, data requirements to support the platform analytical tools, and international road safety reporting requirements. Furthermore, the development of DRIVER enabled the government to also review the institutional arrangement of road safety management as a whole. DRIVER became the catalyst for government institutions to coordinate with each other in relation to road safety data. The World Bank assisted by conducting capacity and needs assessments, recommending minimum indicators, and sharing best practice in terms of road safety data. The result of these discussions is the current iteration of DRIVER as well as a road map to the improvement of road safety data in the country.

The establishment of a formal institutional arrangement played a key role in the success of DRIVER in the Philippines. The DOTr took the lead in implementing DRIVER in the Philippines. Under the DOTr, there are several user groups acting as Data Providers and Data Users. Data Providers such as the PNP and local government units provide crash-related data and upload them into the database system. The DOTr is also piloting a link between DRIVER and the injury surveillance system of the DOH to collect reliable fatality and injury data. Data Users such as road engineers, transport planners, police, researchers, and advocates take advantage of DRIVER's analytical modules and custom report generation features. Finally, the Department of Science and Technology (DOST) provides technical support and maintains the servers where DRIVER is hosted. This institutional arrangement enabled multiple stakeholders in road safety to collect reliable and accurate data and effectively share and analyze data for meaningful decisions in road safety.

In September 2016, DRIVER 1.0 was officially unveiled with the official government domain name, roadsafety.gov.ph. It was initially hosted in physical servers in the Department of Science and Technology – Advanced Science and Technology Institute (DOST-ASTI) and was later on migrated to the DOST's OpenStack cloud platform. Policy and legal instruments such as data sharing agreements have been signed by the government, one of which is for the police to use DRIVER in their regular operations. To formally launch DRIVER, the DOTr led the National Road Safety Idea Hack in March 2017, inviting local developer talents to solve big road safety challenges using the new large, open crash datasets made available through DRIVER. More than 450 developers, road safety advocates, and government officials participated in the hack-a-thon, and a series of workshops on road safety advocacy,data and technology to improve safety were held.

To date, the DOTr and PNP with the support of the World Bank and other local partners have conducted trainings for 4,060 individuals coming from various sectors such as the police, local governments, research, and the academia. During each training, each police precinct and participating local government unit were provided with their own account and corresponding access to DRIVER. They were trained on basic principles of road safety and how to encode and analyze data in DRIVER. Training materials were also provided to all of the participants. Starting 2019, stakeholders such as the traffic police started using DRIVER to record crashes throughout the country. And even amidst the pandemic, training has continued throughout the country (Fig. 6).

Trainings were also organized for DPWH engineers where participants were taught to use DRIVER data to conduct a preliminary assessment on a chosen site and to

**Fig. 6.** DRIVER trainings with the police in the Philippines

conduct a road safety diagnosis based on DRIVER and International Road Assessment Program (iRAP) assessments (Fig. 7).

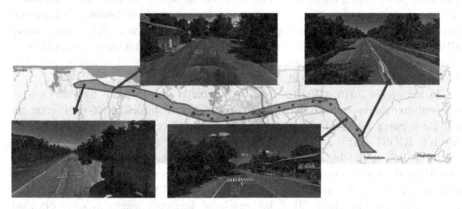

**Fig. 7.** Sample cross sections of urban and rural segments and their respective locations using the DRIVER crash database system

Currently, an enhanced version of DRIVER, called DRIVER 2.0, has been rolled-out for use by the Philippines government. Enhancements to the platform such as DRIVER integration with the iRAP Toolkit Application Programming Interface (API), integration with the iRAP Star-Rating GIS shapefiles, and integration with Mapillary were introduced. Updated DRIVER guidelines and training materials were also prepared and published. At the same time, activities which aim to strengthen the institutional arrangement on road safety data collection and management are organized and implemented. This includes continued capacity-building and the execution of legal and policy instruments to support the sustained collaboration among government agencies.

## 4 Driver Case Studies

Following the success of DRIVER in the Philippines, multiple countries have used DRIVER for different purposes and at different scales to address challenges in crash data.

In Lao People's Democratic Republic (PDR), DRIVER was first piloted in the capital city of Vientiane in 2017. DRIVER was used by the Vientiane Traffic Police and is now being scaled out at the national level for Lao PDR. Before DRIVER, traffic police were collecting crashes manually on paper and had no crash database system. Now, DRIVER is implemented as a national database system. To date, four DRIVER workshops for the police and the transport department have already been held and three years of crash data in Vientiane have been entered into the platform. Currently, the data is used to investigate high-risk locations in the city and to develop evidence-based interventions (Fig. 8).

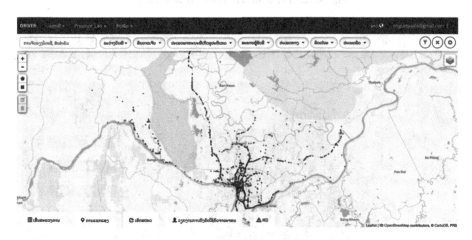

**Fig. 8.** DRIVER in Vientiane, Lao PDR, 2021

In Sao Paulo, Brazil, DRIVER is used by the city's Traffic Engineering Company to share official and anonymized crash data with the general public. In turn, the public can use the data to propose locations for road safety interventions (Fig. 9).

In Fortaleza, Brazil, DRIVER is being used by the city's traffic agency to link data from different police bodies and the health sector to consolidate a single database for road crashes in the city. The city is also using DRIVER to monitor the impact of different engineering and enforcement road safety interventions. The anonymized data is also available for the public to visualize and download (Fig. 10).

Meanwhile, in Malawi, the Ministry of Transport and Public Works and the Directorate for Road Traffic and Safety Services have recently received training in collecting and analyzing crash data in DRIVER. They are currently testing the web based and the android app as a means to collect crashes and analyze the impact of interventions.

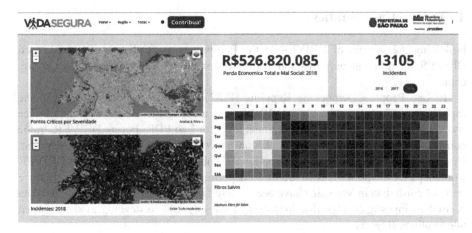

**Fig. 9.** DRIVER in Sao Paulo, Brazil, 2021

**Fig. 10.** DRIVER in Fortaleza, Brazil, 2021

## 5 Driver Experience in the Asia-Pacific Road Safety Observatory (APRSO)

DRIVER is aligned with the objectives and goals of the APRSO, specifically in strengthening road safety data collection, management, sharing and analysis within Asia and the Pacific.

The APRSO, being a collaborative effort among development organizations and member countries, has been established out of the growing demand in the region to improve road safety in general and road safety data in particular. The APRSO also aims to collect, manage, and analyze road safety data across member countries to support informed activities and programs in road safety. Data that supports cross-country comparison and evaluation of regional trends will enable the observatory to provide adequate and effective technical and policy assistance to its member countries, to efficiently allocate resources and funding, to engage civil society, international development organizations, and private sector, and ultimately to have targeted and effective interventions in road safety. This regional platform will also facilitate monitoring of progress towards the Sustainable Development Goals (SDG) 2030 road safety targets, and on the UN Voluntary Road Safety Targets.

Data at the regional level will also enable countries to cooperate and coordinate with each other, to harmonize and standardize data definitions and collection methodologies, and to foster health competition among countries. With these aims, the APRSO was officially launched in February 2020 at the Stockholm High Level Ministerial Road Safety Conference. Following the official launch, member countries have met on different regional APRSO workshops and shared their challenges with regards to data. The Asian Development Bank currently hosts the Technical Secretariat for APRSO, coordinating data collection efforts and providing technical assistance, in collaboration with other partner organizations, including the World Bank, WHO, UNESCAP, FIA, ITF and others.

It is in this context that DRIVER becomes a tool not just for individual cities and countries but also internationally within the APRSO. Currently, an APRSO DRIVER platform is being piloted to analyze its potential for cross-country comparison of crash data and safety performance data.

DRIVER includes APIs that enable both the importing and exporting of crash records. This means that member countries through their own individual crash database systems can voluntarily upload their data into the APRSO DRIVER platform which can then be used for cross-country comparison, benchmarking, identification of regional strategy and programs, among others. Apart from this, through DRIVER, the APRSO is able to promote the need for more reliable and accurate data in road safety in the region.

## 6 Conclusions

DRIVER was first developed for low resources countries, but it has proven an attractive and useful tool globally, thanks to its flexibility and ability to link with other road safety tools.

Implementing DRIVER in multiple countries brought forward interesting insights regarding road crash data. First of all, it is clear that there is a growing need for better road crash data and database systems around the world. DRIVER, being open-source, enables countries to improve the collection of crash data without the need to procure proprietary systems.

Secondly, whether it is implementing a road safety management system or conducting capacity-building, it is crucial that activities are responsive to a country's local needs and context. This requires collaboration and consultation with all relevant stakeholders including target users and participants, funders, technical staff, and high-level authorities. Addressing local needs and context also means engaging with local experts, hiring local consultants, knowing how different ministries work together, and conducting capacity-building when needed (e.g. computing skills). This also requires that DRIVER is continuously improved, updated, and contextualized to the needs of its users.

Thirdly, in implementing DRIVER, it is important to build capacity and expertise in every aspect of the system. This includes not just for the front-end users of DRIVER (i.e. encoding of records or analyzing data) but also the software and technical side (i.e. troubleshooting, fixing of bugs, implementing enhancements). Transferring the knowledge and expertise of the developers to the government is necessary in order to ensure the sustainability of the platform. This transfer of knowledge can be achieved through training of developers and development of technical manuals. In addition to knowing every aspect of the system, it is also important that stakeholders know the principles in road safety and how collecting reliable and accurate data can lead to evidence-based interventions.

Fourthly, improving the database system is not enough to improve data collection in a country. The implementation of DRIVER provides an opportunity to reach agreements between stakeholders to share data, to adequately fund data collection, and allocate sufficient human resources to this task, all of which are necessary for a sustainable improvement in road crash data. As important as DRIVER is a robust and well-defined institutional arrangement that will allow individual government agencies to collect and share data with each other. The implementation of DRIVER in different countries also required standardizing definitions of crash data elements, assessing capacities and resources, reviewing legal and policy instruments, and clarifying the mandate of each government agency in road safety. Crash investigation trainings were also essential in developing the skills of the police officers collecting the crash data.

Finally, implementing DRIVER at the country-level is vastly different from implementing it in the context of the APRSO. In the APRSO, DRIVER becomes a platform for comparing the performance of one country from another and monitoring the road safety situation of an entire region. This warrants close coordination and consultation among member countries not just to standardize data definitions and collection methods but more critically, to develop a technical and governance framework that will facilitate the seamless and easy sharing of data among countries. The APRSO has been leading these discussions as well as sharing best practice in data and promoting DRIVER especially to countries who plan to improve their data. DRIVER, being an open-source platform, can easily be modified to link to individual database systems and is flexible enough to respond to the varying needs and contexts of the member countries.

DRIVER has contributed to raising the awareness on the importance of data but more significantly, to the importance of road safety. The platform has provided the much-needed analytics and visualizations for identifying high risk locations and developing evidence-based interventions. From here on, DRIVER becomes one of the tools that governments can use to more effectively and successfully save lives on the road.

**Acknowledgements.** The DRIVER tool was developed and deployed thanks to support from the following donors: Bloomberg Philanthropies, UK Aid and the World Bank through the Global Road Safety Facility (GRSF); and the Government of Japan through the Quality Infrastructure Investment (QII) Partnership.

# References

Canadian Department of Foreign Affairs, Trade and Development, DOH, PSA, WHO, and United Nations Economic and Social Commission for Asia and the Pacific. Strengthening Civil Registration and Vital Statistics: A Case Study of the Philippines (2014). https://getinthepicture.org/sites/default/files/resources/phl_crvs_case_study_2014.pdf, Accessed 28 May 2021

Philippine Statistics Authority. Philippine Statistical Yearbook (2021). https://www.psa.gov.ph/tags/philippine-statistical-yearbook, Accessed 25 May 2021

Philippine Statistics Authority. Philippine Statistical Yearbook [Online] (2015). https://psa.gov.ph/vital-statistics/table-year, Accessed 25 May 2021

World Health Organization. Global Status Report on Road Safety 2018 (2018). https://www.who.int/publications/i/item/9789241565684, Accessed 25 May 2021

# Road Financing Strategies

# Is a Further Increase in Fuel Levy in Kenya Justified?

Evans Omondi Ochola[1]([⊠]) and Jennaro Boniface Odoki[2]

[1] STC World Bank, Nairobi, Kenya
ocholaevans@gmail.com
[2] IMES Ltd., London, UK
jennarodoki@yahoo.co.uk

**Abstract.** Road transport is the predominant transportation mode in Kenya, contributing to about 9% of the GDP. Increased expenditure in roads has seen road assets become a significant portion of public investments that must be preserved. In Kenya, funds for road maintenance are primarily derived from fuel levy, which is charged per liter of petroleum imported. Despite the recent doubling of fuel levy charge, it is still inadequate for the entire network maintenance needs. This has prompted calls for a further increase in fuel levy in line with the government's "user pay" policy, much to motorists' chagrin. This study seeks to assess whether these calls are justified by comparing the benefits of increasing maintenance expenditure versus the road user cost (RUC) savings.

The study methodology involved undertaking network strategy analysis using the HDM-4 model. The model was used to quantify the optimal network maintenance needs, network performance under different budget scenarios, and the impact of increased maintenance funding on RUCs.

The study revealed that Kenya needs to double its fuel levy from the current US\$ 0.18 per liter in order to meet its entire network maintenance requirements. It also revealed that every dollar invested in road maintenance translates to RUC savings of about US\$ 5.53.

However, fuel levy charging is under threat as road user charging is becoming "politically unpopular". Also, vehicle fleets are switching to alternative fuels, and engine efficiencies are improving, leading to declining fuel consumption hence the need to explore alternative sources of maintenance financing.

**Keywords:** Road user charging · Road user costs · Maintenance needs

## 1 Introduction

Road transport is the predominant mode of transportation in Kenya, where it accounts for over 90% of all freight and passenger traffic (KRB 2019a). So important is the role of road transport in Kenya that it is currently estimated to be contributing about 9% of the Gross Domestic Product (KNBS 2020). Road assets, therefore, represent a critical portion of Kenya's public investments that must be preserved and maintained in good condition. Against this background, improvement in the road network condition has become a priority not only for Kenya, but also for most African countries, with current emphasis being on asset preservation of the capital investments made in the road sector

© The Author(s), under exclusive license to Springer Nature Switzerland AG 2022
A. Akhnoukh et al. (Eds.): IRF 2021, SUCI, pp. 151–167, 2022.
https://doi.org/10.1007/978-3-030-79801-7_11

since the turn of the millennia (Gwilliam 2011). Gwilliam (2011), further notes that most countries in Africa face the quandary of having to marshal adequate resources for road maintenance from fuel levy charges. This is because, the level of those charges in most cases is inadequate to cover the maintenance costs arising from network wear and tear, let alone funding the rehabilitation backlog. This insufficient expenditure in road maintenance, has resulted in deficient road infrastructure, which is estimated to be limiting economic growth by as much as 2% per annum (PIDA 2020).

In Kenya, funds for road maintenance are primarily derived from the Road Maintenance Levy Fund (RMLF) and Transit Tolls (KRB 2019a). RMLF is charged per liter of fuel imported into Kenya, while Transit Tolls are charged on foreign-registered commercial trucks exceeding 2-tonnes that ply Kenyan roads (KRB 2019a). Presently, the RMLF contributes more than 99% of the road maintenance funds, highlighting its significance in meeting Kenya's maintenance needs (KNBS 2020). However, a recent publication by the Kenya Roads Board (KRB) revealed that the current RMLF collections were only able to meet about 70% of the total routine maintenance needs despite the fuel levy charge having been doubled over the last few years (KRB 2019a). There are, therefore, calls for a further increase in the fuel levy charge in a bid to bridge this financing deficit in line with the government's policy of the "user-pay" principle, much to the chagrin of road users.

Calls to increase the fuel levy charge have often elicited heated debate amongst road users who often question whether they are getting value for money from these road user charges. This study seeks to quantify the benefits of savings in road user costs, if any, to be passed on to road users if the fuel levy charge is increased further. To assess this assertion, an analysis was undertaken using a road management assessment tool called Highway Development and Management model (HDM-4). The study assessed the relationships between road network conditions as a function of various funding level scenarios and their impact on road user costs.

The aim of the study was, therefore, to assess whether calls to increase the fuel levy charge in Kenya are justified by answering the following questions: -

i. What is the optimal level of expenditures for road network maintenance needs on Kenya's classified road network?
ii. By what proportion should the fuel levy charge be increased to meet Kenya's entire road maintenance needs?
iii. What would be the accrued road user cost savings that can most likely be derived from improved road conditions, if the fuel levy charge is increased to cover the "optimum" expenditure level?

The relevance of this study to a Kenyan context has been informed by the ballooning road maintenance financing gap and the Kenyan government policy directive, which calls for the adoption of the "user-pay" principle for financing the maintenance of transport infrastructure (MoT 2010). Based on this policy directive, bridging the maintenance financing gap calls for increased road user charges such as the fuel levy charge.

## 2 Study Context

### 2.1 Study Country

As already stated, this study was focused on Kenya, a low-middle income country located in East Africa (See Fig. 1).

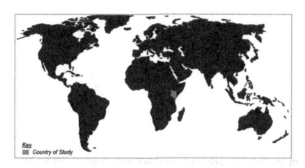

**Fig. 1.** Country of study. Source: Free World Maps (2019)

### 2.2 Kenya's Road Sector

Kenya's road sector has undergone tremendous policy, legal and institutional reforms over the last two decades culminating in the establishment of the Kenya Roads Board (KRB) in 1999 to manage the fuel levy charge and the establishment of road authorities in the year 2007 following the recommendations of a Government Sessional Paper of 2006 on the management of roads (World Bank 2018). The promulgation of a new constitution in Kenya in 2010 further devolved the management of the country's road network into a two-tier system of governance consisting of the National Government and devolved 47 County Governments (KRB 2019a). Under the new constitution, the National Government is in charge of the national trunk roads which are managed by the various road authorities while the County Governments are in charge of the county roads (KRB 2019a).

Of the total classified road network of about 161,821 km, 40,000 km constitutes the national road network, while the remaining 121,821 km constitutes the county road network. Only about 10% of the classified road network is paved with trends in network condition over the last two decades as shown in Fig. 2. The trends reveal that the condition of Kenya's road network has improved considerably over the last decade, with the proportion of the network in the maintainable state (i.e., fair and good condition) increasing by about 21% since the year 2009.

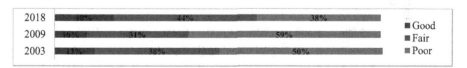

**Fig. 2.** Trends in Kenya's road condition 2003–2018. Source: KRB (2019a)

Figure 3 below shows Kenya's entire classified road network which has been considered in this study:

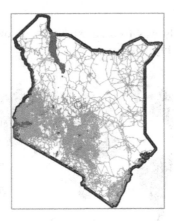

**Fig. 3.** Kenya's classified road network. Source: KRB (2015)

### 2.3 Road Maintenance Financing Status and Challenges

KRB (2019a) noted that maintenance expenditures have remained below the projected maintenance needs, resulting in a continued accumulation of maintenance backlog on the Kenyan road network. This issue had also been identified by Gwilliam (2011) who attributed it to the fact that most countries had a bias towards capital expenditure, resulting in a vicious cycle of low maintenance budgets. In Kenya, the institutional reforms that led to the creation of the Kenya Roads Board has, however, ensured that the capital investments and the maintenance expenditures are handled as separate budget lines with a cross-plot of capital expenditures against maintenance expenditures over the past 5-years yielding a positive correlation of 0.54 (See Fig. 4). The positive correlation indicates that maintenance expenditure is increasing in tandem with capital expenditure, albeit at nearly half the rate. A continued increase in maintenance expenditures along with the capital investments bodes well if the backlog of maintenance is to be reduced going forward (Gwillian et al. 2008).

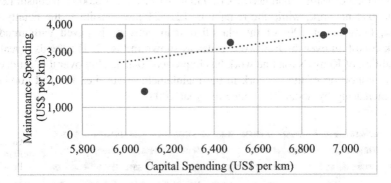

**Fig. 4.** Cross-plot of capital expenditure vs. maintenance expenditure. Source of data: KNBS (2020)

Several other studies have highlighted Africa's road maintenance management and financing challenges. For instance, Foster and Briceño-Garmendia (2010) noted that despite the best attempts of most countries to use fuel levies for road maintenance, the collected amounts were nowhere near enough to meet the maintenance needs. Besides, less than 10% of the road network attracted the minimum traffic flow of 15,000 vehicles per day required to make road infrastructure concessions to the private sector economically viable (Foster and Briceño-Garmendia 2010). A policy paper prepared for the World Bank prepared by Calderón et al. (2018) estimated the infrastructure bottleneck within the Sub-Saharan Africa region to top US$ 93 billion per year over the next decade with only less than half of that amount being provided.

In Kenya, the national transport policy was published by the Ministry of Transport in 2010 in a report dubbed: "*Integrated National Transport Policy: Moving a Working Nation,*" which identified funding for road infrastructure as a significant challenge. The policy recommended the adoption of "user pays" and "polluter pays" principles to generate sufficient revenues to support the development and maintenance of transport infrastructure (MoT 2010). Another recent government publication titled "*National Surface Transport Funding (2016–2025)*" estimated the funding gap for road infrastructure to top US$ 1 billion annually from 2021–2025 and recommended a raft of proposals to bridge this financing gap, including (i) an annual 30% increase in fuel levy charge between the years 2017/18 and 2019/20, after that annual adjustment for inflation; (ii) tolling of high-trafficked viable roads; (iii) levying of charges on driver/vehicle licenses; (iv) levying of a 16% tax on all gross motor vehicle insurance premiums; and, (v) imposing of "polluter pays" on specific vehicle types.

## 2.4 Need for Increased Expenditure in Road Maintenance

Heggie (1995a, b) estimated that low investment in maintenance increased the cost of road transport to African economies to the tune of about US$ 1.2 billion per year or 0.85% of the regional GDPs. The same publication called for more investment in road maintenance since various road maintenance strategies were highly effective with every dollar spent on road maintenance being recouped several times over by reduced vehicle operating costs with annualized benefit/cost ratios varying from 3.4 to 22.1 (Heggie 1995a, b). Thriscutt and Mason (1989), on the other hand, also estimated that each dollar spent on road maintenance reduced vehicle operating costs by between US$ 2 to US$ 3.

Heggie (1995b) further emphasized the need for timely maintenance with a World Bank study noting that for every dollar not invested in preventive road maintenance, road users were wasting US$ 3 on extra transport costs while road authorities were having to spend US$ 4 on reconstruction costs.

# 3 Study Methodology

## 3.1 The HDM-4 Model

The analysis was undertaken using the Highway Development & Management model (HDM-4), a tool for appraisal and analysis of road management investment decisions

(Kerali and Odoki 2009). Its analytical framework is premised on pavement life cycle analysis which is applied to predict road deterioration; road works effects; road user effects; and, socio-economic and environmental impacts (Kerali and Odoki 2009). The applications of HDM-4 in road management include strategy analysis, programme analysis, project analysis, and operational analysis (Kerali and Odoki 2009). Strategy analysis is recommended for high-level network-wide analysis, with operational analysis being preferred for detailed sub-section analysis. This study involved undertaking strategy analysis for the entire classified Kenya's road network. The decision to use the HDM-4 model stemmed from the fact that its sub-models were developed from large-scale field experiments conducted worldwide (Kerali 2001) and that it was recently adapted to reflect the local Kenyan conditions.

## 3.2 The HDM-4 Strategy Analysis

Network strategy analysis was used to determine the maintenance needs and the relationship between network conditions and funding levels. To undertake strategy analysis, the first step involved the development of homogenous representative road sections based on the main attributes that influence road user costs, such as road class, surfacing type, traffic level, road condition, and climate data. Data for the homogenous road sections were then populated using aggregate data which included traffic bands, traffic composition, and ride quality. Suitable traffic growth rates were then applied to predict future network performance.

Basic and economic vehicle fleet data were thereafter compiled with the basic vehicle fleet representing the actual physical vehicle characteristics in Kenya (e.g., no. of wheels, no. of axles, etc.); tire characteristics (e.g., tire type, re-tread cost, etc.); utilization parameters (e.g., annual km driven, working hours, average life, occupancy, etc.); and, loading characteristics (e.g., load factors, operating weights, etc.). On the other hand, economic vehicle fleet characteristics reflect the average unit costs of vehicle resource consumption and included vehicle resources (e.g., new vehicle purchase cost, tire cost, fuel cost, lubricant cost, crew wages, annual overhead costs, etc.); and, time value (e.g., the value of driver and passenger working and non-working time, cost of cargo delay, etc.).

Maintenance standards proposed by the Kenya Roads Board were then defined, taking cognizance of road conditions, traffic levels, etc. The standards described the work type, intervention criteria (responsive or scheduled), after-work effects, and unit costs for maintenance works.

An unconstrained network analysis (i.e., analysis assuming there are no budget constraints) was then undertaken to determine the 10-year maintenance needs of Kenya's classified road network. A comparison of the unconstrained analysis budgetary requirements and that of the expected road fund collections was then used to determine the maintenance financing deficit. Suitable graphs showing the relationship between network condition and funding levels were then plotted, and the impact of various funding levels on network condition assessed. The optimal maintenance financing needs were then established and the corresponding additional funds required to attain such network condition determined.

The analysis scenarios, therefore, involved a base model scenario which assumed there should be no significant change in policy regarding the current fuel levy charge of about US$ 0.18 per liter. The other analysis scenario considered increasing the fuel levy in line with government's policy that road users should pay the full cost of maintaining road infrastructure. A comparison of the impact of these two funding scenarios on future road network performance and savings on road user costs were then undertaken. The savings on road user costs, if any, were used as the basis of justifying whether an increase in fuel levy charge is justifiable.

## 3.3 Data Collection and Processing

Most of the critical data that influences road agency costs and road user costs, such as pavement surfacing types, traffic levels, road condition, and climate zone, was provided by the Kenya Roads Board (KRB). Less critical data for strategic level analysis that was not readily available were estimated either using the concept of Information Quality Level as recommended by Paterson and Scullion (1990) or using HDM-4 default data sets. The primary data sets, formats, and data sources are shown in Table 1.

**Table 1.** Data input, formats, and sources

| Data type | Data | Availed format | Processed format | Source |
|---|---|---|---|---|
| Road network | Inventory, pavement type, pavement strength, road condition, length, climate, etc. | GIS | Attributes extracted using *QGIS* software and then exported to *Excel* and *Access* | KRB |
| Traffic | Traffic composition, traffic volumes, speed-flow types, traffic flow pattern etc. | PDF report | File converted to *Excel* then to *Access* | KRB |
| Vehicle fleet | Vehicle physical characteristics, vehicle utilisation, loading and performance etc. | HDM-4 object files | No processing required | KRB |
| Road works | Construction maintenance standards and unit costs | | | |
| Budgets | Budget projections by work types | PDF report | Excel | Economic survey |

The road inventory data used in the study was the most recent available, having been collected in the year 2018. Since the most critical data inputs were up to date and that the data covered the entire country, the analysis outputs can be considered reasonably accurate for this level of analysis. The overall confidence level in the data used is categorized as being "High."

## 3.4 Study Assumptions

i. This study focused on the impact of increasing the Road Maintenance Levy Fund (RMLF) charge as it currently contributes more than 99% of the road maintenance funds, and ignored Transit Tolls.
ii. It was assumed that about 10% of the total collected RMLF is used to cover the funds' operational and administrative costs, with the remaining 90% being used to pay for the actual maintenance works.
iii. It was further assumed that about 60% of the total RMLF collections should be used to maintain paved roads, with the remaining proportion being utilized for the unpaved network. This ratio was estimated from the Annual Public Roads Programme for 2019/20 (KRB 2020).

# 4 Study Results and Discussions

## 4.1 Assessment of Optimal Road Network Maintenance Needs and Maintenance Financing Gap

HDM-4 allows for the assessment of optimal maintenance needs either through engineering-based or economic efficiency-based optimization methods. The engineering-based optimization method allows for the maximization of road condition improvement, while economic efficiency-based optimization method allows for maximization of the net present value (Morosiuk et al. 2006). For this study, the economic efficiency-based optimization method was used to ensure priority was given to alternatives with the maximum economic return.

Projected road fund collections were derived from past fuel sales trends, given that fuel levy is charged per liter of fuel imported into Kenya. The projections assumed that the impact of a changing vehicle fleet on fuel consumption would be negligible over the next decade (KRB 2019b).

The maintenance financing gap was derived from comparing the unconstrained maintenance needs and the projected road fund collection. The maintenance requirements, projected fuel levy collections, and the maintenance financing gaps are summarized in Fig. 5.

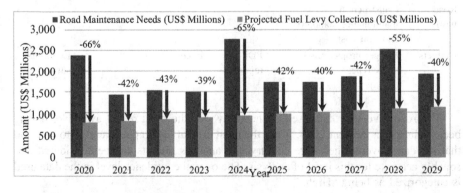

**Fig. 5.** Road maintenance needs, projected fuel levy collections, and financing gaps

From the 10-year analysis, the total projected maintenance needs were estimated to hit US$ 19.7 billion against the projected fuel levy collection of about US$ 10 billion over the same period. This implies that, on average, only about 51% of the required maintenance funds shall be available during the 10-year analysis period, with the annual maintenance financing gap ranging between 39% to 65% over the 10-years.

## 4.2 Impact of Maintenance Under-funding on Network Performance

To assess the impact of maintenance under-funding on network performance, 10-year annual average roughness progression graphs were prepared for both the "unconstrained" and "status quo" funding scenarios. The "unconstrained" budget scenario assumed adequate funds were available to keep the network in an optimum condition. In contrast, the "status quo" funding scenario assumed the current maintenance practices with inadequate maintenance financing continues. The roughness, in this case, represents the quality of the riding surface and is expressed in terms of the international roughness index (IRI). The higher the IRI, the worse the road condition.

The average network roughness for the unconstrained budget scenario analysis stays below 4.2 m/km IRI for the paved network. At the same time, roughness for status quo funding level continues to increase, eventually hitting about 7.3 m/km IRI in the tenth year as inadequate financing worsened the average network condition. Figure 6 shows the trends in average roughness for these two scenarios.

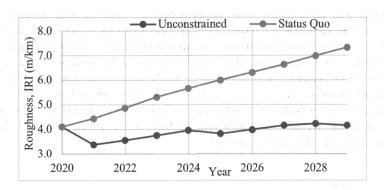

**Fig. 6.** Paved network average roughness by funding scenarios

For the unpaved road network for the unconstrained budget scenario, the average network roughness stays below 19.1 m/km IRI for the unpaved network. In comparison, the average roughness continues to increase for the current funding level, eventually hitting about 19.7 m/km IRI in the tenth year. Again, the "status quo" funding scenario revealed a worsening average network condition due to inadequate financing. Figure 7 shows the trends in average roughness for these two scenarios.

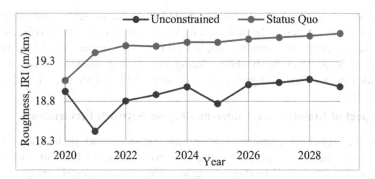

**Fig. 7.** Unpaved network average roughness by funding scenarios

With the current road user charging projected to meet only about half of the total maintenance needs over the next decade, Kenya risks losing the massive investments made in improving the overall road network condition over the past two decades. This problem has also been aptly captured by Calderón et al. (2018), who notes that less than half of the required finances are likely to be available over the next decade if the current financing trends continue.

### 4.3 Relationship Between Network Performance and Funding Levels

The average long-term road network performance was assessed under varying maintenance funding scenarios for paved and unpaved networks. To develop these relationships, the unconstrained maintenance needs and the corresponding average network roughness were first established. The impact of reducing the funding level at various proportions of the unconstrained maintenance needs on the average network roughness was then established. As expected, decreased funding levels led to a deteriorating network condition for both paved and unpaved roads. Figure 8 shows the maintenance funding vis-a-vis average roughness relationship for the paved network with the goodness of fit (i.e., $R^2$) approaching 1.0.

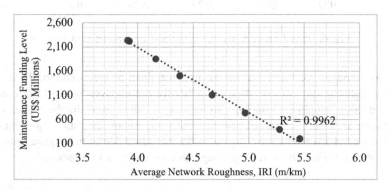

**Fig. 8.** Relationship between maintenance funding and average network roughness for paved roads

Figure 9 shows the maintenance funding versus average roughness relationship for the unpaved network with the goodness of fit (i.e., $R^2$) of about 0.96.

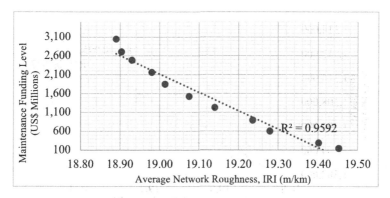

**Fig. 9.** Relationship between maintenance funding and average network roughness for unpaved roads

### 4.4 Impact of Road Condition on Road User Costs (RUCs)

The study then assessed the relationship between network condition and network RUCs. RUC is the summation of vehicle operating costs and travel time costs. Generally, RUCs increase as the road condition worsens as this often translates to low travel speeds, higher fuel consumption and increased wear and tear.

Although the relationship between the RUC and the roughness (IRI) is S-shaped, for simplicity purposes, a straight-line relationship has been assumed for assessing the changes in RUCs. In fact, the near-straight section of the graph conforms to the range within which most roads are considered to be in a maintainable condition (e.g., the roughness of less than 7.0 IRI for paved roads). Figure 8 shows the relationship between RUC and roughness for the paved network, with the straight-line curve ($R^2 = 0.83$) being assumed for simplicity in the analysis (Fig. 10).

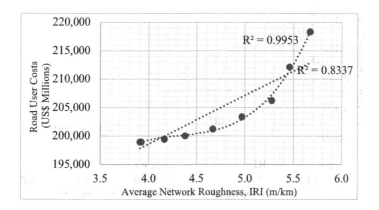

**Fig. 10.** Relationship between road user costs and average network roughness for paved roads

Figure 11 shows a similar relationship for the unpaved network. Again, the straight-line relationship ($R^2 = 0.995$) was adopted for computing RUC changes.

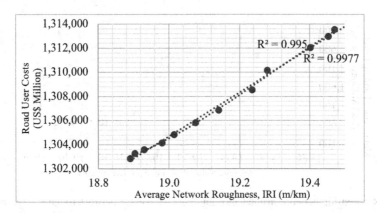

**Fig. 11.** Relationship between road user costs and average network roughness for unpaved roads

### 4.5 Return on Investment Due to Increased Maintenance Expenditure

To assess whether increasing the fuel levy charge is justified, the benefits in terms of savings in RUCs were determined by comparing RUCs from the current funding level and those to be derived from increased funding level. The ratio of RUC savings to that of additional maintenance expenditure was then obtained for the various networks.

#### 4.5.1 Paved Road Network

For the paved network, the established relationship between maintenance funding level and network condition was used to assess the impact of increasing the maintenance expenditure from, say F1 to F2 ($\Delta F$) on a change in network condition from, say R1 to R2 ($\Delta IRI$) as shown in Fig. 12.

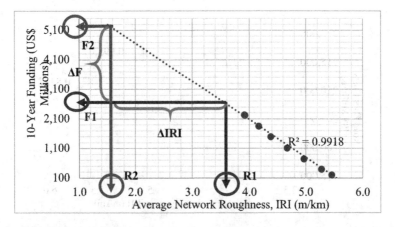

**Fig. 12.** Impact of increasing funding level on paved network condition

As shown in Fig. 12, doubling the projected level of expenditure in road maintenance shall improve the average network condition from an IRI of 3.6 m/km to an IRI of 1.6 m/km over the 10-year analysis period. This change in network condition, $\Delta$IRI, translates to RUC savings ($\Delta$RUC) of about US$ 17,063 million over the 10-year analysis period, as shown in Fig. 13. It is important to note that a network IRI of 1.6 m/km is difficult to achieve in real life given the differences in road classes in the network mix.

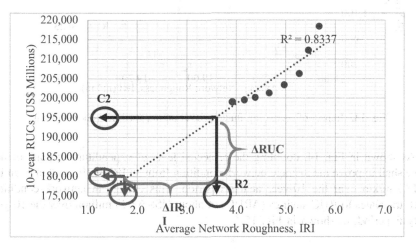

**Fig. 13.** RUC savings due to increased maintenance expenditure on paved roads

The ratio of RUC savings ($\Delta$RUC) to that of the corresponding increased maintenance expenditure ($\Delta$F) yields the return on investment (ROI) for extra spending in maintenance. It shows that every additional dollar invested in the maintenance of paved road network translates to RUC savings of about US$ 6.64.

$$ROI = \frac{\nabla RUC}{\nabla F} = \frac{195{,}359 - 178{,}296}{5{,}143 - 2{,}572} = 6.64 \qquad (1)$$

### 4.5.2 Unpaved Road Network

Similarly, for the unpaved network, the established relationship between maintenance funding level and network condition was used to assess the impact of increasing the maintenance expenditure from, say F1 to F2 ($\Delta$F) on a change in network condition from, say R1 to R2 ($\Delta$IRI) as shown in Fig. 14.

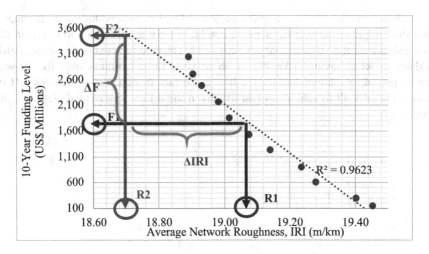

**Fig. 14.** Impact of increasing funding level on unpaved network condition

As shown in Fig. 14, doubling the projected level of expenditure in road maintenance shall improve the average network condition from an IRI of 19.1 m/km to an IRI of 18.7 m/km over the 10-year analysis period. This change in network condition, ΔIRI, translates to RUC savings (ΔRUC) of about US$ 6,623 million over the 10-year analysis period, as shown in Fig. 15.

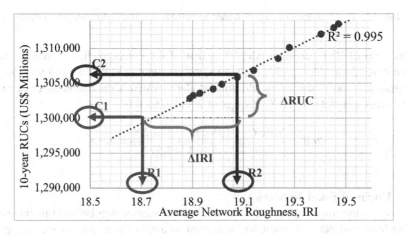

**Fig. 15.** RUC savings due to increased maintenance expenditure on unpaved roads

The ratio of RUC savings ($\Delta$RUC) to that of the corresponding increased maintenance expenditure ($\Delta$F) yields the return on investment (ROI) for extra spending in maintenance. It shows that every additional dollar invested in the maintenance of unpaved road networks translates to RUC savings of about US$ 3.86.

$$ROI = \frac{\nabla RUC}{\nabla F} = \frac{1,306,231 - 1,299,608}{3,429 - 1,714} = 3.86 \tag{2}$$

The combined return in investment for the combined paved and unpaved network translates to about US$ 5.53 for every extra dollar invested in the road network. The savings in RUC due to maintenance supports Heggie's (1995a, b) and Thriscutt and Mason's (1989) findings, thereby justifying the need to allocate more resources to road maintenance.

## 5 Conclusion

This study stemmed from calls in Kenya to increase the fuel levy charge further despite the levy being doubled over the last few years. Much to the chagrin of motorists, the calls are clearly in line with the government's policy directive, which emphasizes the adoption of the "user-pay" principle to generate sufficient revenues to support the maintenance of transport infrastructure (MoT 2010).

The methodology adopted for the study involved undertaking strategy analysis using the HDM-4 model based on the concept of pavement life-cycle analysis. The HDM-4 model was used to quantify the optimal network maintenance needs, network performance under different maintenance budgets, and the benefits of increased funding levels to meet the entire optimal maintenance needs.

The study revealed the following:

- The projected total road maintenance needs of the next decade are about US$ 19.7 billion against a projected road fund collection of about US$ 10 billion, translating to a financing gap of about 49%.
- The current level of fuel levy charging is inadequate and leads to the eventual deterioration of the road network over the next decade. To meet the optimal network maintenance needs of the next decade, the fuel levy charge shall have to be nearly doubled from US$ 0.18 per liter to US$ 0.35 per liter.
- Every additional dollar invested in the maintenance of Kenya's paved road network translates to RUC savings of about US$ 5.53. Paved roads contribute to RUC savings of US$ 6.64 for every dollar invested in maintenance, while unpaved roads contribute to RUC savings of US$ 3.86 for every dollar invested in maintenance. These findings reaffirm that more expenditure on maintenance is important as it reveals that road users bear the cost of poor maintenance directly.

The road user charging in this study was focused on the principle of full cost recovery and the principle of economic efficiency. Additional studies to ensure the principle of equity is observed in fuel levy charging are required. Also vital are studies on social and environmental aspects of the envisioned road user charging.

**Acknowledgements.** This paper is an extract for the final year master's dissertation from the University of Leeds. It presents the findings of an original research without any funding. In writing the paper, we received a great deal of support from a plethora of professionals. Firstly, we would wish to profoundly thank Margaret Ogai and Victor Odula, both of the Kenya Roads Board, who provided the data that we used in this research. We would also like to express our heartfelt gratitude to my dissertation supervisor from the University of Leeds, Jeffrey Turner, who guided me throughout the research.

# References

Calderón, C., Cantú, C., Chuhan-Pole, P.: Infrastructure development in Sub-Saharan Africa - a scorecard. The World Bank, Washington DC (2018)

Foster, V., Briceño-Garmendia, C.M.: Africa's Infrastructure: A Time for Transformation. The International Bank for Reconstruction and Development/The World Bank (2010)

Free World Map (2019). www.freeworldmaps.net

Gwilliam, K.: Africa's Transport Infrastructure: Mainstreaming Maintenance and Management. The International Bank for Reconstruction and Development/The World Bank (2011)

Gwilliam, K., Foster, V., Archondo-Callao, R., Briceño-Garmendia, C., Nogales, A., Sethi, K.: Africa infrastructure country diagnostic: roads in Sub-Saharan Africa. The World Bank, Washington DC (2008)

Heggie, I.G.: Management and financing of roads: an agenda for reform. World Bank technical paper 275. Africa Technical Series. World Bank. Washington D.C (1995a)

Heggie, I.G.: Commercializing Africa's roads: transforming the role of the public sector. Africa region findings & good practice info briefs; no. 32. World Bank, Washington, DC (1995b)

Kenya National Bureau of Statistics (KNBS): Economic survey 2020, Nairobi, Kenya (2020)

Kenya Roads Board (KRB): ARICS road map, Nairobi, Kenya (2015)

Kenya Roads Board (KRB): State of our roads 2018: summary report on road inventory and condition survey and policy implications, Nairobi, Kenya (2019a)

Kenya Roads Board (KRB): Road user charges study and development of a road maintenance funding policy. Draft report, Nairobi, Kenya (2019b)

Kenya Roads Board (KRB): Annual public roads programme for the financial year 2019/20 (2020). https://www.krb.go.ke/interactive-aprp-2019-2020/. Accessed 7 Apr 2021

Kerali, H.R., Odoki, J.B.: The role of HDM-4 in road management: HDM-4 training course. The University of Birmingham, UK, September 2009

Kerali, H.R.: The role of HDM-4 in road management. In: Proceedings, First Road Transportation Technology Transfer Conference in Africa, Ministry of Works, Tanzania, pp. 320–333 (2001)

Morosiuk, G., Riley, M., Toole, T.: Applications guide HDM-4 version 2.0. The HDM Series Volume 2, ISOHDM. PIARC (Paris) and the World Bank (Washington, DC) (2006)

Paterson, W.D.O, Scullion, T.: Information systems for road management: draft guidelines on system design and data issues. Technical paper INU77, Infrastructure and Urban Development Department, World Bank, Washington, D.C. (1990)

Program Infrastructure Development for Africa (PIDA). Overview (2020). https://au.int/en/ie/pida. Accessed 20 Apr 2021

Republic of Kenya - Ministry of Transport (MoT): Integrated national transport policy: moving a working nation. Final draft, Nairobi, Kenya (2010)

Thriscutt, S., Mason, M.: Road deterioration in Sub-Sahara Africa. In: The Road Maintenance Initiative: Building Capacity for Policy Reform, Volume 2. Readings and Case Studies, EDI Seminar Series, 1991, The World Bank, Washington DC (1989)

World Bank: How reforms transformed Kenya's roads (2018). https://www.worldbank.org/en/news/feature/2018/09/18/how-reforms-transformed-kenyas-roads. Accessed 7 Apr 2021

# Bridge and Structural Management

# Aerial Robotic System for Complete Bridge Inspections

Antidio Viguria[✉], Rafael Caballero, Ángel Petrus, Francisco Javier Pérez-Grau, and Miguel Ángel Trujillo

Advanced Center for Aerospace Technologies (FADA-CATEC),
La Rinconada, Sevilla, Spain
aviguria@catec.aero

**Abstract.** Bridge inspections have a large variety of procedures to ensure the safety of its facilities and personnel, and at the same time tightly budget constraints. These procedures involve extensive inspections, most of which should be performed at height and using both cameras and other sensors that require to be in contact with the surfaces being inspected. Then, bridge inspections traditionally require access to specific inspection points using man-lifts, cranes, scaffolds, or rope-access techniques, which increments importantly the costs of these inspections.

This work will present a system formed by two drones that will perform complete inspection operations in bridges in less time, reducing costs, improving quality of the inspection, and increasing safety of operators. The first one is an aerial robot that can obtain pictures of the overall bridge fully autonomously thanks to a GPS-free navigation system. The second one is the AeroX drone platform, a novel solution for inspection of difficult access areas. The AeroX can perform contact inspection due to its robotic contact device, which is equipped with an end-effector. Finally, both drones will be presented with videos of the validation experiments.

**Keywords:** Drones · Aerial robots · GNSS-free navigation ·
Bridge inspections · Visual inspection · Contact inspection

## 1 Introduction

The bridge inspections sector has a large variety of procedures to ensure the safety of its facilities and personnel, and at the same time tightly control costs. These procedures involve extensive inspections, many of which should be performed at height and using sensors that require to be in contact with the surfaces being inspected. Contact inspection is traditionally performed by technicians accessing to the specific inspection points using man-lifts, cranes, scaffolds, or rope-access techniques. Apart from that, only box type bridges allows inner inspections, as it is possible for a person to get inside them. However, it becomes necessary to inspect these types of bridges from outside as well.

There is a strong demand among operators for bridge inspections to develop alternatives to manual inspection. Sensors deployed at selected locations in the bridge,

© The Author(s), under exclusive license to Springer Nature Switzerland AG 2022
A. Akhnoukh et al. (Eds.): IRF 2021, SUCI, pp. 171–186, 2022.
https://doi.org/10.1007/978-3-030-79801-7_12

such as in the following works (Akhondi et al. 2010) and (Savazzi et al. 2013), provide punctual measurements at a high rate, which is suitable for intensive monitoring of few or small critical structures, but it is not ideal for covering/inspecting a full bridge. The use of aerial robots is very promising in order to reduce the cost and time required to perform complete inspections at height in bridges. However, there are two main technological challenges to be solved. First is the autonomous navigation of aerial vehicles without GNSS that will allow the automatization of general visual inspections using drones. Second, it is the use of drones for contact inspection applied to detail measurements on specific spots.

## 2 Aerial Visual Inspection Robot

The AERO-CAM RPAS is the aerial platform used for the visual inspection of the viaduct. The UAV is based on a DJI Matrice 600 Pro customized for the purpose of the mission. The current version of the UAV is the one shown in Fig. 1, where it can be seen the CAD model of the drone and the final implementation of it. As it can be seen, the configuration of the UAV is a 6-motors UAV, so the aircraft is robust against the failure of one engine due to the redundancy achieved by the configuration.

**Fig. 1.** AERO-CAM aerial robot

The aerial robot is being designed as small as possible, considering the weight and size. Moreover, the flight time has been optimized, maximizing its endurance to achieve the required flight times with the needed payload, that in that case are the camera and localization sensors. The AERO-CAM operates in BVLOS (Beyond Visual Line of Sight), and the camera is integrated on top of the aircraft, to allow photos to be taken of the underside of the viaduct deck, as well as to have cleaner pictures of the horizontal surfaces of the pilar and the rest of possible walls. The electronics of the UAV incorporates an electronic cover for dust protection. Due to the expectable lost of GNSS signal under the viaduct or bridge, AERO-CAM integrates a LIDAR underneath the airframe, between the two legs of the landing skid.

Regarding the hardware architecture for the integration of the camera in charge of the visual inspection, two Wi-Fi antennas are integrated onboard the aerial robot. These antennas allow the communication between the main computer, located at the bottom of the aircraft, and the computer of the camera system. The schematic of the hardware integration of the camera and the localization sensors is shown in Fig. 2.

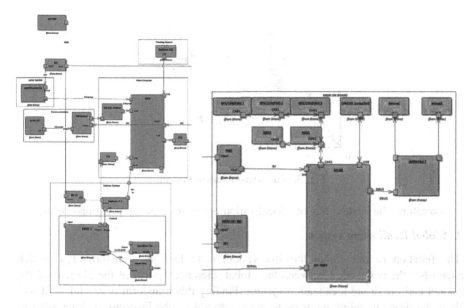

**Fig. 2.** Aerial robot hardware integration relative to the main added components

## 3 GNSS-Free Aerial Robot Navigation

To be able to perform the GNSS-free localization and navigation, the drone is equipped with an Ouster lidar sensor, model OS0 with 128 channels. As mentioned, this lidar sensor is used to perform an in-flight location of the drone movements, but it is also used to locate the drone's reference system with respect to the global inspection reference system. To complement this lidar, the drone has a 9-axis IMU that works at 400 Hz. In addition, the system includes a laser altimeter that allows more precise landings. The processing of all the sensor readings is carried out on an onboard computer, model Intel NUC i7.

In order to locate the aerial robot (or drone) in the same coordinate system than the one used for the location of the detected defects on the bridge, it requires the creation of a previous 3D map using a total station. This map will be a point cloud that identifies a coordinate origin for the entire inspection system. To create this map, operators must ensure that the ENU system is followed (x = east, y = north, z = up). This map can be reused in future inspections of the viaduct.

The drone system has its own localization and navigation algorithm ($T_{LD}$) that has its origin in the take-off point ($L$). Since this location may vary, the complete system requires a second localization system that stablishes the 3D transformation between the initial drone pose and the global reference system ($T_{GL}$), expressed in ENU coordinates at the origin of the map created by the total station ($G$). These transforms can be visualized in the following Figure, where $i$ is and arbitrary time instant (Fig. 3):

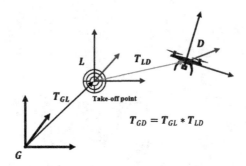

Fig. 3. Full transformation system.

Therefore, the system has two localization processes that are described below:

## A. Global localization system

The function of the global location system is to find the transform $T_{GL}$, which establishes the connection between the global reference system of the 3D map of the viaduct and the drone localization system. Finding this transform is crucial, as it will allow the drone to safely navigate to areas selected by the inspector as interesting to inspect without maintaining the same take-off position between flights. This is especially beneficial for eliminating the total station dependency beyond creating the initial 3D map. In addition, since the viaduct can be found in an inaccessible area, the take-off position may not be replicated between flights or even inspections on different days.

The transform $T_{GL}$ is fixed and will only vary during the flight if another transform with better accuracy has been obtained. Therefore, this global localization system is designed to calculate the transform at the initial instant, just before the drone takes off. During the flight, this system could continue working by calculating the transform between the drone current position and the base map ($T_{GD}$), so that if the accuracy of the transform improves, update it. This last case can be visualized in the following Figure (Fig. 4).

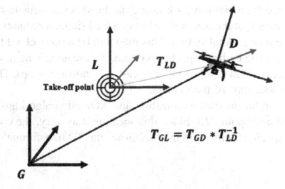

Fig. 4. In-air global localization transformation system.

This localization system does not have to be executed on the onboard computer, and it can be executed on a ground computer, since it does not require the calculation to be instantaneous. In this case, the onboard computer would send the data to the ground computer, which will perform the calculations and send the result back to the drone.

To find the correspondence between the previous 3D map generated by the total station and the data from the onboard sensors, we are working on an algorithm that makes use of the geometric characteristics of the point clouds. Specifically, the algorithm makes use of the characteristics of both clouds extracting FPFH (Fast Point Feature Histogram) features. These features encode the geometric properties of the k nearest neighbors of a given point using the average curvature of the multidimensional histogram around that point. Among its advantages, these features are invariant in position and a certain level of noise. After this feature extraction process, the algorithm applies RANSAC (Random Sample Consensus) to find a first approximation between both inputs. The result is then corrected with the assumptions of the problem and ICP (Iterative Closest Point) is applied to refine the result. The following Figure shows a block diagram of the main steps of the algorithm (Fig. 5):

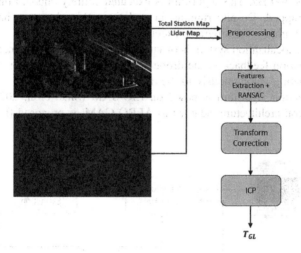

**Fig. 5.** Point clouds matching diagram.

## B. Relative localization system

The function of the relative localization system is to find the transform $T_{LD}$, which describes the motion of drone from its take-off point. This take-off point will be a flat area near the viaduct and is assumed to be flat with respect to the horizon so that the drone can take off safely. This localization is performed only with the onboard sensors and does not require any data beforehand, only the current sensor reading. Of course, it is desirable that this localization is as accurate as possible and, above all, minimize drifts in time as much as possible to avoid a significant divergence between reality and the pose in which the drone thinks it is (Fig. 6).

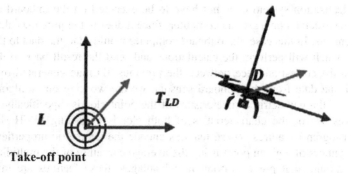

**Fig. 6.** Relative localization transformation system.

The relative localization system makes use of the lidar and the 9-axis IMU to calculate the drone's pose at each instant. The algorithm operates at high frequency in real time, updating the pose at the same frequency as the IMU, which in the case of AERO-CAM is 400 Hz. This algorithm is executed entirely onboard the drone in the Intel NUC equipped. Despite running in real time, this algorithm has the highest processing load among the programs executed by the drone.

The relative localization system is of vital importance to the system, as it is used to provide localization feedback to the drone control algorithm so it can ensure a stable flight by navigating autonomously to the desired target points.

The localization algorithm is based on LIO-SAM (Shan et al. 2020), where the following general architecture, adapted to AERO-CAM, is proposed (Fig. 7):

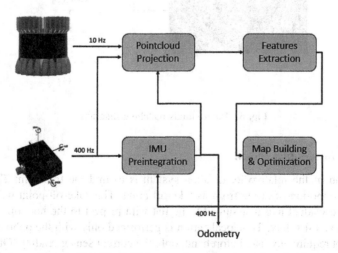

**Fig. 7.** LIDAR+IMU localization general architecture.

This architecture establishes a tightly coupled fusion between the lidar and the IMU, build a factor graph in which the measurements made by the sensors are integrated to build and optimize the map. The IMU pre-integration is based on (Forster et al. 2016). An important clarification it that to integrate the IMU measurements the magnetometer is not used, only the angular velocities and linear accelerations. Since the double integration of an IMU leads to large drift, the architecture proposes the short-term integration of the IMU, correcting its bias thanks to the localization at lower frequency in the built map using the information of the lidar point cloud. In order to process everything in real time, the algorithm discards lidar readings if they are not sufficiently displaced with respect to the previous reading considered (lidar keyframes). In this way, a lot of redundant information that would increase the computational load is discarded. Between lidar keyframes, the IMU readings are integrated, converging in a node of the graph that would be the state of the location at that given instant. The following diagram, simplified from (Shan et al. 2020), shows a schematic of the processing of the sensor readings (Fig. 8):

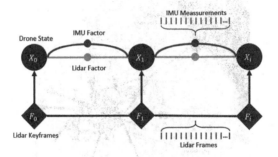

**Fig. 8.** Localization factor graph example.

As already mentioned, the result of all this processing is the location of the drone without GPS with a high frequency (400 Hz) that server the control algorithm to proceed with the AERO-CAM.

## 4 Aerial Contact Inspection Vehicle

The design and development of contact drone, AeroX, was based on requirements driven by the inspection sector end-user needs for giving successful solutions to the industry. The operation of the robotic solution should be robust under nominal working conditions at heights with potential strong wind gusts. Besides, the proposed aerial robotic system should be integrated with the current maintenance operations in many industries and should be easily operated by the personnel that is currently involved in bridges inspections, which often have low training in robotics. The main requirements considered in the design of AeroX can be summarized in the following:

- It should keep a sensor in steady physical contact at a point on the surface where a measurement is going to be taken.
- It should mount a variety of physical contact sensors.
- The aerial platform should have fast reactivity and controllability in order to fly close to obstacles with wind gusts.
- It should be easy to be operated for the personnel currently involved in bridge inspection with low training.
- Its operation should be easily adaptable to the specific inspection procedures currently used in bridge inspections.
- The robotic system should be very robust and reliable for everyday operations in industrial settings.

The proposed aerial robotic system—the AeroX robot—has three main components, see Fig. 9:

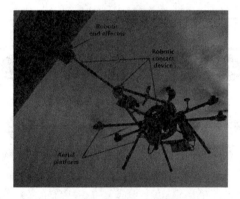

**Fig. 9.** The AeroX robot keeping physical contact during a validation experiment

- The aerial platform. Its design enables applying on the surface the contact forces required for physical inspections. The aerial platform is endowed with tilted rotors, which increases its stability and reactivity to compensate perturbations.
- The robotic contact device. It is responsible for providing the capability of steady contact with surfaces. It is a mechatronic device with six DoF and has the robotic end-effector at its end. Due to its efficient design, surface contact forces are transmitted to the centre of mass (CoM) of the aerial robot which enables their efficient and effective compensation.
- The robotic end-effector. Located at the end of the robotic contact device, it is endowed with wheels for moving on the surface under inspection. It integrates the sensors to be used for inspection and also additional sensors to facilitate operation. A variety of robotic end-effectors can be mounted on the robotic contact device. An end-effector with three wheels, an ultrasonic emitter and a sensor, and a camera is presented in this paper.

The operation of AeroX has two main modes. In the free-flight mode, the pilot guides the aerial robot (in manual or assisted way) to the element to be inspected and moves the robotic contact device to the selected inspection point until the robotic end-effector is in contact with the surface. As soon as the contact has been performed, and so the contact-flight mode starts, the pilot activates the fully-autonomous stabilization mode, which keeps the aerial vehicle steady w.r.t. the surface contact point with uninterrupted contact using only the measurements from the robot internal sensors. The inspector teleoperates the movement of the wheels of the robotic end-effector on the surface. As a result, when the inspector moves the wheels of the end-effector, the surface contact point changes and the aerial vehicle moves to keep its position steady w.r.t. the surface contact point. Hence, the aerial robot follows the end-effector commanded by the inspector.

## 5 Main AeroX Components

This section summarizes the main components of AeroX including the aerial platform, the robotic contact device, and the robotic end-effector. In order to have a better understanding of the overall aerial robot, in Fig. 10 is shown the overall system architecture.

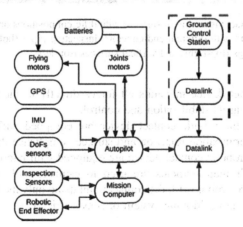

Fig. 10. AeroX UAV hardware architecture

### A. Robotic Contact Device

One of the main components of the proposed aerial robotic system is its novel robotic contact device, whose patent has been granted at the beginning of 2018 (Trujillo et al. 2018). Due to its mechanical design and integration to the aerial platform, all the surface contact forces are transmitted to the CoM of the aerial robot, which enables simplifying the stabilization and control of the aerial vehicle. With the robotic contact device, the aerial robot is capable of absorbing perturbations, keeping a robust and stable contact perpendicular to the contacted surface. The main characteristics of the robotic contact device are:

- Its mechanical design and integration with the aerial platform enable inspecting surfaces with different positions and orientations such as vertical, horizontal and inclined surfaces, see pictures from outdoor experiments in Fig. 11.

**Fig. 11.** AeroX in contact inspection of surfaces with different positions orientations in different outdoor experiments and with different end-effectors: inspecting pipes (bottom-left and bottom-right), horizontal ceilings (top-right) and vertical surfaces (top-left).

- The robotic contact device transmits all forces directly to the CoM of the aerial vehicle, simplifying its stabilization and control.
- All the joints of the robotic contact device are equipped with shock absorbers, which increase perturbation rejection and surface compliance during contact.
- The joints of the robotic contact device are equipped with internal sensors. During contact-flight their measurements are used to estimate and control the relative position of the aerial robot w.r.t. the contact point using only its internal sensors and without any external positioning system or sensor (Fig. 12).

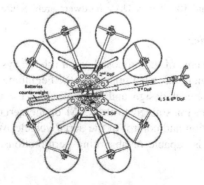

**Fig. 12.** Drawing of AeroX. Its size is 170 × 230 cm.

## B. The Aerial Platform

The controllability and reactivity to fly close to obstacles and withstand wind gusts, manoeuvrability and agility to fly in constrained industrial environments and the robustness for everyday operation have been the main characteristics in the design of the aerial platform.

The proposed aerial vehicle has eight 2 kW motors (KDE5215XF-220), providing a maximum take-off weight of 25 kg. The motors are set such that the robotic contact device can rotate around the vehicle CoM without colliding with the propellers. Each set of motor-propellers is alternatively tilted 30° around the motor boom. The tilting of the motors is fixed and cannot be controlled. The actuation variables that can be controlled are each of the eight motor desired forces. Tilted rotors enable independent control of the six DoF of the vehicle adding the capability to control the linear lateral forces, providing full control of the aerial platform during hovering, highly improving the aircraft controllabilit.

Furthermore, as demonstrated in (Michieletto et al. 2018), six tilted propellers are sufficient to be able to control the six DoF of the vehicle and ensure control of four DoF in case one motor fails.

The adopted configuration with eight motors was selected to overcome the failure of one motor keeping the full robot six DoF controllability. This makes this aerial vehicle the first one of its class, this feature being an important step to its future industrialization.

The aerial platform design and its hardware elements integration have been carefully implemented for distributing the weight in order to keep the CoM of the robot at the geometric center of the aircraft. Besides, the integration of the robotic contact device is such that the surface contact forces are transmitted to the CoM of the aerial vehicle, which improves stability and controllability during contact-flight mode.

## C. The robotic end-effector

The end-effector is the device that is connected to the robotic contact device and it contains the ultrasonic sensor to accurately measure the depth of the cracks, see Fig. 13. The following requirements for the end-effector were identified in coordination with the end users and with the ultrasonic sensor developers:

**Fig. 13.** General view of the AeroX end effector.

- The end-effector must be able to move along the surface of the bridge. The main reason for this requirement is that it is very difficult to position the ultrasonic sensor from the air exactly in the place where the end-user wants to perform the measurement. Even in an autonomous mode, the navigation precision is not enough.
- The end-effector should incorporate one or two cameras in order to allow the ground operator to know exactly where the ultrasonic inspection is going to be performed.
- The end-effector must have all the components included in order to facilitate the installation on the robotic arm, and also, allow exchanging the end-effector if it is required.
- The end-effector should be light and small enough to be carried by the AeroX UAV.
- The end-effector should be designed so the ultrasonic sensor can be operated remotely from the ground.

The end-effector main components can be seen in Fig. 14. They are:

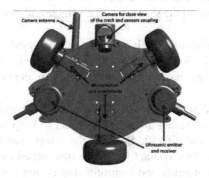

**Fig. 14.** End effector description

- Motion subsystem: a set of three omnidirectional wheels have been used. These three omnidirectional wheels allow us to have full motion capabilities on the end-effector such as moving to any direction on the plane or even rotating around a single point at any time. The wheels also facilitate the positioning of the sensor in the precise place where the measurement is required. Furthermore, this motion subsystem is composed of the necessary electronics and communications systems which allow the operation of the motion subsystem from the ground.
- Sensor subsystem: the sensor subsystem is composed of the ultrasonic sensor, the electronics required to process the information and the actuators required to move the ultrasonic probes (and apply a specific contact force towards the surface).
- Video cameras: two cameras have been installed in the system. One camera is integrated in the end-effector and allows knowing where the ultrasonic measurement is being performed with the required precision. The camera is positioned in a way that the crack will be seen in the centre of the image, being in this way in the middle between both probes. A second camera is used to have a wider view and have an idea of the position of the end-effector. This second camera is actually installed in the robotics arm in order to cover enough surface of the infrastructure and increase the situational awareness of the ground operator.

This designed specific end-effector for proper concrete surfaces applications, such as bridges, have been developed. The result is shown in Fig. 15. This end-effector has 3 omnidirectional motorized wheels distributed in a way the operator can move the tool in every direction, essential for its correct positioning with respect to the crack because it has to be aligned with the centre. Additionally, two micro-motors are in charge of moving the up and down mechanism for placing the ultrasonic transducers against the concrete. Additional sensors have been integrated on the end-effector to be used at the same time of the UT transducers, like a crack width measurement system as shown in Fig. 15. The piezoelectric emitter was mounted on one of the two micro-motors of the end effector and the acoustic-optical microsensor was attached to it using a rigid plastic support. In this way, both devices could be brought in contact with the surface of the concrete by the end-effector during the measurements. This design can be seen with more detailed in Fig. 15.

**Fig. 15.** Images of the integration experiments of the end-effector for concrete cracks characterization. The left images shows different perspectives of the end-effector. The top-right image shows the operator view of the end-effector internal camera aligned with a crack while taking a measurement. The bottom-right image shows the integration of a crack width measurement sensor which could be used at the same time than the ultrasonic transducers for crack depth.

## 6 Experimental Validation

Several experiments have been conducted in order to validate the UAV platforms and associated technologies (GNSS-free navigation and contact inspection). Afterwards, the system was tested on a real viaduct and bridge on operation in Spain, where the general visual inspection and the contact device and the end-effector for depth measurements in concrete were tested.

### A. Validation for Visual Inspections in Viaducts

A number of experiments were performed first to validate both the GNSS-free navigation solution and the general visual inspection procedure using AEROCAM platform. These experiments took place in a viaduct. A total of 360 images were obtained with AERO-CAM platform. The flights were performed under the viaduct in a space between two pillars. The captured photos contain visual information of both pillars and the viaduct deck (see Fig. 16).

**Fig. 16.** Two of the inspected areas

All images are in JPG format and were attempted to be taken at a safe distance to ensure the integrity of the AERO-CAM but still meet the 1mm per pixel density requirement (see Fig. 17).

**Fig. 17.** Example of images of the Álora viaduct dataset

### B. Validation for Bridges using Ultrasonic Sensors for Concrete

For the second set of experiments, these were performed at a bridge in order to test the capability of contacting a surface of the bridge, move along it and take ultrasonic measurements to obtain the depth of the cracks. Unfortunately, there were not visible cracks at the selected bridge and we had to limit the measurements to the concrete surface velocity.

The surface velocity measurement is the first step that has to be performed before the crack measurement. In this way, the velocity of transmission of the ultrasonic waves through the concrete is calculated. If a crack depth wants to be measured, it is needed to have that data before, that typically is taken near the crack. Afterward, the crack is measured using the difference of time measured between the previous one. Furthermore, this means that if the system is able to take a surface velocity measurement, it will validate the crack depth capability because it is using precisely the same tools and procedure.

The views the operator had for controlling the end effector were: a general view used for moving while avoiding obstacles and a close view of the operation of the sensors to check their coupling and the working surface conditions. Thus, the operator navigated from the initial contacted point to the one defined in the operation plan. Once the operator was at that point, he checked the surface with the close view camera and decided to move the sensors against the surface and proceed with the measurement of the surface velocity.

After two measurements on the vertical surface of the pillar, the pilot performed a detachment of the contact, and the ground operator commanded a change of the configuration of the robotic arm to inspect at the top of the UAV in order to touch the first beam of the first span.

After making contact with the beam of the bridge, the operator moved the end-effector along it to reach the point that was planned. Even if the end-effector is easy to be controlled, that operation had to be performed with caution due to the proximity of the wheels to the limits of the beam. However, with just some minutes of training, an operator with no previous training can control the inspection system. After reaching the objective point, the operator took another surface velocity measurement and concluded the operation giving the pilot the end mission instruction. The pilot commanded a detachment of the beam surface, and at the same time, the robotic arm returned to its standard front configuration and proceeded to land.

## 7 Conclusions

A novel solution for complete bridge inspection has been designed, and its validation is presented on this paper. The system is formed by two drones: one for general visual inspection and a second one for specific and detail inspections using ultrasonic contact inspections. The first one is an aerial robot that can obtain pictures of the overall bridge fully autonomously thanks to a non-GPS navigation system. The second one is the AeroX drone platform, a novel solution for inspection of difficult access areas. The AeroX can perform contact inspection due to its robotic contact device, which is

186 A. Viguria et al.

equipped with an end-effector. Also, both drones were validated and results and videos of the validation experiments were presented.

In conclusion, a new opportunity for the inspection of bridges and viaducts has arisen, allowing complete inspection operations in bridges in less time, reducing costs, improving quality of the inspection, and increasing safety of operators.

# References

Akhondi, M., Talevski, A., Carlsen, S., Petersen, S.: Applications of wireless sensor networks in the oil, gas and resources industries. In: Proceedings of 24th IEEE International Conference on Advanced Information Networking and Applications, Perth, Australia, 20–23 April 2010, pp. 941–948 (2010)

Forster, C., Carlone, L., Dellaert, F., Scaramuzza, D.: On-manifold preintegration for real-time visual-inertial odometry. IEEE Trans. Robot. **33**(1), 1–21 (2016)

Michieletto, G., Ryll, M., Franchi, A.: Fundamental actuation properties of multirotors: force-moment decoupling and fail-safe robustness. IEEE Trans. Robot. **34**(3), 702–715 (2018)

Ryll, M., et al.: 6D physical interaction with a fully actuated aerial robot. In: Proceedings of the IEEE International Conference on Robotics and Automation, Singapore, 29 May–3 June 2017, pp. 1–6 (2017)

Savazzi, S., Guardiano, S., Spagnolini, U.: Wireless sensor network modeling and deployment challenges in oil and gas refinery plants. Int. J. Distrib. Sens. Netw. **9**, 383168 (2013)

Shan, T., Englot, B., Meyers, D., Wang, W., Ratti, C., Rus D.: LIO-SAM: tightly-coupled lidar inertial odometry via smoothing and mapping. In: IEEE/RSJ International Conference on Intelligent Robots and Systems (IROS) (2020)

Trujillo, M., et al.: Aeronave con Dispositivo de Contacto. Spain Patent ES 2 614 994 B1 (2018)

# Damage Inspection, Structural Evaluation and Rehabilitation of a Balanced Cantilever Bridge with Center Hinges

Chawalit Tipagornwong[1], Koonnamas Punthutaecha[1],
Peerapat Phutantikul[1], Setthaphong Thongprapha[1],
and Kridayuth Chompooming[2(✉)]

[1] The Department of Rural Roads, Ministry of Transport, Bangkok, Thailand
[2] Thammasat University, Bangkok, Thailand
kridayut@tu.ac.th

**Abstract.** A main objective of the present investigation is to carry out bridge inspection and evaluation, and determination of structural rehabilitation of Phra Pinklao Bridge, having structural configuration of a balanced cantilever bridge with center hinges. The bridge structure is a 3-span, continuous prestressed concrete box girder with a total length of 280 m and a width of 26.6 m supporting 6 traffic lanes. The investigation tasks include 1) bridge visual inspection, 2) measurement of bridge alignment and profile, and joint movements, 3) bridge load test and behaviors measurement, 4) structural analysis and load-rating evaluation, and 5) appropriate bridge rehabilitation plan. Based on the results of bridge inspection, structural damages of the center hinge bearings in a critical state are reported. The damages include permanent deformation of stayed plates and plunger edges of the hinge bearings causing the bearing movement to be restrained and direct impact on serviceability and load-carrying behaviors of the bridge structure. Responses of the bridge structure and identification of any structural damages and deficiency are examined by employing diagnostic load testing under predetermined loading. The results of load-rating evaluation yielding sufficient load-carrying capacities of the existing box girder are reported. According to the investigation carried out, it can be found that appropriate implementation of the planar hinge bearing concept yields a most suitable and effective method for the center hinge bearing replacement of the bridge structure.

**Keywords:** Balanced cantilever bridge · Plunger hinge bearing ·
Prestressed box girder · Bridge damage inspection · Bridge load test ·
Bridge rehabilitation · Planar hinge bearing

© The Author(s), under exclusive license to Springer Nature Switzerland AG 2022
A. Akhnoukh et al. (Eds.): IRF 2021, SUCI, pp. 187–201, 2022.
https://doi.org/10.1007/978-3-030-79801-7_13

188 C. Tipagornwong et al.

# 1 Introduction

Phra Pinklao Bridge, having structural configuration of a balanced cantilever bridge with center hinges and being considered among the oldest prestressed concrete bridges in Thailand, was built in 1971 and opened to traffic in 1973. The bridge structure is a three-span, continuous prestressed concrete box girder with a total length of 83.0 + 114.0 + 83.0 = 280.0 m and a width of 26.6 m supporting 6 traffic lanes. The bridge structure was initially designed according to AASHTO HS20-44 design live load. Figure 1 illustrates the bridge configuration and main components. The box sections from each side are connected at the bridge mid-span employing four center hinges, each made of a metallic plunger-type hinge, installed over girder web with concrete horizontal shoe at bottom slab to ensure continuity of the deck ends across each side of the expansion joint. However, under load transferring, the top and bottom edge of the plunger portion are subjected to stress concentration due to limited (edge/line) contact area between the plunger and socket of the hinge bearings. This incident results in permanent deformation of plunger top and bottom edge and induces vertical gap between the plunger and socket causing excessive vibration and accumulative damage of the hinge bearings as shown in Fig. 2. Regarding serviceability and load-carrying capabilities of the bridge structure, corresponsive measures on structural assessment and remedial actions need to be carried out accordingly. A main objective of the present investigation is to perform bridge inspection and evaluation, and determination of the center hinge bearing replacement for Phra Pinklao Bridge.

# 2 Investigation Procedure

The main tasks of the investigation procedure depicted in Fig. 3 consists of:

1) Review of previous records and reports of bridge repair and maintenance
2) Visual inspection of the bridge structural components including superstructure, substructure, bearings, and expansion joints
3) Measurement of bridge alignment and profile, and movements of bridge bearings and expansion joints under long-term creep and shrinkage effects
4) Bridge load testing and behaviors measurement of the bridge structure under pre-specified load and under normal traffic condition
5) Structural analysis and load-rating evaluation of the existing bridge structure
6) Determination of appropriate method for rehabilitation of the center hinge bearings.

Damage Inspection, Structural Evaluation and Rehabilitation    189

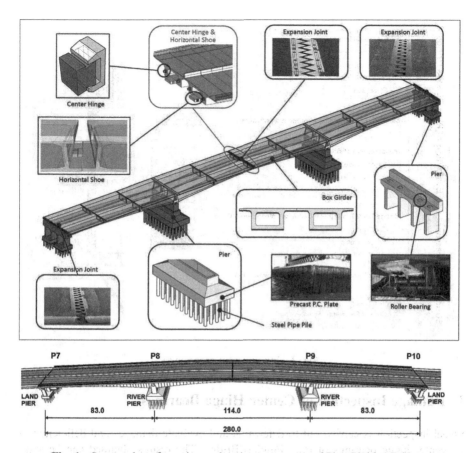

**Fig. 1.** Structural configuration and main components of Phra Pinklao Bridge.

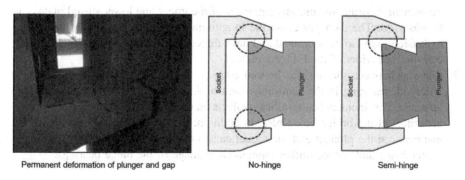

**Fig. 2.** Damage of center hinge bearing and gap due to permanent deformation of the plunger

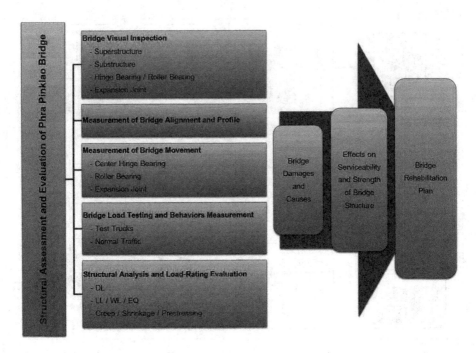

**Fig. 3.** Investigation procedure.

## 3 Damage Inspection of Center Hinge Bearings

Visual inspection is carried out to determine and evaluate damages and deficiency of the center hinge bearings of Phra Pinklao Bridge. The results obtained can be summarized as follows (Fig. 4).

1. Permanent deformation and misalignment of the upper and lower stayed plates can be observed. The damages cause the longitudinal movement of center hinges and expansion joints to be restrained and have direct impact on serviceability and load-carrying behaviors of the bridge structure
2. The gap between the top and bottom edge of the plunger and the socket can be observed, resulting in the semi-hinge and no-hinge condition of the center hinge bearings. This incident has significant effects on overall load-carrying behaviors and capacities of the bridge box girder. In addition, free vertical movement due to the gap between the plunger and the socket causes excessive vibration over the bridge center hinge and induce further cumulative damage of the hinge bearings.

3. Based on the results of bridge inspection, structural damages of the center hinge bearings in a critical state are reported. This necessitates urgent remedial actions for the center hinge bearings so that the bridge structure can be maintained under normal and safe operating condition.

An overall damage inspection of Phra Pinklao Bridge can be illustrated in Fig. 5.

## 4 Measurement of Bridge Alignment and Profile, and Longitudinal Movements

According to the measurement results of bridge geometric surveying, no abnormalities of bridge alignment and profile of the existing structure are found. The results of bridge longitudinal movement are summarized in Fig. 6. Considering the bridge structure on each side of the center hinges, the longitudinal movement of the roller bearings and expansion joints over the land pier and the center expansion joint can be observed with the tendency of movement direction toward the river pier. In conjunction with the results obtained on creep and shrinkage deformation analysis of the bridge structure, it is found that the accumulative movement of the bridge components is likely to be resulted mainly due to creep and shrinkage of the concrete under long-term compression effect. It is anticipated that under the current conditions of the existing structure, the bridge creep nearly approaches a constant state and will have no further effect on the movement of the hinge bearings in the future after replacement.

192  C. Tipagornwong et al.

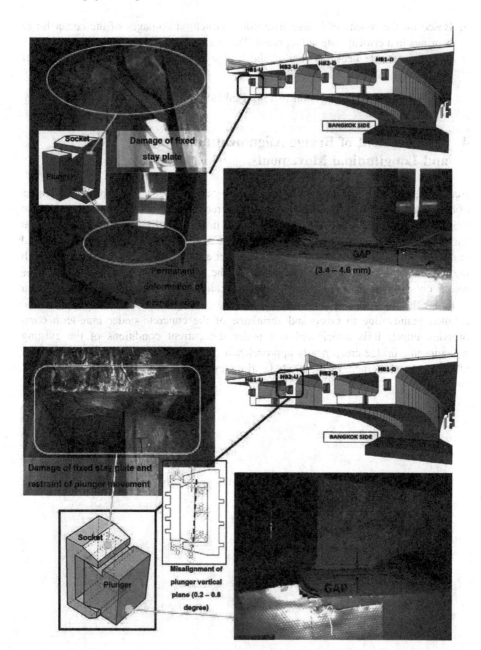

**Fig. 4.** Damage and deterioration of the bridge center hinge bearings.

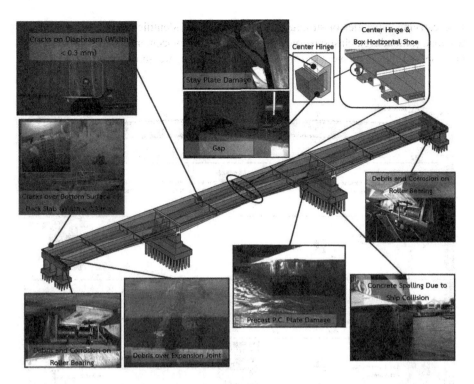

**Fig. 5.** Overall damage inspection of Phra Pinklao Bridge.

## 5 Bridge Load Testing and Behaviors Measurement

A diagnostic load test of Phra Pinklao Bridge, adopting the procedure of the nondestructive load testing of bridges outlined in the Manual for Bridge Evaluation (AASHTO 2008), is carried out to gain physical insight into load-carrying behaviors regarding structural deterioration and susceptible damages that may occur due to bridge aging and center hinge deficiency. Nine 3-axle domestic trucks with the predetermined gross weight of 245 kN per truck as shown in Fig. 7 are employed as test load. The effects of the test trucks are approximately equivalent to 80% of the design live load HS20-44 (AASHTO 2002) based upon the results of bending moments of the box girder.

The installation of measurement sensors to measure structural responses of the box girder is illustrated in Fig. 8. Examples of strain measurement at box girder Sect. 2 and 4 are shown in Fig. 9. The analytical results due to the test trucks obtained from the finite element model indicated by the dashed lines, are provided for comparison. Figure 10 shows the measurement results of the natural frequencies corresponding to bending modes of the bridge structure.

Based on the measurement results of the box girder under the test trucks, the overall structural responses are in normal conditions. No indication regarding any loss of load-carrying capacities or structural deficiency can be identified. Upon removal of the test load (unloading), elastic recovery of strains and displacements of the box girder can be observed. No permanent deformation of the bridge structure is reported.

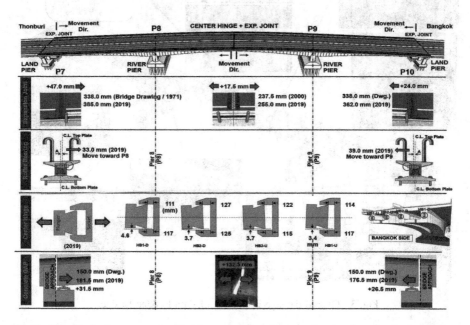

**Fig. 6.** Longitudinal movement of bridge expansion joints, roller bearings, and center hinge bearings

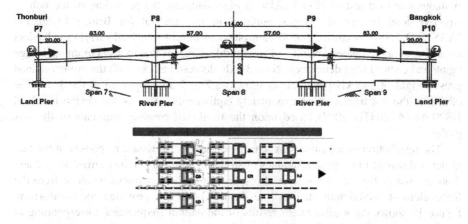

**Fig. 7.** Test trucks employed in diagnostic load test of Phra Pinklao Bridge.

Damage Inspection, Structural Evaluation and Rehabilitation 195

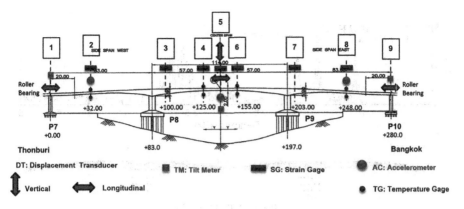

**Fig. 8.** Installation of measurement sensors of the box girder.

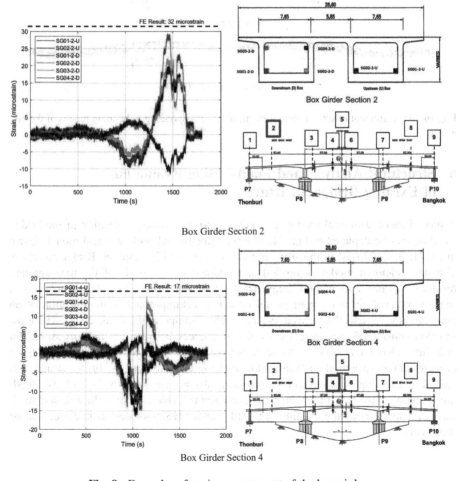

**Fig. 9.** Examples of strain measurement of the box girder.

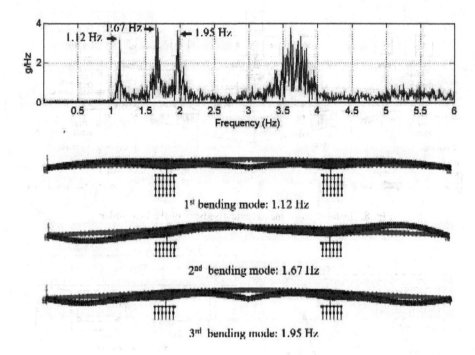

**Fig. 10.** Measurement results of natural frequencies corresponding to bending modes of the box girder.

## 6 Structural Analysis and Load-Rating Evaluation of Existing Bridge Structure

A procedure of structural modeling, analysis and load-rating evaluation of the bridge structure can be depicted in Fig. 11. A three-dimensional finite element model shown in Fig. 12 is employed in the structural analysis of the Phra Pinklao Bridge structure. The finite element model is verified and updated on the basis of the measurement results of strains and natural frequencies under the bridge load test employing a sensitivity-based finite element model updating technique and a minimization of an error function (Friswell and Mottershead 1995, Chompooming et al. 2007). The error function considered is represented by the differences between the field measurement and finite element results of the response parameters. Based on the updated finite element model, structural analysis and load-rating evaluation of the bridge structure is carried out. Examples of structural analysis results are illustrated in Fig. 13. The load-rating results of the box girder can be summarized in Table 1. In overall, sufficient load-carrying capacities of the existing box girder under HS20-44 design live load can be illustrated.

## 7 Determination of Appropriate Replacement Method for Center Hinge Bearings

An appropriate replacement method for center hinge bearings is determined on the basis of the following considerations.

- Current state of long-term deformation due to concrete creep and shrinkage effects of the bridge structure and accumulative damages of the center hinge bearings as the results of stress concentration over the edge or line contact between the plunger and socket
- Serviceability and load-carrying capacities of the bridge structure before and after the center hinge bearing replacement
- Durability and maintenance of the hinge bearings
- Cost and construction period

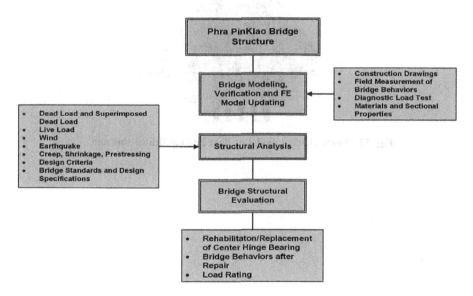

**Fig. 11.** Procedure of structural modeling, analysis and evaluation.

198   C. Tipagornwong et al.

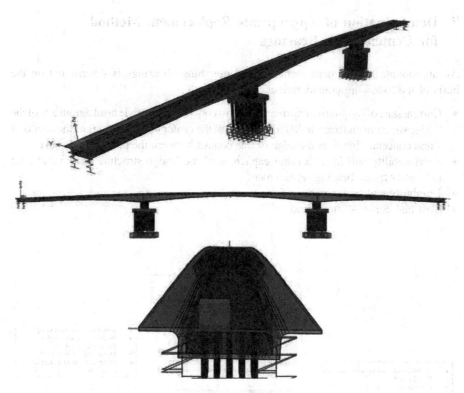

**Fig. 12.** Finite element model of Phra Pinklao Bridge structure.

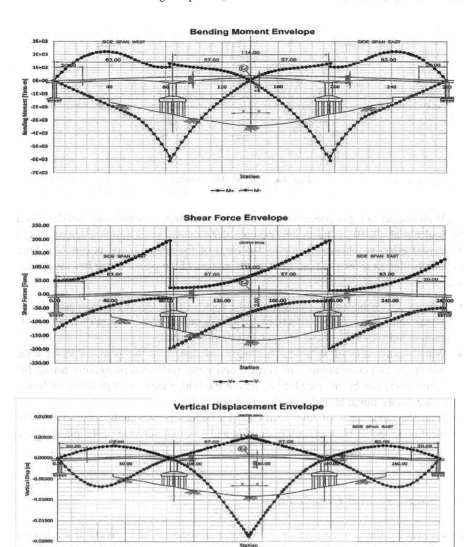

**Fig. 13.** Examples of structural analysis results of the box girder under HS20–44 design live load.

**Table 1.** Inventory rating factors of the box girder under HS20-44 design live load.

| Girder section | Inventory RF | Remark |
|---|---|---|
| +33.5 | 1.63 | Bending |
| +78.0 | 2.30 | Bending |
| +100 | 2.01 | Bending |
| +131 | 2.50 | Shear |

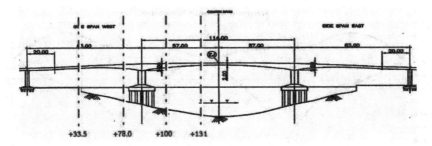

**Fig. 14.** Phra Pinklao Bridge

With respect to these concerns, a planar hinge bearing, as introduced and described by Spuler et al. (2010), is taken into the consideration and in overall it is determined to be a most effective alternative for replacement of the center hinge bearings of Phra Pinklao Bridge as shown in Fig. 14. The advantages of the planar hinge bearings are as follows.

- Contact area between the plunger and the socket of the bearing is increased due to planar contact, hence the effects of stress concentration causing permanent deformation and damages of the plunger can be diminished.
- With a utilization of spherical bearings installed over the planar contact area, continuity and compatibility of vertical and multi-rotational movement across the center hinge can be enhanced. Evenness of the bridge expansion joints on both side of the center hinge is anticipated to be improved.
- With the advantages of high load-carrying capacities and multi-rotational movement of spherical bearings, more durable and low maintenance of the planar hinge bearings after replacement can be expected.

## 8 Summary and Conclusions

In this paper, the implementation of damage inspection and structural evaluation including determination of center hinge bearing replacement for a balanced cantilever bridge is described and illustrated. Responses of the bridge structure and identification of any structural damages and deficiency are examined by employing diagnostic load testing under predetermined loading. The response measurements of the bridge structure are employed in the finite element model verification and updating process considering a sensitivity-based minimization of the error function represented by the differences between the field measurement and finite element results. Bridge load rating is then evaluated on the basis of the updated model under HS20-44 design live load. The results yielding sufficient load-carrying capacities of the existing box girder are reported. A recommendation for replacement of the original edge or line contact hinge bearings with planar hinge bearings is presented. In addition, spherical bearings are employed to enhance load-carrying capabilities and compatibility of bridge vertical and multi-rotational motion across the center hinges. Currently, Phra Pinklao Bridge rehabilitation including replacement of center hinge bearings and expansion joints has been under an ongoing process.

# References

AASHTO: Standard Specifications for Highway Bridges, 17th edn. American Association of State Highway and Transportation Officials, Washington DC (2002)

AASHTO: The Manual for Bridge Evaluation, American Association of State Highway and Transportation Officials, Washington DC (2008)

Chompooming, K., Sirimontree, S., Prasarnklieo, W., Wiriyakowittaya, T.: Finite element model updating of a segmental box girder based on measured responses under load testing. In: Sixth International Workshop on Structural Health Monitoring, Stanford, CA (2007)

Friswell, M.I., Mottershead, J.E.: Finite Element Model Updating in Structural Dynamics. Kluwer Academic Publishers (1995)

Spuler, T., Moor, G., Ghosh, C.: Supporting economical bridge construction – the central hinge bearings of the 2nd Shitalakshya Bridge. In: IABSE-JSCE Joint Conference on Advances in Bridge Engineering-II, Dhaka, Bangladesh (2010)

# Reinforcement of Scoured Pile Group Bridge Foundations with Spun Micro Piles

Apichai Wachiraprakarnpong, Juti Kraikuan,
Jirasak Watcharakornyotin, and Thanwa Wiboonsarun[✉]

Bureau of Bridge Construction, Department of Rural Roads, Bangkok, Thailand
kevin.wib@gmail.com

**Abstract.** This article demonstrates an approach to reinforce the bridge's foundations, of which its safety factor had been reduced to a dangerous level by scouring, with the use of spun micro piles. The bridge was built more than 30 years ago at Dong Pa Lan, Mae Fai Subdistrict, San Sai District, Chiang Mai Province, Thailand. The Department of Rural Roads investigated the existing piles' lengths embedded in the soil using Parallel Seismic Test and then recalculated the capacity of the foundations. The calculation suggested that one of the foundations had a safety factor of 1.04, which was considered extremely dangerous from a geotechnical engineering aspect since a value higher than 3.0 is typical. A collapse would occur if the bridge was subjected to more scour and traffic without reinforcement. The department examined a few solutions to the problem considering limitations on the construction space, necessity of traffic during the repairing process, cost, and safety for the construction, and then decided to reinforce the foundations with spun micro piles. The process began with pressing the piles around the old foundations into the stiff soil layer by using hydraulic jacks. The weight of the superstructure was transferred by temporary steel structures to the new micro piles. The bridge was then lifted a little to pre-load the piles. Finally, concrete columns reinforced with steel tubes were cast to connect the old foundations with the micro piles. The result suggested that the spun micro piles were able to support the weight of the bridge and complied with the design criteria and engineering standards.

**Keywords:** Reinforced concrete bridge · Safety factor · Spun micro pile · Foundation reinforcement · Scouring

## 1 Introduction

Dong Pa Lan Friendship Bridge is located on Rural Road No. 1414 at Mae Fai Subdistrict, San Sai District, Chiang Mai Province, Thailand crossing Ping River according to Fig. 1. The bridge with two traffic lanes was built from reinforced concrete in 1997 and had been opened for traffic for more than 20 years. The geometry of the bridge includes eight spans with two 15-m exterior spans and six 20-m spans for a total of 150 m in length and 10 m in width. The foundations of the bridge were designed as pile groups containing 20 driven piles. The third to sixth foundations were in the channel. The elevation and plan drawings are demonstrated in Fig. 2.

© The Author(s), under exclusive license to Springer Nature Switzerland AG 2022
A. Akhnoukh et al. (Eds.): IRF 2021, SUCI, pp. 202–216, 2022.
https://doi.org/10.1007/978-3-030-79801-7_14

**Fig. 1.** Location of Dong Pa Lan Friendship Bridge (Google Maps 2021).

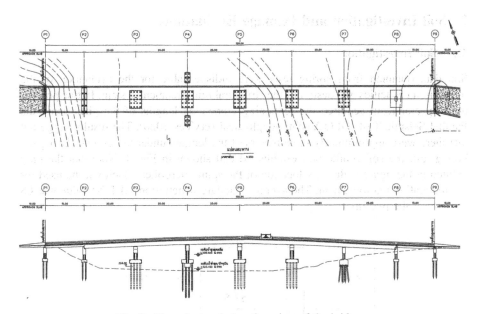

**Fig. 2.** Plan view and elevation view of the bridge.

Ping River has been utilized as sand suction sites for many years resulting in a wider bankfull width, lower bottom level, and more scouring. Department of Rural Roads investigated the bridge and found that almost four meters long of the existing piles' length was already above the soil as illustrated in Fig. 3. This problem caused the piles to be more slender, and it reduced the pile lengths embedded in the soil. If the bridge was still used without repairing, there would be danger to the entire structure and the people commuting on it.

**Fig. 3.** Damage on pile-group bridge foundations.

## 2 Soil Investigation and Damage Evaluations

### 2.1 Soil Investigation

Soil data obtained from boring tests were indispensable for the evaluation of the foundation capacities and design of the structural reinforcement system. In this project, the inspector team acquired the remaining embedded lengths of the piles using the Parallel Seismic Test (GEO Vision geophysical services 2020). The results suggested that there were approximately three meters of the lengths under the ground. Using the boring test, the soil profile was established and shown in Fig. 4. Note that the gray column in the figure is the new location of the spun micropiles which will be used for the foundation reinforcement. The average standard penetration (SPT-N) value was 68 blows/ft at the pile tip.

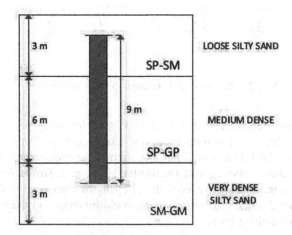

**Fig. 4.** Soil profile obtained from soil boring tests.

## 2.2 Damage Evaluation

From the inspection, all the four foundations located in the channel were scoured, causing more than half of the pile lengths to be above the ground. The maximum scoured pile length was almost four meters, as measured from the bottom of the foundation to the ground. The damage caused a direct consequence that results in only 16.7 tons per pile of the remaining capacity, which was three times less than the designed capacity of 50 tons per pile. As a result, the remaining safety factor approximated was only 1.04. Generally, from a geotechnical perspective, the typical value used is between 2.5 to 3.0, making this case extremely dangerous. Therefore, the department examined several solutions, taking into consideration the limited construction space, the necessity of using the bridge during the construction, cost, and safety.

# 3 Alternatives and Solutions

The first alternative was to demolish the bridge and reconstruct a new one. The department, however, estimated that the cost of the bridge would be around $1,200,000. Also, this method would affect the people who needed to use the bridge because the adjacent route was 10 km away and would cause additional time and cost to people. The other solution was to reinforce the bridge foundations. This method did not require the replacement of the superstructure. Thus, people could still utilize the bridge, and the cost was cheaper than reconstructing a new bridge.

## 3.1 Alternatives

Construction of a new foundation in replacement of the old one is very challenging because the weight of the superstructure was still on it. There are several options, depending on the type of damage and the site characteristics. The U.S. Federal Highway Administration suggests many options and case studies regarding the repair of bridge foundations (U.S. Department of Transportation, The Federal Highway Administration 2016). A typical method used in Thailand is to build a portal frame over the existing foundation. This frame will transfer the load of the superstructure to the two new foundations on a new cap beam (Kumar et al. 2013). Figure 5 shows a foundation already repaired by using the portal frame method.

**Fig. 5.** Portal frame foundation repair.

From Fig. 5, one of the foundations of Dong Pa Lan Friendship Bridge had already been reinforced with a portal frame. However, it was rather expensive because of the two new foundations and the new cap beam. The new structures would obstruct and change the water flow, which would cause more erosion on the river embankment. The cost of this method was estimated to be around $650,000 for this project.

The next popular type of foundation is called "Drilled Shaft" which is a pile boring method. The construction step begins with pre-boring to the designed depth. Steel casings are driven into the hole with a stabilizing solution to prevent the hole from collapse. The reinforcing steel is then placed into the hole ready to cast a reinforced concrete pile. Finally, the casings will be removed (Garder et al. 2012). Figure 6 shows the procedure to construct a drilled shaft.

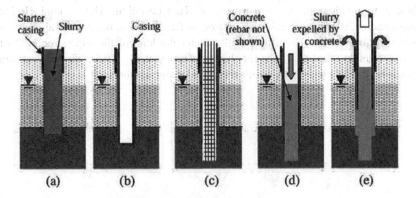

**Fig. 6.** Construction of drilled shafts.

A major advantage of the drilled shaft method is that the pile can be cast to the desired depth no matter how the soil profile is. The structural strength of the pile itself is more flexible than a pre-cast pile since the diameter of the pile, the amount of reinforcing steel, and the strength of the pile materials can be designed. However, the equipment and machine required for boring a drilled shaft are typically tall and large, making this method unsuitable for work under a bridge foundation or in a limited space. Drilled shafts are suitable for constructing a new bridge without the superstructure already standing. As a result, this piling method was not appropriate for the project.

Because of the limitations, the department introduced the idea of using spun micro piles, which had become popular in the reinforcement of residential foundations in the country. Reinforcement of the foundations with the micro piles was very convenient due to their small sizes (less than 1.50 m). This alternative allowed the use of the bridge during the construction process; it prevented the new foundation from obstructing too much water flow, and reduced the construction cost. This type of pile could be driven under the scoured foundation for replacement of the old piles. After that, the weight of the superstructure would be transferred to the micro piles instead. The total cost of the project was only $380,500 with only a one-year construction period, which was cheaper and faster than the other methods mentioned earlier.

## 3.2 Spun Micro Piles

Spun micro piles are made of pre-cast concrete reinforced with pre-tensioned steel. The piles have either rectangular or circular cross-sections with a hollow in the middle as shown in Fig. 7 (Bhumi Siam 2017). The piles are cast inside formworks which rotates around their longitudinal axes faster than the gravitational acceleration. The production causes the concrete to condense more than the normal cast-in-place method. In general, a spun micro pile has an allowable capacity of 20 to 40 tons, and a length varying from 1.0 to 1.50 m. A total pile length can be increased as designed by welding the plates at the ends of the piles (Irawan et al. 2015).

**Fig. 7.** Spun micro piles (Bhumi Siam 2017).

Spun micro piles are very convenient. The hollow cross-sections help reduce the vibration and sound from the driven process. The earth pressure exerting on the pile tips are less because some soil can move through the hollow (Irawan et al. 2015). Transportation of the piles is quick and easy due to their sizes. The related driven equipment is also smaller than a typical driven pile, allowing construction on limited space or an area difficult to reach. Considering these benefits, the department decided to use spun micro piles to reinforce the foundations P3, P5, and P6, which were located underwater (See Fig. 2). Then, the new pile foundations were covered by a reinforced concrete wall to decrease incoming scouring. Figure 8 shows the drawing for the reinforced bridge. The spun micro piles used in this project had a diameter of 24 cm, length of 1.50 m long, and allowable strength of 25 tons per pile. The standard used was Thai Industrial Standard Institute (TISI) No. 397-2534.

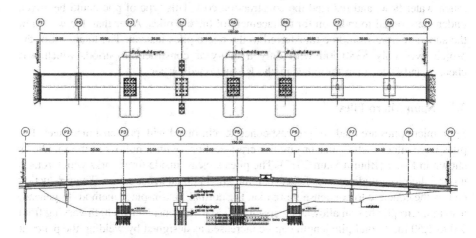

**Fig. 8.** Elevation and plan views of reinforced bridge foundations.

## 4 Design of Spun Micro Piles

The new foundations with spun micro piles were designed to have more capacity than the maximum capacity of the old foundations. This would ensure long-term safety and utilization of the bridge. The capacity of the old foundations was recalculated based on the soil data and the pile characteristics. The structural capacity of the piles themselves was still sufficient as could be assessed from the current bridge condition.

### 4.1 Capacity of Existing Foundations

From the damage to the foundation shown in Fig. 3, the soil around the piles was already scoured for most of the lengths. Obtaining the remaining pile lengths embedded in the ground would incur more expenses and result in a little more skin friction capacity to the foundation. Furthermore, the soil could be scoured more in the future. Therefore, it was safe to assume that the capacity of the pile consisted of the end bearing alone without the skin friction. From the parallel seismic tests and the soil data, the pile capacity was computed to be 16.7 tons per pile. For a foundation with 20 piles, the capacity was 334 tons with a factor of safety of 1.04. Thus, the actual capacity was 347 tons without a safety factor.

## 4.2 Design of Spun Micro Pile Foundations

The department designed the foundations using spun micro piles based only on the end bearing capacity of the piles and without the skin friction. This concept would extend the use of the bridge in case scouring occurs in the future which will reduce the skin friction. Thus, the capacity of a pile can be computed using Eq. 1 (Al-Hashemi 2016).

$$Q_b = q_b A \qquad (1)$$

where

$q_b$    is the end bearing capacity of the pile $\leq 1000\,\text{t/m}^2$
$A$    is the cross-sectional area of the pile

Since after the pile was driven in place, the soil would be compacted in the hollow section and act as a part of the cross-section. Therefore, the full cross-sectional area of 453 cm$^2$ was used. The end bearing capacity at the pile tip level of 10.5 m below the ground was computed from the soil data. The SPT-N value was 68 blows per foot, which limited $q_b$ at 1000 ton/m$^2$. Therefore, the end bearing capacity was 45.3 tons per pile. The value is more than the allowable capacity of the pile of 25 tons, which the factor of safety is 45.3/25.0 = 1.81. Note that the desired safety factor is more than 3.0. Thus, the number of piles designed would be increased to enhance safety. The actual capacity of each pile would be tested again once they had been driven in place.

The number of piles required for the reinforcement of the old foundations can be recalculated from Eq. 2.

$$F.S. = \frac{Q_{old} + NQ_{pile}}{Q_{old}} \geq 3.0 \qquad (2)$$

Thus, $N \geq \frac{2Q_{old}}{Q_{pile}}$.

where

$F.S.$    is the factor of safety
$Q_{old}$    is the capacity
$Q_{pile}$    is the capacity of the spun micro pile
$N$    is the number of pile required

By setting the factor of safety to 3.0, the capacity of the old foundation of 347 tons, and the capacity of the micropile of 25 tons per pile, the minimum number of piles required for a foundation is 28. In this project, the department chose 34 piles per foundation for symmetry. Figure 9 presents the drawings for the reinforced foundation.

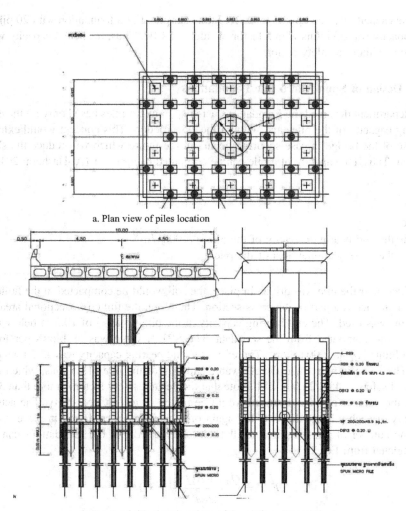

a. Plan view of piles location

b. Front and side elevation views of the reinforced foundation

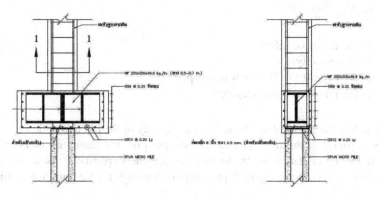

c. Connection between old columns and new spun micro piles

**Fig. 9.** Reinforced foundation drawings

# 5 Construction Procedure

As mentioned previously, the use of spun micro piles allowed the bridge to open for traffic during most of the construction period, except for when the new foundations were preloaded to transfer the weight of the superstructure. This was due to the small size of the micro piles, which caused little influence on the structure during the driving process, and there was no replacement or demolition of the existing foundations. However, there were strict weight limitations on the bridge to ensure safety. The construction process included the following steps:

## 5.1 Preparation of Construction Site

The first step of all was to prepare the construction site. The area boundary was clearly set and signed for unauthorized people not to enter to ensure safety. A storage area, site office, and site utilities were established.

## 5.2 Soil Excavation and Embankment

After the construction site was set, the workers embanked soil and compacted it around each foundation at a time. The embankment helped adjust the working area in order that the construction equipment could access the foundation. This also prevented water to pass through the foundation during the construction. The embankment worked well since the water level in the channel was not too high. Figure 10 shows the working area with an embankment around a foundation.

**Fig. 10.** Adjusted working area around the bridge foundation.

## 5.3 Driving Spun Micro Piles

This step began by marking all the pile locations according to the design in Fig. 6. After that, casing steel tubes, which were slightly larger than the micro piles in size, were driven as presented in Fig. 11. The soil and rock inside the casing tubes were then bored and removed until the hole depth reached 7.0 m. The workers used a PVC tube whose diameter was smaller than the casing tube, air pressure, and water to push the soil and rock out of the hole. This method had been applied from the bored pile procedure to all the micro piles to pass the rock layer at the depth of 4.0 to 5.0 m. Then, the micro piles were connected and driven to the remaining designed depth of 10.5 m.

**Fig. 11.** Spun micro piles and casing tubes.

Generally, a driving hammer is necessary to drive a pile into the ground. In this project, however, the locations of the reinforcing piles were underneath the old foundation, and this condition prevented the hammer from accessing the piles. As a result, the department decided to drive the piles by using hydraulic jacks. The jacks were installed at the head of the piles against a temporary frame connected under the old foundation. The weight of the superstructure helped counterweigh and push the piles into the ground when the jack expanded as shown in Fig. 12.

**Fig. 12.** Use of hydraulic jacks to drive spun micro piles.

## 5.4 In-Situ Test of the Load-Bearing Capacity of the Driven Piles

A hydraulic jack with a pressure gauge was used to recheck the actual load-bearing capacity of the piles. The pressure value in psi was calibrated to the load unit in tons according to ISO 17025:2017 (International Organization for Standardization 2017) at Chulalongkorn University in Thailand in 2018. As mentioned in the previous step, the piles were dropped to a depth of 7.0 m inside the casing tubes. Then, they were pressed to the designed depth of 10.5 m with the hydraulic jacks. After reaching the target depth, the calibrated hydraulic jack was used to measure the actual capacity of each pile. Figure 13 demonstrates the calibration curve showing the load in tons vs. pressure in psi.

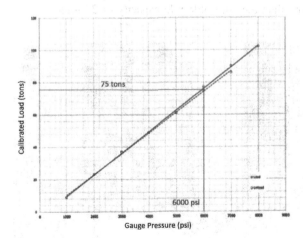

**Fig. 13.** Calibration curve for testing the actual pile capacity (Faculty of Engineering, Chulalongkorn University, Thailand 2018).

The allowable load of a pile was 25 tons with a safety factor of 3.0. Thus, the actual capacity needed to be at least 75 tons. From the tests, the piles moved a little more down into the ground due to the hydraulic jack and stopped moving when the load was stable at 75 tons. According to Fig. 10, the pressure gauge needs to read 6000 psi to obtain this load. Figure 14 shows the pressure gauge reading 6000 psi (red scale).

**Fig. 14.** Pressure gauge reading the target load-bearing capacity.

## 5.5 Transferring the Weight of the Superstructure to the New Piles

After all the spun micro piles were driven and their capacity tested as planned, the construction team built the load transferring structure to transfer the weight of the superstructure to the new piles. The structure consisted of circular steel tubes connected to steel bearings at the tops of the piles and attached to the bottom of the old foundation as shown in Fig. 15. In this project, the circular tubes have an outside diameter of 8 in. and a thickness of 4.5 mm with an ultimate capacity of 31.8 tons per column. This number was more than the allowable pile capacity of 25 tons.

**Fig. 15.** Load-transferring structure.

To transfer the weight to the new piles, hydraulic jacks were installed to lift the superstructure up slightly. After that, the load was transferred and the new foundation connected to the new piles was cast. Note that traffic was not allowed during this process.

### 5.6 Casting New Foundations and Protection Walls

The load of the superstructure was transferred to the micro piles. The new foundations and their protection walls were cast. This was where the old foundations were already reinforced. The new foundations helped connect the original foundations and their old piles to the new micro piles so that they could work together. The walls helped protect the foundations and slow down the scouring rate. Figure 16 shows the final look of the bridge after the construction.

### 5.7 Returning the Site to the Beginning Condition

As outlined in the first step, there was some soil embankment created to adjust the channel in order to prepare the construction space. After the process was complete, the channel was readjusted to the condition before the project. Everything was rechecked to ensure that the bridge would work well for a long time.

**Fig. 16.** The bridge after the reinforcement.

## 6 Conclusions

The Department of Rural Roads has reinforced the foundations of Dong Pa Lan Friendship Bridge, which was located on Rural Road No. 1414 at Mae Fai Subdistrict, San Sai District, Chiang Mai Province, Thailand. The groups of pile foundations were damaged from scouring, which removed most of the soil around the existing piles. From the evaluation before the reinforcement, the factor of safety was decreased to a dangerous value of 1.04. The department considered several alternatives based on many factors: structural strength, long-term durability, limitation on the working space, cost, and the need for the use of the bridge during the construction. Considering all the factors, the department decided to use spun micro piles to build a new foundation that is underneath and attached to the existing structure.

The process started by recalculating the remaining capacity of the old pile foundation and then designing a micro pile system to obtain a desirable safety factor of 3.0. There were three foundations located in the channel that needed to be reinforced. The pile driving process consisted of pre-boring the holes and driving the piles to the design depth using hydraulic jacks. All the piles driven were checked for their actual capacity by pressing them with a hydraulic jack calibrated with the load scale. The results suggested that all the piles achieved the target factor of safety. Then, a load transferring structure was built and used to transfer the weight of the superstructure to the new foundation. During the construction period, the bridge could still be utilized and crossed, making this process non-disruptive for the local people. Therefore, reinforcement of the pile-group bridge foundations with spun micro piles is very cost-effective, convenient, and efficient. This method would be an excellent alternative in repairing old bridges while greatly extending their lifespan.

# References

Bhumi Siam: Spun Micro Piles. Bhumi Siam Supplies Co. Ltd. (2017). https://www.bhumisiam/micro-pile051017/

Irawan, C., Suprobo, P., Raka, P., Djamaluddin, R.: A review of prestressed concrete pile with circular hollow section (Spun pile). Jurnal Teknologi **72**(5) (2015). https://doi.org/10.11113/jt.v72.3950

Kumar, D., et al.: Design of Bridge Component. BML Munjal University, India (2013)

Faculty of Engineering, Chulalongkorn University: Calibration of hydraulic jack report (reference no. CHJ-84/61). Faculty of Engineering, Chulalongkorn University, Thailand (2018)

GEO Vision Geophysical Services: Parallel Seismic Test Method. GEO Vision Geophysical Services Co. Ltd., Corona (2020)

Google Maps: Location of Dong Pa Lan, Mae Fai Subdistrict, San Sai District, Chiang Mai Province, Thailand (2021). maps.google.com

Al-Hashemi, H.M.B.: End Bearing Capacity of Pile Foundations. University of Bahrain (2016). https://doi.org/10.13140/RG.2.2.21744.46080

International Organization for Standardization: General requirements for the competence of testing and calibration laboratories (ISO 17025:2017). International Organization for Standardization, Geneva, Switzerland (2017)

Garder, J., et al.: Development of a Database for Drilled SHAft Foundation Testing (DSHAFT) (2012). Corpus ID: 108019510

Thai Industrial Standard Institute: Standards for Prestress Spun Concrete Piles (TISI No. 397-2534). Government Gazette No. 111 Section 93 (1994)

U.S. Department of Transportation, The Federal Highway Administration: Protocols for the Assessment and Repair of Bridge Foundations. The Federal Highway Administration, McLean (2016)

# Repair of Settled Pile Bent Bridge Foundations with Spun Micro Piles

Apichai Wachiraprakarnpong, Juti Kraikuan, Wichai Yu, and Pawin Ritthiruth[(✉)]

Bureau of Bridge Construction, Department of Rural Roads, Bangkok, Thailand
kevin.wib@gmail.com

**Abstract.** This article demonstrates an approach for reinforcing bridge's foundations, which had been settled from scouring, with the use of spun micro piles to substitute the existing piles. The bridge was built 27 years ago crossing Ranae Canal in Phipun District, Nakhon Si Thammarat Province in the southern region of Thailand. The bridge has eight pile-bent foundations. The Department of Rural Roads investigated the bridge in 2020 and found that two foundations in the middle of the channel settled down by 7 and 10 cm from the initial elevation, causing a shift in the bridge's horizontal alignment. This deflection significantly reduced the capacity of the bridge and would cause danger if it was still opened for traffic. The department examined a few solutions to the problem considering limitations on the construction space, necessity of traffic during the repairing process, cost, and safety during construction, and then decided to reinforce the foundations with spun micro piles. The process began with driving the micro piles around the old foundations and constructing temporary structures to transfer the weight of the bridge to the new piles. The superstructure of the bridge was then lifted using hydraulic jacks against the load-transferring structures to obtain the desired level. The old concrete columns were replaced by the new composite ones. Finally, new reinforced concrete foundations were cast on the micro piles to support the bridge. The result suggested that the spun micro piles were able to support the weight of the bridge and complied with the design criteria and engineering standards.

**Keywords:** Reinforced concrete bridge · Settlement · Spun micro pile · Foundation reinforcement · Scouring

## 1 Introduction

Ranae Bridge is a local bridge located at Ranae Canal, Phipun District, Nakhon Si Thammarat Province, Thailand as shown in Fig. 1. The bridge was built 27 years ago with two traffic lanes. The geometry of the bridge includes seven 10-m-long spans for a total of 70 m long and 10 m wide. The foundations of the bridge were designed as pile bents containing six driven piles for each. The fourth to seventh foundations were in the channel. The elevation and plan drawings are demonstrated in Fig. 2.

© The Author(s), under exclusive license to Springer Nature Switzerland AG 2022
A. Akhnoukh et al. (Eds.): IRF 2021, SUCI, pp. 217–231, 2022.
https://doi.org/10.1007/978-3-030-79801-7_15

218     A. Wachiraprakarnpong et al.

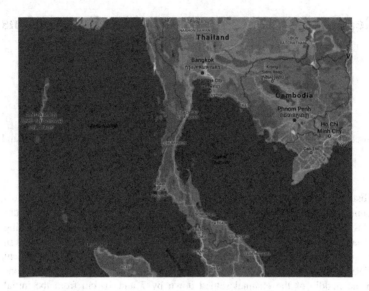

**Fig. 1.** Location of Dong Pa Lan Friendship Bridge (Google Maps 2021).

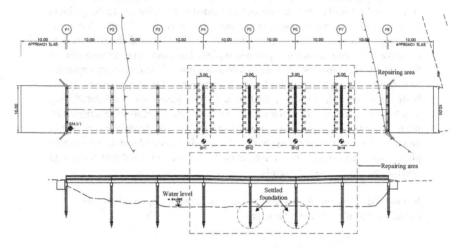

**Fig. 2.** Plan view and elevation view of the bridge.

Ranae canal is located in a valley surrounded by mountains. This geography allows for flash floods during heavy rain, making the bridge foundations more vulnerable from scouring. Department of Rural Roads received a request from the locals and then investigated the bridge. From the initial inspection, the fifth and sixth foundations, P5 and P6, which were in the middle of the channel, settled as illustrated in Fig. 3. Although this bridge is under the jurisdiction of the local authorities, the severity of the damage was beyond the capability of the local administration. If the bridge was still used without repairing, there would be a danger to the structure and the people commuting on it. As a result, the department helped design and repair the settled bridge.

**Fig. 3.** Bridge settlement damage.

## 2 Soil Investigation and Damage Evaluations

### 2.1 Soil Investigation

Soil data obtained from boring tests were indispensable for the evaluation of the foundation capacities and design of the structural reinforcement system. The data helped the designer to calculate the capacity of the existing foundations and design new ones. In this project, there were four boring holes at each of the four piles located in the channel. The soil profile from the soil boring test at the P4 and P5 foundations is shown in Fig. 4. The data informed which location was appropriate for the pile tip. From the profile, the stiff rock layer is at 10.5 m depth. Therefore, the design length of the new piles would be 10.5 m.

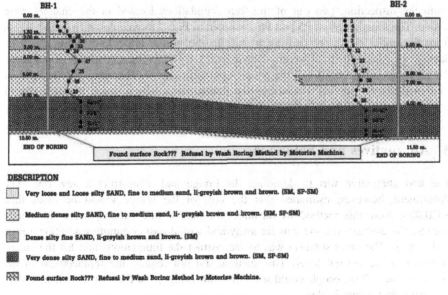

**Fig. 4.** Soil profile obtained from soil boring tests.

Besides the soil profile, the structural capacities of the soil were determined. The capacities of the pile include the skin friction and the end bearing values, which vary by the depth, as presented in Fig. 5. From the plots, the skin friction increases from zero nonlinearly to 15.5 tons per meter at 10.5 m depth. The increase in skin friction is due to the increase in soil pressure. However, when the soil around the piles is scoured out, the capacity decreases. Therefore, to design a new bridge foundation for longer lifespan, the design team wound not utilize the skin friction. On the other hand, the end bearing capacity of soil typically depends on the soil profile itself; stiffer soil or rock layer has more capacity than the weaker soil layer. This portion of soil strength remains similar even though the soil around the piles begins scouring. The maximum end bearing capacity is limited to 1000 tons per square meter. In this project, the team intended to use this maximum value, therefore the new piles were then driven to 10.5 m under the ground.

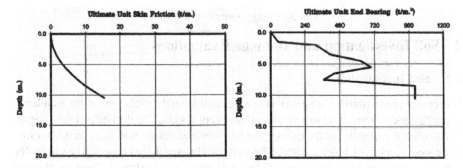

**Fig. 5.** Ultimate skin friction and end bearing capacities of soil.

### 2.2 Damage Evaluation

From the inspection, two out of the four foundations located in the channel were settled. The foundations at P5 and P6 as shown in Fig. 2 moved down by 7 and 10 cm from the initial elevation, causing a shift in the bridge horizontal alignment. This damage weakened the total weight-carrying capacity of the bridge, which would cause danger to the bridge itself and the people commuting on it. Therefore, the department examined several options considering the limited construction space, the necessity of bridge usage during the construction, cost, and safety during construction.

## 3 Alternatives and Solutions

The first alternative was to demolish the bridge and reconstruct a new one. The department, however, estimated that the cost of the bridge would be more than $550,000. Also, this method would affect the people who needed to use the bridge because the alternative route was far away and would cost commuters additional time and money. The other solution was to reconstruct the foundations and lift the superstructure to the desired level. This method did not require the replacement of the superstructure. Thus, people could still utilize the bridge, and the cost was cheaper than reconstructing a new bridge.

## 3.1 Alternatives

Construction of a new foundation in replacement of the old one is very challenging because the weight of the superstructure was still on it. There are several options depending on the type of damage and the site characteristics. The U.S. Federal Highway Administration suggests many options and case studies regarding the repair of bridge foundations (U.S. Department of Transportation 2016). A typical method used in Thailand is to build a portal frame over the existing foundation. This frame will transfer the load of the superstructure to the two new foundations on a new cap beam (Kumar et al. 2013). Figure 6 presents a foundation already repaired by the department using the portal frame method. However, it was rather expensive because of the two new foundations and the new cap beam. The new structures would obstruct and change the water flow, which would cause more erosion on the river embankment. The cost of this method was estimated to be around $480,000 for this project.

The next popular type of foundation is called "Drilled Shaft" which is a pile boring method. The construction step begins with pre-boring to the designed depth. Steel casings are driven into the hole with a stabilizing solution to prevent the hole from collapse. The reinforcing steel is then placed into the hole ready to cast a reinforced concrete pile. Finally, the casings will be removed (Garder et al. 2012). Figure 7 shows the procedure to construct a drilled shaft.

A major advantage of the drilled shaft method is that the pile can be cast to the desired depth no matter how the soil profile is. The structural strength of the pile itself is more flexible than a pre-cast pile since the diameter of the pile, the amount of reinforcing steel, and the strength of the pile materials can be designed. However, the equipment and machine required for boring a drilled shaft are typically tall and large, making this method not suitable for work under a bridge or in a limited space. Drilled shafts are suitable for constructing a new bridge without the superstructure already in place. As a result, this piling method was not appropriate for the project.

**Fig. 6.** Portal frame foundation repair

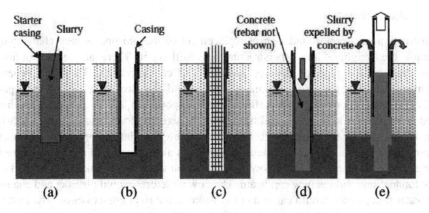

**Fig. 7.** Construction of drilled shafts

Because of the limitations, the department introduced the idea of using spun micro piles which had become popular in the reinforcement of residential foundations in the country. Reinforcement of the foundations with the micro piles was very convenient due to their small sizes; It prevented the new foundation from obstructing too much water flow, and reduced the construction cost. After that, the weight of the superstructure would be transferred to the micro piles instead, and the superstructure would be lifted back to the initial elevation before the damage. The total cost of the project was only $380,500 with only 10 months of the construction period, which was cheaper and faster than the other methods mentioned earlier. Furthermore, this alternative allowed the use of the bridge during some construction process, which will be described later in Sect. 5.

### 3.2 Spun Micro Piles

Spun micro piles are made of pre-cast concrete reinforced with pre-tensioned steel. The piles have either rectangular or circular cross-sections with a hollow in the middle (see Fig. 8). They are cast inside formworks rotating along their longitudinal axes faster than the gravitational acceleration. The production causes the concrete to condense more than the normal cast-in-place method. In general, a spun micropile has an allowable capacity of 20 to 40 tons, and a length varying from 1.0 to 1.50 m. A total pile length can be increased as designed by welding the plates at the ends of the piles (Irawan et al. 2015).

**Fig. 8.** Spun micro piles and casing tubes.

Spun micro piles are very convenient. The hollow cross-sections help reduce the vibration and sound from the driven process. The earth pressure exerting on the pile tips are less because some soil can move through the hollow (Irawan et al. 2015). Transportation of the piles is quick and easy due to their sizes. The related driven equipment is also smaller than a typical driven pile, allowing construction on limited space or an area difficult to reach. Considering these benefits, the department decided to use spun micro piles to reinforce the foundations P4 to P7 in Fig. 2, which were in the channel. Figure 9 shows the drawing for the repaired bridge. The spun micro piles used in this project had a diameter of 24 cm, length of 1.50 m long, and allowable strength of 25 tons per pile. The standard used was Thai Industrial Standard Institute (TISI) No. 397-2534.

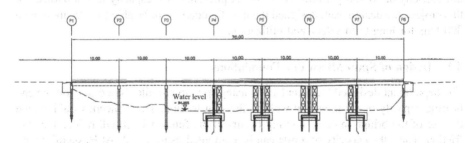

**Fig. 9.** Elevation views of repaired bridge foundations.

## 4 Design of Spun Micro Piles

The new foundations with spun micro piles were designed to have more capacity than the maximum capacity of the old foundations. This would ensure long-term safety and utilization of the bridge. The capacity of the old foundations was recalculated based on the soil data and the pile characteristics. The structural capacity of the piles themselves was still sufficient as could be assessed from the bridge condition at that moment.

## 4.1 Capacity of Existing Foundations

The first step was to evaluate the structural capacity of the piles themselves, which could be recalculated from the geometry and material data. Each foundation had 6 rectangular columns with sizes of 40 cm wide and 5 m long. There was no bracing connecting the columns. The other information in the standard drawing for bridges (Department of Rural Roads 2018) included: concrete with the ultimate strength ($f_c'$) of 350 ksc and reinforcing steel with the yield strength ($f_y$) of 2400 ksc. Assume the minimum steel was 1% of the cross-sectional area of the column. Therefore, the ultimate axial capacity of the column was determined per the ACI 318 standard (American Concrete Institute 2008) in Eq. 1.

$$\phi P_n = 0.80\phi\left[0.85f_c'\left(A_g - A_{st}\right) + f_y A_{st}\right] \tag{1}$$

Where

$\phi P_n$ is the ultimate axial capacity of the column
$A_g$ is the gross cross-sectional area of the column
$A_{st}$ is the total cross-sectional area of the reinforcing steel
$\phi$ is the reduction factor (0.65 for short columns)

The capacity of the pile is the least of the structural capacity of the pile, the allowable strength of the pile, and the soil capacity. From Eq. 1, using all the information mentioned above, the ultimate axial capacity of a column is 265 tons. According to the standard drawing for bridges (Department of Rural Roads 2018), the allowable capacity of a driven pile is 50 tons per pile. From the end bearing capacity shown in Fig. 5 and the cross-sectional area of the pile of 1600 cm$^2$, the existing pile length only needs to be less than 3 m under the ground to obtain 57.6 tons capacity per pile. Therefore, in this project, the capacity of the pile was controlled by the allowable pile capacity of 50 tons per pile. Thus, for six piles, the total capacity was 300 tons. For this project, the design team designed the new foundations to be able to carry more than 300 tons for long-term safety and utilization.

## 4.2 Design of Spun Micro Pile Foundations

The department designed the foundations using spun micro piles based only on the end bearing capacity of the piles and without the skin friction. This concept would extend the use of the bridge in case scouring occurs in the future which will reduce the skin friction. Thus, the capacity of a pile can be computed using Eq. 2 (Al-Hashemi 2016).

$$Q_b = q_b A \tag{2}$$

Where

$q_b$ is the end bearing capacity of the pile $\leq 1000 \text{ t/m}^2$
$A$ is the cross-sectional area of the pile

Since after the pile was driven in place, the soil would be compacted in the hollow section and act as a part of the cross-section. Therefore, the full cross-sectional area of

453 cm$^2$ was used. The end bearing capacity at the pile tip level of 10.5 m below the ground was computed from the soil data. Therefore, the end bearing capacity was 45.3 tons per pile. The value is more than the allowable capacity of the pile of 25 tons. The actual capacity of each pile would be tested again once they had been driven in place. The least number of micro piles required to obtain the capacity of 300 tons is 300/25 = 12 tons. In this project, 16 piles were used to increase safety and symmetry. Figure 10 presents the drawings for the reinforced foundation (unit in meters).

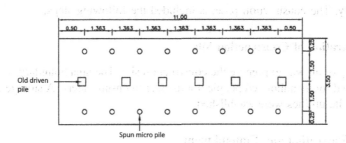

a. Plan view of piles location

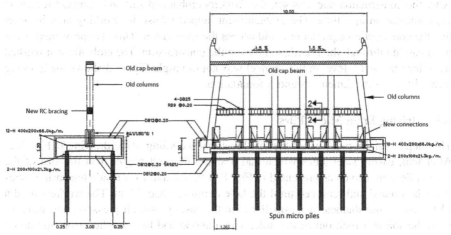

b. Front and side elevation views of the reinforced foundation

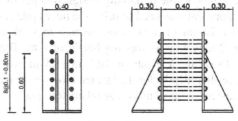

14-Stud Bolt dia.160mm.x0.50mm.
Steel Plate 9mm.

c. Connection between old columns and new spun micro piles

**Fig. 10.** Reinforced foundation drawings.

# 5 Construction Procedure

As mentioned previously, the use of spun micro piles allowed the bridge to open for traffic during some of the construction period; this excludes the process of lifting the settled superstructure, demolishing the old columns and cap beams, and casting new foundations. The small size of the micro piles caused it to have little influence over the structure during the driving process, and there was no replacement or demolition of the existing foundations. However, there were strict weight limitations on the bridge to ensure safety. The construction process included the following steps:

## 5.1 Preparation of Construction Site

The first step of all was to prepare the construction site. The area boundary was clearly set and signed for unauthorized people not to enter to ensure safety. A storage area, site office, and site utilities were established.

## 5.2 Soil Excavation and Embankment

After the construction site was set, the workers embanked soil and compact it around each foundation at a time. The embankment helped adjust the working area in order that the construction equipment could access the foundation. This also prevented water from passing through the foundation during the construction. The embankment worked well since the water level in the channel was not too high. Figure 3 shows the working area with an embankment around a foundation.

## 5.3 Driving Spun Micro Piles

This step began by marking all the pile locations according to the drawing in Fig. 10a. After that, casing steel tubes which had a size a little bit bigger than the size of the micro pile were driven as presented in Fig. 8. The soil and rock inside the casing tubes were then bored and removed until the hole depth reached 7.0 m. The workers used a PVC tube whose diameter was smaller than the casing tube, air pressure, and water to push the soil and rock out of the hole. This method had been applied from the bored pile procedure to all the micro piles to pass the dense sand layer. Then, the micro piles were connected and driven to the remaining designed depth of 10.5 m.

Micro driving equipment was used to drive the micro piles as presented in Fig. 11. The equipment had a height of 3 m, short enough to work under the bridge. The hammer weight was 2 tons with a dropping height of 1 m. There were 7 micro piles welded and driven to obtain a depth of 10.5 m (1.5 m each). After reaching the designed depth, the reading depths for the last ten blows were recorded to calculate the actual driven pile capacity according to a standard formula.

**Fig. 11.** Use of a micro driving equipment to drive spun micro piles.

## 5.4 Lifting the Settled Superstructure to the Desired Elevation

After all the spun micro piles were driven and their blow counts recorded, the construction team built the load transferring structure to transfer the weight of the superstructure to the new piles according to Fig. 10b. The structure consisted of wide flange frames connected to the tops of the piles and connected to the old columns using the connection shown in Fig. 10c. 16 temporary circular steel tubes were used as the load-transferring structure to the wide flange frames. In this project, the circular tubes have an outside diameter of 8 inches and a thickness of 4.5 mm with an ultimate capacity of 31.8 tons per column. This value was more than the allowable pile capacity of 25 tons. The tubes supported 12 hydraulic jack systems, which were connected to the bottom of the superstructure on both sides of the cap beam as shown in Fig. 12. Then, the hydraulic jacks were used to lift the superstructure to the horizontal level before the settlement. This procedure also helped ensure that the new micro piles were able to support the weight of the superstructure since the weight was transferred directly to the piles while being lifted. Note that traffic was not allowed during this process.

**Fig. 12.** Pressure gauge reading the target load-bearing capacity.

### 5.5 Casting New Foundations

At this step, the load of the superstructure was successfully transferred to the micro piles, and the elevation of the bridge was adjusted. The new foundations were cast to cover the wide flange frames with some more reinforcing steel according to the drawing in Fig. 10b. The new foundations helped support the columns and transfer the weight to the new piles instead of the old ones. Note that traffic was not allowed during this process.

### 5.6 Returning the Site to the Beginning Condition

As outlined in the first step, there was some soil embankment done to transform the channel into a construction space. After the process was complete, the channel was returned to the condition before the project. Everything was checked again to ensure that the bridge would work well for a long time. Figure 13 shows the final look of the bridge after the construction.

**Fig. 13.** The bridge after the repair.

## 6 Assessment of Load-Carrying Capacity

### 6.1 Pile Capacity Evaluation Using Blow Counts

Evaluation of the actual capacity of the driven piles can be performed by using the last 10 blows. There are many mathematical models conducted to approximate the driven pile capacity. In this project, Janbu's formula, which was very well-known, was used (Fragaszy et al. 1985). The capacity can be calculated using Eq. 3.

$$\begin{aligned} Q_u &= \tfrac{Wh}{K_u S} \\ K_u &= C_d [1 + \sqrt{1 + \tfrac{\lambda}{C_d}}] \\ C_d &= 0.75 + 0.15 \tfrac{P}{W} \\ \lambda &= \tfrac{WhL}{AES^2} \end{aligned} \qquad (3)$$

where

$Q_u$    is the actual driven pile capacity
W    is the weight of the hammer
h    is the hammer's drop height
S    is the average driven depth from the last 10 blows
P    is the weight of the pile
E    is the modulus of elasticity of concret
A    is the cross-sectional area of the pile
L    is the total length of the pile

In this article, an example of one of the piles will be presented. The depth of the pile driven into the ground during the last 10 blows was recorded during construction. The average depth was 1.7 cm. The hammer had a weight of 2 tons and a drop height of 3

230    A. Wachiraprakarnpong et al.

m. The weight of the pile was 0.73 tons with a cross-sectional area of 453 cm$^2$. Therefore, the capacity of the pile was 63.4 tons. This number was more than the allowable strength of the pile of 25 tons, which has a safety factor of 2.54. As a result, it can be concluded that the pile had sufficient capacity as designed.

### 6.2   Load Test on the Bridge

"Load Test" is a universal method of testing the real-life behaviors of bridges. The test applies known load on bridges and measures the structural responses with several types of sensors. For Ranae Bridge, which has successfully been repaired, The Bureau of Maintenance in the department will conduct a load test on the bridge using two 25-ton trucks. These loads are reasonable for testing a local bridge. Table 1 summarizes the types of sensors which will be installed on the bridge, their locations, and how to use their recorded data to evaluate the behaviors of the bridge.

**Table 1.** Sensor type, data type, sensor location, and data analysis

| Sensor type | Data type (Unit) | Sensor location | Data analysis |
| --- | --- | --- | --- |
| Strain gauge | Strain (m/m) | Middle of the settled span | Compared with the allowable strain calculated from the maximum allowable strength |
| Accelerometer | Acceleration (m/s2) | Both edges of the settled span | Compared with the allowable vibration from the AASHTO standard (AASHTO 2018) |
| Tilt meter | Tilt angle (degree) | Both edges of the settled span | Checked if the bridge has unequal settlements |
| Displacement | Vertical deflection (m) | At supports and midspan of the settled span | Compared with the allowable deflection of beams (ACI 2008) |

## 7   Conclusions

The Department of Rural Roads has repaired the foundations of Ranae Bridge, which was located in Phipun District, Nakhon Sri Thammarat Province, Thailand. Two of the pile-bent foundations were settled from scouring, which caused the superstructure of the bridge to no longer be horizontal. The department considered several alternatives based on many factors: structural strength, long-term durability, limitation on the working space, cost, and the need for bridge usage during the construction. Considering all the factors, the department decided to use spun micro piles to build new foundations to replace the old ones, and lifted the superstructure of the bridge back to the initial elevation.

The process started by calculating the remaining capacity of the old pile foundation and then designing a micro pile system to obtain a desirable safety factor of 2.5 or more. There were 4 foundations located in the channel that needed to be repaired to prevent further damage in the future. The pile driving process consisted of pre-boring the holes and driving the piles to the design depth using micro driving equipment. All the piles driven were monitored with their last ten blows recorded for the calculation of their actual capacity. The results suggested that all the piles achieved the target factor of safety. Then, a load transferring structure was built on the new piles and used to lift the superstructure to the initial elevation. During the construction period, the bridge could still be utilized and crossed, making this process rarely affecting the local people. Therefore, reinforcement of the bridge foundations with spun micro piles is very cost-effective, convenient, and efficient. Furthermore, the department will continue monitoring the bridge and conducting a load test using known truckloads to confirm the repair results.

# References

American Association of State Highway and Transportation Officials: Manual for Bridge Evaluation, Washington, D.C. (2018)

American Concrete Institute: Building code requirements for structural concrete (ACI 318M-08) and commentary (ACI 318R-08). American Concrete Institute, Farmington Hills (2008)

Irawan, C., Suprobo, P., Raka, P., Djamaluddin, R.: A review of prestressed concrete pile with circular hollow section (Spun pile). Jurnal Teknologi **72**(5) (2015). https://doi.org/10.11113/jt.v72.3950

Department of Rural Roads: Standard Drawings for Bridge Construction and Repair No. 1. Department of Rural Roads, Bangkok (2018)

Al-Hashemi, H.M.B.: End Bearing Capacity of Pile Foundations. University of Bahrain (2016). https://doi.org/10.13140/RG.2.2.21744.46080

Google Maps. (2021). Location of Dong Pa Lan, Mae Fai Subdistrict, San Sai District, Chiang Mai Province, Thailand. maps.google.com

Garder, J., et.al.: Development of a Database for Drilled SHAft Foundation Testing (DSHAFT) (2012). Corpus ID: 108019510

Fragaszy, R.J., Higgins, J.D., Lawton, E.: Development of guidelines for construction control of pile driving and estimation of pile capacity. Report WA-RD 68.1. Washington State Transportation Center, Olympia, WN, USA (1985)

Thai Industrial Standard Institute: Standards for Prestress Spun Concrete Piles (TISI No. 397-2534). Government Gazette No. 111 Section 93 (1994)

U.S. Department of Transportation, The Federal Highway Administration: Protocols for the Assessment and Repair of Bridge Foundations. The Federal Highway Administration, McLean (2016)

# Overview of Integral Abutment Bridge Applications in the United States

Amin Akhnoukh[1(✉)], Rajprabhu Thungappa[2], and Rudolf Seracino[2]

[1] East Carolina University, Greenville, NC, USA
akhnoukhal7@ecu.edu
[2] North Carolina State University, Raleigh, NC, USA

**Abstract.** Integral abutment (IA) bridges have been successfully used in constructing highway bridges by different State Departments of Transportation (DOTs) in the United States over the past 60 years. IA bridges main advantages include the absence of construction joints which expedites the bridge construction process, increase bridge redundancy (improve seismic performance), and reduce the need to costly and laborious maintenance.

Despite the afore-mentioned advantages, several DOTs have recently discontinued the IA bridge construction. Other state DOTs have conducted research to investigate the efficiency of IA bridge systems, their advantages, and disadvantages. This paper presents the history of IA bridge construction in the United States, different IA bridge construction techniques including IAs with Frame Abutments, Bank Pad Abutments, Embedded Wall Abutment, and Flexible Support Abutments, relevant advantages, and disadvantages. In addition to listing different maintenance activities required to maintain the IA bridge inventory within the United States. The presented information can be used by bridge design engineers, precast plants, and DOT personnel as a decision support tool to decide on the type of bridge system to be selected for new bridge projects.

**Keywords:** Integral abutment · Bridge decks · Maintenance · Approach slab · Sleeper slab · Expansion joints

## Acronyms

| | |
|---|---|
| IA | Integral Abutment |
| FIAB | Fully Integral Abutment Bridge |
| SIAB | Semi Integral Abutment Bridge |
| FHWA | Federal Highway Administration |
| DOT | Department of Transportation |
| NCDOT | North Carolina Department of Transportation |
| MSE | Mechanically Stabilized Earth Walls |
| MR&R | Maintenance, Repair, and Replacement |

## 1 Introduction and Literature Review

Integral abutment (IA) bridges provide many advantages over conventional bridges during construction and subsequent maintenance activities. IA bridges have been adopted by different State Departments of Transportation (DOTs) since the 1960s.

© The Author(s), under exclusive license to Springer Nature Switzerland AG 2022
A. Akhnoukh et al. (Eds.): IRF 2021, SUCI, pp. 232–240, 2022.
https://doi.org/10.1007/978-3-030-79801-7_16

Different types of IA bridge systems are being used by different State DOTs including spill through, mechanically stabilized earth walls (MSE), and semi-integrated abutment systems. Conventional IA bridge construction components are shown in Fig. 1.

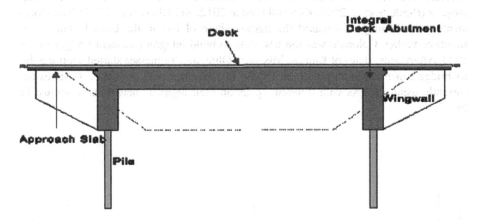

**Fig. 1.** Main construction components of IA bridge

Unlike conventional bridges, IAs do not have expansion joints within the bridge deck or between the bridge deck and supporting abutments. The elimination of joints is advantageous as it reduces initial bridge construction cost, increases structural redundancy of the bridge, and minimize the costly and laborious activities associated with joints maintenance activities (Arsoy et al. 1999; Hassiotis et al. 2006).

The deterioration of joints results in leakage, deterioration of bridge girders, and deterioration of bridge bearings. The consequences of joint deteriorations include the deterioration of the overall bridge condition, lower bridge load rating, potential unsafe rides for bridge users, high maintenance, repair, and replacement (MR&R) costs, and increased bridge projects life cycle cost. Examples of expansion joints failure are shown in Fig. 2 a and b.

**Fig. 2.** a. Expansion joint plate spalling and b. Expansion joint broken plate (Miller and Jahren 2014)

Advantages of IAs include ease of construction, time and cost savings, and high performance under seismic activity, and extreme loading conditions (Arenas et al. 2013). The afore-mentioned advantages and the relationship between the IAB superstructure and substructure performance is explained in detail in relevant research projects (Burdette et al. 2004, Kim and Laman 2012, and Olson et al. 2012). The afore-mentioned advantages increased the market share of IAs in the United States construction market. Colorado was the first state to build integral abutment bridges in the 1920. Afterwards, states of Kansas, Missouri, Ohio, and Tennessee started adopting the IA bridge construction through their DOTs. In a recent survey, a total of 41 states are currently using the IAs with different spans on their highway networks, as shown in Fig. 3.

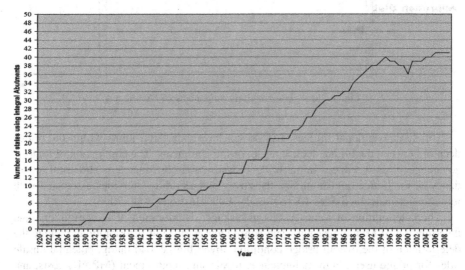

**Fig. 3.** Numbers of integral abutment bridges in the United States up to 2008 (Paraschos and Amde 2011)

Despite their significant advantages, there are problems associated with IA bridge construction. Most of the IA problems arise from the elimination of joints, thus, forcing lateral movement to occur at the abutments. The severity of the problem varies according to the design loads, bridge span and skew, and weather variations. Recently, three states, namely Alaska, Arizona, and Mississippi discontinued the use of IA bridges due to significant service life problems. The following sections of this paper presents different IA bridge systems, design and construction considerations of IA bridges, advantages, and disadvantages of IA bridges.

## 2 Types of Integral Abutment Bridges

Bridge abutments are classified into 4 main types: I) Wall type abutments, II) Stub abutments, III) Integral abutments, and IV) Semi-integral abutment. Integral and semi-integral abutment bridges include single or multi-span bridges that have continuous concrete decks and approach slabs integrated with bridge abutments. In 2010, the United States bridge inventory included more than 9,000 fully integrated abutment bridges (FIAB), and more than 4,000 semi-integrated abutment bridges (SIAB). The main difference between FIAB and SIAB is that FIAB superstructure is fully integrated in the bridge abutment, with a direct load transfer in the absence of any bearings, while SIAB has an integrated deck with the abutment, yet, the girders loads are transferred to the abutment through bearing pads. In IAs, the abutments are rested on flexible foundation that allows lateral movement under specific types of loading as breaking forces, seismic forces, prestress relaxation, shrinkage, creep, and temperature variations. Types of IAs superstructure include steel girders with concrete decks, prestressed, and post-tensioned bridge girders. Spans of IAs ranges extend up to 300 ft. for steel bridges, and 500 to 600 ft. for concrete. Four different types of IAs are adopted on a federal and state level within the United States, as follows:

I. Integral Bridges with Frame Abutment
  In IAs with frame abutment, the bridge deck and girders are rested on high abutment walls forming a frame structure that has an effective load-resistance capability. Due to the high abutments, horizontal earth pressure is increased compared to other types of IAs, as shown in Fig. 4.

**Fig. 4.** IA bridge system with frame abutment construction

II. Integral Bridges with Bank Pad Abutment
  IAs are constructed with bank pad abutment if the construction site at the bridge ends have a high slope, as high walls will not be feasible (due to site topography-imposed restrictions). Bank Pad abutments are constructed on piles that has tendency to move laterally to accommodate lateral movement due to thermal variations. If Bank pads are supported on soil (no piles are used), the Pads should

be flexible-enough to accommodate lateral movement. IAs with Bank Pads are shown in Fig. 5.

**Fig. 5.** IA bridge system with Bank Pad Abutment

III. Integral Bridges with Embedded Wall Abutment
Embedded Wall abutments are full height reinforced concrete integrated abutments similar to Frame Abutments. Wall Abutments are required to have a high stiffness to support excessive earth pressure. Embedded Wall abutments example is shown in Fig. 6.

**Fig. 6.** IA bridge system with Embedded Wall Abutment

IV. Integral Bridges with Flexible Support Abutment
Flexible Support Abutments are a recently developed technique to provide the bridge abutment with sufficient flexibility to efficiently accommodate lateral displacement developed due to secondary lateral loadings including temperature

variations, breaking forces, seismic activities, girders shrinkage and creep, and lateral earth pressure. The IAs are supported on piles embedded on larger diameter sleeves. Sleeves are filled with material that would enable lateral movement as needed. This IA system is shown in Fig. 7.

**Fig. 7.** IA bridge system with flexible support abutment

## 3 Loads on Integral Abutment Bridge Systems

Integral abutment bridges are subjected to dead and live loads (primary loads) similar to other bridge types, in addition to secondary loads due to the absence of joints. Secondary loads include creep, shrinkage, temperature gradient loading, and loads due to differential settlement. Secondary loading is defused through the lateral displacement of bridge abutment resting on flexible foundations. Details of secondary loading, and subsequent effects are presented in the following section:

  I. Shrinkage and Creep
     Shrinkage and creep of bridge superstructure represents a major challenge in the design of IABs. The continuous shrinkage of bridge superstructure components and creep of precast/prestressed girders are usually associated with increased lateral earth pressure. The lack of expansion joints significantly increases the strains added to the rest of the bridge components (Fennema et al. 2005).
  II. Temperature Gradient (Variation)
     Temperature variations result in continuous expansion and contraction of bridge components, mainly bridge superstructure. Temperature variations include linear variations resulting in uniform length changes of bridge deck and girders, or temperature gradient (where temperature effect varies dependent on the bridge location). In conventional bridge construction, where superstructure loading acts on bearings which allow lateral superstructure movement, temperature effects are reduced. However, in IAs, due to the absence of joints and bearings, lateral

movement is applied on bridge abutment and piers. This secondary loading may result in cracked abutment, and/or piles reduced capacity (LaFave et al. 2021)

III. Differential Settlement

The lateral movement of bridge abutment results in the de-stabilization of the earth fill behind the abutment. Due to the repetitive (cyclic) movement of bridge abutment, it was found that the earth fill tends to settle resulting in a cracked approach slab and possible differential settlement resulting in unlevelled roadway-approach slab. The differential settlement results in a poor-quality ride for bridge users.

## 4 General Construction Provisions for Integral Abutment Bridges

The Federal Highway Administration (FHWA) and different State DOTs have conducted research projects to develop design guidelines and recommendations for efficient construction practices. Currently, there is no unified guidelines or specifications for design and construction of IAs in the United States. Different DOTs developed their own guidelines dependent on their research findings, and according to their topography, predominant soil conditions, temperature gradients, local wind, and seismic activities. Research findings for different State DOTs specifies a maximum acceptable bridge skew of $20°$ to $45°$ Few DOTs allow spans of steel girder IAs to extend to 500 ft., while other DOTs allow spans of concrete/prestressed girder bridges to extend to 700 ft.

Most DOTs, except Illinois and North Carolina, requires the weak axis orientation for abutment piles. The weak axis orientation is advantageous as it reduces the pile

**Table 1.** Specific DOTs guidelines for integral abutment bridge construction

| | Maximum Skew (Degree) | Maximum bridge Length (ft) | | Pile orientation (HP piles)- Perpendicular to bridge center line |
|---|---|---|---|---|
| | | Steel girders | Concrete/pre-stressed girders | |
| North Carolina | $20°$ | 300 | 400 | Strong axis |
| Illinois | $30°$ | 310 | 410 | Strong axis |
| Indiana | $30°$ | 500 | 500 | Weak axis |
| Iowa | $45°$ | 300 | 300 | Weak axis |
| New Jersey | $30°$ | 450 | 450 | Weak axis |
| Pennsylvania | $45°$ | 90 | 90 | Weak axis |
| Minnesota | $45°$ | 100 | 100 | Weak axis |
| Utah | NA | NA | NA | NA |
| Vermont | $20°$ | 395 | 695 | Weak axis |
| Virginia | $30°$ | 150 | 250 | Weak axis |

inertia and enables the abutment to perform lateral movement to accommodate displacements resulting from secondary "lateral" loads. Current DOTs specifications provide extensive details regarding permissible abutment heights, wingwall specifications, pile details, drilled shaft, and allowable maximum displacement for the IAs. Table 1 shows major differences for IAs specifications in 10 different State DOTs.

## 5 Limitations of Integral Abutment Bridges

The main impediments to the widespread of IAs is mainly attributed to problems associated with the lack of construction joints within the bridge. Despite the ease of construction, budget savings, and reduced maintenance attained due to the absence of joints, different set of problems evolves which negatively impact the IAs including:

I. IAs are not suitable when excessive lateral expansion and/or contraction are expected due to temperature variations. Lateral variations may result in significant cracking in the bridge superstructure or the supporting abutments, as shown in Fig. 8.

II. Integral abutment design is highly impacted with lateral displacement due to material properties of super structure. Thus, spans of bridge girders are limited compared to other bridge construction systems. In addition, the permissible lateral movement may dictate a specific type of construction material for a given project.

III. IAs are not advantageous in construction sites with subsoils or embankments with relatively low strength.

IV. If IAs are exposed to unexpected lateral movements, supporting piles may develop plastic hinges due to the high stresses generated within the piles. Plastic hinge formation results in reduced axial strength of piles supporting the abutment. Hence, the bridge load rating is substantially reduced.

## 6 Conclusions

IAs have been successfully used in the construction of steel and concrete bridges within for several decades in the United States. Major advantages of IAs include ease of construction, absence of joints and possible joint construction and maintenance-related problems, and high performance against lateral forces. Different types of FIAB are currently available in the United States, to be selected according to site topography, geotechnical conditions, and temperature gradient. There are no unified guidelines for the design and construction of IAs. However, every State DOT has developed its own construction manual dependent on parameters impacting the construction market within the state including temperature variations, anticipated lateral loading due to wind and seismic activities, and soil nature within the state. Despite the design and construction advantages of IAs, multiple states recently discontinued the construction of IAs due to structural problems resulting from the absence of joints. These problems include excessive abutment cracking, super structure cracking due to the inability of the bridge to accommodate lateral movement, and settlement of bridge approach slab resulting in bridge operational problems.

**Acknowledgements.** The authors would like to thank North Carolina Department of Transportation (NCDOT) for funding this research project. The assistance provided by Trey Caroll at NCDOT Structures Management Unit is highly appreciated. Finally, the authors would like to acknowledge Dr. Greg. Lucier at North Carolina State University for his support to the project activities.

# References

Arenas, A., Filz, G., Cousins, T.: Thermal response of integral abutment bridges with mechanically stabilized earth walls. Report No. VCTIR 13-R7, Virginia Center for Transportation Innovation and Research, Virginia Polytechnic Institute & State University, Virginia (2013)

Arsoy, S., Barker, R.M., Duncan, J.M.: The behavior of integral abutment bridges. Virginia Transportation Research Council, Charlottesville (1999)

Burdette, E.G., Ingram, E.E., Tidwell, J.B., Goodpasture, D.W., Deatherage, J.H., Howard, S.C.: Behavior of integral abutment supported by steel H-piles. Transp. Res. Record **1892**(1), 24–28 (2004)

Fennema, J.L., Laman, J.A., Linzell, D.G.: Predicted and measured response of an integral abutment bridge. ASCE J. Bridge Eng. **10**(6), 666–677 (2005)

Hassiotis, S., Khodeir. Y., Roman, E., Dehne, Y.: Evaluation of integral abutments. Final Report, No. FHWA-NJ-2005-025. Department of Transportation and Stevens Institute of Technology, Hoboken (2006)

Kim, W., Laman, J.A.: Seven-year field monitoring of four integral abutment bridges. Eng. Struct. **32**(8), 2247–2257 (2012)

LaFave, J.M., Brambila, G., Kode, U., Liu, G., Fahnestock, L.A.: Field behavior of integral abutment bridges under thermal loading. J. Bridge Eng. **26**(4), 04021013 (2021)

Miller, A.M., Jahren, C.T.: Rapid replacement of bridge deck expansion joints study – phase I. Report, Institute of Transportation, Iowa State University, Ames, Iowa (2014)

Olson, S.M., Holloway, K.P., Buenker, J.M., Long, J.H., Lafave, J.M.: Thermal behavior of IDOT integral abutment design limitations and details. Illinois Center for Transportation Research Rep. No. ICT-12-022. University of Illinois, Urbana-Champaign (2012)

Paraschos, A., Amde, A.M.: A survey on the status of use, problems, and costs associated with integral abutment bridges. In: Better Roads, pp. 1–20 (2011)

# Case Studies in ITS Development

Classification in IDS Development

# Pacemaker Lighting Application to Prevent Traffic Congestion in a V-Shaped Tunnel and Provide Sustainable Operation: A Case Study: Eurasia Tunnel

Aşkın Kaan Kaptan[✉] and Murat Gücüyener

Eurasia Tunnel, Istanbul, Turkey
{kaan.kaptan,murat.gucuyener}@avrasyatuneli.com

**Abstract.** Eurasia Tunnel is a V-Shaped TBM tunnel connecting Asian and European sides of Istanbul by reaching its deepest point at 106 m from sea level with a longitudinal slope of 5% in accordance with the highway standards. Free flow without any traffic congestion along the tunnel is one of key elements for traffic management not only traffic safety, but also good qualify of driving satisfaction in the tunnel. It has been observed that traffic flow speed decreases and light traffic congestions at the middle of the tunnel which cause increase in traffic density at the deepest point of the tunnel and the following uphill section due to sudden brake, wrong gear and/or wrong speed choices of the drivers. This study will discuss the Pacemaker Lighting application started in Eurasia Tunnel in June 2020, -which is a speed regulating moving light reflected on the ceiling of the tunnel by using LED fixtures-. This application aims to guide users for a smooth acceleration and following distance, to avoid undesirable traffic jams, to prevent potential traffic accidents, engine/transmission breakdowns and to decrease level of vehicle emissions. This paper also reports the sustainability provided by the system with the results of the application such as significant speed and efficiency increase in the Pacemaker Lighting Zone, significant decrease in the number of traffic speed decreases and vehicle emissions in 11 months of operation time.

**Keywords:** Eurasia Tunnel · Tunnel lighting · Pacemaker lighting · Traffic congestion · Traffic management · Roadway tunnel · Sustainability

## 1 Introduction

The Eurasia Tunnel Project (Istanbul Strait Road Tube Crossing Project) connects the Asian and European sides of Istanbul via a highway tunnel crossing underneath the seafloor since its commencement of operation in December 2016. 5.4-km section of the Project consists of a double-deck tunnel under the seafloor built with Tunnel Boring Machine technology and connection tunnels. Due to its structure, Eurasia Tunnel reaches its deepest point with a longitudinal slope of 5% in accordance with the PIARC highway standards, and after the deepest point it reaches the other side with a 5% longitudinal slope as well. Even it is not clearly noticed during driving inside the tunnel, due to this slope, the vehicles brake at the deepest point in line with the driver's

© The Author(s), under exclusive license to Springer Nature Switzerland AG 2022
A. Akhnoukh et al. (Eds.): IRF 2021, SUCI, pp. 243–256, 2022.
https://doi.org/10.1007/978-3-030-79801-7_17

behaviour, and drivers may have problems with the correct gear level and speed while moving against the gradient (uphill) (Fig. 1). In 2019 in the light of 3 years of operation experience, it is determined that a decrease in traffic flow speed occurs at the deepest point and uphill zone, therefore temporary queues and increase of traffic density occurs in the tunnel. In line with the applied traffic management plan which aims to prevent congestion within the tunnel, there might be lane closures applied due to this problem and it causes a capacity decrease for the tunnel.

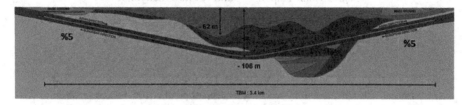

**Fig. 1.** Cross-sectional view of the Eurasia Tunnel

## 2 Literature Survey

Literature research and international examples have shown that a similar situation is also valid for similar, V- shaped- tunnels. This issue has been researched academically in Tokyo Bay Aqua Line Tunnel where a similar situation was experienced, and it was determined that the traffic density occurrence is reduced up to 60% by help of the Pacemaker Lighting LED (Speed Regulator Moving Lighting) application placed at the point where the traffic density increase occurred (Kato et al. 2015). This application consists of blinking LED lights through the tunnel to guide users as it is mentioned in the same study similar to Tomei Expressway. It has been proven by academic data that the same application is successful in the tunnel located on the Gaikan Expressway route with similar features.

Pacemaker Lighting (Speed Regulator Moving Lighting) is also available in the detailed technical report on "Prevention and Mitigation of Tunnel-Related Collisions from 2019R03EN' published by PIARC (World Road Organization) in March 2019 and listed as recommended systems for safer future tunnels. In the same publication, it is stated that the mentioned practice will prevent possible accidents by reducing traffic congestion and speed differences.

Traffic efficiency increase calculations in this study are done according to the proposed formula (Brilon 2000).

$E = q\,V\,T$ via below listed parameters which is applicable with the data gathered in Eurasia Tunnel.

where:

E = traffic efficiency (veh*km/h)
Q = volume (veh/h)
v = travel velocity over an extended section of the freeway (km/h)
T = duration of the time period for analysis of flow (h)

## 3 Specification and Application of Pacemaker Lighting in Eurasia Tunnel

The zones affected due to congestion and speed decrease especially in peak hours are determined in the Europe- Asia direction by several methods such as drivers' observation, CCTV monitoring, speed, and occupancy statistics of the direction in both lanes and traffic management application numbers in the portal area. Pacemaker Lighting application, which starts 500 m before the deepest point in the tunnel in the direction of Europe applied to keep the traffic speed constant and in order to provide a smooth traffic flow and its length is 1.000 m after the deepest point in the direction of Asia which is total of 1.500 m.

White light in the form of an arch with adjustable speed, size and distance synchronously flows in the blue background lighting on the ceiling which is the color of the thematic lighting used in the tunnel. System is established by LED tubes mounted on the cable trays in the tunnel reflects the light to the ceiling in wall washer principle and it is operated synchronously by remote control system. Remote control system is supported with the local control units mounted on the cable trays and the MEP rooms in the tunnel. Main server is located at the Eurasia Tunnel Data Center and integrated with the main SCADA system of the tunnel. Thus, the system is able to be monitored and controlled by the traffic control room operators. Overall system is in patent process with this unique design, application and control features (Figs. 2 and 3).

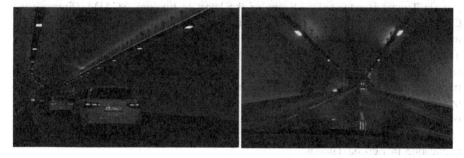

**Fig. 2.** Eurasia Tunnel pacemaker lighting application view from the tunnel and driver's perspective

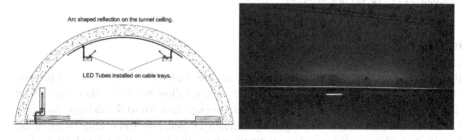

**Fig. 3.** Eurasia Tunnel pacemaker lighting application from the technical drawings and site application

## 4 Main Aims of the Project

The Pacemaker Lighting application is used to regulate the desirable traffic flow speed which is 70 km/h in the Europe-Asia direction. It is operated according to the scenario where 5 meter-length arches will move continuously with 40 m safe following distance. With the help of the system, below listed items are aimed to be obtained.

- Minimize potential accidents by ensuring the safety following distance between vehicles in the tunnel.
- Minimize vehicle breakdowns related with the engine and transmission problems by guiding drivers to adjust themselves more easily to the traffic speed and uphill direction.
- Resolve traffic congestions occurring at the deepest point and the following uphill section thus use the traffic capacity more efficiently.
- Contribute interior air quality positively by decreasing exhaust gases to protect environment by preventing high gear usage or sudden acceleration due to uphill speed reduction.
- Improve driving comfort in compliance with the thematic lighting concept.

## 5 Methodology

Eurasia Tunnel is operated with one of the latest technology SCADA (Supervisory Control and Data Acquisition) Systems to monitor and control integrated systems including energy, lighting, ventilation, fire detection etc. System is also supported with a CCTV (Closed Circuit Television) System including more than 400 cameras and AID (Automatic Incident Detection) System which can detect stopped vehicles, pedestrians, traffic congestions etc. All the incidents and activities are logged in this system based on date, time, and incident type from the beginning of the operation. To avoid any kind of seasonal effect, 11 months periods are selected for the comparison as before (15.06.2019–30.04.2020) and after (15.06.2020–30.04.2021) the Pacemaker Lighting application in Eurasia Tunnel.

Lane speed and occupancy values are gathered from the SCADA System and passing vehicle numbers are received from the Toll Collection System. To avoid any effect due to lane closures applied in the nighttime for maintenance works, data between 21.00–06.00 are excluded.

## 6 Main Results of the Project

Pacemaker Lighting System is operated in the Eastbound Tunnel (Europe-Asia Direction) consist of 5-meter length white arches following each other with 40 m distance which is the recommended safety following distance to the drivers via safety brochure, VMS (Variable Message Sign) messages and FM-Radio announces in the tunnel. According to the accident numbers within the period before and after Pacemaker Lighting application, accident occurrence is realized almost the same respectively 7 and 6, however there are no accidents reported in Pacemaker Lighting Zone

Pacemaker Lighting Application to Prevent Traffic Congestion     247

after the beginning of application which was 5 accidents in that zone before. This can be interpreted that Pacemaker Lighting helps drivers to keep their following distance and prevent accidents in the application area (Table 1).

**Table 1.** Accident numbers comparison for the periods before and after pacemaker lighting

| Period | Accident numbers in the Eastbound tunnel | |
| --- | --- | --- |
| | Eastbound tunnel total | Pacemaker zone and uphill section |
| Before pacemaker (15.06.2019–30.04.2020) | 7 | 5 |
| After pacemaker (15.06.2020–30.04.2021) | 6 | 0 |

Pacemaker Lighting System is operated in the Eastbound Tunnel (Europe-Asia Direction) consist of 5-meter length white arches moves with the speed of 70 km/h to support drivers to keep their speed stable and gears on the correct level. According to the engine and gearbox related vehicle breakdown numbers within the period before and after Pacemaker Lighting application, breakdown occurrence is realized exactly the same with 88 incidents, however breakdown numbers decreased by 4.65% in the Pacemaker Lighting Zone and the rest of the tunnel from the beginning of application. This can be interpreted that Pacemaker Lighting helps drivers to keep their speed and gearbox at the correct and required level to prevent any potential breakdown in the application area and afterwards (Table 2).

**Table 2.** Breakdown numbers comparison for the periods before and after pacemaker lighting

| Period | Breakdown numbers in the Eastbound tunnel | |
| --- | --- | --- |
| | Eastbound tunnel total | Pacemaker zone and uphill section |
| Before pacemaker (15.06.2019–30.04.2020) | 88 | 86 |
| After pacemaker (15.06.2020–30.04.2021) | 88 | 82 |

Eurasia Tunnel has 26 egress stairs on each 200 m which connects two decks of the tunnel for a safety evacuation in any kind of emergency and these 26 zones are monitored in 13 traffic zones (2 egress zones in each zone) via CCTV and AID cameras for traffic situation. Data gathered for this study is based on traffic situation in each traffic zone hourly based between 06.00 and 21.00 to avoid any effect of lane closures due to maintenance works.

According to the hourly average traffic speed values, it is determined that traffic speed is increased in the traffic zones which Pacemaker Lighting is applied –between Zone 5 and Zone 9- for all traffic volume conditions. The same results are determined when it is analysed for different traffic volume conditions in the direction. The detailed average speed volume data for different traffic volumes can be seen on Table 3, Fig. 4 and Fig. 5.

**Table 3.** Zone based average speed values in Eastbound tunnel lane 1

| Traffic | Period | Zone 1 | Zone 2 | Zone 3 | Zone 4 | Zone 5 | Zone 6 | Zone 7 | Zone 8 | Zone 9 | Zone 10 | Zone 11 | Zone 12 | Zone 13 |
|---|---|---|---|---|---|---|---|---|---|---|---|---|---|---|
| All | Before pacemaker | 60.6 | 61.8 | 60.1 | 57.8 | 57.5 | 58.8 | 56.6 | 55.3 | 58.5 | 57.2 | 57.4 | 61.0 | 58.4 |
|  | After pacemaker | 56.0 | 60.5 | 59.9 | 59.3 | 60.7 | 61.1 | 58.7 | 57.2 | 59.6 | 58.8 | 60.6 | 62.8 | 60.4 |
| >500 | Before pacemaker | 58.7 | 60.1 | 58.0 | 54.9 | 54.5 | 56.3 | 53.6 | 52.1 | 56.3 | 54.4 | 55.3 | 58.8 | 56.3 |
|  | After pacemaker | 51.5 | 57.5 | 56.6 | 55.5 | 57.4 | 58.0 | 54.8 | 52.9 | 56.4 | 55.0 | 57.6 | 60.2 | 57.6 |
| >1000 | Before pacemaker | 58.2 | 59.7 | 57.4 | 54.1 | 53.8 | 55.6 | 52.8 | 51.4 | 55.8 | 53.6 | 54.6 | 57.8 | 55.5 |
|  | After pacemaker | 50.9 | 57.0 | 55.9 | 54.4 | 56.5 | 57.2 | 53.9 | 52.1 | 55.7 | 54.4 | 56.7 | 58.7 | 56.5 |
| >1500 | Before pacemaker | 57.4 | 59.2 | 56.9 | 53.6 | 53.1 | 54.5 | 51.9 | 50.7 | 54.8 | 52.8 | 54.0 | 56.8 | 53.5 |
|  | After pacemaker | 50.3 | 56.5 | 55.4 | 53.8 | 55.9 | 56.1 | 53.1 | 51.4 | 54.9 | 53.7 | 56.0 | 57.4 | 54.8 |
| 2000 | Before pacemaker | 57.1 | 59.1 | 56.7 | 53.3 | 52.8 | 54.0 | 51.5 | 50.4 | 54.4 | 52.4 | 53.7 | 56.2 | 52.6 |
|  | After pacemaker | 49.9 | 56.0 | 55.1 | 53.4 | 55.3 | 55.1 | 52.4 | 50.8 | 54.2 | 53.0 | 55.4 | 56.4 | 53.4 |
| >2500 | Before pacemaker | 57.0 | 59.0 | 56.5 | 53.1 | 52.7 | 53.5 | 51.2 | 50.1 | 54.1 | 52.1 | 53.4 | 55.9 | 52.0 |
|  | After pacemaker | 49.0 | 55.3 | 54.5 | 52.9 | 54.7 | 54.2 | 51.8 | 50.2 | 53.7 | 52.6 | 55.0 | 56.0 | 52.7 |

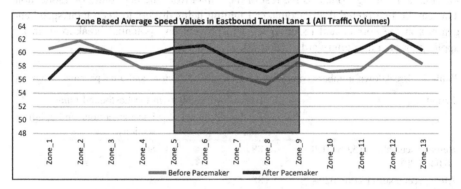

**Fig. 4.** Zone based average speed values in Eastbound tunnel lane 1 (All Traffic Volumes)

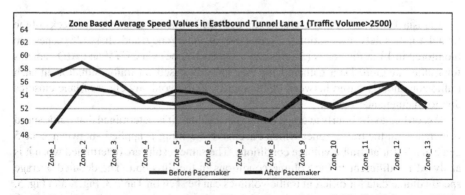

**Fig. 5.** Zone based average speed values in Eastbound tunnel lane 1 (Traffic Volume > 2500)

Same speed increase is also observed for the second lane of the tunnel in that direction which is mostly preferred by the 2$^{nd}$ class (minibus) vehicles and relatively slow driving users as can be seen in Table 4, Fig. 6, and Fig. 7.

**Table 4.** Zone based average speed values in Eastbound tunnel lane 2

| Traffic | Period | Zone 1 | Zone 2 | Zone 3 | Zone 4 | Zone 5 | Zone 6 | Zone 7 | Zone 8 | Zone 9 | Zone 10 | Zone 11 | Zone 12 | Zone 13 |
|---|---|---|---|---|---|---|---|---|---|---|---|---|---|---|
| All | Before pacemaker | 56.7 | 59.5 | 58.0 | 56.3 | 55.8 | 52.9 | 51.4 | 53.0 | 56.1 | 54.2 | 53.6 | 57.1 | 59.1 |
| | After pacemaker | 52.4 | 57.9 | 58.3 | 58.1 | 58.2 | 55.9 | 53.6 | 55.5 | 57.8 | 56.0 | 55.8 | 58.9 | 61.0 |
| >500 | Before pacemaker | 55.2 | 58.5 | 56.7 | 54.7 | 54.1 | 51.1 | 49.6 | 51.2 | 54.6 | 52.5 | 52.0 | 55.4 | 57.3 |
| | After pacemaker | 48.9 | 55.5 | 55.8 | 55.3 | 55.6 | 53.2 | 50.6 | 52.6 | 55.4 | 53.1 | 52.9 | 56.1 | 58.4 |
| >1000 | Before pacemaker | 55.0 | 58.3 | 56.6 | 54.5 | 53.9 | 50.7 | 49.2 | 50.9 | 54.2 | 52.1 | 51.6 | 54.9 | 56.7 |
| | After pacemaker | 48.8 | 55.3 | 55.7 | 55.1 | 55.4 | 52.6 | 50.1 | 52.2 | 54.9 | 52.6 | 52.4 | 55.4 | 57.5 |
| >1500 | Before pacemaker | 54.5 | 57.9 | 56.3 | 54.3 | 53.6 | 49.8 | 48.4 | 50.3 | 53.6 | 51.5 | 51.1 | 54.0 | 55.2 |
| | After pacemaker | 48.4 | 55.0 | 55.5 | 54.8 | 55.0 | 51.7 | 49.4 | 51.6 | 54.3 | 52.0 | 51.8 | 54.4 | 56.4 |
| >2000 | Before pacemaker | 54.3 | 57.9 | 56.3 | 54.1 | 53.5 | 49.4 | 48.1 | 50.0 | 53.3 | 51.2 | 50.8 | 53.5 | 54.5 |
| | After pacemaker | 48.0 | 54.7 | 55.3 | 54.5 | 54.5 | 50.9 | 48.8 | 51.1 | 53.8 | 51.5 | 51.3 | 53.6 | 55.4 |
| >2500 | Before pacemaker | 54.1 | 57.8 | 56.1 | 54.0 | 53.4 | 49.0 | 47.8 | 49.9 | 53.1 | 51.0 | 50.6 | 53.2 | 54.1 |
| | After pacemaker | 47.4 | 54.2 | 55.0 | 54.2 | 54.2 | 50.2 | 48.3 | 50.8 | 53.4 | 51.2 | 50.9 | 53.2 | 55.1 |

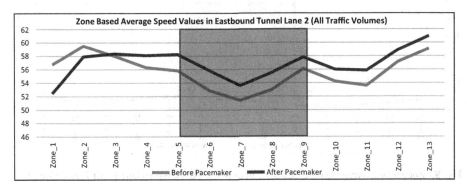

**Fig. 6.** Zone based average speed values in Eastbound tunnel lane 2 (All Traffic Volumes)

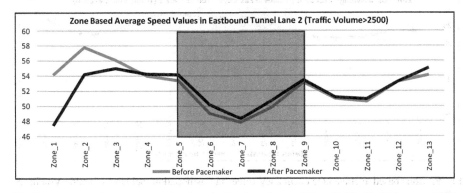

**Fig. 7.** Zone based average speed values in Eastbound tunnel lane 2 (Traffic Volume > 2500)

According to the hourly average lane occupancy values, it is determined that lane occupancy is decreased in the traffic zones which Pacemaker Lighting is applied for all traffic volume conditions due to the traffic speed increase. However, it is determined lane occupancy values are increased in the direction especially for higher traffic volumes. This can be interpreted that the Pacemaker Lighting helps decreasing the congestion in lower traffic volumes and helps increasing the traffic capacity in higher traffic volumes. The details can be seen on Table 5, Fig. 8, and Fig. 9.

**Table 5.** Zone based average occupancy values in Eastbound tunnel lane 1

| Traffic | Period | Zone 1 | Zone 2 | Zone 3 | Zone 4 | Zone 5 | Zone 6 | Zone 7 | Zone 8 | Zone 9 | Zone 10 | Zone 11 | Zone 12 | Zone 13 |
|---|---|---|---|---|---|---|---|---|---|---|---|---|---|---|
| All | Before pacemaker | 14.7 | 13.7 | 13.2 | 13.3 | 13.2 | 14.4 | 15.1 | 15.4 | 15.9 | 15.6 | 14.3 | 14.4 | 17.6 |
|  | After pacemaker | 12.6 | 11.9 | 11.1 | 11.1 | 10.9 | 12.1 | 12.7 | 13.0 | 13.6 | 12.9 | 11.8 | 11.1 | 15.0 |
| >500 | Before pacemaker | 18.5 | 17.3 | 16.7 | 16.9 | 16.8 | 18.3 | 19.2 | 19.5 | 20.2 | 19.8 | 17.7 | 18.3 | 22.1 |
|  | After pacemaker | 17.5 | 16.6 | 15.5 | 15.5 | 15.4 | 17.1 | 17.9 | 18.1 | 19.1 | 18.1 | 16.4 | 15.6 | 20.3 |
| >1000 | Before pacemaker | 21.1 | 19.9 | 19.2 | 19.4 | 19.4 | 21.2 | 22.2 | 22.4 | 23.3 | 22.8 | 20.2 | 21.2 | 25.1 |
|  | After pacemaker | 21.5 | 20.7 | 19.4 | 19.4 | 19.3 | 21.5 | 22.5 | 22.7 | 23.9 | 22.6 | 20.5 | 19.7 | 24.2 |
| >1500 | Before pacemaker | 26.8 | 26.0 | 25.1 | 25.5 | 25.7 | 28.3 | 29.5 | 29.4 | 30.7 | 29.9 | 26.1 | 28.1 | 31.9 |
|  | After pacemaker | 26.9 | 26.3 | 24.7 | 24.7 | 24.9 | 28.0 | 29.0 | 28.9 | 30.4 | 28.6 | 26.0 | 25.4 | 29.7 |
| >2000 | Before pacemaker | 29.6 | 29.0 | 28.1 | 28.6 | 29.0 | 32.1 | 33.1 | 32.9 | 34.4 | 33.4 | 28.9 | 31.7 | 35.3 |
|  | After pacemaker | 30.6 | 30.3 | 28.5 | 28.6 | 28.9 | 32.8 | 33.6 | 33.3 | 35.1 | 33.0 | 30.0 | 29.5 | 33.9 |
| >2500 | Before pacemaker | 31.8 | 31.3 | 30.5 | 31.1 | 31.4 | 34.8 | 35.8 | 35.6 | 37.4 | 36.2 | 31.2 | 34.2 | 38.1 |
|  | After pacemaker | 33.8 | 33.7 | 31.9 | 31.9 | 32.5 | 36.9 | 37.6 | 37.2 | 39.0 | 36.5 | 33.1 | 32.8 | 37.2 |

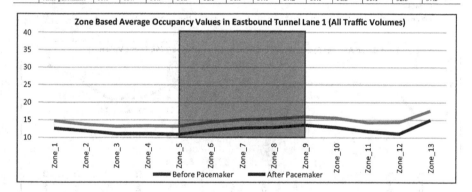

**Fig. 8.** Zone based average lane occupancy values in Eastbound tunnel lane 1 (All Traffic Volumes)

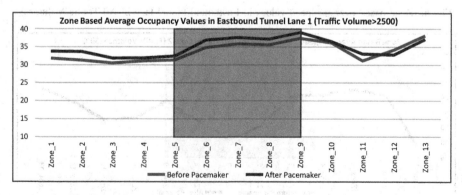

**Fig. 9.** Zone based average lane occupancy values in Eastbound tunnel lane 1 (Traffic Volume > 2500)

Same lane occupancy decrease is also observed for the second lane of the tunnel in that direction for the lower traffic volumes however it can be again observed that Pacemaker Lighting has a role of increasing the traffic capacity with the higher traffic volumes as can be seen in Table 6, Fig. 10, and Figure 11.

**Table 6.** Zone based average occupancy values in Eastbound tunnel lane 2

| Traffic | Period | Zone 1 | Zone 2 | Zone 3 | Zone 4 | Zone 5 | Zone 6 | Zone 7 | Zone 8 | Zone 9 | Zone 10 | Zone 11 | Zone 12 | Zone 13 |
|---|---|---|---|---|---|---|---|---|---|---|---|---|---|---|
| All | Before pacemaker | 14.0 | 14.9 | 14.7 | 15.1 | 14.9 | 16.9 | 16.4 | 15.8 | 15.2 | 15.1 | 15.4 | 15.3 | 16.9 |
|  | After pacemaker | 12.3 | 13.2 | 12.7 | 12.9 | 13.0 | 14.9 | 14.3 | 13.7 | 13.5 | 13.0 | 13.2 | 12.4 | 13.6 |
| >500 | Before pacemaker | 17.3 | 18.5 | 18.2 | 18.7 | 18.4 | 20.9 | 20.3 | 19.5 | 18.9 | 18.7 | 18.5 | 18.9 | 21.1 |
|  | After pacemaker | 16.7 | 18.0 | 17.4 | 17.7 | 17.8 | 20.4 | 19.5 | 18.7 | 18.5 | 17.7 | 17.7 | 16.9 | 18.5 |
| >1000 | Before pacemaker | 19.3 | 20.5 | 20.2 | 20.7 | 20.4 | 23.2 | 22.5 | 21.6 | 20.9 | 20.7 | 20.2 | 20.9 | 23.6 |
|  | After pacemaker | 19.6 | 21.0 | 20.2 | 20.7 | 20.8 | 23.9 | 22.8 | 21.8 | 21.6 | 20.7 | 20.6 | 19.8 | 22.0 |
| >1500 | Before pacemaker | 23.1 | 24.3 | 23.9 | 24.6 | 24.2 | 27.8 | 26.9 | 25.8 | 25.1 | 24.9 | 23.7 | 25.0 | 29.1 |
|  | After pacemaker | 23.1 | 24.5 | 23.6 | 24.2 | 24.4 | 28.3 | 26.9 | 25.7 | 25.4 | 24.4 | 24.0 | 23.4 | 26.7 |
| >2000 | Before pacemaker | 25.1 | 26.1 | 25.7 | 26.4 | 26.1 | 30.2 | 29.1 | 27.8 | 27.2 | 27.0 | 25.5 | 27.1 | 31.9 |
|  | After pacemaker | 25.5 | 27.0 | 25.9 | 26.6 | 27.0 | 31.5 | 29.7 | 28.4 | 28.2 | 27.0 | 26.5 | 26.0 | 30.2 |
| >2500 | Before pacemaker | 26.7 | 27.6 | 27.2 | 27.9 | 27.6 | 31.9 | 30.8 | 29.4 | 28.9 | 28.7 | 26.9 | 28.7 | 34.4 |
|  | After pacemaker | 28.4 | 28.6 | 27.2 | 27.9 | 29.1 | 34.5 | 31.9 | 30.2 | 29.8 | 28.9 | 28.7 | 28.9 | 36.6 |

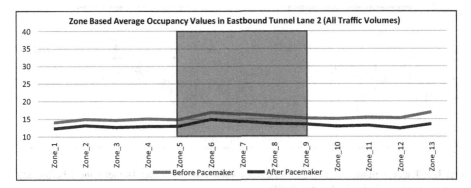

**Fig. 10.** Zone based average lane occupancy values in Eastbound tunnel lane 2 (All Traffic Volumes)

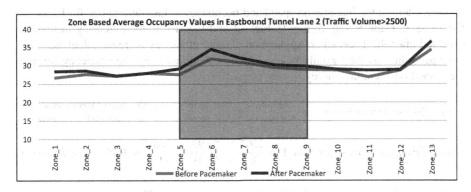

**Fig. 11.** Zone based average lane occupancy values in Eastbound tunnel lane 2 (Traffic Volume > 2500)

According to the data, the efficieny of the Pacemaker Lighting is calculated and analysed for higher traffic volumes which is over 2500 vehicles per hour.

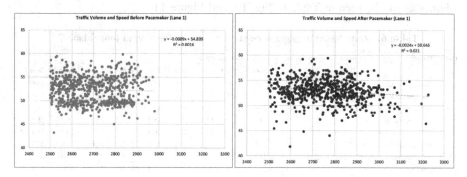

**Fig. 12.** Traffic volume and average speed values in Eastbound tunnel lane 1 before and after pacemaker (Traffic Volume > 2500)

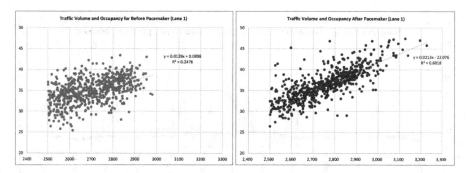

**Fig. 13.** Traffic volume and average occupancy values in Eastbound Tunnel Lane 1 before and after pacemaker (Traffic Volume > 2500)

According to the increase on speed and occupancy values determined for Lane 1 at the Table 3, Table 5, Fig. 5, Fig. 9, Fig. 12 and Fig. 13, traffic efficiency increase is expected after the Pacemaker Application. Traffic efficiency increase calculations are done according to the formula below which is modified from the efficiency formula mentioned on literature survey.

$$\text{Traffic Efficiency Increase for Lane 1} = \frac{Avg.\ volume \times Avg.\ Speed \times Avg.\ Occupancy\ (After\ Pacemaker)}{Avg.\ volume \times Avg.\ Speed \times Avg.\ Occupancy\ (Before\ Pacemaker)}$$

$$\text{Traffic Efficiency Increase for Lane 1} = \frac{2762 \times 52.93 \times 36.64}{2702 \times 52.31 \times 34.99} = 1.083\ (8.3\%\ increase)$$

Traffic efficiency increase for Lane 1 is calculated as 8.3%.

Pacemaker Lighting Application to Prevent Traffic Congestion 253

According to the data, the efficiency of the Pacemaker Lighting is calculated and analyzed for higher traffic volumes which is over 2500 vehicles per hour.

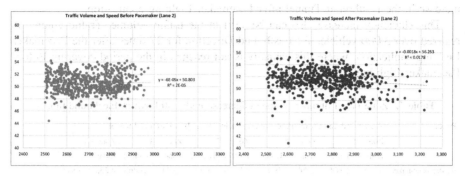

**Fig. 14.** Traffic volume and average speed values in Eastbound Tunnel Lane 2 before and after pacemaker (Traffic Volume > 2500)

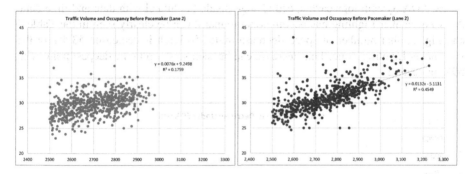

**Fig. 15.** Traffic volume and average occupancy values in Eastbound Tunnel Lane 1 before and after pacemaker (Traffic Volume > 2500)

According to the increase on speed and occupancy values determined for Lane 2 at the Table 4, Table 6, Fig. 7, Fig. 11, Fig. 14 and Fig. 15, traffic efficiency increase is expected after the Pacemaker Application. Traffic efficiency increase calculations are done according to the formula below.

$$\text{Traffic Efficiency Increase for Lane 2} = \frac{\text{Avg. volume} \times \text{Avg. Speed} \times \text{Avg. Occupancy (After Pacemaker)}}{\text{Avg. volume} \times \text{Avg. Speed} \times \text{Avg. Occupancy (Before Pacemaker)}}$$

$$\text{Traffic Efficiency Increase for Lane 2} = \frac{2762 \times 51.38 \times 31.09}{2702 \times 50.64 \times 29.69} = 1.086 \ (8.6\% \ increase)$$

*Average Pacemaker Efficiency Increase for Eastbound Tunnel* (*Lane* 1 *and Lane* 2)
$= 8.45\%$

According to the determined results, the efficiency of the Pacemaker Lighting is calculated and analyzed for the higher traffic volumes over 2500 vehicles per hour for both Lane 1 and Lane 2. Results shows that there is an average of 8.45% efficiency increase in the Eastbound Tunnel after Pacemaker Lighting application.

This efficiency increase also can be seen in the traffic volume. Traffic volume is realized over 3.000 vehicles for Eastbound Direction for 27 times after the Pacemaker Lighting application, where this capacity is not reached in the period covering Before Pacemaker Lighting.

Table 7. Zone based number of speed decrease under 50 km/h in Eastbound tunnel

| Period | Zone based number of speed decrease under 50 km/h in Eastbound tunnel ||||||||||
|---|---|---|---|---|---|---|---|---|---|---|
| | Zone - 5 || Zone - 6 || Zone - 7 || Zone - 8 || Zone - 9 ||
| | Lane 1 | Lane 2 | Lane 1 | Lane 2 | Lane 1 | Lane 2 | Lane 1 | Lane 2 | Lane 1 | Lane 2 |
| Before pacemaker | 4,268 | 1,864 | 5,299 | 28,785 | 14,248 | 48,475 | 23,245 | 21,463 | 4,233 | 4,915 |
| After pacemaker | 1,926 | 1,727 | 3,537 | 8,757 | 5,076 | 25,773 | 12,128 | 8,931 | 2,442 | 2,647 |

SCADA System logs each speed change inside the tunnel with date and time data. After Pacemaker Lighting has become operational it can be clearly seen in the Table 7 and Fig. 16 that number of speed decrease under 50 km/h is decreased 53.4% in the Pacemaker Lighting Zone. This decrease can be seen so sharp in the deepest point with 69% in Lane 2.

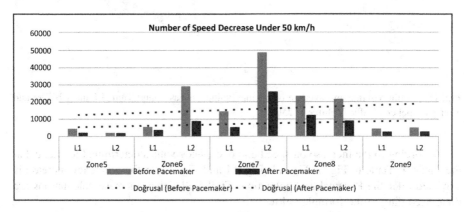

**Fig. 16.** Comparison for number of speed decrease under 50 km/h before and after pacemaker lighting

Table 8. Interior air quality analysis in Eastbound tunnel (ug/m$^3$)

| Period | Interior air quality analysis in Eastbound tunnel (ug/m$^3$) |||
|---|---|---|---|
| | CO | NO$_2$ | PM |
| Before pacemaker (15.06.2019–30.04.2020) | 12155.05 | 894.52 | 98.36 |
| After pacemaker (15.06.2020–30.04.2021) | 10651.04 | 826.03 | 103.47 |

Interior Air Quality is analysed in Eurasia Tunnel monthly based according to the day with the highest traffic volume for CO (Carbon Monoxide), $NO_2$ (Nitrogen Dioxide) and PM (Particulate Matter) via the data gathered from the sensors inside the tunnel before the ventilation shaft. According to the hourly averages for determined days of each month (April 2020 and April 2021 data is excluded due to traffic volume decrease in Covid-19 Period). CO and $NO_2$ values are decreased 12.37% and 7.66% respectively after the Pacemaker Lighting application. However, PM value increased 5.19%, which can be explained by change in condition of PM from settled to ground to the suspended in the air due to increase of vehicle speed in the tunnel. Also, it should be noted that PM value in Eurasia Tunnel is 4% of the tunnel closure threshold level according to PIARC requirements. Detailed information can be examined on Table 8.

**Table 9.** Summary of analysis for pacemaker lighting application

| Target item | Summary of pacemaker lighting targets and results | |
|---|---|---|
| Accidents | No accidents reported in the period | |
| Vehicle breakdowns | 4.65% decreased | |
| | Lane-1 | Lane-2 |
| Speed | 1.18% increased | 1.46% increased |
| Occupancy | 4.71% increased | 4.72% increased |
| Lane traffic efficiency | 8.3% increased | 8.6% increased |
| Eastbound tunnel traffic efficiency | 8.45% increased | |
| | Deepest point | Pacemaker |
| Traffic speed decreases under 50 km/h | 69% decreased | 53.4% decreased |
| | CO | $NO_2$ |
| Exhaust emission gases | 12.37% decreased | 7.66% decreased |

# 7 Conclusion

Pacemaker Lighting System commenced in Eurasia Tunnel established all its targets according to the results in this study. There are no accidents reported in the period after Pacemaker Lighting is functional and breakdown numbers sourced by engine and transmission are decreased according to the incident reports. Efficiency of the tunnel which Pacemaker Lighting is increased by 8.45% by increasing both speed and occupancy of both lanes at the same time. On the other hand, potential traffic congestions are prevented 53.4% in the Pacemaker Zone and up to 69% in the deepest point. Pacemaker Lighting System also helps decreasing exhaust gases such as CO and $NO_2$ 12.37% and 7.66% respectively to protect environment.

In the light of all the facts mentioned above in this study, Pacemaker Lighting is a complete system and good practice for tunnel operation sustainability (Table 9).

# 8 Recommendations for Future Studies

Pacemaker Lighting in Eurasia Tunnel is used with a constant speed of 70 km/h till now to ease driver's adaptation process since it is a new technology used for the first time in Turkey. The effect on the efficiency and the other related factors can be studied when the system is used in adaptive mode which means changing the speed of the lighting according to the average speed and the traffic volume inside the tunnel.

Also, the effect of the Pacemaker Lighting on accident and breakdown numbers can be followed in the longterm.

# References

Brilon, W.: Traffic flow analysis beyond traditional methods. In: Proceedings of the 4th International Symposium on Highway Capacity, pp. 26-41, TB Circular E-C018, Transportation Research Board, Washington D.C (2000)

Kato, H., Nakagawa, H., Hashimoto, D., Inaba, Y.: Measures to prevent congestion on the Tokya Bay Aqua Line by using pacemaker light. In: 22nd ITS World Congress, Bordeaux, France, 5–9 October 2015

Prevention and Mitigation of Tunnel-Related Collisions Technical Committee D5 Tunnel Operations, PIARC 2019R03EN (2019)

# Real Time Multi Object Detection & Tracking on Urban Cameras

Rifkat Minnikhanov[1(✉)], Maria Dagaeva[1,2], Timur Aslyamov[1],
Tikhon Bolshakov[1], and Emil Faizrakhmanov[1]

[1] "Road Safety" State Company, Kazan, Russia
its.center.kzn@gmail.com
[2] Kazan National Research Technical University - KAI named
after A. N. Tupolev, Kazan, Russia

**Abstract.** Processing video from urban video surveillance cameras requires the use of algorithms for multi-object detection and tracking on video in real-time. However, existing computer vision algorithms require the use of powerful equipment and are not sufficiently optimized to process multiple video streams simultaneously. This article proposes an approach to using the tracker in conjunction with the YoloV4 object detector for real-time video processing on medium-power equipment. Paper also presents the solution for difficulties that arise during work with optical flow. The results of the comparison of the accuracy and speed of image processing of the applied approach with such trackers as IOU17, SORT, KCF, and MOSSEE are also presented.

**Keywords:** Deep learning · Object detection & tracking · Video surveillance · Computer vision

## 1 Introduction and Problem Definition

Currently, one of the most relevant areas in the field of intelligent transport systems (ITS) is the task of detecting and tracking objects in real-time on video streams of closed-circuit television systems (CCTV). Existing algorithms for detecting and tracking objects in the video image allow to determine the position of the object, its trajectory and class with high accuracy. However, when processing video images from multiple video surveillance cameras in a data centre, there is a problem associated with the need to accommodate many expensive graphics processor units (GPU). The development of our solution for object detection and tracking allows real-time analysis of video images on medium-power equipment with sufficient accuracy that allows using of video surveillance cameras as an optimal and operational source of information.

## 2 Related Works

Today, most existing tracking solutions follow the tracking-by-detection paradigm. The paradigm involves splitting the tracking process into 2 stages: detecting all objects in the image (frame) and linking the corresponding detected objects to form a trajectory.

© The Author(s), under exclusive license to Springer Nature Switzerland AG 2022
A. Akhnoukh et al. (Eds.): IRF 2021, SUCI, pp. 257–268, 2022.
https://doi.org/10.1007/978-3-030-79801-7_18

258    R. Minnikhanov et al.

According to the features of the application, existing trackers can be divided as follows:

1. Trackers that require a marked-up training dataset (Braso and Leal-Taixe 2020; Fang et al. 2018; Sadeghian et al. 2017; Leal-Taixe et al. 2014). This feature creates difficulties in obtaining a marked-up dataset, which makes their use difficult.
2. Trackers that perform processing with insufficient FPS (an indicator of video processing, denoted as the number of frames processed per second), which makes real-time processing impossible (Bergmann et al. 2019; Karthik et al. 2020).
3. Trackers that perform independent tracking of individual objects, without the use of a detector on each frame (Chu et al. 2019; Bolme et al. 2010). This group of trackers has a problem related to the insufficient quality of tracking objects.
4. Trackers that provide acceptable tracking quality and image processing speed (hereinafter referred to as FPS), provided that an effective detector is used (Bewley et al. 2016; Bchinski et al. 2017).

Although, the problem, outlined in the article, should be solved by trackers from the 4-th group, to date, there are no sufficiently effective, pre-trained detectors for detecting objects on video from surveillance cameras in the public domain. The exception is the YoloV4 detector (Bochkovskiy et al. 2020), which is the best in FPS/accuracy ratio. It is capable of providing real-time image processing on Nvidia GeForce 2070 graphics cards with an input image size of 320*320 pixels. However, its use on medium-power video cards (GeForce 960, 1050 Ti, 1060, 1070 and 1080) is not possible due to the low FPS rate. At the same time, the use of a lightweight version of this detector – YoloV4 – tiny (Bochkovskiy et al. 2020) is not sufficiently effective in detecting objects. The Table 1 shows a comparison of the FPS output of YoloV4 on different GPUs.

**Table 1.**  Comparison of YoloV4 FPS on different GPUs

| Nvidia GPU | Framework | An input image size | FPS |
|---|---|---|---|
| GeForce 850M | OpenCV 4.4.0 for Windows | 416 * 416 | 5.4 |
| GeForce 1050 Ti | | 416 * 416 | 14.3 |
| GeForce 1050 Ti | | 608 * 608 | 9.1 |
| GeForce 1060 Ti | | 416 * 416 | 20.4 |
| GeForce 1060 Ti | | 608 * 608 | 16.1 |
| Jetson Nano | TensorRT FP16 | 416 * 416 | 3.9 |
| Jetson Nano | | 608 * 608 | 1.9 |
| GeForce 1080 Ti | TensorRT FP32 | 416 * 416 | 27.3 |
| GeForce 1080 Ti | | 608 * 608 | 18.2 |
| Tesla V100 16 Gb | OpenCV 4.5.0 – pre for Linux | 416 * 416 | 62.5 |
| Tesla V100 16 Gb | | 608 * 608 | 37 |

## 3 Proposed Solution

In this paper, we propose a new tracker for tracking multiple objects based on the pyramidal implementation of the allowed optical flow KLT (Bouguet 2001).

Our tracker independently predicts the position of the object bounding box several frames ahead, and every few frames the YoloV4 detector is used (with the size of the input image - 608*608 pixels). This is necessary to correct the predicted values of the object frame position and track new objects. For effective prediction, the tracker uses statistical characteristics of the distribution of the calculated allowed optical flow KLT to predict the current position of the frame from its previous location. The association of objects between frames is carried out by solving the assignment problem with a greedy algorithm according to the following principle: if the centroid of a newfound object falls into one of the existing frames, then it is considered that it is the same object. If the centroid of the newfound object falls within more than one of the existing frames, then it is associated with the one whose distance to the centroid is less. In this case, an object is considered lost if the detector has not confirmed its existence a certain number of times in a row.

### 3.1 Predicting the Position of the Object Frame with the Tracker

To predict the position of an object's frame, it is enough to know the position of the object in the current frame and the contour or mask of the object in the previous frame. Also, given that the object is defined by a bounding rectangle rather than a contour, it is sufficient to determine how much the frame's centroid must be offset for the frame to continue to reflect the object's position. The displacement of the centroid can be determined using a statistical characteristic of the distribution of displacements of points of the object (mathematical expectation, mode or median of the optical flow of points of the object). However, there are some problems when calculating the optical flow.

1. Fast detectors do not return the object mask, but its bounding box, which does not allow you to simply accept new frame coordinates after calculating the optical flow for the coordinates of points in the current frame, for a number of reasons: some points fall on a fixed background; the object mask may be partially covered by a fixed background object and for overlapping points; the optical flow is cleared incorrectly.

    This problem was solved in the following way. The absolute majority of points for which the zero optical flux is calculated will belong either to the background or to a static occluder (an object that overlaps a moving object). Points associated with the background or static occluder appear gradually in the object frame and, having a sufficiently large area, retain their position relative to the object in neighbouring frames. Excluding "standing points" (points with the near-zero optical flow) from consideration, we will most likely exclude all points related to the background or static occluder. An example of excluding points can be seen in the Fig. 1.

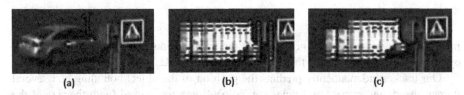

**Fig. 1.** (a) – the tracked object (car) moving behind a static occluder (road sign) indicating its direction of movement; (b) - display of the optical flow of a number of points inside the frame; (c) - display of the optical flow of points, except for those whose optical flow was calculated incorrectly, or for the second or more consecutive times turned out to be near-zero.

2. Calculating the optical flow over all points of an object is a very expensive operation if there are many objects or they are large.

The problem of slow calculation of the optical flow is eliminated by selecting a representative sample of points belonging to the object frame. It was assumed that the frames returned by the detector are well centred relative to the object, that is, the centroid of the frame is close to the centroid of the object and if the object is solid, then the points closer to the centre of the frame will be more likely to fall on the object, and not on the background. Obtaining a representative sample was carried out by creating a "template for placing control points" on a hypothetical object that has a side size of one length. Such a template is a tensor of size H*W*2, where H is the number of points in the sample in height, W is the number of points in the sample in width. The tensor contains the coordinates of the sample control points for the hypothetical object under consideration. The value of the tensor elements is determined by the following formulas:

$$\begin{cases} template_{i,j,0} = 0.5 + \dfrac{10^{\left(0.5-\frac{1}{W}\right)\cdot\left(j-\frac{W}{2}\right)}}{1+0.25\left(10^{\left(0.5-\frac{1}{W}\right)\cdot\left(W-1-\frac{W}{2}\right)}-10^{\left(0.5-\frac{1}{W}\right)\cdot\left(W-2-\frac{W}{2}\right)}\right)}, j > \frac{W}{2} \\ template_{i,j,0} = 0.5 - \dfrac{10^{\left(0.5-\frac{1}{W}\right)\cdot\left(\frac{W}{2}-j\right)}}{1+0.25\left(10^{\left(0.5-\frac{1}{W}\right)\cdot\left(W-1-\frac{W}{2}\right)}-10^{\left(0.5-\frac{1}{W}\right)\cdot\left(W-2-\frac{W}{2}\right)}\right)}, j < \frac{W}{2} \\ template_{i,j,1} = 0.5 + \dfrac{10^{\left(0.5-\frac{1}{H}\right)\cdot\left(\frac{H}{2}-i\right)}}{1+0.25\left(10^{\left(0.5-\frac{1}{H}\right)\cdot\left(H-1-\frac{H}{2}\right)}-10^{\left(0.5-\frac{1}{H}\right)\cdot\left(H-2-\frac{H}{2}\right)}\right)}, i > \frac{H}{2} \\ template_{i,j,1} = 0.5 - \dfrac{10^{\left(0.5-\frac{1}{H}\right)\cdot\left(i-\frac{H}{2}\right)}}{1+0.25\left(10^{\left(0.5-\frac{1}{H}\right)\cdot\left(H-1-\frac{H}{2}\right)}-10^{\left(0.5-\frac{1}{H}\right)\cdot\left(H-2-\frac{H}{2}\right)}\right)}, j < \frac{H}{2} \\ template_{i,j,0} = 0.5, j = \frac{W}{2} \\ template_{i,j,1} = 0.5, i = \frac{H}{2} \end{cases} \quad (1)$$

Next, a similar pattern is applied to the frame of the object by adding to it the offset of the object relative to the coordinates (0, 0), as well as multiplying the corresponding coordinates by the width and height of the object. An example of such a template superimposed on an object with a size of 64*128 pixels is shown in the Fig. 2.

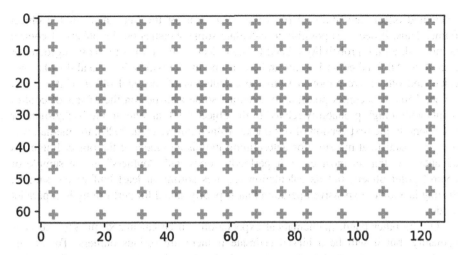

**Fig. 2.** "Template for placing control points" on objects defined by a 12*12*2 tensor superimposed on a hypothetical object with a size of 64*128 pixels

To determine whether the object is moving or not, we calculate the proportion of control points that are standing for the first time, from all control points that are not standing for the second or more times. To do this, the optical flow of each object is calculated at the points corresponding to the previously described pattern superimposed on the object frame.

The condition that the object is worth will be:

$$\frac{\sum stop_{mask\,s,n} - \sum (stop_{mask\,s,n} \overset{\circ}{} stop_{mask\,s-1,n})}{H*W - \sum (stop_{mask\,s,n} \overset{\circ}{} stop_{mask\,s-1,n}) - \sum st_{s,n}} > 0.75 \qquad (2)$$

Taking into account that:

- $st_{s,n}$ - binary matrix H*W of the correctness of calculating the optical flow for the n-th object on the s-th frame, in which 1-means that the optical flow is calculated incorrectly, and 0-that it is calculated normally;
- $stop\_mask_{s,n}$ - binary matrix H*W of immobility of control points for the n-th object on the s-th frame, in which 1 is the optical flow for the point near zero, 0 is significantly different from zero;

The coefficient 0.75 was determined empirically.

The matrix $st_{s,n}$ is determined automatically when calculating the offsets of control points by the optical flow. The matrix $stop\_mask_{s,n}$ is filled in by comparing the coordinate differences of all control points for which the optical flow was calculated with a certain threshold, which is calculated based on the size of the object frame. If it was determined that the object is stationary, then the frame offset for it is not calculated. In addition, all elements of the matrix $stop\_mask_{s,n}$ for this object are set to 0, so that on the next frame, if this object remains stationary, there is no "parasitic" jitter for its frame. All fixed control points and control points for which the optical flow has not calculated offsets are removed from the calculation of the final offset of the object frame. Thus, it becomes possible to calculate some statistics on the offsets of control points with a high probability related to the object itself, and not to the background or occluders. All fixed control points and control points for which the optical flow has not calculated offsets are removed from the calculation of the final offset of the object frame. Thus, it becomes possible to calculate some statistics on the offsets of control points with a high probability related to the object itself, and not to the background or occluders. In the first version of our tracker (Makhmutova et al. 2020), the median was used as a statistical measure for determining the displacement of the object frame. Its advantage is that in most cases it perfectly copes with "outliers" in the sample of control point offsets, but its calculation requires sorting at least half of the sample. Sorting is a rather expensive operation that is poorly suited for performing it in parallel mode.

On the other hand, mathematical expectation can be calculated quickly and conveniently, but it will be a biased estimate if there are serious outliers. To fix this problem, you can recalculate the mathematical expectation after removing the control points from consideration, the offset of which differs from the already calculated mathematical expectation by a certain coefficient that depends on the standard deviation. Moreover, such an operation can be carried out iteratively the required number of times, which will consistently reduce the offset of the mathematical expectation. The visual effect of the iterative recalculation of the mathematical expectations can be seen in the Fig. 3.

**Fig. 3.** Comparison of different ways to remove outliers. (a) - displaying offsets of control points without any filtering; (b) – filtering outliers with a cut-off of ±1.5 σ; (c) – filtering outliers with a cut-off of ±1.1 σ; (d) - filtering outliers with 2 cut-offs of ±1.5 σ with a recalculation of the expectation value.

At each iteration, the corresponding binary mask korr with size H*W is calculated for the points:

$$korr_{c,s,n,i,j} = \begin{cases} 1 * korr_{c-1,s,n,i,j}, & |x_{s,n,i,j} - m_{c-1,s,n}| < K * \sigma_{c-1,s,n} \\ 0 * korr_{c-1,s,n,i,j}, & |x_{s,n,i,j} - m_{c-1,s,n}| > K * \sigma_{c-1,s,n} \end{cases} \tag{3}$$

where:

- $x_{s,n,i,j}$ is the offset of the control point (i, j) for the n-th object on frame s;
- $c$ - the outlier elimination iteration number;
- $m_{c,s,n}$ is the expected value of the control point offsets on iteration $c$ for the n-th object on frame s;
- $\sigma_{c,s,n}$ is the standard deviation of the control point offsets on iteration $c$ for the n-th object on frame s.

The matrix $korr_0$ is initially initialized as:

$$korro_0 = \overline{stop_{mask}} \circ st \tag{4}$$

The expectation value $m_{c,\,s,\,n}$ and the standard deviation $\sigma_{c,\,s,\,n}$, in this case, are determined by the formulas:

$$m_{c,s,n} = \frac{\sum x_{s,n}^{\circ} korr_{c,s,n}}{\sum korr_{c,s,n}} \tag{5}$$

$$\sigma_{c,s,n} = \frac{\sum_{i=1,j=0}^{H,W} (x_{s,n,i,j} * korr_{c,s,n,i,j} - m_{c,s,n,i,j})^2}{\sum korr_{c,s,n}} \tag{6}$$

The expectation value, together with the application of this method of eliminating outliers, is a fairly accurate and easily calculated measure that was used to determine the final offset of the object frame from a set of control point offsets.

3. Optical flow for the "hidden" points are calculated has been grossly inadequate, and the fact of such calculations can often be very difficult to detect automatically.

The optical flow is calculated inadequately if the control point has disappeared behind the occluder or has gone beyond the image. With the gradual overlap of the object, static occluder offset checkpoints before occluders computed optical flow is much less than the real displacement of the object frame. If the occluder has small dimensions relative to the object, these differences do not make a significant contribution to the final frame displacements, being compensated by the displacements of the remaining control points. At the same time, the frame begins to lag behind the object horizontally if a significant part of the columns of the control point placement template is blocked, and vertically if a significant part of the rows is blocked. We solved this problem by adding artificial acceleration to objects horizontally if a certain percentage of the columns of the object's control point placement pattern is significantly overlapped, and vertically if a certain percentage of rows are significantly overlapped.

The problem of inadequate calculation of the optical flow for control points located near the image boundaries can be solved by tracking how much the width and height of the object frame have decreased due to going beyond the image boundaries relative to the original ones. Then, according to the calculated shares, exclude from consideration the corresponding shares of rows and columns from the corresponding sides of the template for placing control points of the object. In addition, if the frame of an object has reduced its area due to going beyond the image boundary, reduce the coefficient before σ for this object, as well as increase the number of iterations of recalculating the mathematical expectation with clipping by the standard deviation by one.

### 3.2 Parameters of the Tracker

By applying all the above methods together, you can significantly improve the efficiency of mathematical expectation as an estimate of the displacement of the object frame with a slight increase in the time of its calculation. We have empirically selected the following parameters for our tracker:

- Percentage of standing points from the total number of points except for standing the second or more times to recognize the object as standing – 75%
- number of points in the template for placing control points in width and height – 12*12
- Number of iterations to recalculate expectation with sigma - 2
- The coefficient in front of σ for drop emissions – 2
- The rate of acceleration of objects horizontally due to the overlapping of a considerable number of columns in the pattern of placement of control points – 1.6
- The rate of acceleration of objects vertically due to the overlapping of a considerable number of lines in the pattern of placement of control points – 1.1
- The share of the points standing in the column/row placement pattern of control points, to make a column/line has been blocked – 0.6
- The proportion of overlapped columns/rows to start the working mechanism of acceleration – 0.2
- The multiplier factor σ for discarding feature points, the part of the framework which went beyond image – 0.85 (Fig. 4)

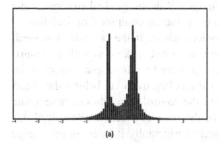 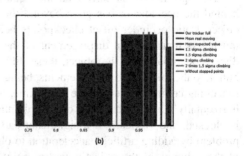

**Fig. 4.** Estimates of the distribution of control point offsets relative to the real displacement of the object frame: (a) - a histogram of the distributions of control point offsets relative to the real displacement of the object frame; (b) - various estimates of the real displacement of the object frame based on the distributions of control point offsets.

# 4 Evaluation

The effectiveness of our tracker was evaluated on 3 datasets: training data 2DMOT2015 (Leal-Taixe et al. 2015), training data MOT16 (Milan et al. 2016) and our own dataset consisting of various videos from surveillance cameras of the city of Kazan. Our own dataset is a set of 5 videos from CCTV Kazan city roads. Their description is presented in the Table 2.

**Table 2.** Description of the dataset for trackers evaluation.

| Videofile's name | FPS | Resolution | Length (Duration) | Tracks | Number of bounding boxes | Density | Description of the video |
|---|---|---|---|---|---|---|---|
| Bulak | 30 | 1280 * 720 | 1 980 (1:06) | 132 | 34 183 | 17,3 | Footage of busy intersection in the centre of Kazan, clear weather |
| Bulak_accident | 30 | 1280 * 720 | 1 483 (0:49) | 89 | 26 926 | 18,2 | Footage of an accident at one of the intersections with rainy weather |
| Bulak_night | 30 | 1280 * 720 | 460 (0:15) | 24 | 3 391 | 7,5 | Footage at one of the intersections in the centre of Kazan at night with defocusing of the camera |
| Kamala | 30 | 1280 * 720 | 2 010 (1:07) | 64 | 24 935 | 12,4 | Footage where there are a large number of static occluders |
| Nesmelova | 30 | 1280 * 720 | 1 285 (0:42) | 57 | 17 230 | 13,4 | Footage where there is a significant difference in light |
| Total | – | – | 7 208 | 366 | 106 665 | 14, 8 | – |

The sets provided by the authors were used as the detection set for the MOT Challenge sets. For our dataset, we used detections made using yolov4 with a 608x608 window.

For comparison with our tracker, 3 trackers were selected: SORT (Bewley et al. 2016), IOU17 (Bchinski et al. 2017), KCF MOSSE. For SORT and IOU17, implementations provided by the authors of these trackers were used as part of the official publication of the results of the MOT Challenge performed by these trackers. For KCF and MOSSE, implementations from OpenCV 4.4.0 were used.

The main key point is the maximum reduction in the cost of processing video from a single camera, this can be achieved by increasing the intervals between the use of the detector while maintaining an acceptable detection efficiency of objects. Also, in the case of increasing the intervals by reducing the FPS of the video stream, there is an extremely undesirable loss of information. Therefore, one of the points of our study was a comparative analysis of the behaviour of trackers with a decrease in the frequency of use of the detector. The comparison results are shown in the Fig. 5.

It is worth noting that for all trackers except IOU17, the position of the frames for objects between detector applications was determined based on tracker predictions. IOU17 does not have a mechanism for predicting the position of the object frame on the next frame, so the only way available for it to reduce the frequency of using the detector is to reduce the FPS of the video itself, by discarding all frames between detections. For this reason, the evaluation results for this tracker are somewhat inflated compared to the rest.

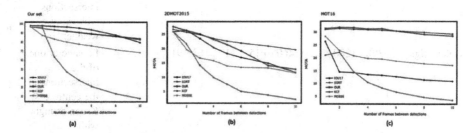

**Fig. 5.** The dependence of the Multi-Object Tracking Accuracy (MOTA) on the frequency of use of the detector: (a) - for our dataset; (b) - for the 2dMOT2015 dataset; (b) - for the MOT2016 dataset.

The results show that our tracker is slightly behind in efficiency and speed SORT and IOU17, when applying the detector on each frame, greatly surpasses them in effectiveness in reducing the frequency of application of the detector 3 or more times, allowing us to confidently move objects by reducing the frequency of application of the detector to 5 times. When compared with KCF, our tracker shows similar efficiency, while reducing the frequency of use of the detector by 4–5 times.

The effect of reducing the frequency of detector application on the final performance of the detector and tracker bundle was also measured. As a detector, YoloV4 was used with a window of 608x608 pixels on hardware with GPU: NVIDIA GeForce 1050ti 4 GB, CPU: Intel Core i5 8300H and 8 GB RAM. The results are shown in the

Fig. 6. The figure shows that with this configuration of equipment, real-time video processing is achieved only when the frequency of using the detector is reduced by 4 times for SORT and IOU17, 5 times for our tracker by 7–8 times for MOSSE. For KCF, real-time processing cannot be achieved. At the same time, the only tracker that provides sufficient efficiency of tracking objects in real-time is the tracker with the suggested approach.

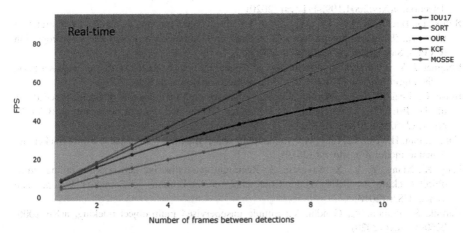

**Fig. 6.** The dependence of the FPS of the detector and tracker bundles on the decrease in the frequency of use of the detector.

## 5 Conclusions

The essence of the work was a comparative analysis of existing trackers with the developed tracker algorithm to achieve real-time video processing on medium-power equipment and with acceptable object detection accuracy. The evaluation results showed that the presented algorithm of the tracker, under the mandatory condition of working in real-time, has a MOTA indicator close to the values of the trackers, which have a much lower FPS indicator. Thus, the tracker is balanced between the acceptable accuracy of object detection and the number of processed frames per second. The tracker provides reliable tracking of objects at intervals of up to 5 frames without the use of a detector, and, when working in conjunction with the YoloV4 detector, which detects once every 5 frames on input 608x608 images, produces 25 FPS on an GeForce 1050 Ti GPU and an Intel Core processor i5 8300H. The use of this tracker allows not only to provide tracking of objects in the video in real-time on relatively weak GPU, but also allows to process several video streams on one copy of YoloV4 on powerful ones: the detector performs object detection for one of the cameras, while for the rest it tracks objects without detections. Experiments show that using a similar technique, one copy of YoloV4 can serve up to 4–5 trackers on Nvidia Tesla V100 16 Gb.

# References

Bchinski, E., Eiselein, V., Sikora T.: High-speed tracking-by-detection without using image information. In: 14-th IEEE Conference on Advanced Video and Signal Based Surveillance (AVSS), Lecce, Italy (2017)

Bergmann, P., Meinhardt, T., Leal-Taixe, L.: Tracking without Bells and Whistles. IEEE/CVF International Conference on Computer Vision (ICCV) (2019)

Bewley, A., Ge, Z., Ott, L., Ramox, F., Upcroft B.: Simple online and realtime tracking. In: IEEE International Conference on Image Processing (ICIP), Phoenix, USA (2016)

Bochkovskiy, A., Wang, C., Liao, H.M.: YOLOv4: optimal speed and accuracy of object detection, arXiv:2004.10934v1 [cs] (2020)

Bolme, D., Beveridge, J.R., Draper, B., Lui, Y.: Visual object tracking using adaptive correlation filters. In: IEEE Computer Society Conference on Computer Vision and Pattern Recognition (CVPR), San Francisco, California (2010)

Bouguet, J.-Y.: Pyramidal implementation of the affine Lucas Kanade feature tracker description of the algorithm. Intel Corporation, Microprocessor Research Labs (2001)

Braso, G., Leal-Taixe, L.: Learning a neural solver for multiple object tracking. In: Proceedings of the IEEE/CVF Conference on Computer Vision and Pattern Recognition (CVPR), pp. 6247–6257 (2020)

Chu, P., Fan, H., Tan, C., Ling, H.: Online multi-object tracking with instance-aware tracker and dynamic model refreshment (2019)

Fang, K., Xiang, Y., Li. X., Savarese, S.: Recurrent autoregressive networks for online multi-object tracking. In: IEEE Winter Conference on Applications of Computer Vision, Lake Tahoe, USA (2018)

Karthik, S., Prabhu, A., Gandhi, V.: Simple unsupervised multi-object tracking, arXiv:2006.02609v1 [cs] (2020)

Leal-Taixe, L., Milan, A., Reid, I., Roth, S., Schindler, K.: MOTChallenge 2015: towards a benchmark for multi-target tracking, arXiv:1504.01942 [cs] (2015)

Makhmutova, A., Anikin, I.V., Dagaeva, M.: Object tracking method for videomonitoring in intelligent transport systems. In: Proceedings of International Russian Automation Conference, RusAutoCon 2020, pp. 535–540 (2020)

Milan, A., Leal-Taixe, L., Reid, I., Roth, S., Scindler, K.: MOT16: a benchmark for multi-object tracking, arXiv:1603.00831 [cs] (2016)

Sadeghian, A., Alahi. A., Savarese. S.: Tracking the untrackable: learning to track multiple cues with long-term dependencies. In: IEEE International Conference on Computer Vision (ICCV), Venice, Italy (2017)

Leal-Taixe, L., Fenzi. M., Kuznetsova, A., Rosenhahn, B., Savarese. S.: Learning an image-based motion context for multiple people tracking. In: IEEE Conference on Computer Vision and Pattern Recognition (CVPR) (2014)

# Effective Advanced Warning for Connected Safety Applications - Supplementing Automated Driving Systems for Improved Vehicle Reaction

Gregory M. Baumgardner[1(✉)], Rama Krishna Boyapati[1], and Amudha Varshini Kamaraj[2]

[1] Battelle Memorial Institute, Columbus, OH, USA
baumgardner@battelle.org
[2] University of Wisconsin-Madison, Madison, USA

**Abstract.** Given the rapid pace of modern technological advancements, the public should expect and demand measurable improvements to highway safety. Yet, it is not so clear how much improvement may be anticipated. Government organizations such as the U.S. Department of Transportation (U.S. DOT) and National Highway Traffic Safety Administration (NHTSA) have already spent decades and millions of dollars researching proper markings, alerting systems, and safety distances to help reduce collisions and other incidents on public roadways. While clearly this effort has had great impact, there are limiting factors that continually constrain the ability of traditional methods to significantly reduce the number of collisions. Such factors include driver behavior aspects such as reaction time, sudden maneuvers, and traffic violations, plus infrastructure aspects such as malfunctioning signals, inadequate signage, and non-standard road design. As increased numbers of connected and automated vehicles (CAV) are introduced into the traffic stream, and advanced safety applications are continually improving, the industry envisions a major decline in incidents across the board. This paper details the limiting factors to why a sizable reduction of incidents is not possible with conventional resources and introduces the framework for adding advanced warnings into connected safety applications in existing vehicles, such as red-light violation warning (RLVW), to achieve measurable results. Further, the paper then applies this same model for use within automated driving systems (ADS). More than just a technological examination, this paper also predicts the expected impact to roadway incidents.

**Keywords:** Connected vehicles · Automated vehicles · Reaction time · Red-light violation · Red-light run · RLVW · Permissive yellow · Restricted yellow · Clearance interval · Advanced warning · Collision avoidance

## 1 Introduction

The goal of any transportation safety application is to decrease the number of crashes. There have been numerous efforts throughout history to produce effective external warning systems such as the flashing lights and gates used in the highway-rail grade

© The Author(s), under exclusive license to Springer Nature Switzerland AG 2022
A. Akhnoukh et al. (Eds.): IRF 2021, SUCI, pp. 269–284, 2022.
https://doi.org/10.1007/978-3-030-79801-7_19

270  G. M. Baumgardner et al.

crossings. Undoubtedly, these have helped many drivers avoid collisions with trains. Likewise, the traffic engineering disciplines that established sophisticated signal timing solutions, such as yellow- and red-clearance intervals, contributed to significant reduction of collisions due to red-light violations. However, many transportation agencies report slower rates of decline than expected (Federal Railway Administration, n.d.), especially given how far technology has advanced inside and around the vehicle. While many promises have been made suggesting that the advent of Connected and Automated Vehicle (CAV) technology will mitigate the majority of incidents by reducing or eliminating all human driver error. However, there still are significant physical and technical constraints before this promise becomes reality. Much of the research in autonomous and robotic technology focuses on the effectiveness and coordination of sensors (Tian, et al. 2007) as well as calculated movements designed to avoid collisions (Urmson 2006). Clearly, the research and development evolving within this discipline is necessary and worthwhile to ensure built-in collision avoidance. However, one tried-and-true method for protecting passengers and the driver is to simply stop the vehicle. This paper explores how the existing policies of dilemma zone management using yellow- and red-clearance intervals at a signalized intersection are used to decrease the number of red-light violations, also known as red-light running (RLR). Moreover, this paper engineers a generally applicable solution for CAV systems to directly increase the number of stops prior to the intersection by sending a directed alert to a vehicle ahead of the signal change.

## 1.1  Problem Statement

Envision the normal operation of a signalized intersection with a programmed permissive yellow-clearance interval (say $Y_P$) design that eliminates the so-called dilemma zone. As a result, drivers may instinctively know whether they should go through or stop during the yellow warning phase. Additionally, the intersection has a programmed all-way red-clearance interval (say $R_t$) that allows for the last vehicles passing through the preceding yellow phase to proceed prior to starting the green phase in the opposite direction. For the sake of argument, let us assume an optimal timing design for all phases based on the intersection characteristics. Yet, despite the efforts of the traffic engineers, RLRs are still prevalent at some intersections. Some of the major contributing factors include distracted driving, speeding, limited line of sight, etc. If there exists some number RLR violations (say $N_{RLR}$) of vehicles attempting to cross during a red-light while traffic is incoming from the adjacent direction, then it can be argued that a successful safety application for this intersection must reduce the amount of violations by some predictable number (say $\Delta_{RLR}$) for it to be useful in this context. Furthermore, if the value of lives saved due to $\Delta_c$ outweighs the implementation and operations costs for the system, then it can be said that the system is commercially viable. Since this reduction $\Delta_c$ is dependent on the initial number of incidents $N_c$ at the intersection, it is more useful to measure a percentage decrease ($\delta$) of RLR instances. That is:

$$\delta = 100\% \times \left(\frac{\Delta_{RLR}}{N_{RLR}}\right), \textit{for any } \Delta_{RLR} \leq N_{RLR} \tag{1}$$

The challenge, then, is to create a supplemental red-light warning system that would produce a predictable $\delta$, especially when considering autonomy. It is self-evident that the introduction of automated driving systems (ADS) is beneficial to reducing RLR since distracted driving, indecision, and human-error would be all but eliminated. However, even a fully automated and bug-free robotic system is subject to the physical constraints of the vehicle, namely the distance required to stop the vehicle under safe braking conditions so not to injure the passengers or lose control. Furthermore, while ADS sensor inputs can quickly detect the transition from a green-light to a yellow-light, there are effective strategies that directly impact the decision to stop or go which do not require this input at all. Careful drivers closely monitor a *stale* green-light in order to prepare for action if and when it turns to yellow. The approach of the Connected Vehicle (CV) red-light violation warning (RLVW) application is a similar approach because the traffic signal controller (TSC) relays timing information to the approaching vehicle which can then determine when the yellow-light transition will occur, and ultimately whether the vehicle should stop or go. With lessons learned from developing CV-based driver warning applications, including RLVW and rail-crossing violation warning (RCVW) (Baumgardner et al. 2020), and using well-established industry research and practices, this paper measures how effective the advanced warning techniques may be in reducing RLR overall and particularly with ADS systems.

## 2 Methods

The conventional method for determining yellow- and red- clearance intervals is based on the work done by Denos Gazis, Robert Herman, and Alexei A. Maradudin in 1960 (Gazis et al. 1960). Their kinematics-based solution argues that a dilemma zone may be eliminated by extending the yellow-light phase to match the *critical distance* it would take to stop prior to entering the intersection. Mats Järlström best illustrates their use of kinematics for intersection clearance in his article on the extended kinematic equation, depicted here in Fig. 1 (Järlström 2020).

The critical distance ($x_c$), which represents the minimum distance required to stop the vehicle, is formulated based on a constant velocity ($v$) and the maximum deceleration value ($a_{max}$) for a vehicle to stop in a safe and comfortable manner:

$$x_c = vt_r + \frac{v^2}{2a_{max}} \tag{2}$$

The reaction, or more appropriately called perception-reaction time, denoted by $t_r$ is the time it takes for the vehicle driver to understand the situation and decide to either apply brakes or continue through the intersection. During this time, the vehicle continues to move toward the intersection at the given velocity, decreasing the available stopping distance. From this basic kinematics study, Gazis et. al. was able to calculate a specific minimum timing model for restrictive yellow lights, meaning the vehicle must stop prior to the change to red:

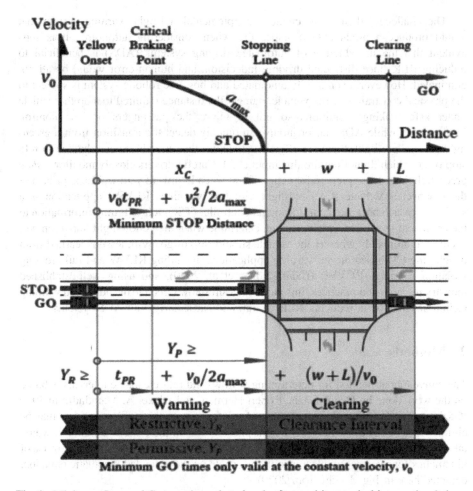

**Fig. 1.** Minimum Stop and Go equations plotted and referenced in a typical intersection [taken from (Järlström 2020)]

$$Y_R \geq \frac{x_c + (w+L)}{v} \quad (3)$$

The $(w+L)$ denotes the fixed width of the intersection plus the length of the vehicle. Substituting the critical distance, a formal restrictive yellow must be:

$$Y_R \geq t_r + \frac{v}{2a_{max}} + \frac{(w+L)}{v} \quad (4)$$

Additionally, many traffic engineers split this model into a permissive yellow-clearance interval and a red-clearance interval in which the vehicle entering at red has the time it takes to cross the entirety of the intersection before other traffic may move:

$$Y_P \geq t_r + \frac{v}{2a_{max}} \tag{5}$$

$$R_t = \frac{(w+L)}{v} \tag{6}$$

## 2.1 Assumptions

The primary assumption made in this paper is that a change in interval timing works to reduce RLR, which in turn impacts the number of collisions at the intersection. The National Cooperative Highway Research Program (NCHRP) in the United States cites numerous studies and concludes that intersections with insufficient clearance intervals can see a reduction of 36 to 50% violations simply by correcting the timing (NCHRP 2012). This utilizes a simplified approach that applies the TSC timing information available over CV communications in order to create a virtual set of clearance intervals preceding the visible signal state change, both based on the kinematic equation, but specific to the approaching vehicle's telemetry. Therefore, the results assume a linear consistency between corrections of physical timings to a virtual counterpart. For purposes of the analysis for both baseline and treatment conditions, a four-legged urban intersection with two lanes of a standard 3.65 $m$ width per each direction was used, plus one 1.8 m wide crosswalk immediately following each stopping line. The speed limit is kept consistent on all four directions.

The kinematic parameters are modeled from previous studies and industry standards. For example, the Policy on Geometric Design of Highways and Streets produced by the American Association of State Highway and Transportation Officials (AASHTO) and more commonly known as the "Greenbook", demonstrates that most drivers decelerate at a rate exceeding 4.5 m/s$^2$ when confronted with the need to stop for an unexpected object in the roadway (AASHTO 2018). Additionally, 90% of all drivers decelerate at rates greater than 3.4 m/s$^2$, which is within the driver's ability to stay within the current lane and maintain steering control during the braking maneuver, even on wet surfaces (AASHTO 2018). Therefore, this comfortable deceleration threshold is recommended for determining stopping sight distance (AASHTO 2018) (Federal Highway Administration 2009). Implicit in this choice is the assessment that most vehicle braking systems and the tire-pavement friction levels on most roadways can support at least 3.4 m/s$^2$ deceleration.

Previous studies suggest a typical reaction time of 0.6 to 1.7 s for an expected braking event (Johansson and Rumar 1971) (Massachusetts Institute of Technology 1935) (Normann 1953). An additional second may be taken for unexpected braking events. Without prior knowledge of the signal timing, reacting to a signal state change would be considered an unexpected braking event. The general consensus, outlined by the Greenbook, suggests a reaction time of 2.5 s to be sufficient for most braking events (AASHTO 2018). Both recent research (NCHRP 1997) and studies documented in the literature show that this braking reaction time for stopping sight situations encompasses the capabilities of most drivers, including older drivers. In fact, this recommended

274 G. M. Baumgardner et al.

design criteria for 2.5 s reaction time exceeds the 90th percentile of all drivers and was used for modeling stopping sight distances in the Greenbook (AASHTO 2018).

## 2.2 Data Sources and Data Cleaning

Much of the traffic crash data obtained for predictive purposes comes from industry-wide standards and practices. However, in order to model the effectiveness of the proposed virtual clearance interval, one must consider the effectiveness of existing clearance times on RLR. To study this, this paper uses data from a simulator study that investigated the effect of wireless telephone use on driving performance (The Human Factors and Statistical Modeling Lab 2013). The study was conducted at the University of Iowa National Advanced Driving Simulator (NADS). During the study, drivers drove through three signalized intersections while engaged in one of three secondary task conditions – using a handheld device, using a headset device, and using a handsfree device for receiving and dialing calls that were initiated before the driver arrives at the intersection. The scenario was designed such that when drivers approached a signalized intersection, they are presented with a dilemma of whether to stop or go. During the approach to the signalized intersection, the traffic signal changes from green to yellow at one of two pre-determined timings. One of these timings is at 3.00 s which is expected to invoke a "go" response from the driver while the second is 3.75 s and is expected to elicit a "stop" response. However, these timed triggers are not exact and fluctuated around these values. The yellow light then remains yellow for 4 s, then transitions to red for 5 s before turning into a green light again. All intersection segments were equivalent with similar ambient traffic at each event. The data was collected at 240 Hz i.e., 240 frames per second.

A total of 1157 observations and 17 variables were collected from the study. Each observation describes the approach to the signalized intersection. Prior to analyzing the data, data from practice runs were removed. Next, there were several incomplete segments during a given test run and these were also removed from the data. Following this, some contradictions where the time frame value shows that the driver made a stop beyond the stop line, but the distance indicates otherwise were also removed. There were also cases where the yellow signal time was either less than 3.9 s or over 4 s thus making shorter or longer yellow signal times. The observations with shorter or longer signals were also eliminated from the data. Thus, the final dataset consisted of 671 observations (Rahman et al. 2018). Only a subset of the 17 variables available were used for the data processing and analysis presented in this paper. These variables are shown in Table 1.

# 3 RLR Reduction

Generally speaking, the more data or information that is available, the greater the safety improvement in automotive scenarios. One example is using a flashing warning on a static "traffic light ahead" sign that provides advanced notification of the prospective signal state ahead of the visible indication. These prove to be valuable input into driver decision making. However, the problem with fixed signage is the possibility that

## Table 1. Variables from the simulator study

| Variable | Description |
| --- | --- |
| 1. Run name | Indicates test run number, cell phone test scenario, treatment conditions, and order of treatment condition |
| 2. Green to yellow | Number of frames indicating when the traffic signal changes from green to yellow |
| 3. Yellow to red | Number of frames indicating when the traffic signal changes from yellow to red |
| 4. Red to green | Number of frames indicating when the traffic signal changes from red to green |
| 5. First stop frame | Number of frames indicating when vehicle first stops or when the vehicle velocity goes to zero |
| 6. Distance from stop line | Distance from the stop line when the velocity is zero |
| 7. Velocity at green to yellow | Vehicle's velocity when the light turns from green to yellow |
| 8. Frame at stop line | Number of frames indicating when the participant reaches the stop line |

drivers may be unaware of the sign due to other distractions, or the view of the sign may be obstructed by passing motorists. One benefit of CV technology is the ability to bring dynamic signage inside a vehicle and fine-tune it to the specific driving environment in which the vehicle is operating. This would include, for example, the approach velocity ($v$), the width of the intersection ($w$) and the length of the vehicle ($L$), all of which are needed to know the appropriate red-clearance interval, and all of which would be available within a proper CV application.

A CV system relies on information sharing via a wireless communication channel. By sending a signal, phase, and timing (SPaT) message from the TSC, along with a basic map geometry identifying lane positions, and an accurate position of the vehicle, the system can determine a precise distance to the end of the approach lane, plus the remaining time that the light will remain green. All things being equal, a computing system should then know if the vehicle would be able to stop prior to the stop line, or if the vehicle would make it through the intersection during a properly timed yellow- or red-clearance interval, thus removing any notion of a dilemma zone. As long as the vehicle's location remains outside the critical distance when the green phase ends, then the system assumes the driver will stop in time and no additional alerting is required. Likewise, if the vehicle will be in the safe stopping distance threshold when the light changes, then the driver would normally decide to proceed through the clearance intervals, and therefore no alert is necessary. However, if the system predicts that the vehicle will be in the critical distance when the yellow phase begins, then it will issue an early warning to the driver to slow down.

For a manually driven automobile, a perfectly timed signal change may be broken down into three distinct distances, as depicted in Fig. 2.

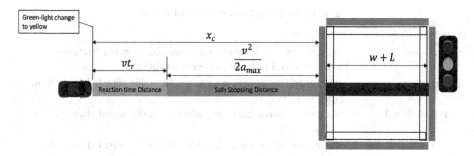

**Fig. 2.** The perfectly timed signal change

One important assumption is that the same vehicle approaching the same intersection at the same velocity will create the same critical distance ($x_c$) and clearing distance ($w + L$). Therefore, if the timing of the situation is consistently perfect, then the resulting number of RLR incidents will remain unchanged. That is to say, the percentage change ($\delta$) is zero. This seems like an obvious statement but also proves to be the key design consideration as it induces a positive $\delta$ under these conditions is by increasing the amount of stopping distance between the vehicle and the stop line. Further, the only way this can be accomplished is for the RLVW application to generate an alert ahead of the signal change in order to induce the intended braking reaction earlier. This advanced warning distance (AWD) constitutes a temporal shift of the necessary critical distance required in order to incur a portion of the perception-reaction time ahead of the signal change. The AWD, then, translates into some cushion distance near the stop line, allowing for more area to brake, as depicted in Fig. 3.

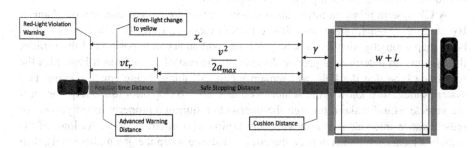

**Fig. 3.** Ideal state using RLVW advanced warning

The ideal case shown here assumes that the driver's reaction to the alert is consistent with their reaction to the signal change, or at least within the modeled 2.5 s. A well-designed user interface that only alerts when the red-light violation would occur is necessary to ensure a consistent reaction. However, given the typical reaction times are well under the model assumption, it is fairly safe to assume the same 2.5 s perception-reaction time. Additionally, the ideal case would be that the driver also decelerates at the expected safe rate of 3.4 m/s$^2$, as in the baseline condition. Under these ideal circumstances, the cushion distance ($\gamma$) would be precisely equal to the advanced warning distance, and the net effect of the advance warning should be similar, if not the same, as applying the permissive yellow-clearance interval, thus leading to approximately 36 to 50% less RLR events. However, this is clearly an upper-bound since the driver may recognize that the distance to the stop line is sufficient to warrant a more controlled braking, and therefore reduce the deceleration rate. One mitigation technique for ensuring that the vehicle is slowing fast enough to ensure a positive $\gamma$ is continually calculating $x_c$ as the vehicle approaches the intersection, using up-to-date velocity and deceleration, and issuing the RLVW accordingly. Only a measured deceleration action that would produce a safe stopping distance less than the remaining distance to the stop bar should extinguish the alert.

## 3.1 Determining an Appropriate Advanced Warning

The effectiveness of advanced warning will depend on whether the alert issued will elicit the same reaction as the signal change. In other words, the driver must consider the issue urgent enough to initiate braking within the 2.5 s perception-reaction time. Alerting too far ahead would serve only to create a nuisance to the driver, who would ultimately learn to ignore the RLVW. Using multiple alerts with escalated urgencies proves useful in eliminating distractions while keeping critical warnings to the minimum required (Campbell et al. 2007). Therefore, an appropriate advanced warning distance would be based on actual driving models. For example, the Greenbook suggests 0.5 s for modeling reaction times of modern control systems (AASHTO 2018). Based on this suggestion, an RLVW alert sent 2 s ahead of the signal change should ensure that the nominal driver would have a total of 2.5 s to react, including 2 s prior to the signal change and 0.5 s after. Another way to look at this would be that issuing advanced warning 2 s ahead of the signal change for a driver with a worst-case reaction time of 2.5 s effectively turns that driver into a best-case 0.5 s reaction. Therefore, the total reaction time consists of the sum of both the advanced warning time (say $t_\alpha$) plus the post-yellow effective reaction time (say $t_\beta$). For human drivers, the sum of these must always be the nominal case of 2.5 s.

$$t_r = t_\alpha + t_\beta = 2.5 \tag{7}$$

Recall from Fig. 3 that the cushion distance ($\gamma$) is bounded by this AWD ($vt_\alpha$) since the alert is issued in advance and while the velocity is constant. Thus, the cushion distance can be formulated as:

$$\gamma \leq vt_\alpha \tag{8}$$

Because the advanced warning time is suggested to be 2 s ahead based on the 90th percentile human reaction, the cushion distance will always be near double the vehicle speed given in $m/s$. This is likely too far in advance and causes $\gamma$ to be bigger than the width of the intersection at most speeds, which could skew the experimental results. This paper, therefore, also considers backing off the advanced warning time in increments of 0.5 s. However, as the time taken to react to the advanced warning approaches

278     G. M. Baumgardner et al.

the nominal 2.5 s, the resultant cushion distance approaches zero, which is to be expected. Table 2 shows the variability in the cushion distance based on the sample of advanced warning times.

**Table 2.** Maximum cushion distance for 15 m/s

| | Advanced warning time (s) | Nominal human reaction time (s) | Post-yellow effective reaction time (s) | Velocity (m/s) | Cushion distance (m) |
|---|---|---|---|---|---|
| | $t_\alpha$ | $t_r$ | $t_\beta$ | $v$ | $\gamma$ |
| No advanced warning | 0.0 | 2.5 | 2.5 | 15 | 0 |
| With advanced warning | 0.5 | 2.5 | 2.0 | 15 | 7.5 |
| | 1.0 | 2.5 | 1.5 | 15 | 15.0 |
| | 1.5 | 2.5 | 1.0 | 15 | 22.5 |
| | 2.0 | 2.5 | 0.5 | 15 | 30.0 |

If the nominal case produces some number of RLR instances across a given critical distance, then it can be assumed that a proportional change in the number of instances would follow a proportional change in the actual distance.

$$\delta \propto \frac{\gamma}{x_c} \tag{9}$$

Substituting the upper bound for $\gamma$ found in Eq. (8), the ideal model for the reduction in RLR instances using advanced RLVW can be derived.

$$\delta \leq \frac{t_\alpha}{Y_p} \tag{10}$$

Equation (10) shows that the bigger the advanced warning time, the more impactful RLVW would be to the number of RLR events. Likewise, as the permissive yellow-clearance time increases, it is likely to reduce the impact of RLVW. The modeled $Y_P$ comes from nominal reaction time, which is 2.5 s, plus the time required for the safe stopping distance, which increases in direct proportion to velocity or as deceleration reduces. Fixing the deceleration to the nominal 3.4 m/s$^2$ yields some predictable upper bounds for $\delta$, based on sampled approach speeds (v) from the data set, and a few different modeled advanced warning times ($t_\alpha$). As shown in Fig. 4, the absolute best-case scenario for reduction of incidents is similar to the typical reduction of RLR following a properly timed $Y_P$, which is around 35 to 50%, depending on the approach speed.

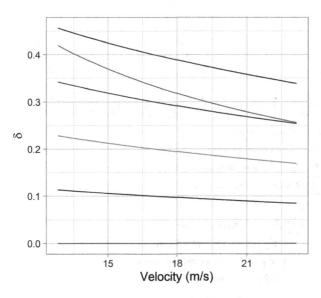

**Fig. 4.** Expected % reduction in RLR ($\delta$) by velocity

## 3.2 Sources of Error

Unfortunately, everything comes at a cost. To this point, while using a CV-based RLVW application makes the decision to stop or go as pre-determined as possible, the development of the early warning system introduces new sources of error:

1. The access time to retrieve the SPaT information ($t_S$)
2. The communication time to pass the SPaT to the vehicle ($t_C$)
3. The time to obtain a valid geo-location of the vehicle ($t_G$)
4. The processing time to calculate the distance and timing ($t_P$)
5. The positional error due to inaccuracies in the geo-location equip ($E_P$)
6. The vehicle length uncertainty ($E_L$)
7. The antenna placement uncertainty ($E_A$)
8. The vehicle speed uncertainty ($E_V$)

The first four of these sources are errors measured as latency of a task which, like reaction time, incites a distance error that varies based on the vehicle speed. The remaining four sources of error are fixed measurements. As shown in Fig. 5, the net effect of all of these error sources increases the amount of distance needed for the advanced warning.

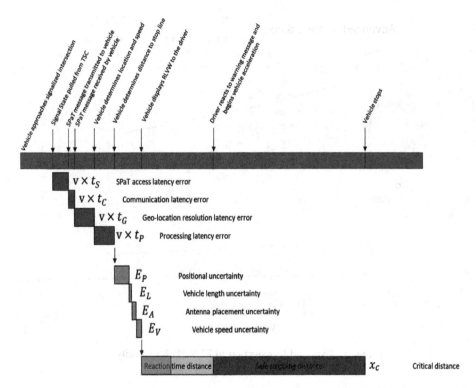

**Fig. 5.** The net effect of error sources on the advanced warning

By definition of the RLVW application, SPaT messages must be sent out 10 times per second (10 Hz), or once every 100 ms. This time would encompass both $t_S$ and $t_C$. Existing economical positioning solutions are also capable of reporting position at 10 Hz or higher. A multi-core processing unit would be able to simultaneously receive the SPaT and position information and make the decision to stop or go based on those inputs. Further, if each core operates at 1 GHz, then that processing can do tens of millions of instructions within 10 ms. Therefore, it can be assumed that the total latency errors take up to a total of 210 ms. Whereas length measurements for the vehicle and antenna positions may be at worst 10 cm in error, the positional uncertainty for the solution itself may be upwards of a meter for the most economical augmented system, although many generally perform better. Likewise, testing uncovers up to another meter of error due to speed uncertainty. Therefore, the total static error may be estimated at around 2.02 m.

Considering these sources of error as a shortening of $\gamma$, a revamped version of Eq. (10) can be generated.

$$\delta \leq \frac{(2.5 - t_\beta - t_S - t_C - t_G - t_P)}{Y_p} - \frac{(E_P + E_L + E_A + E_V)}{vY_p} \qquad (11)$$

This produces a much less idealistic and more converging model for $\delta$, in which the well-designed advanced warning alert should lead to around 36% reduction of RLR, depending on the speed (see Fig. 6).

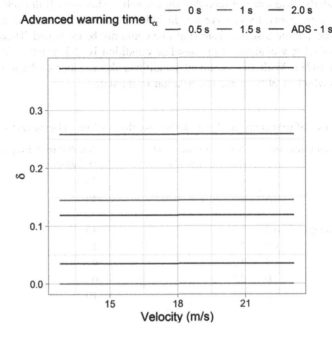

**Fig. 6.** Expected % reduction in RLR ($\delta$) by velocity including the error sources

### 3.3 Number of Reductions in RLRs

Recall that one of the aims of this paper is to investigate whether a red-light warning system would produce a predictable number of reductions in RLR violations i.e., $\delta_c$ from Eq. (1). This is investigated using one baseline condition with no advanced warning and two treatment conditions consisting of scenarios with an advanced warning in a manually driven vehicle and one with an ADS (see Table 3). First, the number of red-light violations is extracted from the simulator study data presented in Sect. 2.2. From the data, drivers who stopped beyond the stop line when the light turned red and drivers who did not stop for the red light were considered to have committed RLR violations (Jahangiri et al. 2016).

To determine the reduction in the number of RLR violations, first, the cushion distance, $\gamma$ from Eq. (8) is estimated based on the advanced warning treatment conditions where the advanced warning ranges from 0.5–2.0 s. Note that the 49 RLR violations in the baseline where the vehicle stopped beyond the stop line has a measure of how far from the stop line the vehicle came to a complete stop (variable 6 from Table 1). A comparison of the cushion distance and the distance from the stop line can indicate if the vehicle could have stopped ahead of the stop line given an advanced

282 G. M. Baumgardner et al.

warning thereby avoiding a RLR violation. As shown in Table 3, a reduction in the number of RLR violation is observed from all treatment conditions where a 1.5 s advanced warning eliminates the number of RLR violations that stopped beyond the stop line. Note that, this does not indicate a complete elimination of RLR violations. In the simulator study, data was not recorded for the vehicle behavior if the vehicle failed to completely stop. Thus, these were not included because a comparison between a cushion distance and the distance from stop line could not be estimated. Therefore, the total number of RLR violations in the baseline condition is 255 which is 38% of the observed data (671). With the treatment conditions the number of RLR violations reduce to 206 which is 30% of the observation in the data.

**Table 3.** Number of reductions in RLR violations for the baseline and treatment conditions

| Advanced warning time (s) | | Nominal perception-reaction time (s) | Number of red-light running violations |
|---|---|---|---|
| $t_\alpha$ | | $t_r$ | $N_{RLR}$ |
| No advanced warning | 2.5 | 2.5 | 49 |
| Advanced warning condition | 0.5 | 2.5 | 10 |
| | 1.0 | 2.5 | 1 |
| | 1.5 | 2.5 | 0 |
| | 2.0 | 2.5 | 0 |
| ADS condition | 0.5 | 0.5 | 10 |

In Table 3, the second treatment condition is that of an ADS vehicle. The difference in the ADS case is that human indecision is eliminated. In other words, the amount of time it takes for the ADS to respond to a yellow light (i.e., $t_\beta$) is bound only by the time it takes for the control and perception systems to read the situation and engage the brakes. Studies conducted by Khoury et al. suggested that the modern perception and control systems used by an ADS can reduce the reaction time from 2.5 s to 0.5 s (Khoury et al. 2019). Note that 0.5 s is equivalent to the best-case post-yellow effective reaction time. In addition, an ADS system can also be tuned to always brake at a prescribed $a_{max}$. Thus, the reduction in the RLR cases for the ADS case are just as effective as the advanced warning condition but are achieved at a lower perception-reaction time.

## 4  Conclusions

Historically, RLR violations create safety challenges for motorized and non-motorized travelers at signalized intersections in urban and rural areas. This paper explored the impact of a CV safety application implemented in conjunction with the traditional traffic signal timing practices in reducing RLR violations at signalized intersections. The proposed CV safety application generates an advanced signal change warning to the drivers. Study findings indicated that a 1 s advanced warning could significantly reduce RLR violations by creating additional cushion distance at the end of the stopping maneuver. Future research should include formal simulation testing in attempt to validate the predictions made in this paper and to refine the advanced warning time for the most common approach scenarios.

An ADS-equipped vehicle utilizes sensors to help the system make decisions. While sometimes forgotten or glanced over, adding CV technologies as additional input for the autonomous system enhances its ability. This is critically important for cases such as the signal state change problem because visual and electromagnetic sensors are incapable of predicting when a green light would be turning yellow. Therefore, RLVW is still a very practical solution for vehicle autonomy. The analysis findings indicated that a half second advanced warning could be equally effective in lowering RLR cases similar to the human drivers' scenario but with a much more effective perception-reaction time.

Not factoring in the cost of the ADS itself, the cost of adding CV technology to infrastructure comes in at a rough order of magnitude of around ten thousand U.S. dollars for each signalized intersection. The in-vehicle radio and processing technology would be at least one quarter of that cost, but with the potential volume in the millions. Therefore, a fair estimate of the total cost of applying RLVW to ADS is in the three billion to five billion U.S. dollar range. Considering governmental infrastructure budgets are significantly higher on a yearly basis, and the continual potential savings on damages and loss of life, deploying RLVW seems to be well worth the time and money.

**Acknowledgements.** I thank my partner authors for all the time and effort they put into this study. In particular, I express my deepest thanks for the great research done by Varshini. We thank her graduate school, the University of Wisconsin, for training her well, and our colleagues, especially Dr. Alex Noble and Martha Eddy, who made this work possible.

# References

American Association of State Highway and Transportation Officials: A Policy on Geometric Design of Highways and Streets. American Association of State Highway and Transportation Officials, Washington, D.C. (2018)

Baumgardner, G., Hoekstra-Atwood, L., Prendez, D.M.: Concepts of connected vehicle applications: interface lessons learned from a rail crossing implementation. In: AutomotiveUI '20: 12th International Conference on Automotive User Interfaces and Interactive Vehicular Applications, pp. 280–290. ACM (2020)

Campbell, J.L., Richard, C., Brown, J.L., McCallum, M.: Crash Warning System Interfaces: Human Factors Insights and Lessons Learned. National Highway Traffic Safety Administration, Washington, D.C. (2007)

Federal Highway Administration: Manual on Uniform Traffic Control Devices for Streets and Highways. Federal Highway Administration, Washington, D.C. (2009)

Federal Rail Administration: 3.01 - Accident Trends - Summary Statistics. Retrieved from Office of Safety Analysis. https://safetydata.fra.dot.gov/OfficeofSafety/publicsite/summary.aspx

Federal Railway Administration: 1.12 - Ten Year Accident/Incident Overview. Retrieved from Office of Safety Analysis. https://safetydata.fra.dot.gov/OfficeofSafety/publicsite/summary. aspx

Gazis, D., Herman, R., Maradudin, A.A.: The problem of the amber signal light in traffic flow. Oper. Res. **8**, 112–132 (1960)

Jahangiri, A., Rakha, H., Dingus, T.A.: Red-light running violation prediction using observation and simulator data. Accid. Anal. Prev. 316–328 (2016)

Järlström, M.: An Extended Kinematic Equation, Beaverton, Oregon, USA (2020)

Johansson, G., Rumar, K.: Driver's brake reaction times. Human Factors **13**(1), 23–27 (1971)

Khoury, J., Amine, K., Abi Saad, R.:, An initial investigation of the effects of a fully automated vehicle fleet on geometric design. J. Adv. Transp. 1–10 (2019)

Massachusetts Institute of Technology: Report of the Massachusetts Highway Accident Survey. CWA and ERA Project, Cambridge, MA (1935)

National Cooperative Highway Research Program: NCHRP Report 400: Determination of Stopping Sight Distance. National Academy of Sciences, Transportation Research Board, Washington, D.C. (1997)

National Cooperative Highway Research Program: NCHRP Report 731: Guidelines for Timing Yellow and All-Red Intervals at Signalized Intersections. National Academy of Sciences, Transportation Research Board, Washington, D.C. (2012)

Normann, O.: Braking distances of vehicles from high speeds. In: Proceedings of the Thirty-Second Annual Meeting of the Highway Research Board, pp. 421–436. Highway Research Board, Washington, D.C. (1953)

Rahman, Z., et al.: Evaluation of cell phone induced driver behavior at a type II dilemma zone. Cogent Eng. **5**(1) (2018)

The Human Factors and Statistical Modeling Lab: 2014 TRB Data Contest. Seattle, WA (2013). https://depts.washington.edu/hfsm/upload.php. Accessed 13 July (2021)

Tian, J., Gao, M., Lu, E.: Dynamic collision avoidance path planning for mobile robot based on multi-sensor data fusion by support vector machine. In: 2007 International Conference on Mechatronics and Automation, pp. 2779–2783. IEEE, Harbin (2007)

Urmson, C.: Driving beyond stopping distance constraints. In: 2006 IEEE/RSJ International Conference on Intelligent Robots and Systems, pp. 1189–1194. IEEE, Beijing (2006)

# Traffic Planning and Forecasting - 1

# A Case Study of Rural Freight Transport – Two Regions in North Carolina

Daniel J. Findley[1]([⊠]), Steven A. Bert[1], George List[2], Peter Coclanis[3], and Dana Magliola[4]

[1] Institute for Transportation Research and Education, North Carolina State University, Raleigh, NC, USA
daniel_findley@ncsu.edu
[2] Department of Civil, Construction, and Environmental Engineering, North Carolina State University, Raleigh, NC, USA
[3] University of North Carolina at Chapel Hill, Chapel Hill, NC, USA
[4] North Carolina Department of Transportation, Raleigh, NC, USA

**Abstract.** Rural and urban transportation needs are known to be different. Our objective was to understand how infrastructure investments could help rural areas of North Carolina with economic development. For many planners and economic developers, freight growth is seen as the physical manifestation of a strong economy but planning for freight is challenging. For us to understand the rural needs better, we conducted a socio-economic study, an economic analysis, and two workshops. The workshops focused on two areas of the state, the deep southwest and the far northeast. These regions are among the most economically depressed and economic development through enhanced freight infrastructure could be very helpful.

Due to the mountainous terrain in the deep southwest, the resiliency and reliability of the state's transportation system is critical. Even more important than building new capacity, is ensuring the functionality of existing highway, rail, and aviation assets. The region does not act as one unit, but instead as a collection of many microeconomies that transcend county and state boundaries. The proximity of the far northeast to Norfolk and the eastern seaboard provides a comparative advantage for marine industries. The area is well-positioned to capitalize on economic growth related to a nearby metropolitan area in Virginia. Highway, waterway, and other transportation networks that connect the region's business and population centers to the Norfolk region can facilitate economic growth.

**Keywords:** Freight · Rural · Investment · Economic development · Transportation · Infrastructure

## 1 Introduction

This paper describes a project that studied the freight transportation needs of rural North Carolina, which are very different from those of the urban areas. The effort included a socio-economic study, an economic input-output analysis, and two major workshops, one held in Sylva, NC in the deep southwest and another in Hertford, NC

© The Author(s), under exclusive license to Springer Nature Switzerland AG 2022
A. Akhnoukh et al. (Eds.): IRF 2021, SUCI, pp. 287–302, 2022.
https://doi.org/10.1007/978-3-030-79801-7_20

in the far northeast. The objective was to understand the rural transportation needs using these two areas as case studies with an aim to help understand how to couple infrastructure investments with economic development for rural areas.

The effort built on two prior studies. One was a report prepared by the Global Research Institute (UNC GRI 2011). It focused on the socio-political-economic trends and the role played by educational institutions in facilitating economic growth. The other was the Seven Portals Study (and implicitly, its antecedent, the Statewide Logistics Plan) (List and Foyle 2011; List et al. 2008). This second study identified ways that regional economic growth could be aided by infrastructure investments and many of its rural recommendations were implemented (e.g., improvements to the Western Carolina Regional Airport and highway network improvements in the north eastern region). Arguably, the centerpieces of the current project were the workshops held in the southwest and northeast. They elicited input about infrastructure needs. Follow-up communications and web-meetings ensued. The inputs received served as the backbone for our findings.

Periodically updating perceptions about freight transport needs is important. The state has limited resources for making infrastructure (transportation-related) investments. So, choosing the right investments in the right places and times is critical. As the Seven Portals study found (List and Foyle 2011), infrastructure investments can be a significant catalyst in the state's economic development; acting as supply push; although, to be effective, such investments must be coordinated with demand pull from the business sector. The "demand pull" is the interest in growth espoused by companies and people that are already located in or might be encouraged to choose these areas. The "supply push" is investment in transportation and possibly other forms of infrastructure (such as information technology) to help reduce transport costs, improve accessibility, and enhance those economic activities.

## 2 Prior Studies

Many previous research efforts have aimed to address the issue of tying infrastructure investment to economic development. Although a lack of consistency exists in the definition of "freight-intensive" or "freight-dependent" industries (Shin et al. 2015n), many industries make heavy use of freight transport. Examples include agriculture, manufacturing, retail, forestry, construction, activities related to energy extraction and mining, as well as transportation (WisDOT n.d.). Many of these industries are prevalent in rural areas. Four economic sectors - services (37%), government (16%), retail and wholesale trade (14%), and manufacturing (11%) - constitute 80% of rural employment (USDOT and USDA 2008). Other rural industries are transport dependent even though they may not be freight-intensive or freight- dependent. One example is the tourist industry, which is heavily dependent upon delivery of supplies. Another is the production of vaccines. It is critically dependent on the availability of transport services with global reach so that medicine can be delivered to critical locations in a timely manner.

Finding more effective, efficient solutions to rural America's transportation needs is an ongoing process that requires the hard work of researchers, elected and appointed policy makers, business leaders, and non-profit advocacy groups (Kidder 2006). Local needs assessments, such as those undertaken by the NC Rural Center, the Southwestern Commission, and the NC East Alliance, shed light on rural infrastructure priorities and include information about local economic conditions and activities that can be supported by transportation infrastructure. These assessments provide a basis from which solutions can be uniquely tailored to specific localities, involving substantial stakeholder input; however, funding for needs assessments is often limited.

It is vital to North Carolina's economy to ensure adequate transportation infrastructure is in place that can facilitate the movement of interstate freight and cargo. In 2015, approximately $765 billion of cargo, weighing nearly 430 million tons, was transported using North Carolina highways (Cambridge Systematics 2017). Rural interstates are major freight corridors, and maintenance of those network links is critical to the overall economic health of the state. There is a nuance here which is important to note. Many state-level "rural transport needs" studies have focused on inter-urban infrastructure investments needed to facilitate transport between urban centers. While important, these studies contain little insight into the transport needs or dynamics of freight movement for the rural areas themselves.

Since the Great Recession of 2007–2009, while urban areas have seen strong gains and rebounds, rural areas have continued to decline (Khanna 2016). One of the central issues is a lack of interstate cooperation. Often, states compete with one another, investing in redundant infrastructure projects to the detriment of rural communities (Khanna 2016). The United States has 11 megaregions with economic activities that transcend state lines (Rockefeller Foundation 2008). Being able to organize investments around economic corridors, instead of competing based on state boundaries, would better suit the needs of rural communities.

# 3 Socioeconomic Attributes of the Study Areas

Understanding the socioeconomic characteristics of the study areas is important in formulating infrastructure investment ideas. We needed to understand the status of these regions, their characteristics, and their business climates. This section summarizes the findings. The investigation placed a special emphasis on the northeastern and southwestern portions of the state. As might be expected, we found that these regions are challenged, economically, and they are heavily dependent upon small, private sector businesses. Unemployment is high and earnings are low relative to averages for the state. Poverty is more common than for the state overall. The statistics cited are mostly from years since 2010, but there is no reason to believe the implications are not still valid.

The deep southwest, sometimes called the Western Prosperity Zone, is one of the eight statewide planning regions established by the N.C. General Assembly. People of European descent migrated into this 13-county region throughout the 19th century. In 2017, an estimated 727,000 people—or seven of every 100 North Carolinians—lived here. The region has historically had an economy dominated by small farmers and merchants. The region's output in 2015 was $23 billion. This means the region generated $5 of every $100 in statewide economic activity. In 2016, the region had some

18,000 private businesses with employees. Those firms accounted for 8% of the state's businesses. The region also had 64,000 single-person establishments. Of all the businesses, 8% were owned by persons of color in 2012. In contrast, persons of color accounted for 15% of the population. The region was home to, on average, 7% of the state's civilian labor force from 2013–17; of those people, 5%, on average, were unemployed. Some 83% of regional employees work in the broad service sector. The largest concentration of employees is in the health care and social assistance super-sector (47,000), followed by the retail trade (39,000), accommodation and food services (37,000), and manufacturing (27,000) super-sectors. The typical working person (age 25+) residing in the area had, on average, annual labor earnings of $30,700 from 2013–17, an amount 12% less than the statewide figure. On average, 15% of the region's population lived in a household with an income below the federal poverty level, with another 22% had an income no greater than twice that level. In all, 37% of all residents were poor or near poor.

The far northeast, sometimes called the Northeast Prosperity Zone, is a largely rural and sparsely populated, 17-county region. As of 2017, an estimated 541,000 people— or five percent of North Carolinians—lived here. Once one of the more prosperous areas of the state, today the region contributes modestly to the state's economy. Noteworthy trends include the following: It is economically depressed. On average, 21% of the region's population lives in households with incomes below the federal poverty level, 22% live in households with incomes no greater than twice that level; and 43% are poor or near poor. Moreover, it has high unemployment. From 2013–2107, 9% of the area's civilian labor force was unemployed. Thirteen counties had unemployment rates above the statewide rate. The labor force also has low earnings. In the period 2013-17, the typical working person (age 25+) earned $31,500, 10% less than the statewide average. The largest concentration of employees is in the health care and social assistance super-sector (30,934), followed by the retail trade (25,577), accommodation and food services (21,358), and manufacturing (16,977) super-sectors. Finally, the area is experiencing deep poverty. An average of 9% of the area's residents —some 48,000 individuals in all—live in households with incomes no greater than half the poverty level.

## 4 Findings

The study's findings were heavily derived from workshops we conducted with local business leaders, economic development professionals, and transportation planners. We engaged in discussions about the current status of economic activity, opportunities for economic growth for existing industries, and new industries that might be developed. They both produced thoughtful ideas about how the North Carolina Department of Transportation (NCDOT) could help foster economic growth in their region. The ideas shared by the workshop participants helped shape the suggestions for infrastructure investments. Those findings are presented here.

### 4.1 The Deep Southwest

The Appalachian Mountains cover North Carolina's southwestern region, offering an intricate network of springs, streams, waterfalls, rivers, and points of high elevation. Certain areas of the Blue Ridge or the Great Smoky Mountains receive up to 90 inches per year of rainfall, outpaced only by the Pacific Northwest (Dykeman et al. 2019). In this region, sudden rainfall brings rapid rises in stream water, which often result in destructive floods and debris flows (Dykeman et al. 2019). Storms in the region that trigger hundreds of debris flows occur about every nine years and those that generate thousands occur about every 25 years (Wooten et al. 2016). In February 2019, for example, landslides closed four of the major arteries in North Carolina's southwestern region, including I-40 in both directions (Marusak and Price, 2019). Additionally, in August 2019, more than 7,600 tons of soil, rock, and tree debris, caused the US Forest Service to close the Nantahala River and the North Carolina Department of Transportation to shut down U.S. 19/74 near Bryson City.

Due to the terrain in the region, the resiliency, reliability, and robustness of the transportation system is critical. The system is tested during times of landslides and construction, when multi-hour detours may result from service disruptions, particularly for freight trucks which may not be able to travel on the same alternative routes as passenger vehicles. According to one participant's observation, there may only be one grocery store in the community with only one convenient route for access. Any disruption in this primary route can lead to substantial delay for product delivery due to a long truck detour.

Over the course of the workshop, the concept of regional identity was also discussed. Workshop attendees were quick to demonstrate that the region does not act as one unit, but instead as a collection of many micro-economies. However, when needed and mutually beneficial, these counties can and do coordinate to improve conditions, increase quality of life, and enhance economic opportunities. For instance, two counties have a mutual aid agreement in which they share resources to assist one another in natural or man-made disasters, including traffic crashes that occur in one county but are served by the other. This collaborative approach mirrors the administrative practices of the NCDOT in the region. Attendees discussed how highway networks influence their commuting patterns. They explained that highway connections between states enabled those living in the region to reach employment locations in other states as well as connecting out-of-state residents to employment centers in North Carolina. Transportation infrastructure planning and investment must take these dynamics into account.

In addition to their similarities, workshop attendees helped demonstrate unique business characteristics found within counties in the region, which included some counties focusing on heavy industry, another in healthcare and services, in addition to several niche businesses that are unique to specific counties - a tax preparation software provider, which is the second largest employer in the county, a construction company, and a university. Additionally, Harrah's Cherokee Valley River Casino was widely discussed as an economic engine for the region. "Human freight is our biggest market," one workshop attendee quipped. "We bus people in from all over to visit the casino." Harrah's Cherokee Casinos received approximately 5.6 million guests in 2018 with 4.4

million guests at Harrah's Cherokee Casino Resort in Jackson County and 1.2 million guests in Harrah's Cherokee Valley River Casino & Hotel in Cherokee County. In addition to Harrah's, several eco-tourism attractors exist in the region including: the Appalachian Mountains, the Nantahala River, Great Smoky Mountain National Park, Pisgah National Forest, among others. Broadband internet access was also discussed during the workshop. With the rise of shared workspaces, short-term transient office rentals, and telecommuting, improved internet access would undoubtedly benefit all industries in the region.

The map in Fig. 1 shows workforce linkages in southwestern North Carolina. Blue arrows indicate a "secondary workflow" or the highest level of workforce commuting for residents of a given county (second only to commuting within the county itself). Tertiary and quaternary workflows indicate the third and fourth highest levels of workforce commuting for residents of a given county. The numbers inside parentheses indicate total county employment. The other numbers indicate the number of workers who are employed in another county.

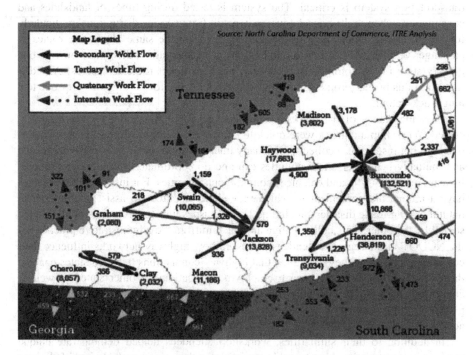

**Fig. 1.** Southwestern North Carolina County Employment Linkages

During the workshop, attendees discussed business opportunities that could benefit the region, citing destination retirement, tourism, and other industries. Many of the counties in southwestern NC have increasing population rates for individuals in their retirement years (60+) and preparing for that could provide an economic opportunity.

Attendees also discussed the importance of securing the region's existing transportation network, rather than focusing on new capacity. They emphasized NCDOT's role in highway maintenance, citing landslides and other events. Attendees also discussed the importance of viewing the individual roads in that region as part of a collective system that needed to be preserved. In other words, a roadway alteration in one community could impact the entire regional network (for better or worse).

Due to the terrain in the region, the resiliency, reliability, and robustness of the transportation system is critical. The transportation system is also tested with the addition of new traffic signals. Though new signals may benefit a local community, they can create travel time slow-downs that negatively impact the region. Workshop attendees voiced a shared concern about maintaining reasonable travel times. They specifically discussed corridor conflict points in the context of traffic signals. According to those in attendance, there are currently no effective regional measures that can be taken to consider and evaluate the regional impact of traffic signal installations or other access-management concerns. Attendees explained the special importance of land use decisions in the mountains due to severe topographic constraints to accessibility and transportation choice. Attendees explained that implementing state-level policies to strengthen access-management in the region (i.e. minimizing or managing the number of conflict points that exist along a corridor) would be invaluable to protecting the area's economic stability.

Reliable travel times are important to community members in the region who travel to neighboring counties for work, school, and healthcare. Additionally, changes in expected travel times may attract or deter travelers to the region. One attendee demonstrated how small changes in travel times affect the region's prosperity. She used an online map trip suggestion to show how a hypothetical driver traveling from Asheville to Atlanta may select a route either through South Carolina or southwestern North Carolina. A difference in two minutes of travel time likely impacts a driver's decision to travel through southwestern North Carolina, Tennessee, Georgia, or South Carolina.

Maintaining a lower travel time through the corridor could potentially give hundreds or thousands of drivers an incentive to pass through the region, thus supporting businesses in the area's economy. For example, as individuals travel through, they may make planned or unplanned stops to purchase fuel, meals, or lodging at locations in the region. Maintaining these economic benefits would require strong state leadership to enforce policies that would benefit the region overall. Attendees acknowledged that locally elected officials are beholden to their communities and therefore have a stronger incentive to put their locality's needs first, even if it comes at the detriment of the region. Attendees reiterated that without an access-management policy intervention, the region's economic livelihood is at risk. Inland ports located in northern South Carolina and northern Georgia affect businesses in North Carolina's southwestern region. The inland ports shorten the supply chain for many manufacturers, processors, and distributors in the region.

The current growth of the retirement-age population is one of the most significant demographic trends in the history of the United States. This demographic has increased steadily since the 1960s, but is projected to more than double from 46 million today to more than 98 million by 2060 (Mather et al. 2015). North Carolina's southwestern counties are experiencing this trend as well. According to U.S. Census data, southwestern county residents above age 60 have increased from 50,000 to 70,000 from 2000 to 2017. In addition to the national trend of an aging population, North Carolina's southwestern counties are experiencing a flatlining population of people 20 to 45 years old. Workshop attendees noted that above and beyond national trends of aging, the southwestern region of North Carolina is attracting retirees from out of state to stay and live. One attendee suggested that the slower pace of life, the lower cost of living, and the mountainous scenery were all factors that were drawing people from Atlanta and other locations inside and outside of North Carolina's borders to settle in the region during retirement. Workshop attendees also mentioned the need to retain a younger workforce through IT, internet advances, and broadband projects.

Ideas from the Rural Freight Workshop coalesced into four primary focus areas to guide economic development in the southwest region, including: (1) System Resiliency and Competitiveness, (2) Demographic Opportunities, (3) Regional Identity, and (4) Regional Industries. Workshop attendees also identified actions that could be taken to support each of these focus areas as well as input from the research team to identify potentially "game-changing" investments that could generate substantial economic activity in the region.

Roadway improvements were high on the list, relating to both freight and the general aging population in the region. This includes widening shoulders and straightening curves to meet safety standards, as well as improving the design and load-bearing capacities of weight-restricted bridges. One participant also discussed safety considerations central to an aging population, suggesting to install raised pavement markers on the roads to help those with night vision issues while driving. The workshop helped lead to key-takeaways, which are summarized below:

- Due to the terrain and hydrology in the region, the resiliency and reliability of the state's transportation system is critical. Even more important than building new capacity, is ensuring the functionality of existing highway, rail, and aviation assets.
- The region does not act as one unit, but instead as a collection of many micro-economies that transcend county and state boundaries. Actions should be taken to strengthen the cross-pollination of business activities between North Carolina and its neighboring states.
- In years past, Advantage West operated to enhance the economic well-being and long- term prosperity of North Carolina's southwestern region with state-support. Currently, the MountainWest Partnership operates to advance economic development priorities of the region; however, it does so without state-support. Issuing renewed support to the MountainWest Partnership, which can serve as an important catalyst for new business and as a critical resource for existing businesses, would greatly benefit the region.
- North Carolina's southwestern counties are experiencing a flatlining population of people 20–45 years old. IT and internet advances, such as the broadband projects being undertaken in Macon County are required to retain a younger workforce. The Southwestern Commission Council of Governments conducted a broadband assessment that can be used to strategically increase broadband access in the region.

- Commercial paddling in the Nantahala and Pisgah National Forest areas is estimated to support 446 full-time jobs and $10 million in employee earnings, annually (Maples and Bradley 2017). The region can continue to capitalize on its natural scenery and outdoor activities to attract visitors from within the state and beyond.
- State-level policies that strengthen access-management in the region (i.e. minimizing or managing the number of conflict points that exist along a corridor) would be invaluable to protecting the area's economic stability. A difference in two minutes of travel time greatly influences a driver's decision to travel through southwestern North Carolina, Tennessee, Georgia, or South Carolina (supporting economies in these areas along the way).
- Improving air and rail access to Harrah's Cherokee Casinos could help increase tourism in the region. Western Carolina Regional Airport may need to relocate its runway to enable regular air carrier service, and railroad right-of-way could be purchased or reactivated to promote increased traffic to the casino as well as rail tourism in the region.

### 4.2 The Far Northeast

Economic development opportunities in northeastern North Carolina are intricately linked to Hampton Roads, Virginia. The Hampton Roads harbor area in southern Virginia has the largest concentration of military bases and government facilities of any metropolitan area in the world, yielding a gross domestic product of $94.86 billion in 2018 (CEAP 2017). Hampton Roads comprises a collection of cities, counties, and towns on the Virginia Peninsula and an extended combined statistical area (CSA) that includes the Elizabeth City, NC micropolitan statistical area and Kill Devil Hills, NC micropolitan statistical area. The harbor at Hampton Roads is essential to its growth and two deep water channels branch out from the harbor, the southern of which is linked with the coastal inlets of North Carolina through the Atlantic Intracoastal Waterway (Tikkanen et al. N.D.). These water connections are key enablers for marine freight. Additionally, the harbor area exists within the Foreign Trade Zone 20 Service Area which offers special procedures to encourage business activities by reducing, eliminating, or delaying duties (Tikkanen et al. N.D.).

During the workshop, participants spoke about Hampton Roads' success and strategies that could be implemented to harness the area's economic activity for growth in North Carolina. One participant explained, "Hampton Roads is one of our primary economic development assets. They keep expanding the ports and have nowhere to move but South." For example, a barge builder recently relocated from Hampton Roads to the Perquimans County Commerce Center. Another participant said economic development opportunities are originating in Hampton Roads, "coming across the border, landing in Currituck County and expanding across North Carolina." To meet these opportunities, a participant spoke about the increasing importance of improved highway infrastructure in the region. Currituck County is experiencing a clustering of housing and development occurring along US-168. The participant voiced the need for future I-87 and limited-access highways connecting US-168 to the future interstate to serve the needs of new residents and businesses looking to locate in Currituck County. He also discussed the possibility of altering the alignment of I-87 to better meet the growth needs of the region.

Echoing these sentiments, another participant spoke about the role of the Virginia ports. This person saw the ports as an economic driver that helped support businesses in Pasquotank and other counties in the region. She spoke about the importance of connecting North Carolina businesses to the ports via future I-87. Other participants built upon the discussion regarding the importance of Virginia's ports and future I-87, mentioning the value of having an interstate in the region for business recruitment purposes. "An interstate designation is a primary site selection criterion [for recruiting businesses]," an attendee explained.

During the workshop, attendees discussed specific strategies that could be implemented to promote industry in the region. These included plans for being an international leader in offshore wind, bolstering boat-building and marine industries, improving seafood production through aquaculture, advancing freight movement through barging, and continuing to be a leader in agriculture including implementing and expanding value-added opportunities. In addition to topics discussed during the workshop, insights from the Seven Portals Study (List and Foyle 2011) highlight opportunities for industries in the region.

The workshop attendees discussed offshore wind manufacturing, supply-chain component distribution, and energy production as invaluable growth industries in northeast North Carolina. The region's navigable waterways, Foreign Trade Zone (FTZ) locations, and proximity to the Atlantic Ocean offer a comparative advantage for offshore-wind manufacturing, distribution, and other marine industries. Additionally, the coast of North Carolina features wind corridors that are favorable for offshore wind energy production (Stearns et al. 2015). Heavy and oversized components are requisite for the offshore wind supply chain. Only a limited number of heavy-lift boats and other vessels are currently equipped to handle the weight and height requirements to install wind turbines and even fewer vessels can install state-of-the-art turbines in transitional depths of 30 to 60 m (USDOE 2016). Barges are used for pile-driving at the site as well as transporting parts (USDOE 2016). The parts can be transported in pieces or put together from the manufacturing site depending on the part size (USDOE 2016).

Workshop attendees discussed multiple opportunities the region has with the offshore wind industry. Attendees explained that the offshore wind supply chain would enable the region to strategically advance boat-building, barging, and manufacturing activities to meet the needs of the offshore wind market. Expanding marine industries in the region was discussed. The participants identified barging, boat-building, and coastal industries as primary opportunities for growth. According to Department of Agriculture statistics, North Carolina ranks first nationally in the production of sweet potatoes, and second in hogs, pigs, and turkeys (NC East Alliance N.D.). The state ranks third overall for cucumbers sold for pickles, trout sold, and poultry and egg products (NC East Alliance 2019).

Outside of its livestock and agricultural strength, N.C.'s northeastern region also has assets that serve to support and grow agricultural output. The state has the facilities to support food processing, ample municipal services support, a variety of logistical support features such as access to trucking companies, large refrigeration facilities and easy access to market through an extensive four-lane highway network (Cambridge Systematics 2019). Much of the state's production in hogs, turkeys, and other poultry is centered in and around the region. There are more than 160 facilities involved in food manufacturing. Total employment in North Carolina's food industry sector exceeds 20,000 people, or 5% of the total workforce (NC East Alliance 2019). Major food-processing employers in the Region include many nationally and internationally recognized companies such as Mt. Olive Pickles, Carolina Turkeys, The Cheesecake Factory Bakery, and Sara Lee Bakeries (NC East Alliance 2019). As the demand for seafood has increased, technology has made it possible to grow food in coastal marine waters and the open ocean. Aquaculture is a method used to produce food and other commercial products, restore habitat and replenish wild stocks, and rebuild populations of threatened and endangered species. It is breeding, raising, and harvesting fish, shellfish, and aquatic plants. U.S. aquaculture is an environmentally responsible source of food and commercial products that helps to create healthier habitats and is used to rebuild stocks of threatened or endangered species. During the workshop, participants discussed the importance of value-added agriculture and saw an increasing role for value-added agriculture in the region. According to the U.S. Department of Agriculture, value-added agricultural products can be described as using raw agricultural outputs to create a distinct product or intentionally altering the production of a good to enhance its value. Examples of the former include a change in the physical state or form of the product, such as milling wheat into flour or making strawberries into jam. Examples of the latter may include the production of a product in a manner that enhances its value, such as production of organic agriculture.

Participants also discussed ways to improve existing transportation networks in North Carolina's northeastern region including highway, rail, air, ferry, barge, and pipeline, to provide increased access with the ports in Virginia. Figure 2 shows the future I-87 corridor, the existing US-158 and US-168 corridors, as well as the numerous waterways that are invaluable to the region's economic competitiveness. Discussion centered around measures that could be undertaken to capitalize on growth that extends across the Virginia-North Carolina state border. The need for full implementation of the future I-87 corridor, which improves connectivity to Hampton Roads, was also discussed. Future I-87 would supersede existing US-64 and US-17 (Regional Transportation Alliance 2019). Participants also mentioned the need for limited access highways to tie into future I-87, which is a priority for NCDOT and is the subject of a recent grant application for completing the corridor, so that industries in northeastern North Carolina could fully access the ports as well as provide housing opportunities for individuals working in the greater Hampton Roads region.

In addition to highway improvements, workshop participants also spoke about opportunities to bolster transportation via the region's waterways. The waterways in northeastern North Carolina have historically served as transportation corridors. They continue to be utilized daily for recreational and some commercial transport. Workshop attendees recommended prioritizing funding to help expand upon marine highways in the region. For example, Washington County's Comprehensive Transportation Plan discusses the role of the Roanoke River and the Albemarle Sound (NCDOT 2015).

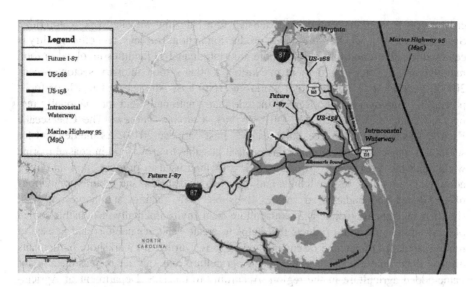

**Fig. 2.** Terrestrial and marine highway assets discussed by participants during the workshop

- Roanoke River: Begins in Roanoke, Virginia and flows 400 miles to its ending point in the Albemarle Sound, near the town of Plymouth. This deep-water river can accommodate barge traffic.
- Albemarle Sound: Protected from the Atlantic Ocean by the Outer Banks, the sound extends east from Washington County for about 50 miles. A vital link in the Intracoastal Waterway, the Albemarle Sound connects with the Chesapeake Bay via canals. Barge traffic travels this route all the way to the Atlantic Ocean.

Workshop participants specifically discussed the importance of barge transportation investment during the workshop. They spoke about a need to update barge infrastructure for transporting freight containers and oversized supply-chain components. Barge investments have been made elsewhere in the United States. For example, the US Department of Transportation's Maritime Administration (MARAD) recently awarded $1.8 million to the James River Barge Lines for the construction of an additional barge to expand service from Hampton Roads to the Port of Richmond. The barge is expected to carry approximately 170 containers per trip (Cambridge Systematics 2019).

Workshop participants also spoke about the general importance of using North Carolina's marine highways to allow increased access to its two deep-water ports and four river ports and continue the development of maritime-dependent industries. The Port of Wilmington is served by CSX and Wilmington Terminal Railroad (WTRY), handles containers, bulk and break-bulk cargo, and features refrigerated storage capabilities (Morley 2019). The Port of Morehead City is served by Norfolk Southern (NS) (on the North Carolina Railroad) and Coastal Carolina Railway (CLNA), has nine berths with approximately 5,500 feet of wharf and handles both breakbulk and bulk cargo at its existing facilities (Morley 2019). Radio Island, which is part of the Port of Morehead City, is located across the Newport River from the port and includes approximately 150 acres of land suitable for port industrial development (Morley 2019).

Edenton is on the Chowan River and handles fertilizers, forest products (i.e., lumber, logs, and wood chips), slag, primary iron and steel products, primary non-ferrous metal products, fabricated metal products, and waste/scrap (Morley 2019). The Knobbs Creek Deepwater Barge Port is in Elizabeth City. The port has a water depth of 30 feet with direct access to shipping traffic lanes through the Albemarle Sound into the Atlantic Ocean (Morley 2019).

Participants also spoke about the importance of shallow draft channels and inlets for keeping North Carolina's seafood industry alive as well as keeping the ferry channels open. In North Carolina, the U.S. Army Corps of Engineers maintains shallow draft projects with dredging depths of less than 20 feet. This includes 10 inlets and 14 inland waterways that are a part of the Atlantic Intracoastal Waterway. Of these channels there are three that directly impact the successful operation of ferry routes, Hatteras Inlet/Rollinson Channel, Silver Lake/Big Foot Slough, and Stumpy Point Bay. Workshop participants shared a common view that the waterways and marine infrastructure in the region were tremendous assets for commerce. However, many attendants felt the resources were being underutilized.

These ideas coalesced into three primary focus areas to guide economic development in the Northeastern region, including: (1) Transportation Upgrades and Redevelopment, (2) Workforce Opportunities, and (3) Regional Identity and Industries.

The workshop helped lead to key takeaways, which are summarized below:

- The region is well-positioned to capitalize on economic growth related to Hampton Roads, Virginia. Highway, waterway, and other transportation networks that connect business and population centers in North Carolina to Hampton Roads can facilitate growth in North Carolina.
- The region's proximity to deep-water channels and the eastern seaboard provides a comparative advantage for marine industries.
- Military personnel stationed in Hampton Roads, Virginia often seek employment elsewhere after fulfilling their service obligations. Connecting military personnel with civilian occupations offers a potential growth opportunity for the region.
- Several industries are poised to benefit from transportation investments. Offshore wind, boat-building, seafood production, barging, and agriculture were potential growth industries discussed during the workshop.
- A state-supported economic development entity can serve as an important catalyst in the Northeast. Additional resources for an organization such as the NC East Alliance could help accelerate business growth in the region.
- Waterways are not currently designated as transportation infrastructure, which makes acquiring grants for dredging, channeling, or other marine transportation network improvements difficult. If the region's waterways were able to receive this infrastructure designation, it would help for economic development grants and programs.

## 5 Conclusions

For many planners and economic developers, freight growth is seen as the physical manifestation of a growing economy (American Planning Association 2016). Ideas from the Southwestern Rural Freight Workshop coalesced into four primary focus areas to guide economic development in the southwest region, including: (1) System Resiliency and Competitiveness, (2) Demographic Opportunities, (3) Regional Identity, and (4) Regional Industries. Workshop attendees also identified actions that could be taken to support each of these focus areas as well as input from the research team to identify potentially "game-changing" investments that could generate substantial economic activity in the region. The Southwestern North Carolina workshop helped lead to key-takeaways, which are summarized below:

- Due to the terrain and hydrology in the region, the resiliency and reliability of the state's transportation system is critical. Even more important than building new capacity, is ensuring the functionality of existing highway, rail, and aviation assets.
- The region does not act as one unit, but instead as a collection of many microeconomies that transcend county and state boundaries. Actions should be taken to strengthen the cross-pollination of business activities between North Carolina and its neighboring states.
- In years past, Advantage West operated to enhance the economic well-being and long-term prosperity of North Carolina's southwestern region with state-support. Currently, the MountainWest Partnership operates to advance economic development priorities of the region; however, it does so without state-support. Issuing renewed support to the MountainWest Partnership, which can serve as an important catalyst for new business and as a critical resource for existing businesses, would greatly benefit the region.
- North Carolina's southwestern counties are experiencing a flatlining population of 20–45 year-olds. IT and internet advances, such as the broadband projects being undertaken in Macon County are required to retain a younger workforce. The Southwestern Commission Council of Governments conducted a broadband assessment that can be used to strategically increase broadband access in the region.
- Commercial paddling in the Nantahala and Pisgah National Forest areas is estimated to support 446 full-time jobs and $10 million in employee earnings, annually (Maples and Bradley 2017). The region can continue to capitalize on its natural scenery and outdoor activities to attract visitors from within the state and beyond.
- State-level policies that strengthen access-management in the region (i.e. minimizing or managing the number of conflict points that exist along a corridor) would be invaluable to protect the area's economic stability. A difference in two minutes of travel time greatly influences a driver's decision to travel through southwestern North Carolina, Tennessee, Georgia, or South Carolina (supporting economies in these areas along the way).
- Improving air and rail access to Harrah's Cherokee casinos could help increase tourism in the region. Western Carolina Regional Airport may need to relocate its

runway to enable regular air carrier service, and railroad right-of-way could be purchased or reactivated to promote increased traffic to the casinos as well as rail tourism in the region.

Ideas from the Northeastern workshop coalesced into three primary focus areas to guide economic development in the Northeastern region, including: (1) Transportation Upgrades and Redevelopment, (2) Workforce Opportunities, and (3) Regional Identity and Industries. The workshop helped lead to key- takeaways, which are summarized below:

- The Northeastern region is well-positioned to capitalize on economic growth related to Hampton Roads, Virginia. Highway, waterway, and other transportation networks that connect business and population centers in North Carolina to Hampton Roads can facilitate growth in North Carolina.
- The Northeastern region's proximity to deep-water channels and the eastern seaboard provides a comparative advantage for marine industries.
- Military personnel stationed in Hampton Roads, Virginia often seek employment elsewhere after fulfilling their service obligations. Connecting military personnel with civilian occupations offers a potential growth opportunity for the Northeastern region.
- Many industries are poised to benefit from transportation investments. Offshore wind, boat-building, seafood production, barging, and agriculture were potential growth industries discussed during the workshop.
- A state-supported economic development entity can serve as an important catalyst in the Northeast. Additional resources for an organization such as the NC East Alliance could help accelerate business growth in the region.

**Acknowledgements.** This study was funded and supported by the North Carolina Department of Transportation Research and Development Unit through NCDOT Research Project Number: 2019-17: "Rural Freight Transportation Needs". We would like to thank the stakeholders from the deep southwest and far northeast that helped to organize and participated in the workshops. Without their input, it would have been far more difficult to identify investment ideas that would have economic payoffs.

# References

American Planning Association (2016). APA Policy Guide on Freight. https://www.planning.org/policy/guides/adopted/freight/

Dykeman, W., et al.: Appalachian Mountains. Encyclopedia Britannica (2019). https://www.britannica.com/place/Appalachian-Mountains

Eastern North Carolina Regional Freight Mobility Plan: Final Report. Cambridge Systematics, October 2019. https://static1.squarespace.com/static/5d9e2a35e701446cb2693f61/t/5f980eb7c5934c1be179869b/1603800774836/Eastern+NC+Regional+Freight+Mobility+Plan+%28Final+2020%29.pdf

Khanna, P.: A new map for America. The New York Times **15** (2016)

List, G.F., Foyle, R.S., Canipe, H., Cameron, H., Stromberg, E.: Statewide logistics plan for North Carolina – an investigation of the issues with recommendations for action. North Carolina Office of State Budget and Management. May 13, 2008. Raleigh, NC (2008)

List, G.F., Foyle, R.S.: Seven portals study: an investigation of how economic development can be encouraged in North Carolina through infrastructure investment. North Carolina Department of Transportation. Report FHWA/NC/2010-34-7. December 2011. Raleigh, NC (2011)

Maples, J., Bradley, M.: Economic impact of non-commercial paddling and preliminary economic impact estimates of commercial paddling in the Nantahala and Pisgah national forests. August 2017. https://static1.squarespace.com/static/54aabb14e4b01142027654ee/t/59d545dcd2b857af3a8f1af5/1507149284387/OA_NPNF_PaddleStudy.pdf

Marusak, J., Price, M.: 4 rock slides in 4 days blockmajor highways and other roads in NC mountains. Charlotte Observer (2019). https://www.charlotteobserver.com/news/state/northcarolina/article226817774.html

Mather, M., Jacobsen, L., Pollard, K.: Population bulletin: aging in the United States, December 2015. https://www.prb.org/wp-content/uploads/2019/07/population-bulletin-2015-70-2-aging-us.pdf

NCDOT: Washington County Comprehensive Transportation Plan. NCDOT, August 2015. https://connect.ncdot.gov/projects/planning/TPBCTP/Washington%20County/REPORT.pdf

NC East Alliance: Value Added Agriculture. NC East Alliance (2019). https://www.nceast.org/economy-and-employers/value-added-agriculture/

Regional Transportation Alliance Interstate 87 (2019). http://letsgetmoving.org/priorities/congestion-relief/interstate-87/

Shin, H., Bapna, S., Farkas, S., Bonaparte, I.: Measuring the economic contribution of the freight industry to the Maryland economy. Morgan State University, May 2015. https://www.roads.maryland.gov/OPR_Research/MD-15-SHA-MSU-3-5_Measuring-the-Economic-Contribution-of-the-Freight-Industry_Report.pdf

Stearns, B.: Offshore Wind Energy Development in North Carolina. University of North Carolina at Wilmington, May 2015. https://uncw.edu/mcop/documents/mcop2015-finalreport.pdf

Wooten, R.M., Witt, A.C., Miniat, C.F., Hales, T.C., Aldred, J.L.: Frequency and magnitude of selected historical landslide events in the Southern Appalachian highlands of North Carolina and Virginia: relationships to rainfall, geological and ecohydrological controls, and effects. In: Greenberg, C., Collins, B. (eds.) Natural Disturbances and Historic Range of Variation, vol. 32, pp. 203–262. Springer, Cham (2016). https://doi.org/10.1007/978-3-319-21527-3_9

# Traffic Delay Evaluation and Simulating Sustainable Solutions for Adjacent Signalized Roundabouts

Mohamad Yaman Fares[1(✉)] and Muamer Abuzwidah[2]

[1] Khalifa University, Abu Dhabi, UAE
100059654@ku.ac.ae
[2] University of Sharjah, Sharjah, UAE
mabuzwidah@sharjah.ac.ae

**Abstract.** Roundabouts are proven to be an excellent solution to enhance traffic safety in comparison with intersections. However, multilane roundabouts with more than four approaches do not yet have any clear recommendations for the traffic signal installation neither the type of solution that can be applied to improve the Level of Service (LOS). This paper aims to evaluate signalized roundabouts that have multilane with more than four approaches to help to find sustainable traffic congestion solutions. The PTV-VISSIM software was used to evaluate the current operational performance and simulate multiple solutions for improvement. Traffic and geometric data of two adjacent signalized roundabouts (Al-Kuwait roundabout and Government square) in Sharjah City were used for this study. The two roundabouts are connected through a one-kilometer road.

The results showed that the current LOS of both roundabouts is "F," and there is no significant effect on the operational performance from each roundabout on the other. Moreover, the suggested solution showed LOS "D" with a 58% reduction in vehicle delay and a 60% reduction in stop delay for Al-Kuwait roundabout. While the reduction on the Government square was 62% and 65% for the vehicle delay and in stop delay respectively with a Level of Service "D" Overall, there is an indication that the suggested solution can reduce traffic delay significantly at this type of roundabouts; however, further research is highly recommended to estimate the economic, environmental, and safety benefits of the suggested solution.

**Keywords:** Signalized roundabouts · Traffic delay · Simulation · Traffic congestions · Sustainable solutions

## 1 Introduction

Roundabouts are generally circular-shaped intersections characterized by circulation movement around a central island and yield on entry. Roundabouts are used throughout the world because of their proven safety over regular intersections (De Brabander and Vereeck 2006; Gross et al. 2013; Robinson et al. 2000; Retting et al. 2001). A city like Sharjah in The United Arab Emirates (UAE) has many roundabouts located in sensitive

© The Author(s), under exclusive license to Springer Nature Switzerland AG 2022
A. Akhnoukh et al. (Eds.): IRF 2021, SUCI, pp. 303–314, 2022.
https://doi.org/10.1007/978-3-030-79801-7_21

urban areas such as the downtown, where they witness a colossal traffic demand due to high population and the city's continuous growth. Even after installing the traffic signals to enhance safety and organize the movement of the traffic movement that is characterized to be directional (e.g., morning movement is toward Dubai for work purposes and afternoon is the opposite way), the driver still faces a very long delay.

Researchers in past works tried to improve the operational performance of multi-lane roundabouts by applying alternative solutions and evaluating them using traffic simulation software such as RODEL, VISSIM, SIDRA, and others. Leite et al. (2020) tried to improve traffic operation at a three-lane roundabout by changing speed, acceleration, and minimum gap distance to investigate if this can reduce the waiting time. The authors used SUMO software and found that by imposing a maximum speed reduction on the two innermost routes within the roundabout, it is possible to improve throughput significantly. And they suggested doing this in the roundabout by rough pavement or speed bumps. A similar result by Martin-Gasulla et al.'s (2016) showed a significant improvement due to an increase in the proportion of long gaps on the conflicting flow. Another study (Osei et al. 2021) showed that by signalizing the roundabout, there would be a capacity increase by 50% in some cases, and both delay and queue length reduces significantly. Several other studies (Akcelik 2006; An et al. 2017; Hummer et al. 2014; Natalizio 2005) reported that signalization of roundabouts could significantly benefit the operational performance.

This paper aims to evaluate the traffic delay in the signalized roundabouts and simulate possible sustainable solutions to help reduce travel time, emissions, and improving safety at this type of roundabouts. Such a research paper is needed for two reasons: First, suggest a solution for this type of roundabouts, which will help reduce travel time and improve the driving experience. The second is to help fill the subject gap by experimenting with a new roundabout configuration that is complicated and uncovered in the known design manuals. The Highway Capacity Manual (HCM 2010) explains the capacity concept, operational analysis, design analysis, and other areas only for one and two lanes' roundabouts. While for Three-lane roundabouts and more, the manual suggests using alternative tools since no data is available. On the other hand, The NCHRP report 672 (National Academies of Sciences, Engineering, and Medicine 2010) has some models for the conventional four-legged multilane round-about (three or four lanes), but it also lacks in other areas.

The paper starts with an introduction that contains a brief background about the published research that used traffic simulation software to predict the multilane roundabout performance and will focus on the PTV VISSIM related papers since this software is used in the study. After that, the methodology section will explain the study area that contains two connected roundabouts, traffic demand, simulated alternative solution detail, and microscopic network modeling. The third step is to show the result of the work and to discuss it. Finally, the conclusion will present a summary of work, findings, and recommendations for future research.

## 2 Methodology

In this paper, a case study from Sharjah, United Arab Emirates, was used to analyze the operational performance of linked multilane signalized roundabouts and suggest a possible alternative solution to enhance the traffic conditions by using PTV VISSIM software (PTV Planung 2020). The two signalized roundabouts are:

1. Al Kuwait roundabout.
2. Government square.

### 2.1 Study Area

In this area in Sharjah city, UAE, four significant roundabouts are connected, and they form a square. Al Kuwait roundabout and Government square are linked through a two-direction three lanes road (around one kilometer of length), and this road separates between Al Musalla and Al Manakh zone. Residential areas surround the studied roundabouts with low and medium-raised buildings. Furthermore, there are government buildings located nearby each of the roundabouts. The reasons for choosing these roundabouts are:

- The roundabouts are controlled by traffic signals.
- Both roundabouts have a complicated geometry shape, with more than four approaches.
- Pedestrians' movement is limited, so it does not affect the operational analysis.
- Both roundabouts are connected to major arterial roads.

**Fig. 1.** Spatial position and an aerial image of the studied roundabouts in Sharjah city, UAE (Retrieved from Google Earth)

### 2.1.1 Al Kuwait Roundabout

Al Kuwait roundabout is circular shaped with three lanes, five approaches that have two or three lanes, and right direction movement filter for each approach as shown in Fig. 1 right side. These five approaches have a static traffic signal light with a red – green – flashing green – amber (three seconds) system. In addition to that five-second layer, traffic signal lights appear inside the roundabout. As retrieved from Sharjah RTA, the roundabout is working with 17 programs for the traffic signal depending on the time of the day, and in off-peak hours it turns to be a non-signalized roundabout (between 23:00 – 05:00 works with flashing mode). During the maximum peak hour, the traffic signal works with program eight (total lap time 258 s). Finally, Al Kuwait roundabout is connected to an arterial road, "Al Wahda road" -shown in Fig. 1 with blue color, a bustling road that contributes to the significant demand on the roundabout (Fig. 2).

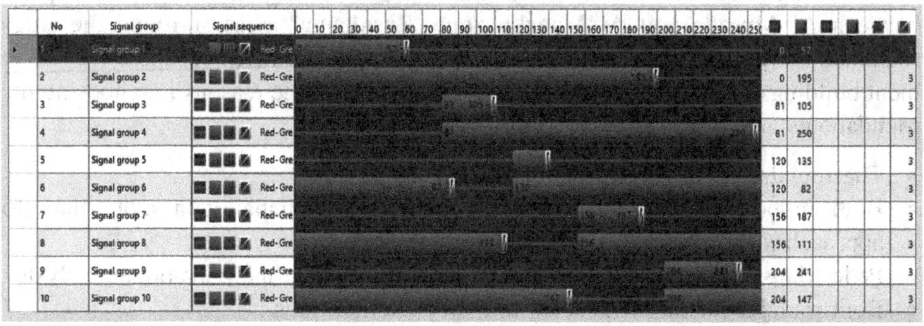

**Fig. 2.** Phase diagram for the traffic signal in Al Kuwait Roundabout

### 2.1.2 Government Square

Government square has an elapsed shape with four lanes, six approaches that have two or three lanes, and some of the approaches have a filter for the right movement as shown in Fig. 1 left side. Furthermore, a static traffic signal light with a red – green - flashing green - amber (three seconds) system on each of those six approaches. And inside the roundabout, a second layer with six traffic signal lights appear. As retrieved from Sharjah RTA, the square is working with 13 different setups for the traffic signals depending on the day, and in off-peak hours, it turns to be a non-signalized roundabout (between 23:00 to 06:00 works with flashing mode). During the maximum peak hour, the traffic signal uses the sixth program (total lap time 259 s). Also, Government square is connected to an arterial road called "Al Rolla" -showed with cyan color- (Fig. 3).

### 2.2 Traffic Demand

Traffic volumes and turning movements were obtained with the help of The Road and Transportation Authority in Sharjah. A full 24-h traffic count for the volumes was done, and as a result, the peak hour was found to be between 13:00 – 14:00 for both roundabouts. The following schematic diagrams and matrices show the distribution of the traffic demand during the peak hour (Figs. 4, 5 and Tables 1, 2).

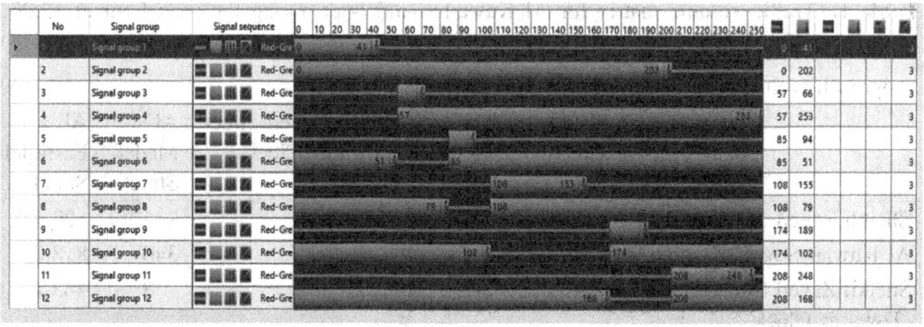

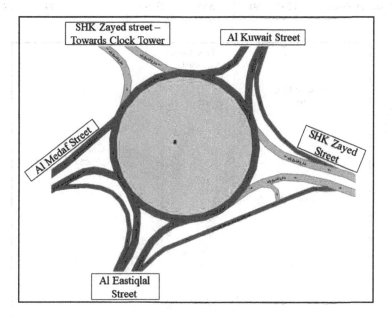

Fig. 3. Phase diagram for the traffic signal in Government square

Fig. 4. Al Kuwait roundabout schematic shape

## 2.3 Alternative Solution

Depending on the traffic demand data and the operational performance, several alternative solutions were tested to improve the Level of Service "LOS" of the roundabout. Starting with the Al Kuwait roundabout, the alternative suggests building a two-lane per direction tunnel that connects the road coming from Flying square to the road that goes to Government square as shown in Fig. 6-a; also, the approach from the Flying square will have an exit to the clock tower road, and this tunnel will be one-directional with two lanes. In addition to these tunnels, a modification was done for the traffic signal limiting the time duration for the approaches with the tunnel and adding this time to other segments.

**Table 1.** Traffic volume matrix for Al Kuwait roundabout during the peak hour at 13:00 to 14:00

| Approach (vehicle/h) | | Exit (vehicle/h) | | | | |
|---|---|---|---|---|---|---|
| | | Sheikh Zayed Street | Al Kuwait Street | Sheikh Zayed Street (Towards Clock Tower) | Al Medaf Street | Al Eastiqlal Street |
| Sheikh Zayed Street | 2323 | 11 | 0 | 1448 | 835 | 29 |
| Al Kuwait Street | 1608 | 414 | 0 | 9 | 340 | 818 |
| Sheikh Zayed Street (Towards Clock Tower) | 1518 | 934 | 81 | 32 | 8 | 463 |
| Al Medaf Street | 1168 | 630 | 479 | 34 | 25 | 0 |
| Al Eastiqlal Street | 1240 | 50 | 555 | 505 | 94 | 36 |

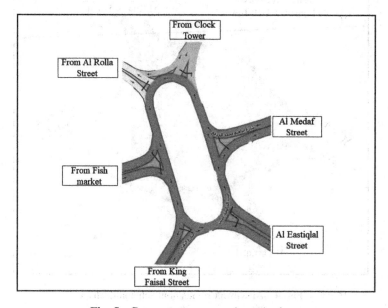

**Fig. 5.** Government square schematic shape

**Table 2.** Traffic volume matrix for Government square during the peak hour at 13:00 to 14:00

| Approach (vehicle/h) | | Exit (vehicle/h) | | | | | |
|---|---|---|---|---|---|---|---|
| | | Al Medaf Street | From Clock Tower | From Al Rolla Street | From Fish market | From King Faisal Street | Al Eastiqlal Street |
| Al Medaf Street | 965 | 15 | 209 | 206 | 177 | 254 | 104 |
| From Clock Tower | 1726 | 286 | 56 | 108 | 336 | 513 | 427 |
| From Al Rolla Street | 784 | 120 | 165 | 2 | 101 | 184 | 212 |
| From Fish market | 800 | 175 | 173 | 100 | 3 | 147 | 202 |
| From King Faisal Street | 1432 | 251 | 375 | 294 | 216 | 53 | 243 |
| Al Eastiqlal Street | 1567 | 227 | 433 | 339 | 250 | 300 | 18 |

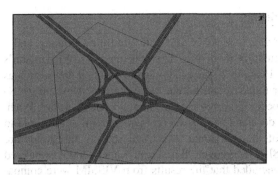 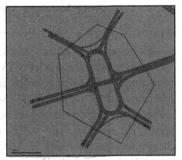

**Fig. 6.** a) To the left, Alternative solution for Al Kuwait roundabout, b) To the right, alternative solution for Government square.

On the other hand, for Government Square, a two-lane bi-directional tunnel, the road from Al Kuwait roundabout to the road that to market, as shown in Fig. 6-b, is to be constructed. In addition to this tunnel, a modification was done for the traffic signal by reducing the time to half to be 130 s for the traffic signal, then the approaches that have a tunnel was redistributed (Figs. 7 and 8).

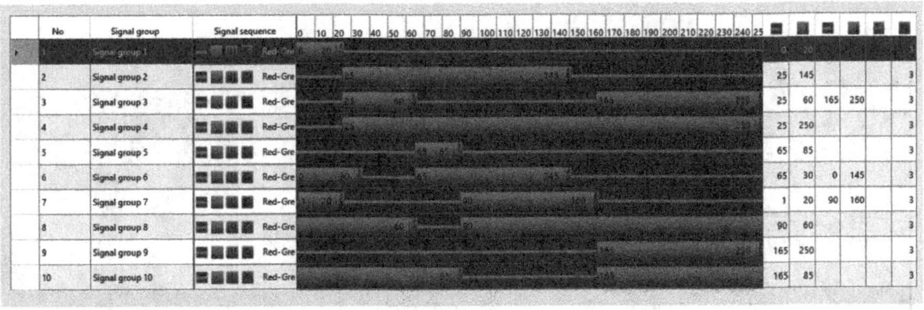

**Fig. 7.** Alternative solution new phase diagram for the traffic signal in Al Kuwait Roundabout

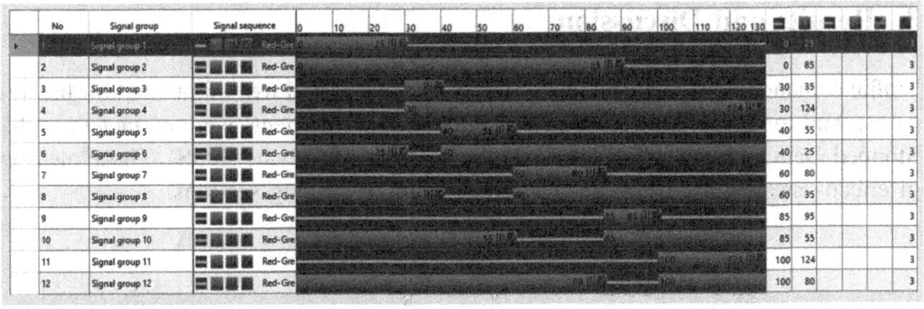

**Fig. 8.** Alternative solution new phase diagram for the traffic signal in Government square

## 2.4 Microscopic Network Modeling

PTV Planung Transport Verkehr AG develops VISSIM software from Germany, and it is specialized in microscopic simulation modeling. The software uses Wiedemann's 1974 driver behavior model (PTV Planung 2020) to simulate interrupted flows, such as roundabouts. This driver behavior model assumes that a driver accelerates and decelerates according to the vehicle's speed of the vehicle traveling in front of his/her. When and how the drivers accelerate or decelerate are based on their perception threshold to speed and spacing. The values of the individual speed and spacing thresholds are distributed stochastically in VISSIM. Moreover, Bared and Edara (2005) simulated roundabouts with VISSIM and concluded that the results from VISSIM were comparable with the U.S. field data.

In this study, three models were constructed for the network using VISSIM. The first one is a model for the two roundabouts case shown in Fig. 9-a. The other one is for Government square case shown in Fig. 9-b, and the reason to build this model is to discover the effect of an adjacent roundabout on the other. The last model was with the addition of the alternative solution. The simulation lasted for 4500 s. The first 900 s were used to warm up the network, and then from 900–4500 s (which is the peak hour), the resulted data were extracted every 15 min.

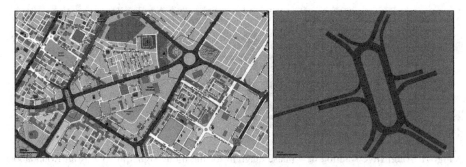

**Fig. 9.** PTV VISSIM model for the study area, a) Both roundabouts, b) Government square

## 3 Results and Discussion

Traffic demand and signal timings for the peak-hour period were simulated for the three cases. And to assess the effectiveness, different measurements were extracted, such as Stopped Delay, Average Vehicle Delay, and Level of Service (LOS). These measures of effectiveness will be explained in detail in the following sections.

## 3.1 Average Vehicle Delay and Stopped Delay

The Average Vehicle Delay (AVD), or in other words, the Delay of a vehicle in leaving a travel time measurement, can be calculated by subtracting the theoretical from the actual travel time (Elefteriadou 2014). The stopped Delay (S.D.) is the total time in seconds when the vehicle is not moving except the stops at P.T. points and parking areas (Elefteriadou 2014). Figure 10 shows that both delay variables in Al Kuwait roundabout have been reduced noticeably after introducing the alternative solution; for example, by average, the AVD decreased from 99.8 s per vehicle to 42.65, which is around 57%, and for S.D., it reduced by 59.5%. While, for the Government square as shown in Fig. 11 it can be seen that the AVD reduced from 102.17 s per vehicle to 38.57, which a 62% reduction can present. Also, for S.D., it showed improvement by at least 65%. Moreover, in Fig. 12, it can be observed that for this study case, there is no significant difference in operational performance for these roundabouts from being adjacent to each other.

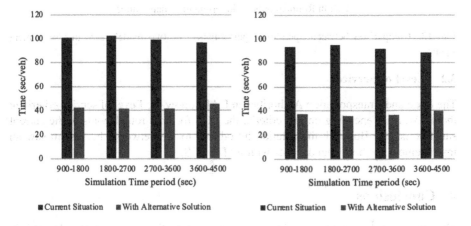

**Fig. 10.** Al Kuwait roundabout analysis result, a) Average Vehicle Delay, b) Stopped Delay

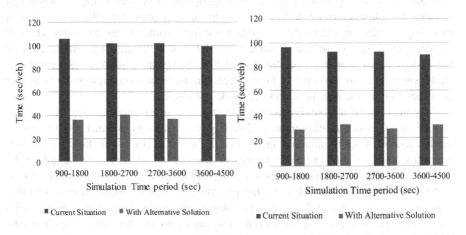

**Fig. 11.** Government square analysis result, a) Average Vehicle Delay, b) Stopped Delay

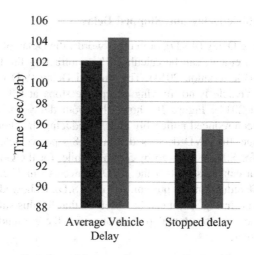

**Fig. 12.** Comparison between operational performance for two model cases in Government square

### 3.2 Level of Service

The Road and Transportation Authorities in UAE follow the Level of service criteria in the HCM, and depending on this criteria, the LOS for both roundabouts in the current situation is "F". But with the solution introduced to the network, the system saw an improvement for both roundabouts with a "D" LOS.

## 4 Conclusions

This paper evaluated traffic delay for adjacent signalized roundabouts located in Sharjah city, UAE, by simulating the existing operation condition and the suggested sustainable solutions using VISSIM software. It was proven in the literature that installing a traffic signal at the standard (four-legs) roundabouts can improve their capacity. However, complicated shaped roundabouts with multilane (three lanes or more) do not have a proper procedure to install traffic signals. And this was clear in this study where there are two layers of traffic lights in the existing condition: the first layer is at the approaches, and the second is inside the circle.

The analysis of this study showed that this system resulted in LOS "F" for the studied roundabouts. Furthermore, in The Highway Capacity Manual (2010), one of the limitations of the roundabout analysis could be the adjacent roundabout when it is connected through a direct link to the nearby roundabout. However, the length of the links was not clearly explained in the literature, although it is one of the critical factors, and the result of this study showed that the effect of both roundabouts on each other was insignificant. This finding can contribute to the microsimulation exercises with other software by separating the linked study area into smaller zones, which can be done in some other software programs that analyze only a single intersection at a time.

The last finding showed that the suggested solution could significantly reduce the average delay and stopped delay. However, the most crucial point was that the tunnels were not enough to make the difference by themselves, and therefore changing the traffic signal timings could also help reduce the delay significantly in both roundabouts. So, this can be considered as a quick partial solution to improve the LOS in this type of roundabout. Moreover, using dynamic traffic signal systems could improve the operational performance of the roundabouts.

The study results and recommendations show that the adopted alternative solution can be applied in different regions, but this should be done with caution and after further testing. However, the results of the solution could vary from one region to another depending on the travel behavior and traffic intensity; for example, since the UAE is considered a car-dependent society where the active travel and transit travel modes are minimum, similar countries with comparable characteristics would benefit significantly from the study.

Overall, further investigation is needed to determine the proper distance between the adjacent roundabouts where they might affect the performance of each other. As well as the estimation of the economic, environmental, and safety benefits of the suggested solutions.

**Acknowledgements.** We want to thank the Road and Transportation Authority (RTA) in Sharjah for collecting the traffic data and sharing the traffic signal timing system for the studied area.

# References

Akcelik, R.: Analysis of roundabout metering signals. In: The 25th AITPM 2006 National Conference, Melbourne (2006)

An, H.K., Yue, W.L., Stazic, B.: Estimation of vehicle queuing lengths at metering roundabouts. J. Traffic Transp. Eng. (English Ed.) 4(6), 545–554 (2017)

Bared, J.G., Edara, P.K.: Simulated capacity of roundabouts and impact of roundabout within a progressed signalized road. In: TRB National Roundabout Conference, Vail (2005)

De Brabander, B., Vereeck, L.: Safety effects of roundabouts in Flanders: signal type, speed limits, and vulnerable road users. Safety effects of roundabouts in Flanders: signal type, speed limits, and vulnerable road users. And vulnerable road users. Accid. Anal. Prev. 39, 591–599 (2006). https://doi.org/10.1016/j.aap.2006.10.004

Elefteriadou, L.: An Introduction to Traffic Flow Theory. Springer, New York (2014). https://doi.org/10.1007/978-1-4614-8435-6

Gross, F., Lyon, C., Persaud, B., Srinivasan, R.: Safety effectiveness of converting signalized intersections to roundabouts. Accid. Anal. Prev. 50, 234–241 (2013). https://doi.org/10.1016/j.aap.2012.04.012

HCM 2010: highway capacity manual. Transportation Research Board, Washington, D.C (2010)

Hummer, J.E., Milazzo, I., Joseph, S., et al.: The potential for metering to help roundabouts manage peak period demands in the U.SUS. In: Transportation Research Board 93rd Annual Meeting, Washington DC (2014)

Leite, B., Azevedo, P., Leixo, R., Rossetti, R.J.F.: Simulating a three-lane roundabout using SUMO. In: Martins, A.L., Ferreira, J.C., Kocian, A. (eds.) INTSYS 2019. LNICSSITE, vol. 310, pp. 18–31. Springer, Cham (2020). https://doi.org/10.1007/978-3-030-38822-5_2

Martin-Gasulla, M., Garcia, A., Moreno, A.T., et al.: Capacity and operational improvements of metering roundabouts in Spain. Transp. Res. Procedia **15**, 295–307 (2016)

Natalizio, E.: Roundabouts with metering signals. In: ITE Annual Meeting and Exhibit, Melbourne (2005)

National Academies of Sciences, Engineering, and Medicine. Roundabouts: An Informational Guide. 2nd Ed. The National Academies Press, Washington, DC (2010). https://doi.org/10.17226/22914

Osei, K.K., Adams, C.A., Ackaah, W., Oliver-Commey, Y.: Signalization options to improve capacity and delay at roundabouts through microsimulation approach: a case study on arterial roadways in Ghana. J. Traffic Transp. Eng. (English Ed.) **8**, 70–82 (2021). https://doi.org/10.1016/j.jtte.2019.06.003

PTV Planung: VISSIM 13.0 User's Manual. PTV AG, Karlsruhe (2020)

Retting, R.A., Persaud, B.N., Garder, P.E.: Crash and injury reduction following installation of roundabouts in the United States. Am. J. Public Health **91**(4), 628–631 (2001)

Robinson, B.W., et al.: Roundabouts: an informational guide. U.S. Department of Transportation, Federal Highway Administration, Report FWHARD-00-067, Washington (2000)

**Safe Roads by Design - 1**

# A Comparative Evaluation of the Safety Performance of Median Barriers on Rural Highways; A Case-Study

Victoria Gitelman[1(✉)] and Etti Doveh[2]

[1] Transportation Research Institute, Technion - Israel Institute of Technology, Haifa, Israel
trivica@technion.ac.il

[2] Technion Statistical Laboratory, Technion - Israel Institute of Technology, Haifa, Israel

**Abstract.** Safety barriers provide forgiving roadsides for highways while their test performance should comply with uniform norms, e.g. EN1317. When selecting a barrier type for a median, the dilemma frequently compares between in-situ concrete and other barriers. In-situ concrete barriers require less space for their installation and rare maintenance, while steel guardrails can deflect, but require a wider median and higher maintenance, with associated crash risks during repair roadworks. This study compared the safety performance of highways with various median barrier types, in Israel, aiming to provide a background for more effective use of safety barriers. The study examined four barrier types: Step-shaped in-situ, pre-cast concrete and steel guardrails, as new barrier types (which satisfy the EN1317), and old NJ-shaped in-situ barriers. The database included 558 km of non-urban Israeli highways. Negative-binomial regression models were fitted for predicting crashes, and differences in crash expectancy were evaluated for various barrier types, while controlling for traffic volumes and other road infrastructure characteristics. The results showed that, on dual-carriageway roads, sections with Step-barriers had a better safety level relative to those with NJ-barriers and, sometimes, to other barrier types. On motorways, the safety level of sections with Step-barriers was better or similar to that with NJ-barriers, with no difference compared to the pre-cast barriers, while sections with steel guardrails had a better safety level, particularly at traffic volumes over 40,000 vehicles. The economic evaluations showed benefits for replacing the old NJ-barrier by a Step-barrier, on dual-carriageway roads, and by steel barriers, on motorways.

**Keywords:** Safety barriers · Median · In-situ concrete barriers · Rural highways · Safety performance · Economic evaluation

## 1 Introduction

Safety barriers provide forgiving roadsides for highways (AASHTO 2011; CEDR 2012). Due to the need for a common approach to the qualification of road safety devices, over two decades ago, uniform rules were introduced for testing road safety barriers, which are currently defined, in Europe, by the European Norms EN1317 (CEN 2010) and in the USA, by MASH (2016). A safety barrier that satisfies the

---

© The Author(s), under exclusive license to Springer Nature Switzerland AG 2022
A. Akhnoukh et al. (Eds.): IRF 2021, SUCI, pp. 317–334, 2022.
https://doi.org/10.1007/978-3-030-79801-7_22

requirements of the EN1317 (or a similar standard in the USA), belongs to "new" barrier types, while barriers which were not tested but are still present on the roadsides are referred to as "old" barriers. In Israel, the use of new barriers is obligatory in all new road construction projects and in road upgrading and major maintenance projects, since 2003 (Gitelman et al. 2014). When a barrier is selected for installation, its characteristics should fit the site conditions as defined by the national guidelines (MOT 2020), which refer to the barrier's performance classes - containment level, impact severity level and the space needed for its deformation (working width, vehicle intrusion) as defined by CEN (2010).

When selecting a barrier type for a median of a non-urban highway, the dilemma frequently compares between in-situ concrete barriers and other barrier types. In-situ concrete barriers do not move during a vehicle collision and usually require less space for their installation and rare maintenance. Other barrier types, mainly steel guardrails, that deflect during a collision, require a wider median and higher maintenance, with associated traffic flow interruptions and crash risks during repair roadworks (Steinauer et al. 2004; Williams 2007). A counter-argument typically refers to possibly higher crash severity in collisions with concrete versus steel barriers (e.g., Martin et al. 2001).

In the past, New Jersey (NJ) slip-formed concrete barriers were mostly applied on medians of non-urban highways in Israel, the type resembling those developed in the USA in the middle of the twenties' century. Such old barriers are still present on many road sections. New safety barriers that were approved for use in the country are mainly of three types: steel guardrails, pre-cast concrete and in-situ concrete barriers. Among the new barriers, a Step-shaped in-situ barrier was introduced, however, there were long debates concerning its acceptability and safety performance relative to other barrier types. Furthermore, empirical findings with regard to the safety performance of new barriers relative to the old ones are not frequent yet in the research literature (Gitelman et al. 2014; Cafiso et al. 2017), thus raising questions on the safety benefits and feasibility of replacing old barriers by the new ones.

Therefore, in this study, empirical data were collected on the installation of various barrier types on non-urban (rural) highways, in Israel, aiming to compare the safety performance of highways with various median barrier types, and to provide a background for more effective use of safety barriers, in the future. The study examined four barrier types: Step-shaped in-situ, pre-cast concrete and steel guardrails, as new barrier types (which satisfy the EN1317), and NJ-shaped in-situ barriers as an old barrier type. Figure 1 provides examples of such barriers, installed on Israeli highways.

a – Step-shaped in-situ    b – Pre-cast concrete    c – Steel guardrail    d – NJ-shaped in-situ

**Fig. 1.** Examples of median barrier types on non-urban Israeli highways.

## 2 Previous Research on Safety Impacts of Various Barrier Types and the Study Motivation

In-situ concrete barriers, due to their rigidity, are associated with greater energy absorbed by the vehicle which collides with the barrier (Ray and McGinnis 1997). This may lead to a higher crash severity, particularly when a small vehicle is involved. Elvik et al. (2009) summarized research findings from various countries (published before 2001) and reported that the installation of a median barrier on multi-lane divided roads in general led to a significant reduction in both fatal and injury crashes, to the extent of 30%-40%. However, concrete barriers on medians were associated with an increase in fatal crashes, yet, not significant, while for steel barriers a significant reduction in injury crashes was found. More recent studies conducted in various states of the USA (Tarko et al. 2008; Hu and Donell 2010; Zou et al. 2014) also indicated a higher severity of single-vehicle crashes with concrete median barriers compared to steel (or cable) barriers. Nevertheless, concrete barriers are in use in the USA, where they are considered as more appropriate for narrow medians, particularly along roadways with high traffic volumes (AASHTO 2011).

Martin et al. (2001) collected data on vehicle collisions with steel and concrete barriers on the French motorway network and found that the share of injury crashes out of the total cases was 1.7 times higher in collisions with concrete barriers. An updated study - Martin et al. (2013), considered barrier crashes over a period of fifteen years and found lower injury levels in crashes with steel guardrails but also indicated a need for concrete barriers at sites with narrow working widths and for preventing cross-median crashes with very serious consequences. In an Israeli study, detailed data on crashes involving barriers on rural roads were collected from the police archives and then matched with the national trauma registry (Gitelman et al. 2006). The police data showed a higher share of severe crashes (fatal and serious) in collisions with concrete versus steel barriers: 20% and 12%, respectively. However, among the casualties hospitalized following the collisions, a higher severity of injury was associated with steel barrier crashes, in terms of injury severity score, length of hospital stay and the frequency of multi-trauma injuries.

Steinauer et al. (2004) conducted a study that compared the whole-life economic costs associated with the use of concrete versus steel barriers on motorway central reserves, in Germany. The study considered barriers' installation and repair costs, crash costs, crash under-reporting by the police, and time losses caused by congested traffic during repairs. The results showed that at high traffic volumes the time losses caused by barrier-repair works are so considerable that the higher investment costs of a concrete barrier are offset during the average service life and, thus, installation of concrete barriers is preferable when daily traffic volume exceeds 65,000 vehicles.

A similar study in the UK (Williams 2007) examined median barrier crashes and the relative whole-life costs of steel versus concrete barriers, based on data collected for the M25 sphere highway network. The crash data showed that the numbers of serious casualties per kilometer were comparable between two barrier types, while concrete barriers resulted in a lower rate of total crashes per kilometer than steel barriers. The whole-life calculations that accounted for barrier installation and maintenance costs as well as for crash costs, repairs, traffic management and traffic delay costs due to roadworks, showed lowest values for concrete barriers with higher containment levels such as the Dutch Step-barrier with an H2-level. As a result, concrete barriers with higher containment levels were advised as preferable on medians of roads with daily traffic of over 25,000 vehicles (IAN 2005).

Most studies of concrete barriers referred to NJ-shaped ones that were originally developed in the USA and were associated with a tendency for small cars to overturn (Jehu and Pearson 1977). McDevitt (2000) overviewed the development of concrete barriers with NJ profile in the USA and noted that the key parameter was the distance between the ground and the point of breaking of the barrier's slope. A high breaking point led to over-rising of small vehicles during the impact with a NJ-barrier that tended to roll-over during a return to the ground, particularly in a collision at high speed. To prevent a roll-over of small vehicles in collisions with concrete barriers, a Step-shaped profile was suggested (Verweij 2000). The Step in-situ barrier was developed in the Netherlands, in the nineties, and passed impact tests in accordance with the requirements of the draft EN1317. Since then, this barrier is in use in the Netherlands, Belgium, Finland and other countries.

In Israel, the Step-barrier was approved for use in 2006 (with a performance level of H2-W1-B, following the test reports from 1995 and a later MIRA approval concerning the compliance with EN1317–5). For a decade this new barrier-model was attacked with statements that its impact severity level should be questioned due to the change in the barrier test conditions in a later version of the EN1317 compared to the draft one. However, such statements were not supported by empirical findings, whereas the evaluations from the UK (as mentioned above) actually promised safety benefits from the use of Step in-situ barriers compared to other barrier types, on highway medians. Moreover, in spite of the indications in research that in-situ concrete barriers, in general, may be less forgiving in collisions with small cars, many countries continue to use them, at least on narrow medians and/or on heavily-trafficked roads (EUPAVE 2012).

Concerning the safety impacts of new barrier types, the research findings are scarce. In Israel, Gitelman et al. (2014) investigated the safety effectiveness of upgrading old barriers by new (steel) barriers; they found safety benefits for dual-carriageway roads whereas a reduction in severe single-vehicle crashes and in total injury crashes was estimated using a multivariate analysis. On sections with new barriers installed, a 15% decrease in injury crashes was indicated, while benefit-cost ratios associated with crash savings ranged in 1.1–2.5. In Italy, Cafiso et al. (2017), Cafiso and D'Agostino (2017) evaluated crash modification factors associated with new (steel) barriers on motorways and found strong reductions, to the extent of 70%, in run-off-road crashes, and smaller reductions, of 17%, in total crashes. In the Clearinghouse of crash modification factors managed in the USA, for a countermeasure of "improve guardrail", an average reduction of 25% was reported for run-off-road crashes and 18% for total crashes (FHWA 2013).

Recognizing the research gaps concerning the safety impacts of various barrier types and respective practical questions with regard to the necessity of replacement of the old barriers by new ones, this study aspired to contribute to existing knowledge,

A Comparative Evaluation of the Safety Performance of Median Barriers   321

by means of examining the safety level of highway sections with a Step-barrier and other barrier types, and, more generally, by comparing the safety impacts of new barriers with the old in-situ barrier.

## 3 The Study Database

The information on the inventory of safety barriers on the non-urban road network was derived from the road survey conducted by the National Transport Infrastructure Company (NTIC), in 2013. First, the information on the presence of Step-barriers was extracted, whereas the locations and the installation terms were verified through contacts with road project managers and the barriers' suppliers. We found that over 92% of Step-barriers were present on medians and that they were installed under the reconstruction of a divided road (with replacing an old barrier) or when a two-lane road was upgraded to a dual-carriageway. Due to the high containment level (H2), Step-barriers were installed on medians of motorways and dual-carriageway highways. Both road types are divided rural roads, where motorways are with controlled access only (no at-grade junctions), better design standards (wider lanes and shoulders, bigger radii) and higher speed limits. These empirical data led to the study's framework - a consideration of the safety level of highway sections with Step-barriers on medians, on motorways and dual-carriageway roads, as treatment groups, relatively to comparison-group sections, i.e. the same road classes but with other median barrier types. The comparison sections were selected based on the NTIC survey, using the road type and the barrier type required. For the study, eight groups of highway sections were defined as follows:

1. dual-carriageway road sections with NJ in-situ barriers (the old type) on medians;
2. dual-carriageway road sections with pre-cast concrete barriers on medians;
3. dual-carriageway road sections with steel barriers on medians;
4. motorway sections with NJ in-situ barriers (the old type) on medians;
5. motorway sections with pre-cast concrete barriers on medians;
6. motorway sections with steel barriers on medians;
7. dual-carriageway road sections with Step in-situ barriers on medians;
8. motorway sections with Step in-situ barriers on medians.

Most barriers, except for groups 1 and 4, are new ones (satisfy the EN1317 demands). Groups 1–3 served as comparison-groups for the treatment group 7; groups 4–6 - as comparison-groups for the treatment group 8. The total length of highway sections in the treatment group was 78 km, the total length of comparison-group sections - 480 km, including 340 km with old NJ-barriers, 31 km with new pre-cast concrete barriers and 109 km with new steel barriers.

Highway sections from all groups were divided into study units of 0.5 km in length. For each unit, the number of crashes was assigned based on the Central Bureau of Statistics' (CBS) crash files (in three years, corresponding to the road survey). The analysis included four crash types: injury crashes, total crashes, severe (fatal and serious together) and single-vehicle crashes, as is common in international practice (Elvik et al. 2009). The single-vehicle crashes included "leaving the road", "overturn" and "collision with fixed object" types, which in sum reflect the run-off-road crashes, in

Israel. The total crashes were a sum of data from two crash files, which are collected by the Israeli police: an "injury crash" file, with cases investigated by the police, and a "general with casualties" file, with cases reported to the police but not investigated. Among the crashes, the collisions with median barriers versus roadside barriers could not be separated, hence, the crash data characterized the safety level of road units, in general. Concerning the crash types examined in the study, it should be noted that safety barriers are not primarily intended to prevent crashes from occurring but to reduce their consequences. However, when non-injury crashes are not reported (e.g. in Israel), a lower injury level leads to a lower number of crashes. Run-off-road crashes are commonly examined with regard to the impact of safety barriers but also total crashes (Elvik et al. 2009; FHWA 2013; Cafiso and D'Agostino 2017); when applicable, both crash counts and their severity are examined. Therefore, in this this study, four crash types were considered to compare the impacts of various barrier types.

To ascertain the effect of safety barriers on crashes, the impacts of other potential confounders should be accounted for, such as the level of exposure (traffic volume), section length and road infrastructure characteristics. For each unit, the infrastructure characteristics were produced based on the NTIC road survey. In addition to road type and median barrier type (which served for selecting the study sections), such characteristics included: width of carriageway, the number of lanes, speed limits, type of road-building, widths of the right (external) and the left (internal) shoulders, presence of roadside barriers. The characteristics were produced as average values of both travel directions (as for all features, high correlations between the values in both travel directions were found). The values of traffic volumes on the study units were extracted from the CBS files, in the form of the average annual daily traffic (AADT), and then classified according to three categories: low, medium, high. Such categories were defined for each road type, motorways and dual-carriageway roads, according to the common levels of traffic volumes. Table 1 presents descriptive statistics of the study database, including the definitions of categories and variables that served in the study analyses.

## 4 Methods of Analysis

Using the data, explanatory models were fitted to crash numbers on the study units. Each model estimates the expected annual number of crashes on a road unit accounting for traffic volumes and road infrastructure characteristics. The effect of a median barrier type is assessed having excluded the impacts of other traffic and road characteristics. Separate models were adjusted to various crash types, on both road types, where treatment and comparison-group sections were used in the same model. As crashes are subject to over-dispersion, negative binomial regression models were applied for the modeling.

First, the form of relationship between crashes and each explanatory variable was explored using generalized additive models (with the *mgcv* library of *R* software). The shape of each relationship was examined and, consequently, a linear relation was kept or a quadratic component was added, on the next step of fitting a parametric model. The adjusted model has the form of:

# A Comparative Evaluation of the Safety Performance of Median Barriers 323

**Table 1.** Descriptive statistics of the study road sections

a - Characteristics of road section groups

| Characteristics of groups (gr) | 1 | 2 | 3 | 4 | 5 | 6 | 7 | 8 |
|---|---|---|---|---|---|---|---|---|
| No of sections | 433 | 37 | 190 | 192 | 23 | 19 | 85 | 65 |
| Total length, km | 239.1 | 19.1 | 99.7 | 100.7 | 11.6 | 9.5 | 44.3 | 33.7 |
| Subdivision according to traffic volume level* (V_level): | | | | | | | | |
| 1 low | 12% | - | 47% | 20% | - | - | 32% | 49% |
| 2 medium | 46% | 95% | 35% | 41% | 9% | 21% | 38% | 17% |
| 3 high | 42% | 5% | 17% | 39% | 91% | 79% | 31% | 34% |
| Total crash numbers, in 3 years: | | | | | | | | |
| injury crashes | 660 | 39 | 173 | 253 | 67 | 27 | 52 | 113 |
| total crashes | 1781 | 109 | 763 | 1019 | 396 | 288 | 105 | 679 |
| severe crashes | 113 | 5 | 29 | 48 | 7 | 9 | 7 | 23 |
| single-vehicle crashes | 182 | 12 | 59 | 105 | 14 | 2 | 12 | 33 |

b – Unit characteristics (on the whole dataset, N=1044)

| Unit characteristics | Variable** | Values and shares according to categories |
|---|---|---|
| Width of carriageway, m | cwRL | mean: 7.6, min: 3.1, max: 14.3 |
| Number of lanes on carriageway | NlaneRL | mean: 2.1, min: 2, max: 4 |
| Speed limit, km/h (categories) | slimit | 50-70 (14%), 80 (4%), 90 (64%), 100 (11%), 110 (7%) |
| Type of road building (categories) | rb | 0 filling (75%), 1 cutting or excavation (25%) |
| Width of external shoulders# (categories) | shERL | 1 narrow (6%), 1.5 narrow-medium (3%), 2 medium (15%), 2.5 medium-wide (20%), 3 wide (56%) |
| Width of internal shoulders# (categories) | shIRL | 1 narrow (30%), 1.5 narrow-medium (10%), 2 medium (21%), 2.5 medium-wide (11%), 3 wide (28%) |
| Presence of barriers on roadsides, % of section length | barLR | mean: 84%, min: 0, max: 100% |
| Presence of concrete barriers on roadsides, % of section length | barLRc | mean: 4%, min: 0, max: 100% |

Notes: *On dual-carriageway roads, AADT: 1 – up to 20,000 vehicles; 2 – 20,000-40,000; 3 – over 40,000. On motorways, AADT: 1 – up to 40,000 vehicles; 2 – 40,000-80,000; 3 – over 80,000. **RL means an average of both travel directions (right and left). # Shoulder width as defined by the NTIC survey: narrow – up to 1.5 m; medium – between 1.5-2.5 m; wide – over 2.5 m.

$$E\{A\} = L \exp \sum [\beta_i x_i] \tag{1}$$

where $E\{A\}$ is the expectancy of crash numbers on the unit; $L$ – the unit length; $x_i$ - unit characteristics, $\beta_i$ - model coefficients. The models were fitted using the GLIMMIX procedure of SAS, with all the variables. The significance of impacts was estimated using *Type III tests* of fixed effects. Based on the models, *post hoc* pairwise comparisons were performed to ascertain differences in crash expectancy on road units with various barrier types. If $A_{gr}$ denotes the expected number of crashes in one group (gr) and $A_{-gr}$ - the expected number of crashes in another group (_gr), then the ratio between the crash numbers in both groups is estimated as:

$$A_{gr}/A_{-gr} = e^{Estimate} \tag{2}$$

where: $e$ – the exponent and *Estimate* values are evaluated using the models. When the *Estimate* value is positive and significant (Adj p < 0.05), this means that the expected crash number in the first group (gr) is higher than in the second group (_gr); when the *Estimate* value is negative and significant, the expected number of crashes in the first group is lower than in the second group. For the p-values, Bonferroni multiple comparison adjustments were applied (Bland and Altman 1995).

Furthermore, based on the models, crash savings were estimated for a case when the old NJ-barrier is replaced by a new barrier type, on the median. The benefit-cost

324 V. Gitelman and E. Doveh

ratio (BCR) evaluations were performed for typical road and traffic conditions, where the crash costs saved during the barrier's length of life were compared to the costs of barrier's installation and maintenance. The method of BCR evaluation of safety features is given, e.g., in Yannis et al. (2008). The evaluation framework was 25 years, with 7% interest rate. The barrier costs were defined based on the NTIC price list, as follows:

- the installation costs for 1 m barrier length: 340, 630 and 350 New Israeli Shekels (NIS; 1 Euro $\sim$ 4.5 NIS), for Step in-situ, pre-cast and steel barriers, respectively (all barriers are double-sided, with H2-level);
- the annual maintenance costs of the new barriers were 0%, 1% and 5%, respectively.

Average crash costs came from the local guidelines for appraisal of transport projects (Nohal Prat 2012) and were estimated as 244,000 NIS for an injury crash, on rural roads (at 2012 prices).

## 5 Results

### 5.1 Safety Performance of Road Sections with Various Median Barrier Types, On Dual-Carriageway Roads

Table 2 presents an example of the explanatory model fitted to injury crashes on dual-carriageway road sections. In this model, a significant impact on crashes (according to *Type III tests* of fixed effects) was found for the section group and the level of traffic volume (p < 0.001), widths of external and internal shoulders (p < 0.01), presence of roadside barriers (p < 0.05), and a close to significant impact – for width of carriageway (p = 0.11). The model shows that comparison-groups 1, 3 (with old NJ and new steel barriers) are associated with higher crash numbers than sections with Step-barriers, on medians; that lower traffic volumes lead to a decrease in crashes; that a wider carriageway and wider external shoulders increase the crash risk, while wider internal shoulders reduce the crash risk.

Table 3 presents the results of pairwise comparisons between the crash expectancy in various groups of road sections, given different traffic levels. Only significant differences in the expectancy of crashes are worth consideration. For injury crashes, the differences show that at all levels of traffic, the comparison sections with old NJ-barriers (group 1) are associated with higher crash numbers than sections with Step-barriers (group 7). In addition, under low traffic volumes, group 1 has higher crash numbers than group 3 (with new steel barriers). Hence, the analysis of injury crashes indicated that road sections with Step-barriers on medians were safer than those with old NJ-barriers, whereas no significant difference was found between the safety level of road sections with Step-barriers compared to other new barrier types.

Furthermore, Table 3 shows the results of comparisons between the crash expectancy on sections with various barrier types, based on the models fitted to single-vehicle and total crashes. (For severe crashes, no model results were significant due to low crash numbers.) Regarding single-vehicle crashes, no significant differences (with

A Comparative Evaluation of the Safety Performance of Median Barriers 325

**Table 2.** Explanatory model fitted to injury crashes on dual-carriageway road sections

| Variables: categories | Estimate | Standard error | t value | Pr > \|t\| |
|---|---|---|---|---|
| Intercept | −2.668 | 1.628 | −1.64 | 0.102 |
| gr: 1 | 0.858 | 0.322 | 2.67 | 0.008 |
| gr: 2 | 0.276 | 0.804 | 0.34 | 0.732 |
| gr: 3 | 0.777 | 0.363 | 2.14 | 0.032 |
| gr: 7 | 0 | | | |
| V_level: 1 | −1.152 | 0.501 | −2.3 | 0.022 |
| V_level: 2 | −0.153 | 0.404 | −0.38 | 0.706 |
| V_level: 3 | 0 | | | |
| gr*V_level: 1 1 | 0.485 | 0.537 | 0.9 | 0.367 |
| gr*V_level: 1 2 | −0.204 | 0.411 | −0.5 | 0.620 |
| gr*V_level: 1 3 | 0 | | | |
| gr*V_level: 2 2 | 0.100 | 0.868 | 0.12 | 0.908 |
| gr*V_level: 2 3 | 0 | | | |
| gr*V_level: 3 1 | −0.455 | 0.564 | −0.81 | 0.419 |
| gr*V_level: 3 2 | −0.406 | 0.462 | −0.88 | 0.380 |
| gr*V_level: 3 3 | 0 | | | |
| gr*V_level: 7 1 | 0 | | | |
| gr*V_level: 7 2 | 0 | | | |
| gr*V_level: 7 3 | 0 | | | |
| year: 2011 | 0.082 | 0.103 | 0.8 | 0.426 |
| year: 2012 | 0.064 | 0.102 | 0.62 | 0.533 |
| year: 2013 | 0 | | | |
| NlaneRL: 2 | −0.321 | 0.968 | −0.33 | 0.741 |
| NlaneRL: 2.5 | −0.578 | 1.010 | −0.57 | 0.568 |
| NlaneRL: 3 | −1.253 | 0.994 | −1.26 | 0.208 |
| NlaneRL: 4 | 0 | | | |
| slimit: 50_70 | −0.038 | 0.122 | −0.31 | 0.759 |
| slimit: 80 | 0.120 | 0.209 | 0.57 | 0.567 |
| slimit: 90 | 0 | | | |
| rb: 0 | −0.026 | 0.099 | −0.27 | 0.790 |
| rb: 1 | 0 | | | |
| log_cwRL | 0.794 | 0.493 | 1.61 | 0.108 |
| log_shERL | 2.592 | 0.857 | 3.03 | 0.003 |
| log_shERL*log_shERL | −1.598 | 0.565 | −2.83 | 0.005 |
| log_shIRL | −0.458 | 0.125 | −3.67 | 0.000 |
| log_barLR | 0.560 | 0.258 | 2.17 | 0.030 |
| log_barLR*log_barLR | 0.193 | 0.139 | 1.39 | 0.164 |
| log_barLRc | −0.908 | 0.532 | −1.7 | 0.088 |
| log_barLRc*log_barLRc | −0.386 | 0.181 | −2.14 | 0.033 |
| Scale | 1.126 | 0.136 | | |

Notes: Fit statistics: -2 Log Likelihood = 3549.51; AIC = 3605.51;
Pearson Chi-Square/DF = 1.01. *log* indicates logarithm

326 V. Gitelman and E. Doveh

**Table 3.** Results of pair comparisons between crash expectancy in section groups, on dual-carriageway roads

| Traffic volume level | Section group | | Injury crashes | | | Single-vehicle crashes | | | Total crashes | | |
|---|---|---|---|---|---|---|---|---|---|---|---|
| | gr | _gr | Estimate | St. error | Adj P | Estimate | St. error | Adj P | Estimate | St. error | Adj P |
| Low | 1 | 3 | 1.02 | 0.30 | 0.002 | 0.33 | 0.42 | 1.0 | 0.29 | 0.27 | 0.847 |
| Low | 1 | 7 | 1.34 | 0.43 | 0.006 | 2.24 | 1.06 | 0.105 | 0.53 | 0.36 | 0.404 |
| Low | 3 | 7 | 0.32 | 0.45 | 1.0 | 1.92 | 1.06 | 0.213 | 0.25 | 0.35 | 1.0 |
| Medium | 1 | 2 | 0.28 | 0.22 | 1.0 | 0.11 | 0.35 | 1.0 | 0.12 | 0.21 | 1.0 |
| Medium | 1 | 3 | 0.28 | 0.18 | 0.740 | 0.00 | 0.29 | 1.0 | 0.50 | 0.19 | 0.040 |
| Medium | 1 | 7 | 0.65 | 0.26 | 0.063 | 1.27 | 0.62 | 0.239 | 1.15 | 0.26 | <.0001 |
| Medium | 2 | 3 | 0.00 | 0.25 | 1.0 | −0.11 | 0.40 | 1.0 | 0.38 | 0.24 | 0.729 |
| Medium | 2 | 7 | 0.38 | 0.32 | 1.0 | 1.16 | 0.68 | 0.540 | 1.02 | 0.31 | 0.007 |
| Medium | 3 | 7 | 0.37 | 0.29 | 1.0 | 1.26 | 0.65 | 0.316 | 0.65 | 0.29 | 0.167 |
| High | 1 | 2 | 0.58 | 0.74 | 1.0 | 13.89 | 974.11 | 1.0 | 0.59 | 0.72 | 1.0 |
| High | 1 | 3 | 0.08 | 0.21 | 1.0 | −0.07 | 0.33 | 1.0 | −0.98 | 0.21 | <.0001 |
| High | 1 | 7 | 0.86 | 0.32 | 0.046 | 0.22 | 0.46 | 1.0 | 0.36 | 0.30 | 1.0 |
| High | 2 | 3 | -0.50 | 0.77 | 1.0 | −13.96 | 974.11 | 1.0 | −1.57 | 0.74 | 0.212 |
| High | 2 | 7 | 0.28 | 0.80 | 1.0 | −13.67 | 974.11 | 1.0 | −0.23 | 0.78 | 1.0 |
| High | 3 | 7 | 0.78 | 0.36 | 0.195 | 0.29 | 0.53 | 1.0 | 1.34 | 0.34 | 0.001 |

*Adj p < 0.05*) were found between various section groups. However, a number of trends (with *Adj p < 0.25*) can be pointed out: at low traffic volumes, the crash expectancy is lower on sections with Step-barriers compared to groups 1, 3 (with old NJ and new steel barriers); at medium traffic volumes, the crash expectancy is lower on sections with Step-barriers compared to group 1. In total crashes, the safety level of sections with Step-barriers was better than that of sections with old NJ-barriers and with new pre-cast concrete barriers, under medium traffic volumes, and was better than safety level of sections with new steel barriers, under high traffic volumes. The results of comparisons between the sections with old NJ barriers (group 1) and new steel barriers (group 3) were mixed indicating a better safety level of group 3 under medium traffic volumes and worse – under high traffic volumes.

## 5.2 Safety Performance of Road Sections with Various Median Barrier Types, On Motorways

In the explanatory model adjusted to injury crashes on motorways (Table 4), a significant effect on crashes was found for traffic volume level and speed limits (p < 0.001), the interaction between the section group and traffic level (p < 0.01), number of lanes (p < 0.05) and width of internal shoulders (p < 0.01). The model indicated that, in general, in comparison-groups 4 and 6 (with old NJ and new steel

A Comparative Evaluation of the Safety Performance of Median Barriers     327

barriers) the crash risk was lower than in the treatment group 8 (with Step-barriers); that lower traffic volume was associated with a decrease in crashes; that sections with lower speed limits were associated with higher crash risk compared to those with 110 km/h speed limit, and wider internal shoulders – with lower crash risk.

Table 5 shows the relative crash expectancy on section groups with various barrier types, based on the models fitted for motorways. For injury crashes, the significant differences indicated that, at low traffic volumes, more crashes are expected in group 4 (with old NJ-barriers) compared to group 8 (with Step-barriers); however, at high traffic volumes, fewer crashes are expected in group 6 (with new steel barriers) than in group 8 (with Step-barriers). In addition, the observed trends (with $Adj\ p < 0.25$) indicated that, at high traffic volumes, sections with old NJ-barriers and new pre-cast concrete barriers (groups 4, 5) are associated with higher crash expectancy than sections with new steel barriers (group 6).

Similarly, for single-vehicle crashes, the significant differences showed that, at low traffic volumes, sections with old NJ-barriers (group 4) were associated with higher crash numbers than sections with Step-barriers (group 8), whereas, at high traffic volumes, sections with new steel barriers (group 6) were safer than sections with all other barrier types (the differences were significant for NJ- and Step-barriers; a trend with $Adj\ p < 0.25$ was found for pre-cast barriers). Regarding total crashes, the results were mixed indicating a better safety level of sections with Step-barriers compared to sections with NJ-barriers, at low traffic volumes, but an opposed finding (worse safety level of sections with Step-barriers versus those with NJ-barriers) at medium traffic volumes. Another significant difference indicated that, under medium traffic volumes, fewer crashes are expected on sections with new steel barriers than on sections with Step-barriers. (For severe crashes no significant results were obtained for motorways.)

In summary, on motorways, at low traffic volumes (below 40,000 vehicles per day), a better safety level was found for sections with Step-barriers compared to those with old NJ-barriers, while at higher traffic volumes, sections with new steel barriers showed better safety performance related to sections with all other barrier types.

## 5.3    Economic Benefits of Replacing the Old NJ-Barriers by Other Barrier Types

Using the models developed, annual crash savings were estimated for typical road sections when an old NJ-barrier on median is replaced by a new barrier type, and then BCRs were evaluated for the barrier's length of life. Based on engineering examinations of the road layouts it was assumed that new barriers can be installed on existing medians and thus no additional investments are required for changing the road layout due to barrier's replacement. Concerning barrier repairs following vehicle collisions, the local practice indicated that usually a traffic lane closing is not applied due to sufficient space of internal shoulders and medians on existing roads. When a lane closing is required, the roadworks take place at night or in hours of low traffic. Thus, unlike previous studies (Steinauer et al. 2004; Williams 2007), the economic values of time losses during barrier-repair works were negligible, for all barrier types, and hence were not quantified in the current study. In addition, traffic delay costs due to crashes

## 328 V. Gitelman and E. Doveh

**Table 4.** Explanatory model fitted to injury crashes on motorway sections

| Variables: categories | Estimate | Standard error | t value | Pr > \|t\| |
|---|---|---|---|---|
| Intercept | 2.537 | 2.767 | 0.92 | 0.359 |
| gr: 4 | −0.487 | 0.254 | −1.92 | 0.056 |
| gr: 5 | −0.155 | 0.403 | −0.39 | 0.700 |
| gr: 6 | −1.193 | 0.375 | −3.18 | 0.002 |
| gr: 8 | 0.000 | | | |
| V_level: 1 | −2.567 | 0.511 | −5.02 | < .0001 |
| V_level: 2 | −0.819 | 0.433 | −1.89 | 0.059 |
| V_level: 3 | 0.000 | | | |
| gr*V_level: 4 1 | 1.846 | 0.502 | 3.68 | 0.000 |
| gr*V_level: 4 2 | 0.379 | 0.503 | 0.75 | 0.451 |
| gr*V_level: 4 3 | 0.000 | | | |
| gr*V_level: 5 2 | −12.4 | 656.4 | −0.02 | 0.985 |
| gr*V_level: 5 3 | 0.000 | | | |
| gr*V_level: 6 2 | −13.1 | 463.4 | −0.03 | 0.978 |
| gr*V_level: 6 3 | 0.000 | | | |
| gr*V_level: 8 1 | 0.000 | | | |
| gr*V_level: 8 2 | 0.000 | | | |
| gr*V_level: 8 3 | 0.000 | | | |
| year: 2011 | 0.253 | 0.154 | 1.65 | 0.100 |
| year: 2012 | −0.036 | 0.159 | −0.23 | 0.820 |
| year: 2013 | 0.000 | | | |
| NlaneRL: 2 | 0.337 | 1.299 | 0.26 | 0.795 |
| NlaneRL: 2.5 | 1.140 | 1.232 | 0.92 | 0.355 |
| NlaneRL: 3 | 1.066 | 1.179 | 0.9 | 0.366 |
| NlaneRL: 3.5 | 1.853 | 1.197 | 1.55 | 0.122 |
| NlaneRL: 4 | 0.000 | | | |
| slimit: 50_70 | 0.773 | 0.395 | 1.96 | 0.050 |
| slimit: 90 | 0.879 | 0.234 | 3.75 | 0.000 |
| slimit: 100 | 0.894 | 0.249 | 3.6 | 0.000 |
| slimit: 110 | 0.000 | | | |
| rb: 0 | −0.158 | 0.193 | −0.82 | 0.415 |
| rb: 1 | 0.000 | | | |
| log_cwRL | −1.382 | 0.960 | −1.44 | 0.150 |
| log_shERL | 0.879 | 0.960 | 0.92 | 0.360 |
| log_shERL*log_shERL | −0.927 | 0.712 | −1.3 | 0.193 |
| log_shIRL | −2.013 | 0.718 | −2.8 | 0.005 |
| log_shIRL*log_shIRL | 1.677 | 0.617 | 2.72 | 0.007 |
| log_barLR | −0.714 | 0.452 | −1.58 | 0.114 |
| log_barLR*log_barLR | −0.522 | 0.358 | −1.46 | 0.145 |
| log_barLRc | −0.037 | 0.121 | −0.31 | 0.759 |
| Scale | 1.107 | 0.189 | | |

Notes: Fit statistics: -2 Log Likelihood = 1537.11; AIC = 1595.11;
Pearson Chi-Square/DF = 1.05. *log* indicates logarithm

**Table 5.** Results of pair comparisons between crash expectancy in section groups, on motorways

| Traffic volume level | Group of sections | | Injury crashes | | | Single-vehicle crashes | | | Total crashes | | |
|---|---|---|---|---|---|---|---|---|---|---|---|
| | $gr$ | $\_gr$ | Estimate | St. error | Adj P | Estimate | St. error | Adj P | Estimate | St. error | Adj P |
| Low | 4 | 8 | 1.36 | 0.46 | 0.003 | 1.61 | 0.71 | 0.023 | 1.81 | 0.42 | < .0001 |
| Medium | 4 | 5 | 12.5 | 656.4 | 1.0 | 11.7 | 644.6 | 1.0 | 12.2 | 387.2 | 1.0 |
| Medium | 4 | 6 | 14.2 | 463.4 | 1.0 | 13.4 | 455.5 | 1.0 | 1.57 | 0.74 | 0.209 |
| Medium | 4 | 8 | −0.11 | 0.44 | 1.0 | 0.81 | 0.70 | 1.0 | −1.20 | 0.44 | 0.039 |
| Medium | 5 | 6 | 1.70 | 803.5 | 1.0 | 1.70 | 789.3 | 1.0 | −10.7 | 387.2 | 1.0 |
| Medium | 5 | 8 | −12.6 | 656.4 | 1.0 | −10.8 | 644.6 | 1.0 | −13.4 | 387.2 | 1.0 |
| Medium | 6 | 8 | −14.3 | 463.4 | 1.0 | −12.5 | 455.5 | 1.0 | −2.77 | 0.84 | 0.006 |
| High | 4 | 5 | −0.33 | 0.35 | 1.0 | 0.74 | 0.51 | 0.896 | −0.31 | 0.37 | 1.0 |
| High | 4 | 6 | 0.71 | 0.35 | 0.248 | 3.39 | 1.09 | 0.011 | 0.40 | 0.35 | 1.0 |
| High | 4 | 8 | −0.49 | 0.25 | 0.333 | 0.00 | 0.36 | 1.0 | −0.21 | 0.30 | 1.0 |
| High | 5 | 6 | 1.04 | 0.46 | 0.153 | 2.65 | 1.19 | 0.155 | 0.71 | 0.48 | 0.814 |
| High | 5 | 8 | −0.16 | 0.40 | 1.0 | −0.74 | 0.57 | 1.0 | 0.10 | 0.46 | 1.0 |
| High | 6 | 8 | −1.19 | 0.37 | 0.009 | −3.39 | 1.11 | 0.014 | −0.61 | 0.39 | 0.741 |

were already accounted for in the crash costs as those were estimated following the local guidelines for assessment of transport projects (Nohal Prat 2012).

Table 6 provides examples of evaluation results, for dual-carriageway and motorway road sections of 1 km in length, when savings in injury crashes are considered. It shows the annual number of saved crashes, the present value of crash costs saved over the barrier's service life and the BCR estimated for various barrier types. The results are presented only for the cases which are applicable for barrier effects' comparisons, based on the modeling results (see Tables 3, 5); values based on significant differences between various barrier types are highlighted. The results indicated that, for a dual-carriageway road section, replacing the old NJ-barrier by a Step-barrier, on the median, is feasible at all levels of traffic, and preferable to other barrier types. In addition, for this road type, safety benefits exceed the costs when the old barrier is replaced by a new steel barrier, at traffic volumes below 40,000, or by a new pre-cast barrier, at traffic volumes over 40,000 vehicles a day. For a motorway section, the feasibility of replacing the NJ-barrier was found for new steel barriers, at medium and high traffic levels, over 40,000 vehicles a day. For Step-barriers, positive crash savings and economic benefits from the replacement of old NJ-barriers on medians were relevant for low traffic volumes only, up to 40,000 vehicles a day; for pre-cast concrete barriers, the replacement of old NJ-barriers is feasible for medium traffic volumes (between 40–80 thousand vehicles a day).

**Table 6.** Results of economic evaluation of 1 km road section, when an old NJ-barrier is replaced by a new barrier type, on the median (based on injury crashes saved)

| New barrier type | On dual-carriageway roads | | | On motorways | | |
|---|---|---|---|---|---|---|
| | Traffic volume level: | | | Traffic volume level: | | |
| | Low | Medium | High | Low | Medium | High |
| a - Number of crashes saved annually | | | | | | |
| Pre-cast concrete | na | 0.20 | 0.53 | na | 0.33 | $-^{\#}$ |
| Steel | $0.39^{*}$ | 0.21 | 0.09 | na | 0.33 | 0.26 |
| Step in-situ | $0.45^{*}$ | $0.40^{*}$ | $0.69^{*}$ | $0.18^{*}$ | $-^{\#}$ | $-^{\#}$ |
| b - Present value of crash costs saved over 25 years (thousand NIS) | | | | | | |
| Pre-cast concrete | na | 578 | 1,501 | na | 933 | – |
| Steel | $1,117$ | 587 | 265 | na | 933 | 734 |
| Step in-situ | $1,290$ | $1,143$ | $1,960$ | 524 | – | – |
| c - BCR of old barrier's replacement | | | | | | |
| Pre-cast concrete | na | 0.8 | 2.1 | na | 1.3 | – |
| Steel | $2.0$ | 1.1 | 0.5 | na | 1.7 | 1.3 |
| Step in-situ | $3.8$ | $3.4$ | $5.8$ | $1.5$ | – | – |

Notes: na – not applicable (using the barrier effects' comparisons); $^{*}$estimates based on significant effects; $^{\#}$negative values obtained, i.e. crash increase is expected following the old barrier's replacement. NIS – New Israeli Shekel.

## 6 Discussion and Conclusions

This study developed statistical models for estimating the safety level of non-urban highway sections with various types of safety barriers on the medians, such as the old NJ-shaped in-situ barrier and the new barrier types (which satisfy the EN1317 requirements) - a Step-shaped in-situ barrier, pre-cast concrete barriers and steel guardrails. The study compared the safety performance of highway sections with various barrier types aiming to examine the safety impacts of in-situ concrete versus other barriers as well as the new versus old barrier types and, in particular, to examine the differences in safety performance of road sections with the new Step-shaped in-situ barriers relatively to other barrier types. Such safety comparisons can be relevant for practical questions raised by the road authorities, for example, regarding the urgency of replacement of the old barriers by new ones or when a certain barrier is preferred for installation, among various new barrier types. Specifically, a question concerning the safety impacts of the Step-shaped in-situ barrier had been raised for a while, as it is a new barrier type (i.e. tested in accordance with the new norms) but still an in-situ concrete barrier that may be less forgiving in collisions with small vehicles.

In the study, differences in crash expectancy were evaluated for sections with various barrier types, while controlling for traffic volumes and other road infrastructure characteristics. The results indicated that, in general, the safety level of highway sections with Step in-situ barriers did not fall below the safety level of sections with other barrier types. On dual-carriageway roads, the sections with Step-barriers had a better safety level relative to those with old NJ-barriers and, sometimes, to other (new) barrier types. On motorways, the safety level of sections with Step-barriers was better or similar to that with the old NJ-barriers, with no difference compared to the (new) pre-cast barriers, while sections with (new) steel guardrails had a better safety level, particularly at higher traffic volumes. The findings were stronger in the analysis of injury crashes and indicated similar trends in single-vehicle crashes. Overall, the study findings did not support the statements on higher crash injury associated with Step in-situ barriers related to other new barrier types. At the same time, the study results did not suggest a bulk usage of Step in-situ barriers on highway medians because safety benefits were also observed for other barrier types. Thus, the current policy of using in-situ Step-barriers on mostly narrow medians of non-urban highways can be continued.

In Israel, the national guidelines (MOT 2020) define which barrier type should be selected for installation, depending on the type of road, traffic volume and composition, e.g. share of trucks and buses, and roadside conditions. The barrier characteristics, in this context, are given in terms of the containment level, working width and vehicle intrusion, as defined by CEN (2010). Concerning the occupant risk values, barriers with impact severity levels A and B are allowed for use in general, while a C-level barrier can be installed as an exception only, on a short road section, when a suitable A- or B-level barrier is unavailable among the approved barrier models. Currently, tens of barrier models, including steel, pre-cast and in-situ barriers, with various containment levels, are approved for use in Israel. (To note, the Step in-situ barrier is approved in Israel with a B severity level, that allows its general use). The national guidelines do not restrict the use of in-situ concrete versus steel or pre-cast barriers, once they belong to the list of approved safety devices in the country (see more details in Gitelman et al. 2014). However, a general preference for more forgiving (A-level) barriers is suggested, when applicable. In practice, the road authorities select the barrier models for a new installation or for upgrading of existing barriers accounting for various considerations, e.g. initial costs and maintenance, road layout limitations, reducing the number of barrier transitions, etc., while the barriers should satisfy the demands on the performance classes (containment level, working width and vehicle intrusion) as defined in the project. Recent developments show that various barrier types are currently applied on Israeli roads, without a tendency to a bulk usage of in-situ concrete barriers, and that the current list of approved barrier models provides solutions for the majority of road conditions (where various barrier performance classes are required).

The economic evaluations showed benefits for replacing the old NJ-barrier by a Step-barrier, on the medians of dual-carriageway roads, for all levels of traffic, and on motorways, for lower traffic volumes, below 40,000 vehicles per day. Additionally, the replacement of old NJ-barriers by new steel barriers was found to be beneficial on motorways with traffic volumes over 40,000 vehicles per day. The safety level of highway sections with old NJ-barriers was generally lower related to sections with new barrier types, thus, supporting the transition to the new generation of safety barriers, in line with previous research (Gitelman et al. 2014; Cafiso et al. 2017). Albeit, one should remember that a positive safety impact of the new barriers can be related also to complementary infrastructure improvements of the road sections where the barriers were replaced.

In contrast to previous studies that found economic benefits of using in-situ concrete barriers mainly on medians of heavily-trafficked motorways (Steinauer et al. 2004; Williams 2007), this study obtained somewhat different results. Indeed, for dual-carriageway highways in Israel, the installation of the new in-situ barriers was found to be beneficial, in line with the previous research. However, for motorways with high traffic volumes higher economic benefits were demonstrated for steel median barriers. Such differences in findings may stem from a distinction in the evaluation frameworks applied in this compared to other studies, e.g. the relative safety benefits from a replacement of the old barriers by the new ones that were considered by the current study versus a direct comparison between the whole-life costs of in-situ concrete and steel barriers that was applied in other studies. In addition, differences exist in the costs of barrier installation and maintenance works, in Israel compared to other countries.

The current study limitations lie in the relatively short length of road sections with the new barriers, which were found on the non-urban road network, at the time of the study analyses, which might lead to a lack of significant results in some cases. Repeated studies with larger data samples may contribute to a better understanding of the relationships between various barrier types and safety performance of road sections. Additionally, more detailed examinations of collisions with safety barriers, as associated with various barrier types, both on medians and on roadsides, would be useful in future research.

In summary, this study aspired to extend the empirical background for a more effective use of safety barriers on the rural road network. Indeed, some useful insights were attained, e.g. concerning the safety benefits of the transition to new barrier types and providing support for the use of the new in-situ concrete barrier on highway medians. However, the complexity of the relationship between the barrier types and the road conditions was exhibited as well, indicating that the barrier choice is not simple and not similar in all conditions and that a variety of factors should be considered in order to show a safety-related preference of a certain barrier type.

**Acknowledgements.** Appreciation is extended to the National Transport Infrastructure Company of Israel that commissioned this research. The authors would like to thank Mss. Fany Pesahov and Eng. Roby Carmel for their great assistance in the preparation of the study database.

# References

AASHTO: Roadside Design Guide. The American Association of State Highway and Transportation Officials, Washington, DC (2011)

Bland, J.M., Altman, D.G.: Multiple significant tests: the Bonferroni method. BMJ **1995**(310), 170 (1995)

Cafiso, S., D'Agostino, C., Persaud, B.: Investigating the influence on safety of retrofitting Italian motorways with barriers meeting a new EU standard. Traffic Inj. Prev. **18**(3), 324–329 (2017)

Cafiso, S., D'Agostino, C.: Evaluating the safety benefit of retrofitting motorways section with barriers meeting a new EU standard: Comparison of observational before-after methodologies. J. Traffic Transp. Eng. **4**(6), 555–563 (2017)

CEDR: Forgiving roadsides design guide. Conference of European Directors of Roads (2012)

CEN: EN 1317-2 Road restraint systems – Part 2: Performance classes, impact test acceptance criteria and test methods for safety barriers including vehicle parapets. European Committee for Standardization (CEN), Brussels (2010)

Elvik, R., Hoye, A., Vaa, T., Sorensen, M.: The Handbook of Road Safety Measures. Bingley, Emerald (2009)

EUPAVE: Concrete safety barriers: a safe and sustainable choice. European Concrete Paving Association, Brussels (2012)

FHWA: Crash Modification Factors Clearinghouse (2013). http://www.cmfclearinghouse.org/. Accessed 20 Dec 2013

Gitelman, V., Hakkert, S., Peleg, K., Givon, A., Koral, A.: A comparative evaluation of safety performance of concrete versus steel barriers on rural roads. Research Report 3007/2006, Transportation Research Institute, Technion, Haifa, Israel (2006)

Gitelman, V., Carmel, R., Doveh, E., Pesahov, F., Hakkert, S.: An examination of the effectiveness of a new generation of safety barriers. Transport Research Arena 2014, Paris, paper #17466 (2014)

Hu, W., Donnell, E.T.: Median barrier crash severity: some new insights. Accid. Anal. Prev. **42**, 1697–1704 (2010)

IAN: Interim Advice Note 60/05. The introduction of a new Highways Agency policy for the performance requirements for central reserve safety barriers on motorways, UK (2005)

Jehu, V.J., Pearson, L.C.: Impacts of European cars and a passenger coach against shaped concrete barriers. Transport and Road Research Laboratory, Crowthorne, UK (1977)

Martin, J.L., Derrien, Y., Bloch, P., Boissier, G.: Severity of run-off-crashes whether motorway hard shoulders are equipped with a guardrail or not. In: Traffic Safety on three Continents Conference, Moscow (2001)

Martin, J.L., Mintsa-Eya, C., Goubel, C.: Long-term analysis of the impact of longitudinal barriers on motorway safety. Accid. Anal. Prev. **59**, 443–451 (2013)

MASH: Manual for Assessing Safety Hardware. American Association of State Highway and Transportation Officials, Washington, DC (2016)

McDevitt, C.F.: Basics of concrete barriers. Public Roads **63**(5), 10–14 (2000)

MOT: Guidelines for Selection and Installation of Permanent Safety Barriers on Rural Roads. Ministry of Transport, Jerusalem, Israel (2020). (in Hebrew)

Nohal Prat: Guidelines for examining feasibility of transport projects. Ministry of Finance, Ministry of Transport, Jerusalem, Israel (2012). (in Hebrew)

Ray, M.H., McGinnis, R.G.: Synthesis of Highway Practice 244: Guardrail and Median Barrier Crashworthiness. Transportation Research Board, Washington, DC (1997)

Steinauer, B., Kathmann, T., Mayer, G., Becher, T.: Criteria for using concrete safety barriers. Federal Highway Research Institute, BASt V112, Germany (2004)

Tarko, A.P., Villwock, N.M., Blond, N.: Effect of median design on rural freeway safety: flush medians with concrete barriers and depressed medians. Transp. Res. Rec. **2060**, 29–37 (2008)

Verweij, A.: Developing H4-safety barriers for Dutch motorways, Road Safety on Three Continents Conference, Pretoria, VTI-konferens CD-ROM (2000)

Williams, G.L.: Whole life cost-benefit analysis for median safety barriers. Published Project Report PPR 279, TRL Limited, UK (2007)

Yannis, G., Gitelman, V., Papadimitriou, E., Hakkert, A.S., Winkelbauer, M.: Testing a framework for the assessment of road safety measures. Transp. Rev. **28**(3), 281–301 (2008)

Zou, Y., Tarko, A.P., Chen, E., Romero, M.A.: Effectiveness of cable barriers, guardrails, and concrete barrier walls in reducing the risk of injury. Accid. Anal. Prev. **72**, 55–65 (2014)

# Ascendi's Safety Barriers Upgrading Program

Telma Silva[(✉)] and João Neves

Ascendi IGI S.A., Porto, Portugal
{tsilva, jneves}@ascendi.pt

**Abstract.** Over the years, there have been some changes regarding road vehicle restraint systems, especially concerning the requirements and the selection criteria for its implementation.

In Portugal, the guidance manual that set the requirements for the safety barriers in accordance with the EN1317 was published in 2010. At that time, most of the Ascendi's network was already constructed and under operation, being the safety barriers applied based on the national requirements available at the time of the project.

In order to evaluate the network compliance with the current established requirements, Ascendi carried out an upgrading program (Ascendi's Safety Barriers Upgrading Program) that started in 2018 and will run until 2023.

This paper will cover the Upgrading Program that was established, going through the challenges of its implementation that included the development and certification of a brand-new safety barrier for bridges with a 30 cm curb in cooperation with Road Steel Engineering, S.L.

**Keywords:** Roadside safety · EN1317 · Road vehicle restraint systems · Safety barrier containment level · Safety barrier development and certification

## 1 Introduction

The 868 km of the Ascendi's road concessions were awarded between 1999 and 2007. The company was responsible for the design, construction and operation of this network and the road design considered the national requirements for safety barriers available at the time developed by *Infraestruturas de Portugal, S.A.* guidance. These technical requirements were mainly based on the French Regulation *"Dispositifs de Retenue des Véhicules"* of 1988.

In December 2007, the European Standard for Vehicle Restraint Systems (EN 1317) was published. The document set the testing and performance standards for the roadside restraining systems products without specifying any guidance regarding the product selection criteria – this was to be set by each Member State.

Later, in 2010, the Portuguese Road Institute (InIR) published a draft guidance manual *"Sistemas de Retenção Rodoviários – Manual de Aplicação"* that finally set the requirements for the selection criteria of safety barriers in the Portuguese roads in accordance with the EN 1317.

Going through these changes, Ascendi decided to develop a program to assess if the safety barriers installed on its network met the necessary requirements and proceed with their upgrading whenever necessary.

© The Author(s), under exclusive license to Springer Nature Switzerland AG 2022
A. Akhnoukh et al. (Eds.): IRF 2021, SUCI, pp. 335–344, 2022.
https://doi.org/10.1007/978-3-030-79801-7_23

## 2 Safety Barriers Upgrading Program

### 2.1 Establishing the Upgrading Program

To evaluate the network compliance with best practices regarding safety barriers, in 2018 and 2019, Ascendi carried out an external audit using InIR's Guidance Manual as reference. Although in a draft version, the InIR document was considered as the most suitable guidance for the selection of vehicle restraint systems on new and operating roads in Portugal.

The audit report recommended the upgrade of safety barriers containment level according to the 4 intervention priorities described below.

Intervention Priorities.

- 1st Priority:
  **Median Barriers** – Where carriageways have difference in level greater than 1 m, as shown in Fig. 1. These cases must all be revised given the probability and the seriousness of its consequences in case of an accident occurring due to the crossing of the median barriers with invasion of the opposite carriageway.

**Fig. 1.** 1st Priority (Median Barriers) - Carriageways with difference in level greater than 1 m.

- 2nd Priority:
  **Bridges** – Upgrading the containment level on the deck and the transitions to mainline barrier, as shown in Fig. 2 and 3, respectively.

Ascendi's Safety Barriers Upgrading Program 337

**Fig. 2.** 2$^{nd}$ Priority (Bridges) - Upgrading the containment level on the bridge deck.

**Fig. 3.** 2$^{nd}$ Priority (Bridges) - Transitions to mainline barrier.

- 3$^{rd}$ Priority:
  **Slopes** – Upgrading the containment level on higher height and/or risky cases (e.g.: slopes close to houses and/or important infrastructures) as shown in Fig. 4.

- 4th Priority:
  **Obstacles and Transitions** – Revision of the protection level near roadside obstacles and transitions between flexible and rigid (concrete) barriers, as shown in Fig. 5 and 6, respectively.

**Fig. 4.** 3rd Priority (Slopes) - Upgrading the containment level on slopes with a higher height and/or risky cases.

**Fig. 5.** 4th Priority (Obstacles and Transitions) - Revision of the protection level near roadside obstacles.

**Fig. 6.** 4th Priority (Obstacles and Transitions) - Transitions between flexible and rigid (concrete) barriers.

### 2.2 The Program

The program covers the four priorities above mentioned and will be implemented in the six road concessions of the Ascendi network. The project will run until 2023 with a current estimated total investment of 17 million euros.

Table 1 shows the total investment already incurred between 2019 and June 2021.

**Table 1.** Program investment.

| Year | Investment |
|------|------------|
| 2019 | € 2 280 248 |
| 2020 | € 5 136 635 |
| 2021 | €[a] 1 824 968 |
| **Total** | **€ 9 241 850** |

[a]Regarding the year 2021, the stipulated budget is € 3 448 005 which means about 53% have already been invested.

At the current date Ascendi has already completed 54% of the Program's budget.

As part of the European Core Road Network, it was possible to apply for a financing support of "CEF Road Safety 2018" program in the Beiras Litoral e Alta concession. The program accepted to support 20% of €2.8m, amounting to €560k.

Figure 7 shows the planning defined for the Program considering all of the six concessions and all the priorities mentioned as is generically. The main work stages for each priority include the following main tasks:

1. Field surveys to characterize the current state of the network's safety barriers;
2. Assessment and identification of all the situations that need an intervention;
3. Tendering and selection of the solution to be implemented;
4. Installation of the safety barrier, according to the solution previously established;
5. Maintenance.

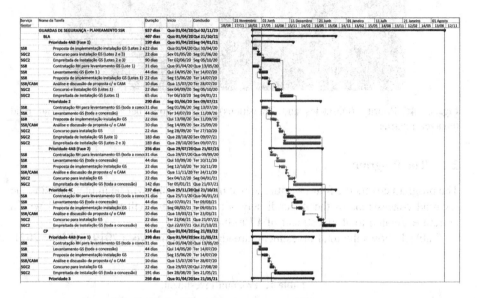

**Fig. 7.** Ascendi's safety barriers upgrading program.

## 2.3 Research and Development

Along the project Ascendi faced some issues that became a challenge to overcome and required innovative solutions.

The main difficulties were related to the 2$^{nd}$ Priority (Bridges), namely:

- No solution market available (no homologation) for safety barriers to be installed over a 30 cm high curb, as presented on Fig. 8.
- The existing system of safety barriers described as "Double U" without homologation $\geq$ unknown containment level;
- Non-existent homologation transitions;
- Non-existent solutions considering the motorcyclist protection system (mandatory in Portugal).

**Fig. 8.** Bridge with a 30 cm high curb.

To overcome these difficulties Ascendi promoted the development and certification of a brand-new safety barrier for application on bridges with a 30 cm curb in cooperation with Road Steel Engineering S.L.

Figure 9 represents the developed system that includes the safety barrier with the motorcyclist protection system.

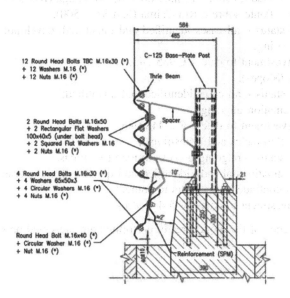

**Fig. 9.** Safety barrier developed for bridges with a 30 cm curb.

This system has already been successfully tested with a containment level H2 and has the certificate of constancy of performance systems according to the procedures defined in the European Standard for Vehicle Restraint Systems (EN 1317) (CE certificated product).

342     T. Silva and J. Neves

The development, testing and certification of transitions between the recent safety barrier applied on the bridges and the mainline barrier is currently being developed and tested.

## 2.4  Implementation

As already mentioned, this Program started in 2018 with some minor interventions and the work continued on the Ascendi's network since then.

Analysing the intervention priorities, the following aspects stand out:

- 1st Priority (Median Barriers):
  - All situations identified were registered only in one concession, namely on Beiras Litoral e Alta Concession;
  - Current status – all cases identified, quantified and solved in 2019;
  - Total investment – € 409 626.
- 2nd Priority (Bridges):
  - Implementation of safety barriers available on market with the homologation for bridges with reduced height curb;
  - Implementation of the new safety barrier developed for application on bridges with a 30 cm curb on two pilot projects:
    - A25 - Viaduto sobre a Linha do Vale do Vouga (Km 30 + 800);
    - A25 – Ponte sobre o Rio Caima (Km 31 + 500).
  - Current status – all cases identified and quantified; development of the transitions ongoing;
  - Total investment to date – € 935 343.
- 3rd Priority (Slopes):
  - Current status – all cases identified and quantified;
  - Implementation ongoing;
  - Total investment to date – € 2 448 086.
- 4th Priority (Obstacles and Transitions):
  - Current status – ongoing process to identify all cases;
  - For the situations already identified and quantified the upgrading work started and will continue in the years to come;
  - Total investment to date – € 5 448 796.

Some evidence of the work already performed are presented on Fig. 10.

**Fig. 10.** Intervention priorities.

Figure 11 summarizes the status of the interventions.

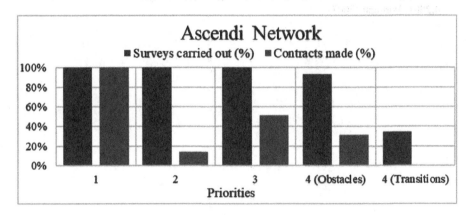

**Fig. 11.** Surveys carried out (%) and contracts made (%), for each priority.

## 3 Conclusions

This program has great potential to improve the safety conditions of Ascendi's network once the modernization of safety barriers will provide higher levels of safety to all its users.

The level of internal and external commitment to the Project leads us to believe that we are on the right track to achieve the goal of improving our level of service and therefore contribute to the decrease of road fatalities and injury accidents in our network.

# References

European Committee for Standardization EN 1317-1 Road Restraint Systems – Part 1: Terminology and general criteria for test methods. Technical Committee CEN/TC 226, Brussels (2007)

European Committee for Standardization. EN 1317-2 Road Restraint Systems – Part 2: Performance classes, impact test acceptance criteria and test methods for safety barriers. Technical Committee CEN/TC 226, Brussels (2007)

LNEC - Laboratório Nacional de Engenharia Civil: Sistemas de Retenção Rodoviários - Manual de Aplicação. This report was prepared by request of the Instituto de Infra-estruturas Rodoviárias, I.P.(InIR), Lisbon (2010)

Portuguese Quality Institute: NP ENV 1317-3 Sistemas de Retenção Rodoviários – Parte 3: Amortecedor de choque. Classes de desempenho, critérios de aceitação do ensaio de choque e métodos de ensaio. Portuguese version of the EN 1317-3:2000. Portugal (2007)

Portuguese Quality Institute. NP ENV 1317-4 Sistemas de Retenção Rodoviários – Parte 4: Classes de desempenho, critérios de aceitação dos ensaios de choque e métodos de ensaio para terminais e transições de barreiras de segurança. Portuguese version of the EN 1317-4:2001. Portugal (2007)

# Effectiveness of Cable Median Barriers in Preventing Cross Median Crashes and Related Casualties in the United States - A Systematic Review

Baraah Qawasmeh[1] and Deogratias Eustace[2]([✉])

[1] Western Michigan University, Kalamazoo, MI, USA
[2] The University of Dayton, Dayton, OH, USA
deo.eustace@udayton.edu

**Abstract.** The purpose of this systematic review is to assess whether installation of cable median barriers (CMBs) in some stretches of freeways in different states have been effective in preventing cross median crashes (CMCs) and related casualties in the United States. As the CMBs get increasingly installed on highways, they have been taunted as the preferred median barriers due to their combination of lower costs of installation compared to other types of barriers such as concrete barriers and their effectiveness in their intended use, i.e., providing median crossover protection by preventing cross median crashes. Several state departments of transportation (DOTs) have been installing CMBs for more than 15 years and some states have evaluated the performance of median cable barriers installed on their freeways. This study evaluated safety evaluation studies from 12 states that have performed the effectiveness studies of CMBs. Ten of these studies were before-after studies and nine of them showed effectiveness of CMBs. Reductions in outcomes across studies ranged from 24% to 93% for fatal and serious injury cross-median crashes and 50% to 91% for total cross-median crashes. However, increasing in outcomes ranged from 18% to 163% for possible injury and property damage only crashes of vehicles that primarily collided with the barriers but did not cross-over to the opposing lanes. Cable barriers have proved to be a valuable safety engineering tool effective in reducing fatalities and serious injuries due to crossover crashes on the states' most heavily traveled roads.

**Keywords:** Cross median crashes · Highway medians · Systematic review · Cable barrier systems · Highway safety

## 1 Introduction

Cross median crashes (CMCs) occur when a vehicle leaves its travel way, enters or crosses the dividing median, and collides with vehicles moving in the opposite direction. CMCs tend to be severe and more likely to occur with higher traffic volumes. CMCs on highways often result in fatalities or severe injuries to occupants of the vehicle or the road users in the opposing lanes (Eustace et al. 2018). Cable median barriers (CMBs) are increasingly being installed on highways due to their lower

© The Author(s), under exclusive license to Springer Nature Switzerland AG 2022
A. Akhnoukh et al. (Eds.): IRF 2021, SUCI, pp. 345–354, 2022.
https://doi.org/10.1007/978-3-030-79801-7_24

installation costs compared to other barriers such as concrete barriers and guardrails. Like other barriers, CMBs are installed to prevent errant vehicles from striking a roadside obstacle, traversing non-recoverable terrain, or colliding with traffic from the opposite direction (Eustace et al. 2018).

Cable barrier systems are life-saving traffic devices ideally suited for use in existing medians to prevent cross-over crashes, absorb collision forces, and reduce the impact on vehicle occupants. Most states that have installed cable barriers in medians report a decrease in CMC fatalities of 90% or more (INDOT 2019). There is no other safety device available that virtually guarantees consistent success in saving lives every year on the interstate system (MNDOT 2012).

According to the American Association of State Highways and Transportation Officials (AASHTO)'s Roadside Design Guide, the primary countermeasure to reduce the opportunity for median crossover crashes is the installation of median barriers (AASHTO 2011). AASHTO's Highway Safety Manual provides estimates that the installation of median barriers results in average reductions of 43% for fatal crashes and 30% for injury crashes (Russo et al. 2016). However, the Highway Safety Manual also indicates that median barriers increase overall crash frequency by approximately 24%, primarily because of the increase in property damage only (PDO) crashes, because of the reduced recovery area for errant vehicles (Russo et al. 2016).

Although the cable median barriers use has increased significantly, cable barriers present possible disadvantages, such as an increase in less severe crashes and the need for frequent maintenance. Given economic considerations, the decision to install a cable median barrier system on a particular freeway segment requires careful examination of the expected frequency of cross-median crashes (CMCs) (as well as all median-related crashes) in the absence of a barrier, as well as the expected frequency of such crashes if such a system were in place.

The frequency of CMCs can be influenced by many factors including traffic volume and median width, which are the two criteria on which AASHTO's Roadside Design Guide bases its recommended guidelines for barrier installation (Russo et al. 2016). Accurate estimates of expected cable barrier safety performance in reductions of severe crashes (as well as increases in less severe crashes) are critical for transportation agencies in determining the most suitable locations for installation, as well as for performing preliminary economic analyses (Russo et al. 2016).

Traditionally, the expected change in safety (in crash frequency and severity) associated with a given countermeasure is presented in the form of a crash modification factor (CMF). CMFs provide a simple multiplicative factor, which estimates the percentage of change in the number of crashes (total crashes or crashes of certain types or severity levels) associated with the installation of a certain countermeasure. For example, a CMF of 0.70 would be associated with a 30% reduction in crashes, while a CMF of 1.30 would be associated with a 30% increase in crashes.

The purpose of this systemic review is to perform a nationwide analysis to determine whether installation of CMBs in some stretches of freeways in different states have been effective in preventing CMCs and related casualties in the United States. This paper analyzed published safety effectiveness of installed median cable barriers studies in various freeways in various states. The earlier review study on the nationwide performance of CMBs was published twelve years ago (Ray et al. 2009).

## 2 Literature Review

Over the years, safety performance evaluation studies of CMBs have been conducted by states that have installed them in their freeways. In 1990s, North Carolina conducted numerous studies on CMCs to develop effective traffic safety strategies. Lynch et al. (1993) assessed cross median crashes by studying crashes that occurred on North Carolina's Interstate Highway System. Then Hunter et al. (2001) analyzed in-service performance evaluation (ISPE) of CMBs in North Carolina, which were installed in 1994 and all these were made of three strands. They developed several models for different types of crashes.

Knuiman et al. (1993) used Highway Safety Information System (HSIS) data from Illinois and Utah to study the relationship between median width and crash rates utilizing log-linear model for the before period (when there were no CMBs). But, for the after period (with CMBs); a study by Sposito and Johnston (1998) concluded a reduced fatality rate but with five times growth in injury crashes with additional number of vehicles hitting the barriers after entering the median. Later, Olsen et al. (2011) used the empirical Bayes (EB) before-after studies method to evaluate the efficiency of cable barriers on Utah's highways.

McClanahan et al. (2004)'s study for the state of Washington calculated the societal benefits of using CMBs at locations having CMCs. They found that CMBs provided a cost-effective solution to prevent such crashes. The cost of CMB installation was calculated to be about $44,000 per mile. Also, the average cost and maintenance cost was found to be $733 per repair and $2570 per mile, respectively. In addition, they found that, the time required between CMB damage and repair was approximately 30% less when compared with W-beam guardrail. A later study by Olson et al. (2013) utilizing data from the state of Washington, which summarized and collected the previous studies of CMBs that were published between 2007 and 2009 reported reduced number of cross-median fatal and injury crashes on Washington's high-speed controlled access highways.

A study by Chandler (2007) showed that CMBs can prevent CMCs from occurring by 95% on Missouri's Interstates. Then Villwock et al. (2009) studied the safety impact of CMBs on rural interstates for eight states. They applied the EB before-after study method that showed that an installation of (low or high tensioned) cable barriers decreases multiple vehicle-opposite direction crashes by 90%. But increases single vehicle crashes by 80%.

Safety performance of CMBs in preventing CMCs on Kentucky Interstates was also assessed by Agent and Pigman (2008). They reported that 392 crashes that were identified as CMCs occurred during the period from 2001 to 2005. Their study found that CMBs are effective in preventing errant vehicles from colliding with traffic from the opposite direction lanes because installing the CMBs only 0.9% of the vehicles were able to breach the barriers and continued into the opposing lanes.

In the state of Texas, an assessment on CMB systems was carried out by Cooner et al. (2009). Texas in 2003 started installing median cable systems almost the same time as Ohio. They maintained that before the year of 2003, CMCs were the main cause of about 96% of fatalities on Interstate highways in Texas. To accelerate projects;

348    B. Qawasmeh and D. Eustace

Texas DOT decided to install high-tension CMBs instead of concrete barriers simply due to economic reasons. Their study evaluated the performance evaluation of various cable barrier systems in Texas by analyzing installation cost, recurring maintenance costs and experiences, crash history before and after implementation, and field performance. They found that a CMB is an attractive option compared to other types of barriers.

The first nationwide safety evaluation of CMBs was conducted by Ray et al. (2009). Their study presented a review of 23 states on the use and effectiveness of CMBs. They found that by 2007, more than 4000 km of CMBs were installed in different states. Their study included 5 of the first pioneering states in initiating the utilization of CMBs in the United States, such as New York, Missouri, Washington, Oregon, North Carolina, and Arizona. Their results showed that using CMBs in depressed medians with moderate slopes had a significant effect on the reduction of CMCs in many states. In addition to that, there was a decrease of about 40 to 100% in the recorded number of severe crashes that occurred in all studied states that installed CMBs. Furthermore, they evaluated the safety effectiveness of CMBs to be more than 88% in preventing the errant vehicles from crossing the median.

In the state of Kansas, Sicking et al. (2009) provided guidelines for CMBs installation along Kansas freeways. By reviewing and analyzing crash database for the period between 2002 to 2006; they found that 115 CMCs and 525 cross median events occurred. They noted that the encroachment rates of CMCs were linearly related to traffic volume. Their findings indicate that CMCs were more frequent in winter months. This means that the warranting criteria of the CMBs may need to be adjusted to accommodate regional climate differences to develop general guidelines on the use of cable median barriers along Kansas freeways. Dissanayake and Galgamuwa (2017) applied cross sectional, case control, and EB methods to estimate CMFs for lane departure countermeasures in Kansas. Their results showed that CMBs have a crash-reduction effect on all lane departure crashes and fatal and injury lane departure crashes on four-lane divided roadway segments.

Alluri et al. (2015) compared the safety performance between the G4 (1S) strong-post W-beam guardrails and CMBs on Florida's freeways. Their comparison was based on the percentages of barrier and median crossovers by vehicle type and crash severity. Their evaluation was done using z-test and odds ratios. Their findings showed that although CMBs resulted into fewer severe injury crashes compared to guardrail barriers; guardrails performed slightly better than cable barriers in terms of barrier and median crossover crashes.

The state of Michigan started installing the CMBs in 2008. By September 2013, approximately 317 miles of high-tension CMBs were installed on the state's freeways (Savolainen et al., 2014). The study by Savolainen et al. (2014) examined the safety performance of CMBs. They found that, there is a significant reduction in fatal and severe crashes after CMB installations. Their study concluded that installation of CMB is an effective approach to reduce CMCs experienced on Michigan's freeways. In addition, Russo et al. (2016) applied the EB before-after method to develop crash modification factors (CMFs) as a function of median width. In addition, they provided important guidance to deal with the expected safety impacts associated with CMB installations. Their results showed that CMB could significantly reduce the frequency of fatal and severe injury crashes.

Eustace and Almothaffar (2018) evaluated the safety effectiveness of Ohio's statewide cable barriers. They utilized the EB before-after study method in their analysis. They studied 41 interstate segments where CMBs are installed. Their study reports that the safety effectiveness of Ohio's statewide cable barriers was found to be 73.9% for total crashes, 80.4% for fatal and injury crashes and 80.1% for fatal, incapacitating, and non-incapacitating injury crashes combined.

For the state of Tennessee, Chimba (2017) evaluated the safety effectiveness of installing CMBs by performing Empirical Bayes (EB) before-after studies. His study reinforced the positive impact of CMBs installation. He found that there was a significant decrease in fatal, incapacitating injury crashes, and fatal and incapacitating injury crashes combined after CMBs installation of about 94%, 92%, and 92%, respectively. Also, he did a direct comparison by using descriptive statistics, his comparison found that statewide fatal crashes were reduced by 82% after the installation of CMB. While incapacitating injury crashes were reduced by 76%. Also, Bryant et al. (2018) developed CMFs for CMBs in Tennessee that respond to the intended benefits in preventing CMCs. They applied comparison group method utilizing CMBs that installed from 2006 to 2010. Their results showed that installation of CMBs in Tennessee significantly reduces fatal and injury crashes but increase minor property damage only (PDO) crashes. For the state of Wyoming Coulter and Ksaibati (2013)'s study summarized the effectiveness of the Wyoming State Highway Safety Plan on crash severity statewide (WSHSP) which included the effectiveness of two types of safety control devices installed on selected sections of the roadway: shoulder rumble strips and CMBs. Analyzing CMBs effectiveness indicated that PDO crashes increased during the study, but there was a significant reduction in severe crashes.

Savolainen et al. (2018)'s study helps in analyzing impacts on traffic crashes, injuries and fatalities and evaluating the efficiency of CMB systems that have been installed in Iowa. Also, their study involved an economic analysis of CMB systems. Results showed that CMB systems have significantly increased PDO crashes, but number of sever crashes across the state have been significantly reduced. So, CMB systems proved a significant return of investment.

A statewide safety evaluation of CMB systems was conducted by Ray et al. (2009). Their study presented a review of 23 states on the use and effectiveness of CMBs. They found that; in 2007, more than 2500 mi of CMBs have been installed. Also, this study included 5 of the most pioneering states in initiating the utilization of CMB in the US, such as New York, Missouri, Washington, Oregon, North Carolina, and Arizona. Their results showed that using CMBs in depressed medians with moderate slopes had a significant effect on the reduction of CMCs in many states. In addition to that, there was a decrease of about 40 to 100% in the recorded number of severe crashes occurred in all states that have installed CMBs. Furthermore, they evaluated the safety effectiveness of CMBs to be more than 88% in preventing the errant vehicles from crossing the median. Although these results are encouraging, infrequent fatal crossover crashes penetrating the cable barrier still demand attention and improved techniques or procedures for selecting or locating cable median barrier will continue to develop. On the

350 B. Qawasmeh and D. Eustace

other hand, this study was initiated in 2007, so a current study should be developed to come up with the most recent statewide safety evaluation of CMBs in the US.

The introductory literature review shows that the use of CMBs throughout the United States is rapidly increasing. Moreover, some of these evaluation studies were comprehensive enough. Some setbacks are due to the small lengths of CMB installations that were studied and some due to simplistic analysis methods used, which may lead to unreliable results due to their inability to control the effect of regression-to-the-mean (RTM) bias.

## 3  Methods

The aim of this systematic review was to perform a review of comprehensive before-and-after crash studies that assessed the safety impacts of CMBs in refereed journal articles, conference proceedings, and research reports that have been published in the United States. A literature search of relevant studies was conducted through standard databases including ScienceDirect, Scopus, MetaPress, ProQuest, Google Scholar, and websites of state departments of transportation. The keywords used in the study include "median cable barriers," "safety effectiveness," and "before-after studies. We selected 25 states to review, based on the title or abstract of the report. After reviewing the full articles, we identified 12 of them that were potentially suitable for inclusion based on the aim of the study, which are Indiana, Texas, Utah, Florida, Washington, Wyoming, Iowa, Ohio, Tennessee, Kansas, Missouri, and Michigan.

## 4  Results

We selected 25 states to review, based on the title or abstract of the report. After reviewing the full articles, we identified 12 of them that were potentially suitable for inclusion which are Indiana, Texas, Utah, Florida, Washington, Wyoming, Iowa, Ohio, Tennessee, Kansas, Missouri, and Michigan. Tables 1, 2 and 3 show a summary of CMB safety evaluation studies conducted in the United States by different states. Reductions in outcomes across studies ranged from 50% to 93% for fatal and serious injury cross-median crashes and 50% to 91% for total cross-median crashes. But an increasing in outcomes ranged from 18% to 163% for possible injury and property damage cross-median crashes.

All state studies attempted to estimate crash modification factors (CMFs) of CMBs associated with their effectiveness in reducing different crash severities. Five states, that is, Missouri, Texas, Florida, Washington, and Wyoming developed their CMFs using simple before-after methods while additional five states of Indiana, Tennessee, Utah, Michigan, and Ohio developed their CMFs using the powerful EB before-after method. Two states, that is, Iowa and Kansas, developed their CMFs using cross-sectional method. Results from states that used the EB before-after evaluation method are more reliable, and their results should be the ones that can be defended with confidence because this methodology is statistically sound and minimizes the bias by taking care of the regression-to-the-mean effect.

# Effectiveness of Cable Median Barriers in Preventing Cross Median Crashes    351

**Table 1.** Summary of statewide safety evaluation of CMBs for total crashes in the United States

| State | Report year | Methodology | CMF* | SE* | CRF %* | Authors |
|---|---|---|---|---|---|---|
| Missouri (MO) | 2007 | Simple before/after | 0.08 | N/A | 92% | Chandler et al. (2007) |
| Indiana (IN) | 2009 | EB before/after | 0.09 | 0.1 | 91% | Villwock, et al. (2009) |
| Texas (TX) | 2009 | Simple before/after | 0.07 | N/A | 93% | Cooner et al. (2009) |
| Utah (UT) | 2011 | EB before/after | 0.38 | 0.104 | 62% | |
| Florida (FL) | 2012 | Simple before/after | 1.378 | 0.108 | −38% | Alluri et al. (2015) |
| Washington (WA) | 2013 | Simple before/after | 0.35 | 0.037 | 65% | Olson et al. (2013) |
| Wyoming (WY) | 2013 | Simple before/after | 0.36 | 0.15 | 64% | Coulter and Ksaibati. (2013) |
| Michigan (MI) | 2016 | EB Before/after | 0.47 | 0.12 | 53% | Savolainen et al. (2014) |
| Tennessee (TN) | 2017 | EB before/after | 0..04 | N/A | 96% | Chimba (2017) |
| Kansas (KS) | 2017 | Regression cross-section | 0.5 | 0.096 | 50% | Dissanayake and Galgamuwa (2017) |
| Iowa (IA) | 2018 | Regression cross-section | 0.38 | N/A | 62% | Savolainen et al. (2018) |
| Ohio (OH) | 2018 | EB before/after | 0.26 | 0.038 | 74% | Eustace and Almothaffar (2018) |

From Table 1 when comparing studies that applied before-after using the empirical Bayes method, we can see that Utah, Missouri, and Ohio have relatively close CMF values (0.38, 0.47, and 0.26; respectively) while Indiana has a CMF of 0.09. When looking closely on the type of crashes used to develop their models, Indiana developed their CMFs using cross-median, frontal and opposing direction sideswipe, and head on crashes while the other states in the same group developed their CMFs using CMCs only.

**Table 2.** Summary of statewide safety evaluation of CMBs for fatal and serious injury crashes in the United States

| State | Report year | Methodology | CMF* | SE* | CRF %* | Authors |
|---|---|---|---|---|---|---|
| Texas (TX) | 2009 | Simple before/after | 0.07 | N/A | 93% | Cooner et al. (2009) |
| Utah (UT) | 2011 | EB before/after | 0.56 | 0.104 | 44% | |
| Florida (FL) | 2012 | Simple before/after | 0.58 | 0.27 | 42% | Alluri et al. (2015) |
| Washington (WA) | 2013 | Simple before/after | 0.34 | 0.124 | 66% | Olson et al. (2013) |
| Wyoming (WY) | 2013 | Simple before/after | 0.21 | 0.14 | 79% | Coulter and Ksaibati. (2013) |
| Michigan (MI) | 2016 | EB Before/after | 0.76 | 0.11 | 24% | Savolainen et al. (2014) |
| Tennessee (TN) | 2017 | EB before/after | 0.07 | N/A | 93% | Chimba (2017) |
| Iowa (IA) | 2018 | Regression cross-section | 0.38 | 0.15 | 62% | Savolainen et al. (2018) |
| Ohio (OH) | 2018 | EB before/after | 0.20 | 0.044 | 80% | Eustace and Almothaffar (2018) |

**Table 3.** Summary of statewide safety evaluation of CMBs for possible injury and property damage only crashes in the United States

| State | Report year | Methodology | CMF* | SE* | CRF%* | Authors |
|---|---|---|---|---|---|---|
| Florida (FL) | 2012 | Simple before/after | 1.881 | 0.22 | −88% | Alluri et al. (2015) |
| Wyoming (WY) | 2013 | Simple before/after | 0.63 | 0.36 | 37% | Coulter and Ksaibati (2013) |
| Michigan (MI) | 2016 | EB Before/after | 2.63 | 0.12 | −163% | Savolainen et al. (2014) |
| Tennessee (TN) | 2017 | Comparison group before/after | 1.18 | N/A | −18% | Chimba (2017) |
| Iowa (IA) | 2018 | Regression cross-section | 2.08 | 0.04 | −108% | Savolainen et al. (2018) |

# 5 Conclusions

Cross-median crashes are among the most hazardous collision types on freeways. The primary countermeasure to reduce such crashes is the installation of a median barrier, with a cable barrier being a widely used alternative. Cable barriers have increasingly been installed in the United States as a median barrier option, primarily because of state agencies looking for a cost-effective means by which to reduce or eliminate CMCs. While the general performance of cable barriers at redirecting or stopping vehicles has been found to be the most effective, several before-and-after studies report that cable barriers reduce fatal and severe injury crashes but increase property damage only (PDO) and minor injury collisions. Cable median barriers are successful in preventing severe median crossover crashes. State studies show that reductions in outcomes across studies ranged from 24% to 93% for fatal and serious injury cross-median crashes and 50% to 91% for total cross-median crashes. However, cable median barriers increase outcomes ranged from 18% to 163% for possible injury and property damage only crashes of vehicles that primarily collide with the barriers but do not cross-over to the opposing lanes where could have resulted into more severe outcomes.

# References

Agent, K.R., Pigman, J.G.: Evaluation of median barrier safety issues. Research Report 30 KTC-08-14/SPR329-06-1F. Kentucky Transportation Center, Lexington, Kentucky (2008)

Alluri, P., Gan, A., Haleem, K., Mauthner, J.: Safety performance of G4 (1S) W-beam guardrails versus cable median barriers on Florida's freeways. J. Transp. Saf. Secur. 7(3), 208–227 (2015)

American Association of State Highways and Transportation Officials (AASHTO). Highway Safety Manual, vol. 1. AASHTO, Washington, D.C (2010)

American Association of State Highway and Transportation Officials: Roadside Design Guide. Task Force for Roadside Safety, AASHTO, Washington, DC (2011)

Bryant, A., Chimba, D., Ruhazwe, E.: Crash modification factors for median cable barriers in Tennessee. In: Proceedings of 97th Annual Meeting of the Transportation Research Board, Washington, DC., 7–11 January 2018 (2018)

Chandler, B.: Research pays off: eliminating cross-median fatalities: statewide installation of median cable barrier in Missouri. TR News (248) (2007)

Chimba, D.: Guidelines for site selection, safety effectiveness evaluation, and crash modification factors of median cable barriers in Tennessee. Project Number: RES2013-28, Final Report, Tennessee Department of Transportation, Nashville, Tennessee (2017)

Cooner, S.A., Rathod, Y.K., Alberson, D.C., Bligh, R.P., Ranft, S.E., Sun, D.: Performance evaluation of cable median barrier systems in Texas. Report FHWA/TX-09/0-5609-1. Texas Department of Transportation, Austin, Texas (2009)

Coulter, Z., Ksaibati, K.: Effectiveness of various safety improvements in reducing crashes on Wyoming roadways (No. MPC 13-262). Mountain Plains Consortium (2013)

Dissanayake, S., Galgamuwa, U.: Estimating crash modification factors for lane departure countermeasures in Kansas. Final Report, Kansas Department of Transportation, Topeka, Kansas (2017)

Eustace, D., Almothaffar, M.: Safety effectiveness evaluation of median cable barriers on freeways in Ohio (No. FHWA/OH-2018-23), Ohio Department of Transportation, Columbus, Ohio (2018)

Hunter, W.W., Richard Stewart, J., Eccles, K.A., Huang, H.F., Council, F.M., Harkey, D.L.: Three-strand cable median barrier in North Carolina: in-service evaluation. Transp. Res. Rec. 1743(1), 97–103 (2001)

Knuiman, M.W., Council, F.M., Reinfurt, D.W.: Association of median width and highway accident rates. Transp. Res. Rec. **1401**, 70–82 (1993)

Lynch, J.M., Crowe Jr, N.C., Rosendahl, J.F.: Interstate across median accident study: a comprehensive study of traffic accidents involving errant vehicles which cross the median divider strips on North Carolina interstate highways. In: 1993 AASHTO Annual Meeting Proceedings. Selected Committee Meeting Papers presented at the 79th Annual Meeting of the American Association of State Highway and Transportation Officials (AASHTO) (1993)

McClanahan, D., Albin, R.B., Milton, J.C.: Washington State cable median barrier in-service study. In: Proceedings of 83rd Annual Meeting of the Transportation Research Board, Washington, DC., 11–15 January 2004 (2004)

Minnesota Department of Transportation (MNDOT): Cable Median Barriers (2012). https://www.dot.state.mn.us/trafficeng/reports/cmbarrier.html. Accessed 10 Feb 2020

Olsen, A.N., Schultz, G.G., Thurgood, D.J., Reese, C.S.: Hierarchical Bayesian modeling for before-and-after studies. In: Proceedings of the 90th Annual Meeting of the Transportation Research Board, Washington, D.C., 23–27 January 2011 (2011)

Olson, D., Sujka, M., Manchas, B.: Cable median barrier program in Washington State (No. WA-RD 812.1). Washington (State). Department of Transportation. Design Policy Research, Olympia, Washington (2013)

Ray, M.H., Silvestri, C., Conron, C.E., Mongiardini, M.: Experience with cable median barriers in the United States: design standards, policies, and performance. J. Transp. Eng. **135**(10), 711–720 (2009)

Russo, B.J., Savolainen, P.T., Gates, T.J.: Development of crash modification factors for installation of high-tension cable median barriers. Transp. Res. Rec. **2588**(1), 116–125 (2016)

Savolainen, P.T., Gates, T.J., Russo, B.J., Kay, J.: Study of high-tension cable barriers on Michigan roadways (2014). http://www.michigan.gov/documents/mdot/RC1612_474931_7.pdf

Savolainen, P.T., Kirsch, T., Hamzeie, R., Johari, M., Nightingale, E.: In-service performance evaluation of median cable barriers in Iowa. Center for Transportation Research and Education, Institute for Transportation, Iowa State University, Aimes, Iowa (2018)

Sicking, D., de Albuquerque, F., Lechtenberg, K., Stolle, C.: Guidelines for implementation of cable median barrier. Transp. Res. Record: J. Transp. Res. Board **2012**, 82–90 (2009)

Sposito, B., Johnston, S.: Three-cable median barrier performance and costs in Oregon. In: Proceedings of the 77th Annual Meeting of the Transportation Research Board, Washington, D.C., 11–15 January 1998 (1998)

Villwock, N.M., Blond, N., Tarko, A.P.: Safety impact of cable barriers on rural inter-states (No. 09-1302). In: Proceedings of the 88th Annual Meeting of the Transportation Research Board, Washington, D.C., 11–15 January 2009 (2009)

# Integrated Transport Planning

# Transforming Infrastructure Projects Using Agile

Nihal Erian[1(✉)] and Brendan Halleman[2]

[1] KemetPro, LLC, Greenville, USA
nserian@kemetpro.com
[2] International Road Federation, Brussels, Belgium

**Abstract.** Traditional project management processes, also referred to as waterfall processes, have been used for decades to manage infrastructure projects. Agile project management, first introduced by the software industry, and known for its adaptability, targets projects delivery using an iterative and/or incremental approach, thus improve project delivery time and cost.

This research paper explores how road infrastructure projects, historically reliant on traditional project management processes, can attain benefits using the Agile set of principles and methods. With current challenges, high uncertainties and ever-changing and evolving customer requirements, there needs to be a more adaptive approach to managing and delivering road infrastructure projects to address changes in a timely and effective manner, while fostering collaboration among stakeholders. Agile can address these challenges by utilizing cross-functional teams which rely on retrospectives and continuous feedback from customers to improve team performance and enhance project outcomes and product delivery. The full benefits of Agile can only be exploited if there is a supporting organizational culture and active leadership participation.

**Keywords:** Road infrastructure projects · Agile project management · Traditional project management · Adaptability · Agile teams · Siloed teams · Retrospectives

## 1 Introduction

Traditional or predictive project management methods have been used for decades to manage Infrastructure Project, which rely on sequential phases and separate teams for the planning and execution of project work. Using these project management methods, the construction industry has focused on delivering projects by meeting scope requirements, however less attention has been paid to the policies and mechanisms required to enhance coordination, avoid costly delays, improve project communications, and ensure effective delivery practices. Additionally, the majority of infrastructure projects stakeholders are not actively involved in the pre-planning and planning phases and, to an extent, during the execution phase of the projects. Infrastructure projects are faced with cost-overruns, schedule delays (Han 2013a), lack of effective data sharing and miscommunication among different teams. These result in conflicts, increased risks and dissatisfaction of stakeholders, and may require project mediation,

© The Author(s), under exclusive license to Springer Nature Switzerland AG 2022
A. Akhnoukh et al. (Eds.): IRF 2021, SUCI, pp. 357–364, 2022.
https://doi.org/10.1007/978-3-030-79801-7_25

and/or litigation. However, due to the importance and critically of infrastructure projects, especially road construction projects, there is little to no tolerance for delays since these projects are highly regarded and valued projects. Their main purpose is to serve the transportation needs of the economy by facilitating the continuous movement of people and goods thus, benefiting the community and impacting the quality of citizens 'daily life. Therefore, there is a need to implement a project management methodology that can continuously and effectively address project stakeholders' needs, enhance decision making, foster better communication and coordination among team members, and enable project plans updates without time-consuming processes of Traditional Project Management (TPM).

Agile Project Management (APM) has been used successfully in the software industry to improve product delivery times, enhance team coordination and stakeholder communication through short and continuous feedback loops. Self-organizing, cross-functional teams which is a main characteristic of APM enhances a team's ability on reaching critical decisions without having to go through the hurdles of the decision-making hierarchy within siloed organizations. Additionally, retrospectives which allow team members to continuously improve team performance enable for faster project delivery and meeting stakeholders' expectations.

## 2 Agile Project Management (APM) Defined

Agile as a movement started in 2001 with the release of the Agile Manifesto (Beck et al. 2001), which includes a set of values and principles that focus on delivering continuous value to customers as the primary goal of work (Denning 2018). Agility is "the ability to both create and respond to change in order to profit in a turbulent business environment. Agility is the ability to balance flexibility and stability" (Highsmith 2002a). It can also be defined as "a project management methodology that relies on short development cycles in order to focus on the continuous and fast improvement of the newly developed product or service" (Mohamed and Moselhi 2019a). In Engineering, Agility refers to the ability of a system to rapidly adapt to market and environmental changes in productive and cost-effective ways (Sharifi et al. 2001). The main priority of Agile is the customer or end-user satisfaction. Therefore, the focus of Agile is doing and not documenting. (The Project Management Institute et al. 2020).

Similarly, Agile project management is defined as the ability to respond in a timely and effective manner to both anticipated and unanticipated changes created by customer and competitors. To be able to respond effectively to changes, the project team needs be innovative, adaptive, has the required knowledge and be accountable. Agile includes several frameworks that have been developed over the years such as Scrum, Scrum of Scrum, Extreme Programming, and Agile Unified Process. With Agile, less planning is done upfront. The project team plans and replans as more information becomes available from review of frequent deliveries. (Agile Practice Guide 2017a). Two main characteristics of Agile are 1) self-organizing, cross-functional teams, which consist of team members with the skills necessary to produce a working product, and 2) retrospectives which help the team to learn from previous work on a product or a process (Agile Practice Guide 2017b). Self-organizing, cross-functional teams are involved in project work from inception to completion. The composition of the team which should include members with all the skills necessary to produce a working product can streamline the delivery process of the project without having to go through the hurdles of siloed teams.

While Traditional Project Management (TPM) can be defined as "the project management techniques predominant today, that is, the method focused on centralized decision making and control within a hierarchical organizational structure. TPM has been proven successful in projects whose solutions can be relatively defined, scoped, and estimated (both time and cost)" (White 2008). It follows a more linear approach, where the majority of the planning is completed upfront. Any changes in the project management plan follows a very definitive, structured, and lengthily process which goes through different departments or entities. Planning is an essential part of TPM and APM, however, the difference lies in how much planning is done upfront and who is involved in the planning process throughout the different stages of the project. Table 1 provides a comparison between APM and TPM.

**Table 1.** Comparison between APM and TPM

| Agile Project Management (APM) | Traditional Project Management (TPM) |
| --- | --- |
| Relies on cross-functional teams | Relies on siloed teams |
| Welcomes and accepts change | Avoids change whenever possible |
| Quick response to change | Slow response to change |
| Proactive response to uncertainties | Reactive response to uncertainties |
| Allows for continued innovation throughout the project | Planning is done in greater detail at the beginning and execution is done according to plan |
| Constant customer interaction | Less customer interaction |
| Retrospectives during project work | Lessons Learned after completion of project |

## 3 Utilizing Agile Project Management for the Enhancement of the Construction of Road Infrastructure

Infrastructure projects especially road projects affect millions of people every day. Road construction is presumed to have an important impact on population and urban development. It affects multiple stakeholders, which can lead to significant changes in adjacent areas. (Khanani et al. 2021). Therefore, the importance of roads as essential economic arteries that connect people and trade routes and the value derived thereof cannot be overemphasized.

A traditional road infrastructure project follows a linear and sequenced process. Each stage of the project is handled by its own team. The majority of the planning is done upfront and in greater detail. An agile project is designed to be more dynamic and adaptive to change. It shares the same characteristic of having deliverables clearly defined with a traditional project. However, the adaptability and dynamic nature of the agile project allows for a quicker reaction to address changes and uncertainties which allows for project work to be completed seamlessly through each project stage.

Due to the criticality of road construction projects, delays are not tolerated, however, they occur often. Some of the main causes of delays are change orders, rework due to error in design, ineffective planning, mistakes in the design documents, and delay of obtaining permits, (Han 2013b).

Agile relies on a set of principles and methods that can be applied to different stages of road construction. These set of principles and methods can help manage the design and construction of different processes and activities. Empowered self-organizing, cross-functional teams, which are at the core of Agile Project Management, can help alienate the hurdles of siloed teams and organizations. Cross-functional teams on road infrastructure projects can include designers, architects, engineers (traffic, civil...etc.), contractors, consultants, subject matter experts and other stakeholders as appropriate and necessary. The composition of the team allows for more innovative solutions to design and construction problems even before the actual start of the construction phase of the project. Highly iterative and incremental processes of Agile Project Management allow the project team to constantly evaluate the evolving product, revisiting plans, reviewing designs and getting immediate and continuous feedback from the stakeholders or customers. This enables the project team to constantly enhance the construction plans which provides greater value to the customer or end user.

Diverse team members can also address changes in a timely manner without the need to go through the structured change request process that TPM supports. These teams can work in shorter and iterative sprints. These sprints "enable the teams to quickly test and adjust ideas, minimizing risk of miscommunication or overdesign.... These principles can be applied across the project life cycle—concept selection, engineering and procurement, and construction and commissioning—to compress schedules and improve productivity while maintaining safety and quality performance." (McKinsey 2020a). During the pre-planning and planning stages, team members can prioritize the features required for the project and what decisions are required. Additionally, these teams have the capability of reducing the time needed to reach a decision due to the availability of individuals responsible for these decisions on the team. This approach can help improve communication and data sharing among team members, reduce the time required to finalize the project plans as well as provide more time for value engineering, which in turn can reduce project costs. A recent report released in 2017 shows that it would be easy for team members to reduce cost by 30 percent by moving to different designs or having the authority to actively participate in the decision-making process regarding those designs. However, if they are given those designs already, rather than being able to influence them, they will have to deliver the more costly project. (McKinsey and IRF 2017).

Figure 1 displays a suggested framework on how to transform the construction management of road infrastructure projects using Agile. It also displays how traditional teams are managed, which provides a comparison between the two methods of managing road infrastructure projects.

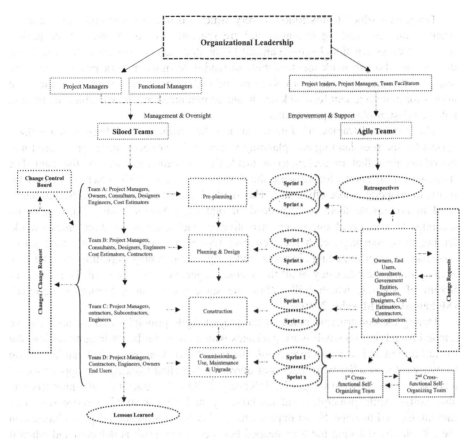

**Fig. 1.** Suggested framework on how to transform the construction management of road infrastructure projects using Agile

## 4 Discussion

The main idea of the framework is to utilize self-organizing cross-functional teams during the different stages of the project. On Agile Projects, project managers, team leaders or team facilitators are tasked to remove any impediments to attain project objectives, facilitate the day-to-day activities of project team as well as taking care of documentation (Highsmith 2002b). Their role is to make sure that the team members are dedicating their full time to project work without having to deal with regular formalities. The planning and decision-making exists within the sphere of the agile team. Therefore, any change or change requests are handled by the team members promptly. Eliminating the structured change request process of traditional projects can save time. The key factor for the success of these processes is the composition of the team, i.e., the team should have the members with the required expertise in their respective areas to be able to address challenges and changes and reach decisions about them in a timely manner.

Traditional siloed teams follow a very structured management and work process, where teams are working in stages and are separate from each other. Agile project teams are cross-functional self-organized teams that include specialists from different disciplines who can work together without having to move a work process from one team to another. The active involvement of end-users or customers with the team allows for more frequent feedback and better communication which enables the team to enhance design plans or procedures.

The suggested framework follows in part the scrum methodology which utilizes sprints for the pre-planning and planning stages of the project. These sprints are time-boxed iterations that are designated to tackle design challenges faced by the team. The framework also allows for utilizing the Scrum of Scrums or Meta Scrum method for teams to cooperate more efficiently. This method can be used when two Scrum teams need to coordinate their work instead of one large Scrum team. "Daily stand-up meetings are conducted among representatives of each team two to three times a week. During these meetings, each representative reports the completed work, next set of work, any current impeding elements, and potential upcoming impediments. The goal is to optimize the efficiency of all the teams. Larger projects may result in a Scrum of Scrum of Scrums, which will follow the same pattern as a Scrum of Scrums." (Mohamed and Moselhi 2019b).

During the construction phase of the project Agile principles and methods can be applied by relying on small work packages which allow for better integration into the overall design and which in turn enable the team to work in sprints that can last for up to two weeks. Smaller work packages allow for "detailed tracking and application of learnings from earlier packages" (McKinsey 2020b). Contractors and Engineers can also utilize Agile principles and methods by making construction processes more measurable and manageable to improve the ability to respond and adapt to changes on the job site, by reducing the time needed between when a risk is detected and when it gets addressed. Utilizing cross-functional teams during the construction stage of the project can help eliminate the silos between the different disciplines, hence effectively manage work packages until completion. The same methodology could be applied to commissioning, usage, maintenance and upgrade of road infrastructure projects. Small, highly efficient cross-functional teams can effectively address changes or challenges that arise during this stage.

Teams working on traditional projects are usually assigned part-time to different projects. However, applying Agile requires the full dedication of project team members. Therefore, they can be fully assigned to a project for weeks and not partially for months or years as with TPM. This can enhance productivity and shorten the time required to complete the project. Meeting twice to three times weekly is highly recommended.

The use of Retrospectives can serve as way of learning from past performances to enhance future performances or as a learning opportunity for all team members. Retrospectives thus enhance accountability and shared responsibility among team members. Retrospectives can be done at specific intervals or when the team feels it is necessary to do so. Retrospectives allow for a continuity of the learning process and performance improvement. Along with continuous interaction with the customers, retrospectives can help the project team to realize the points of strengths or weaknesses in processes or team performance, and work on improving them in a timely manner.

One of the key factors of Agile success in the construction industry is having a well-trained, empowered and highly motivated workforce. Therefore, training is needed to ensure that team members understand what is requested from them. Knowing and understanding the value-driven agile mindset is essential when working on agile projects.

The success of agile teams can never be realized without the support of organizational leadership. The redefined roles and responsibilities of project managers or project leaders on agile projects can be met with resistance from traditionalists. Organizations can utilize change management techniques to a systemized adoption of Agile into road construction projects. Hiring calibers and providing agile training to employees are key success factors for the agile projects to succeed.

## 5 Conclusion

APM is a tried methodology that has been proven successful within the software industry. The construction industry is lagging when it comes to the full adaption of APM due to the fact that TPM has been long used within the industry. In addition, the construction industry is known for its slow adoption of new technologies and techniques. APM can help change the way project team work on road infrastructure projects, due to their importance and impact on a wide spectrum of stakeholders. Companies can successfully transform road construction projects by having a well-trained, empowered, and highly motivated workforce, which increases responsiveness to changes and challenges during the different stages of project, enhances team productivity, reduces cost and time required to deliver the project hence providing the highest value to stakeholders. Organizational leadership support is essential for the implementation of Agile. It is essential for managing the adoption of the Agile methodology and transforming traditional teams to be more dynamic and agile.

## References

Beck, K., et al.: Manifesto for agile software development (2001). http://AgileManifesto.org
Denning, S.: Age of agile, AMACOM, First Edition (2018)
Han, F.: Defining and evaluating agile construction management for reducing time delays in construction, University of New Mexico (2013)
Highsmith, J.: Agile Project Management, 2nd Ed. (2002)
Khanani, R., Adugbila, E., Martinez, J.: The impact of road infrastructure development projects on local communities in peri-urban areas: the case of Kisumu, Kenya and Accra. Ghana. Int. J. Com. WB **4**, 33–53 (2021). https://doi.org/10.1007/s42413-020-00077-4
McKinsey & Company: Agile delivery of capital projects (2020). https://www.mckinsey.com/business-functions/operations/our-insights/agile-delivery-of-capital-projects
McKinsey, IRF: Improving the global delivery of road infrastructure, McKinsey, IRF (2017). https://www.irf.global/mckinsey-irf-diagnostic/

Mohamed, B., Moselhi, O.: A framework for utilization of agile management in construction management, Concordia University, Canada (2019)

Sharafi, H., Colquohoun, G., Barclay, I., Dann, Z.: Agile manufacturing: a management and operational framework. J. Mech. Engrg **21**, 857–869 (2001)

The Project Management Institute, N. A. P. A.: Building an agile federal government: A call to action (2020)

The Project Management Institute. Agile practice guide (2017)

White, K.R.J.: Agile project management: a mandate for the changing business environment. Paper presented at PMI® Global Congress 2008—North America, Denver, CO. Newtown Square, PA: Project Management Institute (2008)

# Savings Potential in Highway Planning, Construction and Maintenance Using BIM - German Experience with PPP

Veit Appelt[1,2,3]

[1] The Technical University of Dresden, Saxony, Germany
Veit.Appelt@apluss.de
[2] The Technical University of Munich, Bavaria, Germany
[3] A+S Consult GmbH, Dresden, Germany

**Abstract.** The digitalization of infrastructure planning in Germany is the current focus of an extensive infrastructure investment program running until 2030. Building Information Modeling technology has been supported by the Federal Ministry of Transport since 2015 and is mandatory for all newly started major projects since 2020.

BIM is a process for creating and managing information on a construction project across the project lifecycle. One of the key outputs of this process is the Building Information Model, the digital description of every aspect of the built asset.

The highway A7 is one of the main arterial roads in Germany. This will be widened in sections to 6 lanes. In the northern German state of Lower Saxony, a section was put out to tender as a Public-Private-Partnership-project (PPP project). The contract for approx. 60 km of construction planning, construction and maintenance over 30 years was awarded to a consortium of Eurovia (VINCI Autoroutes) at a cost of around €1 billion. The project will be planned, executed, and maintained using digital tools using the BIM framework.

This report explains the entire value chain and highlights the use of a uniform, fully parametric BIM overall model. It shows how digital twin, design model and construction model are used on the construction site. The temporal processing, as well as quantity and cost accounting are dynamically adapted to the requirements of logistics, and material deliveries. The construction site reports progress on a weekly basis, while management at headquarters always has an overview of budget and time.

**Keywords:** BIM · Planning · Construction · Maintenance · Public-private-partnership · AI · Big data

## 1 Short Description of the Overall Project or Task

The project presented here was awarded the title of buildingSMART Champion 2021 in the field of construction execution in the buildingSMART international competition in Germany.

---

© The Author(s), under exclusive license to Springer Nature Switzerland AG 2022
A. Akhnoukh et al. (Eds.): IRF 2021, SUCI, pp. 365–378, 2022.
https://doi.org/10.1007/978-3-030-79801-7_26

The A7 federal highway is the longest highway in Germany and one of the most important north-south connections between Scandinavia and central and southern Europe.

The A7 project comprises the six-lane expansion of a previously four-lane section over a total length of 60 km, as well as the operation and maintenance of the project route over a period of 30 years (see Fig. 1).

**Fig. 1.** A7 construction site

In addition to the six-lane expansion, the infrastructure project also includes 170 structures, including 2 major bridges, 9 interchanges, 12 parking and toilet facilities and 2 refueling and rest areas, as well as around 40,000 square meters of noise protection walls and embankments. In addition, there are rainwater retention basins and drainage lines as well as environmental protection and landscaping measures. The rerouting of the federal road B248 is also part of the project (Table 1).

**Table 1.** A7 project data

| Project length | Project duration | Financial volume | Construction costs (expansion) |
|---|---|---|---|
| 60 km | 30 years | Approx 1 billion € across 30 years | Approx 330 million € |

Compiled by A+S Consult GmbH (2021)

## 2 Digitization of the German Construction Industry

Both the economy and society are in the middle of a profound digital transformation. The buildings and infra-structure that will be needed in the future require a great deal of technical expertise and much closer interaction between all players and stakeholders, as they are becoming increasingly complex. Digitalization can make a significant contribution in this area, as well as enabling innovations. It is both a tool and a method with which added value can be created for our built environment.

During the construction of a building, there are many interfaces that are affected by digitalization. It is not only about the interaction between planners and contractors. It is also about the fact that within a company, different interfaces between people or people and machines or machines among themselves are being redesigned. In addition, there are the supply chains and all the companies and craftsmen involved in the construction, the greater use of prefabrication and serial production concepts, and finally, of course, the clients. Digitization can improve each individual process stage and facilitate the transition to the next process stage. However, only the linking - if possible, of all process stages and participants - makes the creation of a construction truly efficient. Only then can the potential of digital value creation be realized.

The prerequisite for successful cooperation is consistent information flows, data, and processes. For this, we need reliable, generally accessible standards. However, cooperation, integrated work processes and partnership are at least as important. That is why we are talking about a real cultural change in the world of construction today.

## 3 Benefits of Building Information Modeling

Building Information Modelling (BIM) is already highly relevant for the construction industry. An EU directive from 2014 recommends the use of BIM as a criterion for awarding public contracts. The step-by-step plan "Digital Planning and Construction" of the German Federal Ministry of Transport and Digital Infrastructure of 15 December 2015 takes up this directive and prescribes the mandatory use of BIM in public infrastructure projects in Germany from 2020. In the UK, the Netherlands, Denmark, Finland and Norway corresponding regulations are already in force.

BIM is a holistic planning and execution method in the construction industry. The aim is to carry out all planning and construction processes on the basis of a digital building information model. The components of a building are not only described geometrically, but also supplemented by semantic information (material, manufacturer's designation, etc.). Unlike traditional project management, which is characterised by media discontinuities, the BIM application aggregates more and more information over the entire life cycle of a building. Through the linking of previously decentralised data sources such as scheduling (4D) and costing (5D) with the model, all project participants have access to all the information on the basis of which further simulations can be performed and analyses can be carried out. The greatest benefit of model-based processing can only be achieved if clear project-specific definitions are made right from the start with regard to the geometric and semantic detailing of the models and components.

368 V. Appelt

In current large-scale projects, problems arise due to an inability to maintain an overview of the overall context and insufficient coordination of those involved in the project. As a result, the above-mentioned goals cannot be met. Digitisation in the sense of the BIM methodology offers solutions here:

- Digital process structuring: Supports the integrative cooperation of those involved in the project. This is intended to achieve an increase in effectiveness and efficiency.
- Consistency and freedom from redundancy: The use of coordinated technical models enables contradiction-free information; combining geometric and alphanumeric information such as component lists, costs and quantities in one shared model ensures that all stakeholders are working with the same information. This forms the basis for consolidated data (single source of truth).
- Decision support: Coordination models provide an overall view of the project. Several aspects can be taken into account in the decision-making process.

The goal of the project is targeted, holistic project management and a positive monetary and temporal orientation of the Public-private-Partnership-project (PPP project) on the planning, execution and operating side through the use of Big Open-BIM. These goals were implemented in an overall process for the infrastructure, which, in addition to planning and construction, also implemented the model-based project controlling and subsequently the maintenance of the built infrastructure. In addition to the transformation of planning-specific BIM subject models and GIS-specific work-flows (geographical information system) into the construction execution, artificial intelligence applications and automation processes were applied in the project. The workflows and use cases of BIM-based planning and construction were used in the ongoing construction execution, validated, regulated accordingly and finally standardized. In the process, the A7 project is to be planned, built and operated holistically in a fully networked team, using BIM standards.

BIM offers a variety of advantages for the decision-making and management processes within the project compared to conventional construction. In contrast to conventional 2D processes, the use of BIM in the project allows most of the technical drawings, such as horizontal and vertical sections, to be derived directly from the model and, thus, ensures consistency.

The collision check between the various sub-models, which were combined to form an overall model, made it possible to identify and resolve conflicts between the individual specialist plannings and executing trades at an early stage. The possibility to identify problems in advance of construction with the help of the model leads to a reduction in the number of subsequent changes and therefore to a reduction in the costs for the project. If planning changes nevertheless occur, a high increase in efficiency can be expected with the BIM-based process. It was no longer necessary to manually retrace all affected plan records, which was a process fraught with errors; instead, changes could be made on the uniform data source of the model and automatically passed on to the corresponding specialised planning. With the help of BIM, such project changes could not only be communicated to all parties involved, but also documented and justified.

Innovative methods of geoinformatics and geodesy with drones, GIS integration, machine processing of point clouds, highly parameterized and networked specialist

Savings Potential in Highway Planning, Construction and Maintenance 369

objects, as well as automated evaluations in a BIM overall model all have roles to play in this digital process. BIM in building construction differs from BIM in linear infrastructure, which is why the term Infrastructure Information Management (IIM) is used in Germany. The focus shifts to distributed heterogeneous linear planning with essential data and information reference.

The BIM model is based on a uniform and standardized parametric and object structure of the digital model with the components:

- As-built model
- Planning model
- Construction model (which will be transferred to the operation model)

The BIM model is hosted in the cloud, fully versioned, and is available centrally and equally to all participants. The parametric and technical model-oriented modeling is based on repetition-free data (single source of truth). All elements and structures of the existing route and the planning status were modeled and attributed without having to resort to further documents. This increased efficiency both in the holistic processing and later in controlling up through the finished as-built model. This served the holistic processing and later the controlling up to the finished (construction?) and are verified by photogrammetric images taken by drones and are the basis for billing.

## 4 Description of the Digital Performance Report in an Overall Model within the Software Ecosystem

With digital performance reporting in a construction status model, the construction site could be viewed digitally by all participants and management, and the performance status, as measured terms of quality and timing, could be seen immediately via linked dashboards. In terms of information technology, an increase in data to be processed in parallel was to be expected. This was to be made controllable by creating a database-based model with 3D to 5D properties (3D model, networked 4D logistics and time, 5D quantities and costs/revenues) for the digital twin, planning model, construction execution model, controlling and coordination phases and the transition to maintenance. For the life cycle, all objects were provided with Globally Unique Identifiers (GUID is a 128-bit number used to identify information in computer systems), which can be used to track them in maintenance after planning and construction and allow easy access to the initial data.

For this purpose, digital twin is created as a coordination model for the construction site, taking into account all technical data of the linear infrastructure. The BIM model consists of linear and point-specific models of the underground and above-ground infrastructure from analog or classic as-built and planning data. The technical data of all trades were semantically merged and networked with each other across all trades via logical technical models. The specialist data is integrated directly into the model in the native standard formats, depending on availability and quality. For this purpose, the software ecosystem (see Fig. 2) with interfaces was analyzed in direct dialog with the planners and common workflows were agreed upon.

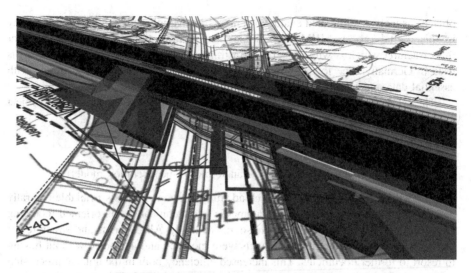

**Fig. 2.** 5D planning model A7 [KorFin® 4.5]

Due to the networking of the specialist models with each other, changes to the BIM model can be detected and integrated automatically. The constant 3D representation consistent with the technical data and the availability of all technical data makes complex planning, visualization, and communication possible in a valid overall context. Due to the highly interconnected overall model, automations in conflict detection can be carried out and achieved with intelligent applications. In the overall model, all underlying sources, such as plans and manufacturer-specific documents, are directly linked to the specialist objects and can therefore be accessed at any time and in the correct version.

Such a dynamically linked overall model, with all dependencies between all specialized planning processes being recorded in real time with all GIS information and all factual data of the specialized objects, is automatically updated with every change in the underlying data. AI algorithms are used to map specific processes, which also enable largely automated coordination of the construction site in a collaboration model (see Fig. 3). The active collaborative involvement and participation in the universally available data repository and the parallel overall model was a challenge for everyone involved. The new workflows were therefore successively implemented in the overall model and concretized and adapted over longer periods of time. The technical planning data was digitized or derived directly from the planning parameters. If possible, the data source was directly linked to the various planning management and software systems.

The versioning of the specialist data for the overall model takes place in a common data environment for the storage of completely revised work statuses. This enables later version comparisons based on consistent overall models with the current specialist data at this status.

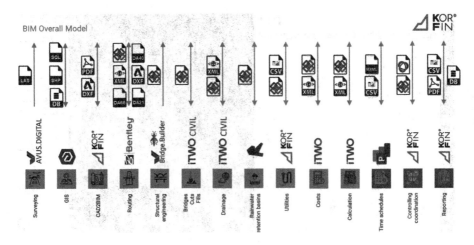

**Fig. 3.** Software ecosystem – software systems

For the coordination and distribution of tasks in the project, communication was implemented directly on the overall model as a virtual model meeting. Various external BIM viewers are available for this purpose, which are maintained centrally and used decentral by all participants. In addition to personal communication, issues are also managed and tracked via this viewer. The basis for this was the standardized BIM Collaboration Format (BCF). In general, the BIM Collaboration Format (BCF) allows different BIM applications to communicate model-based issues with each other by leveraging IFC models that have been previously shared among project collaborators.[1]

The model structure and workflows for business models and business objects as well as the underlying data design have already been developed in other projects and tested in practice. They were initially proposed to the partners, discussed with them, adapted, and transferred into binding common standards.

Another change in processing requires hard time management in the project. Successful changes and parallel use of the BIM model in a self-coordinating approach, while at the same time maintaining stable networking, are essential quality features of the openBIM workflows that were set up.

### 4.1 Implicit Planning Model and Validated Quantity Calculation

To create a digital terrain model (DTM), an aerial survey of the entire route was carried out by drone. These images were used for various purposes. On the one hand, a point cloud was generated for a large-scale inventory, and on the other hand, a digital meshed terrain model was calculated from the point cloud to calculate the necessary cuts and fills for the project (see Fig. 4).

The aerial images serve as historicized documentation and are also incorporated into the BIM model via the textured terrain.

---

[1] https://technical.buildingsmart.org/standards/bcf/.

372　V. Appelt

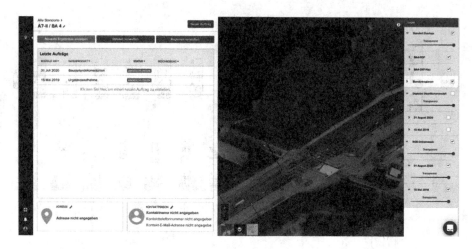

**Fig. 4.** Evaluation of the drone flight and point clouds in AVUS.DIGITAL

In the following, the elevation mesh serves as the basis during the further planning steps for the construction of the individual as built and planning volumes: by meshing to external axes and their transverse, all planning volumes are defined parametrically. This parametric now generates the volumes implicitly. This means that when the routing is changed externally or when the survey is updated, the planning volumes are automatically adjusted.

This workflow gives rise to all TARGET volumes of superstructure and earthwork, such as unbound and bound surface courses, base course, filling and cutting (see Fig. 5).

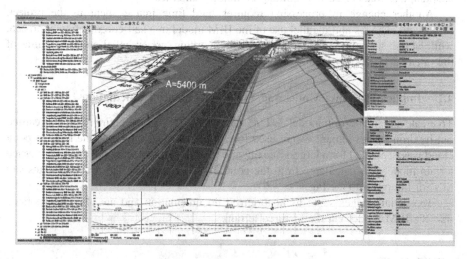

**Fig. 5.** BIM model in KorFin®

In this phase, the results of the modeling can be output as a result with quantity calculation and classification as IFC or CPIXML in a database structure. The CPIXML ("Construction Process Integration") format allows geometric objects for infrastructure projects to be output in 3D for transfer to 4D and 5D BIM processes and coordination processes. A direct continuation in the further workflow pipeline is guaranteed. Since IFC 4 does not have any extensions for implicit geometries of road infrastructure, the conversion into explicit geometry (and normalized attributes) takes place at this step.

The BIM model is extended in the same parametric logic by all other specialist models such as drainage, road equipment and construction subsoil. The merging is carried out including implicit modeling and attribution while maintaining the networking and lifecycle of all specialist objects within and directly in the BIM model.

Pipes and manholes are planned and modeled. However, the specialist model of the drainage results from the transfer with German specific supply line interface and subsequent parameterized component assignment in KorFin®. This closes the gap of the transfer of a node-edge model for the drainage with simultaneous re-engineering of the component usage in the specialist model drainage. Pipes are modeled implicitly; manholes are positioned exactly (see Fig. 6).

**Fig. 6.** Specialist objects of the substrate and superstructure (road) with internal implicit segmentation

Rainwater retention basins are special structures and were modeled in Autodesk® Revit® in addition to the classic design process. In addition to civil engineering elements such as the layered structure of rainwater retention and sedimentation basins, they consist of many structural elements such as stairs, terraces, and baffles. They are viewed like specialized models of bridges and transported via IFC.

### 4.2 External IFC Models

The models of the structural engineering structures and excavations were generated using the BRIDGE BUILDER LOD 200 software developed by VIA IMC (see Fig. 7). The software imports the relevant CAD files of the roadway, as well as the planning documents. Based on this, bridge structure models can be placed in the LOD 200 level of detail and user defined. For example, parameters such as position, angle to the route, component dimensions and material types can be selected. This parametric is used to design the superstructure and substructure.

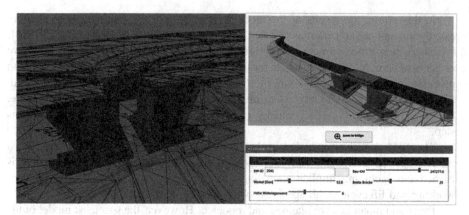

**Fig. 7.** Generating of bridges in Level of Detail 200

The structures created in this way are each provided as a separate model with full attribution as IFC and integrated into the BIM model. For this purpose, the attributes are transferred to the normalized data schema in a fully automated process and verified again. The explicit planning volumes available externally behave exactly like the implicit planning volumes created by KorFin®. They are added directly and loss-free to the entire workflow with full life cycle of the individual specialist objects and networking. IFC works in both directions without any loss of content.

### 4.3 Process Model for Road Construction

The quantities are then immediately incorporated into the 4D workflow with the associated processes (resource deployment, ordering, and transport logistics with quantities and component tracking). The parallel display of the model in real time and with integrative evaluation of partial volumes as well as attached factual database serves the immediate recognition of conflicts and the need for adaptation of the 4D processes to be planned (see Fig. 8). In addition to the display of the earthwork target profiles, the earthwork technology is directly visualized, as well as the transport of the earth quantities and the earth quantity distribution and reinstatement concept implemented including the intermediate storage of earth materials. The 4D planning itself can be done in KorFin® or externally with scheduling tools such as Microsoft Project or Tilos and from several parties including subcontractors.

In both cases, the specialist objects (models) remain correctly networked even in the event of subsequent external changes. The stable lifecycle of the models and the networking are unique selling points of the workflow presented here; GUIDs of the domain specialist objects (planning) and operations are consistently pre-served. Accordingly, the subject object is networked exactly once with the 4D process, optimally by machine (selection parameters), or, alternatively, manually for small adjustments (with feedback to the automatic system).

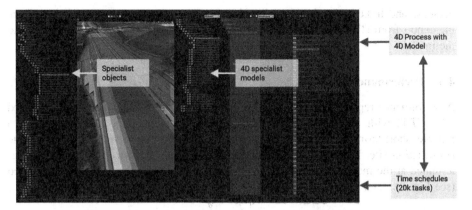

**Fig. 8.** The principle of networking of 3D/5D layers with the 4D layer

### 4.4 Dynamic Segmentation

One innovation is dynamic segmentation, which makes it possible after planning to react to partial volumes that have been realized after planning – later, during construction: For this purpose, no multitude of partial volumes is calculated via a fixed segmentation (e.g. every 5 m) during planning, but the total volume of the planning is left as one volume body (specialist object), which is only divided when necessitated by 4D requirements, and then only internally - invisible to the user. This innovation is the direct further development of implicitly defined intelligent compartment objects in road infrastructure.

### 4.5 5D BIM Model

The result of the combination of all partial planning and partial schedules of the construction sections and various trades therein is a 5D model with the temporal linkage of the specialist projects (frost protection layer, drainage, bridge structures equipment, etc.) with the corresponding processes from the construction schedule (approx. 20,000 processes). Each planning volume knows its construction period. Each construction period has its own separate time behavior and the exact sequence of construction work for the entire planning volume. The dynamic segmentation and the integrative time behavior make it possible to adapt the models and schedules to the actual construction process in a very simple and logically structured way. The construction process and the linked logistics allow the time sequence to be simulated again and again, both in history and in the future and to use AI search for potential savings (opportunity finding). The technology of separate timing with dynamic segmentation is a further unique selling point in the current BIM workflow.

For the connection to bills of quantities and budgets of activities, automatic preparations were installed, which enable the docking of iTWO results. For this purpose, among other things, the Bill of Quantities items are stored in the normalized database and linked with the specialist objects such as the quantity calculation. Since the export from iTWO is not complete or loss-free, the BIM workflow was adapted to the model.

### 4.6 Performance Message

A performance report is required to compare the planned quantities and costs incurred (TARGET) with the actual performance (ACTUAL) and to be able to produce a reliable short-term profit and loss statement, including the operating result of the construction site. In addition to the TARGET (currently known) and ACTUAL (determined at the moment), the budget is checked and, if necessary, a forecast is issued (see Fig. 9).

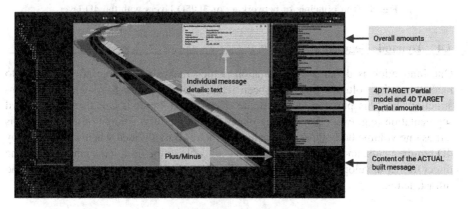

**Fig. 9.** Practical ACTUAL-built message (performance message)

VIA IMC reports the ACTUAL state via the 5D model in the model. This means that the ACTUAL information from the construction site is transferred as a construction model to the BIM model. The benefit results immediately from the automatic comparison of TARGET and ACTUAL including partial quantities and partial costs (quantities and costs of the planning and the actual status with differences).

For this purpose, construction schedules are created as a time-dependent evaluation from the 5D model and output as a dashboard or also as a list if required. On the construction site, the model is updated with performance periods and details of the built station area or quality checks. This updated "partial model" leads to the ACTUAL linking of the construction model by generating specific specialist objects of the construction message. Among other things, the total change in comparison with the last construction message to the last completion is calculated and linked in time.

The performance report is made directly by the site manager via mobile app or from the site computer into the model. The reported status achieved, and the further process can be viewed on site via augmented reality. Through automated point cloud comparison, the statements of the site manager are verified based on a regular weekly verification flight by drone. However, only the form of input changes and not the intelligent data networking among each other.

As a result, a TARGET model, ACTUAL model (green), model of construction delays (red), and a model of services which will have already been performed in the future (yellow) can be output for any past, present, or future date and visualized and evaluated in real time. For each subject object (planning volume), the completion on a past or future date can be evaluated or calculated. This completion can be calculated from the quantities in terms of built lengths, areas or volumes or determined from the estimation of the construction manager (see Fig. 10).

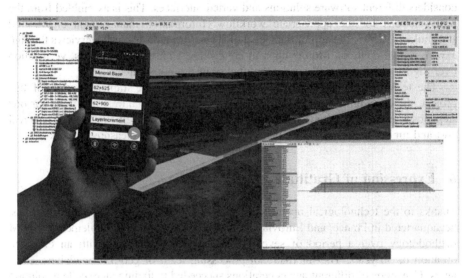

**Fig. 10.** Performance reporting takes place via mobile applications.

Here, the dynamic segmentation plays a role again since the ACTUAL area of completion does not have to correspond to the TARGET area of the planning and a reliable statement on partial volumes must still be determined. The monthly performance reporting based on the comparison between TARGET and ACTUAL in construction in the model supports the budget and risk planning of the construction project.

The as-built model is linked to the structures of the client's SAP software for transfer to maintenance. This means that the client already has access to the key facts of the maintenance in the 30-year period and can integrate the assets into its balance sheet.

## 5 Summary, Conclusions

BIM models are used throughout planning, construction, and maintenance. They are built in 5D and contain the temporal progression of 1) the digital twin (as-built model), 2) the planning model (TARGET) and 3) the construction model (ACTUAL) which is linked in the form of an as-built model to the maintenance model. All information exists only once (single source of truth) and is dynamically updated. Due to the commitment to build and maintain over 30 years, software independent open XML data structures have been implemented and open native and standardized interfaces are preferred. The versioning of the data is done together with the software solutions. On the construction site, all TARGET information is made visible for the construction management and the ACTUAL is mapped via the performance report in the overall model. This means that construction logistics, construction progress, and quantities can be viewed directly, and all management decisions are fact-based. All construction site data, such as quality control results, are also integrated there.

Open interfaces and data exchange formats allow the use of an ecosystem that considers different software solutions and varied interfaces. This is assembled from the best solutions for the appropriate workflow. Information management enables all stakeholders to have the appropriate level of access, from dashboard overviews through expert system controls. Mobile apps in connection with cloud solutions, verification over-flights with drones, all ensure the quality of the statements.

The technical challenges which were confronted in the implementation of this project, the opportunity to be a part of moving the German infrastructure sector one step closer to the future, and the pleasure of seeing envisioned efficiencies become real cost and time savings have combined to make this project a pleasure to work on.

## 6 Expression of Gratitude

Thanks to the technological openness of the international group VINCI Autoroutes, headquartered in France, and Eurovia Germany, it was possible to implement the BIM methodology over a period of several years in a major project with an expected duration of 30 years. The international processing team of geo-informaticians, engineers, IT experts of different age generations succeeded in finding an excellent solution for the project.

The German chapter of buildingSMART has named this the BIM Champion in Construction for 2021, in a tough competition featuring a selection of the best projects. Thanks to the jury of outstanding personalities from the VDI (Association of German Engineers), the German Ministry of the Interior, the construction industry, and scientific institutions for this recognition.

# Demonstrating Connectivity and Exchange of Data Between BIM and Asset Management Systems in Road Infrastructure Asset Management

Sukalpa Biswas[1(✉)], John Proust[1], Tadas Andriejauskas[1],
Alex Wright[1], Carl Van Geem[2], Darko Kokot[3], António Antunes[4],
Vânia Marecos[4], José Barateiro[4], Shubham Bhusari[5],
and Jelena Petrović[6]

[1] TRL Ltd., Wokingham, UK
sbiswas@trl.co.uk
[2] BRRC, Brussels, Belgium
[3] ZAG, Ljubljana, Slovenia
[4] LNEC, Lisbon, Portugal
[5] Royal HaskoningDHV, Amersfoort, Netherlands
[6] BEXEL, Ljubljana, Slovenia

**Abstract.** Road infrastructure asset management is currently facing rapid digital transformation opening up new opportunities for the application of data from both traditional and new data sources, the ability to integrate data across multiple Assets and the potential to improve efficiency in decision making throughout the asset lifecycle. However, there are challenges structuring, integrating, and linking data between different Assets and data sources – especially those data contained separately within asset management and BIM platforms. The CoDEC (Connected Data for Effective Collaboration) project aimed to address these challenges by creating a Data Dictionary to link/integrate static and dynamic data for "key" infrastructure assets (including road pavements, bridges and tunnels). Having proposed such a dictionary, the potential benefits offered for connecting and linking data have been shown in three pilot projects to demonstrate (1) how tunnel monitoring data can enrich BIM models of tunnels (2) how data from bridge sensors can be linked and visualised and (3) how highway data generated at construction in a BIM model can be linked to asset-management GIS for the operational phase. The results from these pilots present the challenges, lessons learned and practical benefits of linked data and semantic web technology to connect BIM to AMS and GIS platforms.

**Keywords:** Asset management · BIM · Data dictionary · Linked data · Ontology · Asset data

## 1 Introduction

Building Information Modelling (BIM) has, in recent years, become firmly established in the infrastructure sector. The process is designed so that asset information can be generated, captured, maintained and used effectively throughout the asset lifecycle. To

---

© The Author(s), under exclusive license to Springer Nature Switzerland AG 2022
A. Akhnoukh et al. (Eds.): IRF 2021, SUCI, pp. 379–392, 2022.
https://doi.org/10.1007/978-3-030-79801-7_27

date the application of BIM has tended to focus on information generated during the construction phase of the asset lifecycle and there are still gaps in applying BIM for the operational phase. On the other hand, well-developed Asset Management Systems (AMS) and processes are in place to support the operational management and maintenance of highway infrastructure. These systems hold considerable amounts of asset information in various formats, and many have been in use for decades by National Road Authorities (NRA). To make BIM more useful in the operational phase, BIM systems need to access and use operational data, typically available within Asset Management Systems (AMS). In contrast, AMS should be able to access and use data from BIM environments to make efficient and informed decisions when maintaining assets. The CEDR Transnational Research Program funded CoDEC project (Connected Data for Effective Collaboration) aimed to address this gap by establishing a standardised process of data integration through a "Data Dictionary" and "Data Ontology" for key infrastructure assets (Roads, Bridge and Tunnel) to facilitate data integration between BIM and AMS. CoDEC has demonstrated the benefit of the data integration process through three pilot projects to demonstrate; (1) how tunnel monitoring data can enrich BIM models of tunnels (2) how data from bridge sensors can be linked and visualised for better risk management and (3) how asset data contained within a BIM model can be linked to data within traditional asset-management GIS to support maintenance decisions. Ultimately, the project aims to assist Road Authorities make efficient and effective use of data to support asset management.

## 2 Asset Management and BIM

Many NRAs use BIM during the design and construction phases of large infrastructure projects, but little further use is made during the remainder of the operational life cycle of the asset (this is more common for pavement assets than for bridges or tunnels). Long-term asset management is still typically carried out in a traditional way using dedicated asset management systems:

- Pavement Management Systems (PMS) contain data about the road network, structural information about the layers of the road, traffic density and current condition (e.g. condition parameters that can be converted into indicators) of individual lengths of pavement
- Bridge Management Systems (BMS) contain bridge inspection data on each component (usually standardized, e.g. following PIARC recommendations (PIARC PIARC 2012)) and data provided by sensors and advanced inspection techniques.
- Tunnel management systems contain data on the tunnel structure itself and the electronic equipment used to support tunnel operation.

BIM is a relatively new concept in highways infrastructure, whereas the above-mentioned systems have been in use for decades to support asset management decisions and hence hold considerable amounts of asset information (data) in various formats. The use of BIM in highways infrastructure is derived from its development in Building industry, where the focus was on design and construction. Therefore, it has not been developed to accommodate the asset management data requirements needs of ageing infrastructure. The disengagement between BIM and Asset Management Systems has contributed to the slow progress in implementing BIM concepts in highways asset management, more importantly, making efficient use of available data.

A significant amount of work has been undertaken to date to define the core requirements for data capture, storage, and maintenance in the BIM environment. For example, Industry Foundation Classes – IFC (ISO 16739-1:2018) have standardised BIM data formats for the BIM Object-Type-Library (OTL) and buildingSMART (buildingSMART 2020) has defined a "Data Dictionary" for BIM data. However, these initiatives have been geared towards construction phase data and do not provide standardised methods to capture and generate data for the operational phase, or integrate data with other AMS. Additionally, recent developments in asset data capture technology (e.g. scanning systems, remote sensing, mobile sensors, IoT, etc.) are now generating data to monitor and manage assets which currently is poorly supported in either BIM or AMS platforms. To make BIM more useful in the operational phase, BIM systems need to access and use operational data, typically available within Asset Management Systems (AMS). In contrast, AMS should be able to access and use data from BIM environments to make efficient and informed decisions when maintaining assets. These systems also need future proofing to support the sharing of data provided by new technologies. The key factor that sustains this disparity is the absence of standardised data formats to allow this exchange of information.

The AM4INFRA (AM4INFRA 2018) and INTERLINK (INTERLINK 2018) research projects, funded by CEDR, have taken the first steps towards a standardised format for data sharing, by developing a European Road Object Type Library (EurOTL), based on the IFC standard. CoDEC has taken these research outcomes further to cover the data used in asset management decision making processes - including data from new technologies - and developing standardised methods to automate the data integration process. Once completed, these methods will be publicly available on CEDR's website.

## 3   Overview of CoDEC

The CoDEC approach draws on a methodically developed framework for data (the Data Dictionary) and translates this into a machine-readable framework (the ontology) to make AMS and BIM data interoperable. This provides a step on the journey to the ultimate goal of making data available seamlessly when and where it is needed across management systems.

CoDEC undertook stakeholder engagement and desktop research to understand the as-is situation, the aspirations of NRAs and the challenges they face, to develop the CoDEC Data Dictionary, followed by a CoDEC Ontology for three key infrastructure assets: Roads, Bridges and Tunnels. Based on the Ontology it has developed, CoDEC has produced a software application (Application Protocol Interface, API) and implemented the developed methods and applications in three demonstration pilot projects. Once completed, the final product of this project will be a final Data Dictionary, set of ontology, and an OpenAPI, all of which are expandable to cater for individual NRA's needs and implementable within their systems and processes. A detailed description of the CoDEC Data Dictionary and CoDEC Data Ontology have been presented in Connected Data for Road Infrastructure Asset Management (Biswas et al. 2021).

## 4 Data Dictionary Structure and Content

The guiding principle to develop a Data Dictionary was focused on its ultimate application: the need to support the management of that asset (including the need for legacy data and new sensors and sensor data). The information from the review and the experience and knowledge of the team in infrastructure asset management (i.e. how Assets are actually managed), was brought together to draw the necessary judgements on: (1) what constitutes "an asset" vs the components of that asset, and (2) the level of detail needed to adequately describe that asset for the purposes of management. Hence the CoDEC Data Dictionary primarily covers three key highway assets (pavements, bridges and tunnels), including both the legacy data from these Assets and the data emerging from new technologies, such as sensors and scanning lasers that offer the potential to transform NRAs' future ability to manage highway assets.

The development of the dictionary built on previous work carried out in AM4INFRA (which developed a Data Dictionary for tunnels and bridges (AM4INFRA 2018)), the Highways England UK-ADMM Data Dictionary (Highways England 2020), the Data Standard for Road Management and Investment in Australia and New Zealand (DSRMI, for tunnels) (Austroads 2019) and ifcRoad (buildingSMART 2020), and the practical expertise of the CoDEC consortium members in asset management and asset data. Having established the design, workshops were held in which the Data Dictionary content was presented and discussed with representatives from CEDR NRAs to validate the approach and the content. An extract from the Data Dictionary is shown in Fig. 1 (the figure is truncated to fit, and as such does not show all fields).

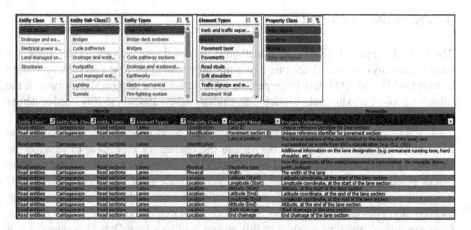

**Fig. 1.** Extract from the CoDEC Data Dictionary showing the entries under the 'Lane' element

## 5 Future Proofing the Data Dictionary

A key aim of CoDEC has been to consider sensors and the data they provide, as these are increasingly used to support infrastructure asset management. Sensors were not considered as 'Assets' in themselves, but rather as separate objects. The Data Dictionary focused on identifying the property sets which would apply across in general to types of sensor. Hence the Data Dictionary would be future-proofed and remain useful in a wide variety of use cases. When considering sensors CoDEC considered it necessary to develop different property sets for sensors that have fixed locations and those that are mobile. This addresses differences in the approach taken to referencing the location of fixed and mobile sensors. In addition, there can be differences in how sensors are defined - for example, one can consider an array (or network) of multiple fixed-location sensors but this does not apply to mobile sensors. Therefore, CoDEC has placed sensors in their own dedicated section of the Data Dictionary, separate from asset entities and elements.

## 6 CoDEC Data Ontology

Whilst the CoDEC Data Dictionary provides a method for creating a domain specific "CoDEC Ontology" in plain English and understandable by asset managers, the CoDEC Data Ontology translates the Data Dictionary content to an ontology modeled with OWL (Web Ontology Language). An Ontology is a framework to describe the interconnectivities between asset data and metadata in a shareable and reusable format. An ontology can be defined as a "formal, explicit specification of a shared conceptualization" (Studer et al. 1998), meaning that concepts, their constraints, and their relationships are encoded in a way that is systematically structured, explicit and machine-readable. This allows ontologies to be used to integrate and retrieve information, obtain semantically enhanced content, and to support knowledge management. However, ontologies are often created in a format which is not easily understandable by the asset managers who work with asset data.

## 7 Methodology for Developing CoDEC Ontology

The CoDEC Ontology has been developed to build on the EUROTL framework (INTERLINK 2018) using "Linked Data" and "Semantic Web" technologies. The Semantic Web helps link datasets that are understandable not only to humans, but also to machines, and "Linked Data" provides best practice for making these links possible. In other words, Linked Data is a set of design principles for sharing machine-readable interlinked data on the Web. CoDEC developed its Ontology using the Resource Description Framework (RDF) Schema and the Ontology Web Language (OWL) which were developed by the World Wide Web Consortium (W3C).

As a first step each Data dictionary concept or relationship was mapped to an existing class or property in EUROTL, as shown in Table 1. Properties are defined either as an object property or data property, meaning a semantic relation between

384     S. Biswas et al.

classes (for objects) or between the class and data (e.g. strings or numbers). Where a mapping was not present in the EUROTL, CoDEC created a new class or property (this ontology was developed using Stanford's Protégé (Musen 2015)). As an example, the Bridge concept already exists in the EUROTL Framework (AM4INFRA: Bridge (Marcovaldi and Biccellari 2018)). However, the concept of a Structural Element (or equivalent) is not found in EUROTL. Hence, a new Structural Element class was created in the CoDEC ontology, as a sub-class of the already existing EUROTL concept EurOTL:PhysicalObject.

**Table 1.** Example of Data Dictionary to ontology mapping

| CoDEC data dictionary | | | CoDEC ontology | | |
|---|---|---|---|---|---|
| Property | Description | Format | Domain | Object/Data property | Range |
| Bridge ID | The unique reference identifier for bridge | String | bridgeID | is-a | Bridge |
| Bridge name | The name of the bridge | String | bridgeID | rdfs:label | xsd:string |
| Environment | Classification of surrounding environment (e.g. Rural/Urban) | String | bridgeID | inEnvironment | xsd:string |
| Region/District/Area | Relevant geographical situation | String | bridgeID | prov: atLocation | eurotl: LocationByIdentifier |
| Owner | Owner of the asset | String | bridgeID | hasOwner | prov:Agent (person or Org.) |

# 8   CoDEC Application Protocol Interface (CoDEC API)

The last step on the process of linking data between different systems was to develop an Open Application Protocol Interface (CoDEC API). Application Programming Interfaces (APIs) are a "set of clearly defined methods of communication subroutine definitions, communication protocols" which help querying data to and from various sources using linked data/ semantic web technology. This helps CoDEC provide CEDR with a practical and systematic approach that can be implemented across Authorities to connect their Asset Data with their BIM Platforms, and vice-versa. The concept of this API is shown in Fig. 2. To manage the complexity of the linked data environment and create a "separate layer" that can be used without interfering with other "layers", CoDEC exposes a set of services (REST Web services and Python services). These services are responsible for communicating with the linked data environment, typically through a set of SPARQL queries and can be used by any application, as long as it has permission to access both services and data.

This layered approach has several advantages, the most critical one being the separation provided by multiple layers, which allows modification of the linked data structures without affecting the normal behavior of external applications, as they just need to know how to call the services (their inputs and outputs). Upon completion, CoDEC will deliver the OpenAPI specification, i.e., description and documentation detailing how services can be called, ensuring this can be used by all NRAs.

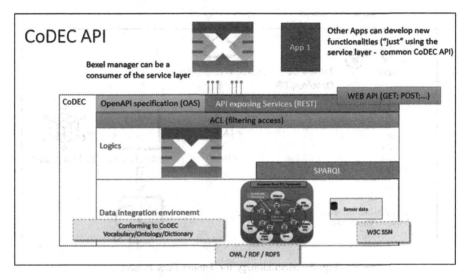

**Fig. 2.** CoDEC API overview

## 9 Demonstrating the Developed Solution Through Real Data

Three pilot projects have been carried out to test the ability to implement the CoDEC approach. These were undertaken with three implementation partners (CEDR NRAs) covering three different asset types: a Road, a Bridge and a Tunnel. Within this paper we will discuss the Road and Tunnel pilot projects. The objectives of the pilot projects were to:

- Show how the CoDEC solution can be successfully implemented across different NRAs for different Asset Types
- Demonstrate how integration of different data sets in one system can improve and help NRA decision making
- Identify potential gaps or challenges, capture those challenges, and refine the solutions if required

## 10 Pilot Project on Tunnel – Integration and 3D Visualisation of Monitoring Data

### 10.1 Concept

This pilot project was carried out on in consultation with Agentschap Wegen & Verkeer (AWV), the Belgian (Flemish) NRA. The pilot aimed to show how the CoDEC approach could support the integration of sensor data within a Tunnel BIM Model, and also support 3D visualisation for maintenance planning. It would hence demonstrate the practical applicability of the CoDEC approach and how the approach could improve the efficiency of Tunnel maintenance. A summary of how the pilot applied the CoDEC approach is shown in Fig. 3.

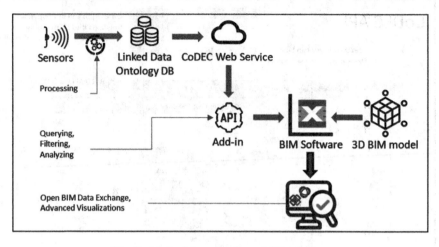

**Fig. 3.** Methodology for Tunnel Pilot Project

## 10.2 Method

A 3D BIM model was provided by Agentschap Wegen & Verkeer (AWV), Belgian (Flemish) NRA. The model included a broad range of categories, families and element types of the Tunnel, along with monitoring sensor data (CO, NO2, temperature, sight distance) collected over a period of one month. CoDEC used Bexel Manager (https://bexelmanager.com/), a BIM platform deploying IFC open BIM format (ISO 16739-1: 2018). The 3D BIM model was imported to Bexel Manager and, using CoDEC the Ontology and API, the sensor data was linked to the corresponding sensors in 3D BIM model within Bexel Manager. Connecting the sensor data to the BIM elements was essentially a 1-1 mapping process fed through a sensor identifier. On the BIM side, this identifier defined the sensors as a BIM element attribute (property), while on the sensor data side it was defined as a data column. This mapping enables an automatic, bi-directional relationship between BIM elements and the related sensor data. The enriched BIM model can be exported using open standard formats such as IFC to other BIM applications that support open standards.

## 10.3 Challenges and Outputs

Having applied CoDEC tools to link sensor data to BIM the pilot project CoDEC addressed the challenge of visualising this data. Sensors represent point locations and are distributed along the length of the Tunnel. However, the imported sensor cannot be shown in the BIM model. Hence, it was a challenge to find an ideal way to visualize imported data. In this case, the elements used for the visualization of imported sensor values are the wall panels that are distributed along the tunnel. Automating the sensor values to align with specific wall panels is one of the key workflows being addressed in the pilot. Ultimately, sensor readings is imported into BIM environment and applied to specific 3D BIM model sensor elements and wall panels to deliver advanced visualisation. For example, Fig. 4 shows the 3D visualization of the sensor data in the BIM Model using Bexel Manager's 3D colour-coded view, with the sensors' values shown in different colours, depending on numerical thresholds.

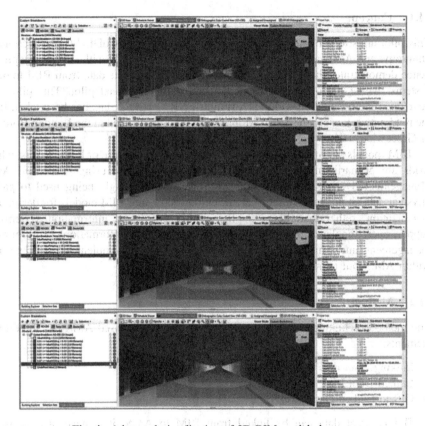

**Fig. 4.** Advanced visualisation of 3D BIM model elements

## 10.4 Benefits

The pilot project is highlighting some of the potential benefits of linking operation data to the BIM model, via the CoDEC API:

- It can support novel automated/improved inspection methods that deliver faster and more accurate inspection of infrastructure, with improved repeatability and reproducibility.
- It supports the development of central information models for data collection and further processing (i.e. a single source of truth).

388 S. Biswas et al.

The advanced 3D visualization tools users to efficiently identify the components of the infrastructure section that require attention. This addresses the key needs expressed by asset managers to have tools to help identify the critical zones and maintenance issues in a reasonable timeframe.

# 11 Pilot Project on Road – Linking BIM and GIS for Road Asset Management

## 11.1 Concept

This pilot aims to demonstrate how the tools developed by CoDEC (Data Dictionary, Ontology, and API) can be applied to integrate data from BIM with AMS. This project hence demonstrates that CoDEC can also be used to integrate data from BIM to other systems, whilst the opposite was demonstrated in the tunnel pilot. This pilot was developed in consultation with FTIA (Finnish NRA) but the data has been taken from the road network using the Smart Mobility Living Lab, located in the London Borough of Greenwich, UK.

BIM models are often created for the design/construct phase of Road Assets, whilst roads are primarily managed during the operational phase using GIS-based Asset Management Systems. Likewise, LiDAR scans – increasingly being used to gather physical data about Road Assets – are often translated into BIM models for design and visualization, rather than being applied within GIS-based information systems to support operational uses. Hence the BIM model often holds information useful for asset management, and which could be used to enrich (and/or complement) the data held within AMS, but is not made available in AMS.

## 11.2 Method

The method of linking data from a BIM model to a GIS based AMS has three main elements:

- Linking asset data from BIM to Linked Database
- Linked Data Base to GIS and
- GIS to Linked Data Base

Before asset data can be linked from BIM to AMS, it is necessary to assign parameters from the detailed geometric representation in a 3D BIM model to the 2D line representation of the road network. Network models are simple by design because they are used to provide insights on network performance as a quick and clear overview. Conversely, BIM models provide a more detailed representation of the network. Converting these complex geometries to simple lines will result in loss of information. In the pilot project, we chose to define the pavement as a set of "slabs" (rectangular units) that together form the road network, and intersect each of them with the lines defining the route of the road. As a result, slabs that do not intersect with the line are excluded. Hence the placement of the lines in the network is important because it can determine if data from slab A or from slab B will be assigned to this network segment.

In the transition, each intersection between network line and slab polygons are stored in a new line dataset that contains information from both data sources.

The positions of the slabs are stored as linked data using the ISO 19148:2021 Linear Referencing ontology, used in the European Road OTL. This ontology provides means for locating objects (Assets) along elements of a network, alignments, or other linear elements. In our case the linear element is an individual slab within the road network. For each slab, the start and end position on the network is determined, by measuring the start and end distance relative to the start of the entire polyline, using the arcpy measureOnLine geometry method. These measurements are stored as 32-bit floating-point (xsd:float) values. Finally, these linear elements are uploaded back into the CoDEC repository using the CoDEC API. This approach enriched the road network model with information from the BIM model using linked data technology and international standards. Figure 5 shows the process of linking data from BIM to GIS model.

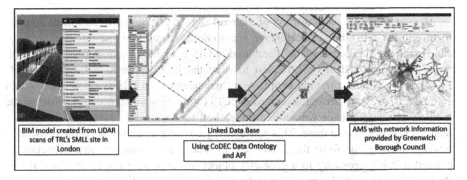

**Fig. 5.** Process of linking of data from the BIM model to GIS

## 11.3 Challenges and Output

The first challenge in the linear referencing method was to store the pavement geometry in the CoDEC linked data repository. The European Road OTL uses the official GeoSPARQL ontology from the Open Geospatial Consortium to represent geospatial location and relationships to objects (Assets). Within the GeoSPARQL standard the property, geo:hasGeometry is used to link a feature with a geometry that represents its spatial extent. This geometry class is defined as geo:Geometry and can contain multiple spatial properties. In this pilot project we use the geo:asWKT property, which is defined to link a geometry with its WKT serialization, to store the convex hull coordinate information from the pavement subsection as a polygon. Figure 6 shows the process CoDEC has taken to convert 3D BIM element to 2D linear referencing system.

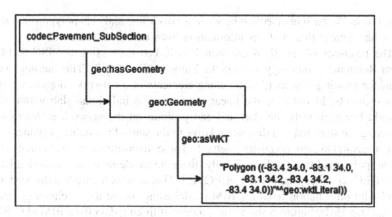

**Fig. 6.** Linear Referencing in CoDEC Ontology

Having delivered the transition of the road slabs from the BIM to a format suitable for use in the GIS, the next challenge was to demonstrate how the information associated with the slabs can be stored and used in GIS-readable format. Again, a SPARQL query was used, with a GET request to store the information as a JSON (ISO/IEC 21778:2017) within a python script. Asset management in GIS is often done within commercial software packages such as ArcGIS (https://www.arcgis.com/), which is the industry standard. Hence we applied the python tool arcpy to convert the JSON to a feature class in a geodatabase. This was done by creating an empty polygon dataset and loading in the slabs from the JSON one by one, while using an arcpy function to convert a WKT geometry to a geometry object readably readable by ArcGIS. It was then possible to view the pavement slabs on a map and visually compare the data with the road network. It was then possible to connect the specific pavement construction information from BIM to the AMS. The connected information is hence available in AMS to use in Asset Management decisions. Figure 7 shows data from BIM model is linked within in GIS based AMS.

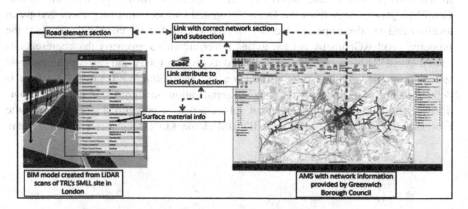

**Fig. 7.** Data from the BIM model linked to GIS

## 11.4 Benefits

This Pilot project demonstrated the use of the CoDEC ontology for successfully linking data between BIM and GIS. This ultimately provides benefits to NRAs such as

- Providing a single source of truth about their Assets
- Having the required data available in the system where Assets are primarily managed
- Providing future-proofing such that data from new technologies (e.g. sensors, digital twins etc.) can be supported within AMS

The pilot also provided experience in the practicalities of applying the CoDEC approach and its implications for further implementation such as: Pavements are linear features but must be modelled in small segments in BIM. There will be a need to determine the optimum size for such segments, and there are many factors influencing the decision – for example, the granularity of the data available to be attached to each segment, the road layout (curvature, length between junctions, complexity etc.), and maybe even constraints on model size; The road alignment should be included as an object in BIM models. Including the alignment as an 'object' (perhaps as connected curvilinear elements) should facilitate the referencing of BIM data to GIS data by effectively including the linear referencing system within the BIM model.

## 12 Conclusion

Data is vitally important to asset managers and supports decisions throughout the asset lifecycle. Although there has been progress integrating BIM into the operational phase of Assets, CoDEC is one of the first projects to consider this from the Asset Management side; creating practical methods to enrich data, data systems, and change our way of working.

Building on the previous research projects, such as AM4INFRA and INTERLINK, CoDEC has applied a methodical approach to develop a framework for data (the data dictionary) and translate this into a machine-readable framework (the ontology) to make AMS and BIM data interoperable. This provides a step on the journey to making data seamlessly available when and where it is needed across data management systems and supports the first steps in the transition from traditional Asset Management to operation via the Digital Twin.

CoDEC has aimed to provide practical and implementable outcomes to NRAs that are also future-proof, by creating a framework that includes data provided by new technologies. Although, Codec has not covered all road infrastructure assets and data types, it gives a structured and practical framework that can be easily expanded to include other asset types and data as required in the future - hence catering for Road Authorities' future needs. The CoDEC methods and applications, once completed, will be available for download from the CoDEC website (www.codec.eu) and CEDR's website (www.cedr.eu).

Although CODEC has successfully developed applications to integrate data from different systems, there is substantial work still to be done in this area. A key recommendation of this research is to encourage collaboration between asset owners (such as NRAs), standardisation bodies (such as ISO and IFC) and the software technology industry to understand the practical needs of asset managers/owners when it comes to data integration, and to build on the outcomes of this project to deliver the tools that will meet these needs.

**Acknowledgements.** The CoDEC project would like to thank all the experts who were involved in stakeholder consultation activities by sharing their knowledge, experience and providing suggestions for the project.

# References

AM4INFRA: Asset Data Dictionary. Deliverable D3.1 (2018)

Austroads. Data Standard for Road Management and Investment in Australia and New Zealand (DSRMI), version 3.0 (2019)

Biswas, et al.: CoDEC: connected data for road infrastructure asset management. In: 30th International Baltic Road Conference (2021)

buildingSMART. Industry Foundation Classes (IFC), IfcROAD (2020)

Highways England. Asset Data Management Manual (ADMM), version 12.0 (2020)

INTERLINK project. Final Project Report – INTERLINK Approach with European Road Object Type Library (2018)

Marcovaldi, E., Biccellari, M.: Asset Data Dictionary, Deliverable D3.1 of the AM4INFRA project (2018)

Musen, M.A.: The Protégé project: a look back and a look forward. AI Matters. Association of Computing Machinery Specific Interest Group in Artificial Intelligence 1(4) (2015). https://doi.org/10.1145/2557001.25757003

PIARC. Inspector accreditation, non-destructive testing and condition assessment for bridges. PIARC Ref.: 2011R07EN (2012). ISBN 2-84060-240-7

Studer, R., Benjamins, R., Fensel, D.: Knowledge engineering: principles and methods. Data Knowl. Eng. **25**(1–2), 161–198 (1998)

# Greening Road Projects

Creating Read Projects

# Road Construction Using Locally Available Materials

Robert D. Friedman[1(✉)] and Ahmed F. Abdelkader[2]

[1] AggreBind Inc., New Haven, CT, USA
info@aggrebind.com, robert@aggrebind.com
[2] AggreBind Inc., Cairo, Egypt

**Abstract.** Road Construction in developing countries can be expensive. Therefore, an effective and a more affordable method is needed. Using local, indigenous, and in-situ materials are becoming necessary as the luxury of constructing asphalt mixing stations with a secured material supply chain for crushed aggregate and more does not always exist. Meanwhile, recently introduced environmental regulations bring additional constraints on traditional road construction practices.

AggreBind's soil stabilization products; RoadMaster RM1/RM2 and Aggre-Bind AGB-WT/BT are unique, patented and patent-pending, water based, cost effective, environmentally friendly, cross-linking polymeric emulsions with or without a proprietary tracer that provides an excellent quality assurance tool for specifiers and contractors alike.

This Paper investigates the influence of using AGB-WT on improving the load-bearing strength and shear-strength parameters, while expanding longevity of stabilization and creating water-resistance in roads constructed using primarily in-situ materials, including sub-soils, sands, mining waste and crushed construction waste.

Reported results have proven that AGB-WT can provide stable and dust-free roads as well as stabilized base-layers for major highways passing the requirement of International Road Specifications. Moreover, it is concluded that the use of AGB-WT in soil stabilization or in stabilizing base-layers in road construction can reduce the cost of road construction by 40% to 60% and increase the in-situ load-bearing capacity by 400% to 600%. Results are documented in AggreBind's Book of Civic Approvals and Independent Test Results.

**Keywords:** Soil stabilization · Dust control · Cross-linking polymer emulsion · In-situ materials · Optimum Moisture Content (OMC) · California Bearing Ratio (CBR) · Water-resistant roads

## 1 Introduction

Roadways are one of the most important signs of a state's/country's economic development and urbanization. The connectivity of adequate transportation stimulates the economy, and facilitates industrial growth, and social developments. Simply stated, good quality roads raise the GNP (gross national products) of a country. On the other hand, unconstructed and dusty roads decrease the air quality index, affect the public health, and damage agricultural crop production.

© The Author(s), under exclusive license to Springer Nature Switzerland AG 2022
A. Akhnoukh et al. (Eds.): IRF 2021, SUCI, pp. 395–404, 2022.
https://doi.org/10.1007/978-3-030-79801-7_28

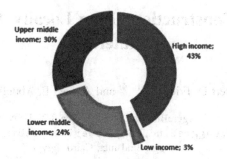

**Fig. 1.** Total Road network distribution by income groups (%).

Figure 1 presents the total road network distribution by income group as reported by IRF in 2018. The data shows the disparity between affluent or developed countries and those that are less so. While the World Factbook of CIA reported in 2013 that the world has 64,285,009 km of unpaved roads.

In traditional soil stabilization and road construction, a wide variety of materials are used as naturally occurring or processed soil, fine and coarse aggregates, and binders like bituminous materials, lime, or cement[4]. Materials selection depends on road specifications and available budget. Moreover, the construction process requires a set of equipment and material processing plants as compactors, motor graders, sub-base station, asphalt plant, pavers, distributors, etc.

There is further pressure upon traditional road construction because of a global shortage of aggregates. Traditional road bases and sub-bases are built with specified materials of varying sizes of aggregate. According to the office of the Environmental Commissioner of Ontario, the quantity of aggregate consumed in an average 1 km of a two-lane road or highway can be 15,000 tons. These constrains were also discussed by Meininger et al. and others.

Moreover, the United Nations Environment Programme (UNEP) concluded that "Sand and gravel represent the highest volume of raw material used on earth after water. Their use greatly exceeds natural renewal rates and the amount being mined is increasing exponentially, mainly, as a result of rapid economic growth in Asia. Negative effects on the environment are unequivocal and are occurring around the world. The problem is now so serious that the existence of river ecosystems is threatened in a number of locations and damage is more severe in small river catchments. The same applies to threats to benthic ecosystems from marine extraction. Therefore, the need to adapt and integrate new ways of building roads and modifying traditional methods of road construction is becoming self-evident.

According to The National Society of Professional Engineers publication, "In a world where construction commodities prices are followed like daily stock quotes, developing a new building material that's cheap and desirable can be a ticket to success".

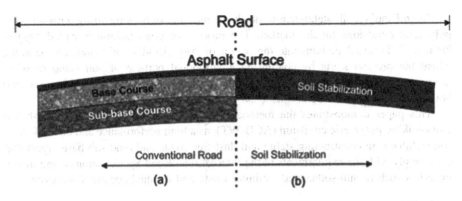

**Fig. 2.** Road design comparison; (a) Mechanical Stabilization (b) Chemical Stabilization

Soil stabilization can be accomplished by several methods which can be classified into two broad categories, namely mechanical stabilization (Fig. 2a), and chemical stabilization. Mechanical stabilization is the process of improving the properties of the soil by changing its gradation, whereas the chemical stabilization (Fig. 2b) is the process of adding a synthetic substance to the soil to react physically or chemically with the soil particles to impart cohesion and hence increase bearing capacity and reduce water absorption.

Recent technology advancements have increased the number of new and green substances used in soil stabilizing purposes. Such non-traditional stabilizers include natural resins, enzymes, surfactants, fibre-reinforcements, salts, synthetic polymers, biopolymers, and more.

Among these materials are polymer emulsions which provide cost effective alternatives compared to other synthetic materials used in soil stabilization, as they are used with locally available soils, eliminating the need for quarried aggregates, reduce the construction time, allow roads to be opened to traffic sooner than other road building methods, and are applied in a straightforward process in which the polymeric emulsion is diluted to the proper concentration to achieve the effective additive volume along with the amount of moisture content desired for the most efficient compaction of the soil.

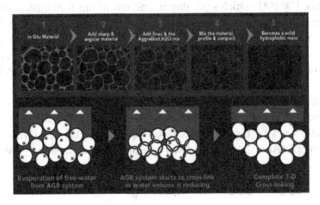

**Fig. 3.** "Aggrebinding©" Process

AggreBind's soil stabilization products are the only long-string cross-linking polymeric emulsions on the market. The process of cross-linking is called "aggre-binding©". Figure 3 demonstrate the action of AggreBind's soil stabilizers on soils, where the process starts by coating each individual particle of soil (sand or other inorganic "waste" material) and ends with a solid hydrophobic mass able to withstand hot and cold temperature extremes ($-57$ °C to 163 °C).

This paper demonstrates the methodology and the results of using water-based cross-linking polymeric emulsion (AGB-WT), as a high performance and cost-effective soil stabilizer, in constructing stable and dust-free roads and road sub-base layers that can comply with the requirements of the International Road Specifications using in-situ materials such as sub-soils, sands, mining waste and crushed construction waste.

## 2 Materials and Methods

### 2.1 Materials

#### 2.1.1 Soil Stabilizer

In this study, AGB-WT, water-based cross-linking polymeric emulsion was used as a soil stabilizer. AGB-WT is patented and manufactured by AggreBind Inc., USA, in conjunction with their global production partners. The AGB-WT emulsion formulation is designed to contain a proprietary tracer for advanced quality assurance inspection. It is classified as an eco-friendly material[21] and received patents in numerous countries including USA; US 9,260,822 B2.

AGB-WT is white in color before curing, colorless after curing and has a solid content of 50% $\pm$ 1% and specific gravity of 1.05. AGB soil stabilizer can also be in colours and AGB-BT is the black grade.

#### 2.1.2 Types of Soils and Nature of the Work

Table 1 identifies the in-situ soils of different sites described in this investigation, including the nature for the work.

The first site (A) was the expansion of the urban road that is the beginning of the International Highway from Kathmandu Nepal through to India. This road carries approximately 100,000 vehicles per day in and out of the Kathmandu area. Work was completed in 2019, https://www.youtube.com/watch?v=GHxquG3sFbI. See Fig. 4.

The second site (B) was the formalization of a small aircraft parking and staging lot at the La Aurora International Airport in Guatemala City, Guatemala Central American. Work was completed in 2014, https://www.youtube.com/watch?v=KnPcTllE1RY. See Fig. 5.

The third site (C) was the road repair of Highway 27 in Waikato New Zealand. This length of road was built on an area with a high-water table. The highway carries 20,000 cars and 4,000 trucks daily. Previous repairs failed quickly, 11 times in 24 months. Work was completed in 2014, https://www.youtube.com/watch?v=_9sWxgkWirs. See Fig. 6.

**Table 1.** Types of in-situ soils and their locations

| Site code | Site name | Country | City | Type of soil | Type of application |
|---|---|---|---|---|---|
| A | Int'l Highway to India | Nepal | Kathmandu | Government Specified | Road |
| B | Airport | Guatemala | Guatemala City | Pumice Sand/Clay | Airplane parking lot |
| C | Road Repair Highway 27 | New Zealand | Wellington | Pulverized Asphalt and Base material | Highway |

(a) Before  (b) After

**Fig. 4.** Site (A) International Highway to India, Nepal

(a) Before  (b) After

**Fig. 5.** Airport, Guatemala

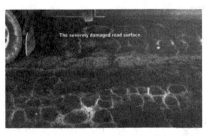 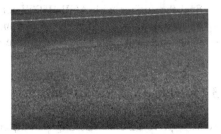

(a) Before  (b) After

**Fig. 6.** Road repair, Highway 27, New Zealand

## 2.2 Test Methods and Application Equipment

Pre-application evaluation and field application monitoring are defined by AggreBind Inc. to produce properly constructed AGB-stabilized roads[23].

### 2.2.1 Soil Characterization

The in-situ soil behavior was characterized using the following test methods:

| - Particle-Size Distribution/Fines Content: | ASTM D6913 |
| - Plasticity Behavior: | ASTM D4318 |
| - Moisture Content: | ASTM D1557 |
| - Optimum Moisture Content (OMC): | ASTM D698 |
| - California Bearing Ratio (CBR): | ASTM D4429 |
| - and Unconfined Compressive Strength (UCS): | ASTM D2166 |

### 2.2.2 Installation Equipment

Regular road-construction equipment is used with in-situ soils including:

- Scarifying/mixing/leveling machinery, as motor-grader,
- Dual-drum compactor,
- Transfer pumps and flow meter,
- and Water truck with a spray bar for topical application.

## 2.3 Installation Procedure

After the pre-installation assessment, each project was treated individually to develop the appropriate installation protocol that suits the in-situ soil and the overall site conditions. However, the general procedure can be summarized as follows:

- Scarify the in-situ soil to the required depth.
- Pre-wet the soil, if required.
- Add/mix additional fines/aggregates into the soil, if required.
- Spray AGB-WT/Water mix over the prepared soil.
- Mix the soil thoroughly.
- Level the surface to desired pitch, camber, and inclines.
- Compact the soil with vibratory compactor.
- Seal the surface with AGB-WT/Water mix.
- Perform Quality Assurance inspection.
- Open road to traffic when the surface seal is dry to the touch.

# 3 Results and Discussion

Table 2 presents both working parameters generated by the pre-installation and the ones used during the stabilization process and road construction. As can be observed, each site has its own characteristics and requirements. OMC and mixed water (pre-

Road Construction Using Locally Available Materials     401

wetting) are the main actions that differentiated the in-situ soils in the three locations. While application rate of the surface-seal and its dilution rate didn't alter throughout the locations.

**Table 2.** Typical characteristics of in-situ soils

| Site code | OMC (cc/200 gm of soil) | Scarify soil depth (mm) | Mixed fines (kg/m$^3$) | Pre-wet mixed water (l/m$^2$) | Mixed larger angular material (kg/m$^3$) | AGB-WT/Water ratio | Application rate of AGB-WT/Water mix (l/m$^3$) | Application rate of surface-seal (L/m$^2$) * |
|---|---|---|---|---|---|---|---|---|
| A | 6 | 150 | 0 | 5 | 0 | 1:5 | 4 | 1 |
| B | 8 | 150 | 0 | 8.8 | 0 | 1:6 | 4 | 1 |
| C | 6 | 150 | 0 | 5 | 0 | 1:5 | 4 | 1 |

*AGB-WT/Water Ratio for Surface-Seal in the three projects was 1:3.

The soils ranged from government specified soil, as a combination of dirt and aggregate, to in-situ soils, amended with pumice fill, to base material and asphalt.

OMC is affected by ambient temperature, humidity, types of soils and moisture content of the soil. The objective of pre-wetting is to bring the residual moisture content up to OMC minus 1%. Generally, a 1:5 dilution rate of AGB-WT:Water closes this gap, and brings the overall moisture content up to OMC. This can be done with moisture test equipment or mechanically, depending on availability of testing equipment.

Results of fines content, gradation of soil size and plasticity index (PI) all fell within acceptable range, as the 'ideal' soil configuration is 35% fines content passing through a #200 sieve, no rocks or stones exceeding 20% of the depth, and a PI of <15. Table 3 presents a typical sieve analysis of soil.

**Table 3.** Typical Sieve Analysis in-situ soil(24)

| Sieve analysis | | | | | |
|---|---|---|---|---|---|
| Sieve | Mesh size | Passing | Sieve | Mesh size | Passing |
| 3" | 75 mm | 100.00 | 10 | 2.00 mm | 54.94 |
| 2" | 50 mm | 87.06 | 20 | 850 μm | 44.27 |
| 1 1/2" | 37.5 mm | 87.06 | 40 | 425 μm | 36.35 |
| 1" | 25 mm | 79.82 | 60 | 250 μm | 30.88 |
| 3/4" | 10.0 mm | 77.35 | 100 | 150 μm | 25.30 |
| 3/8" | 9.5 mm | 71.28 | 140 | 106 μm | 22.71 |
| 4 | 4.75 mm | 64.09 | 200 | 75 μm | 20.77 |

On the other hand, based on independent laboratory testing, adding AGB-WT to soil increased the load bearing strength (CBR) by 400% to 600% (Table 4). It takes up to 28 days to reach maximum hardness, maximum strength, regardless of soils, as is the case of the three projects in Table 1 (Fig. 7).

Figure 8 illustrates the developing of unconfined compressive strength at 14 days of curing. The same as CBR results, the increase in UCS was comparable for each site.

**Table 4.** CBR compaction results at Day 28[24]

| Sample No. | Strikes No. | Compaction H (%) | γd (Lb/pie$^3$) | C (%) | Expansion (%) | C. B. R (%) |
|---|---|---|---|---|---|---|
| 1 | 10 | 6.35 | 114.98 | 89.8 | 0.00 | 53.77 |
| 2 | 25 | 6.35 | 119.95 | 93.7 | 0.00 | 99.36 |
| 3 | 56 | 6.35 | 125.48 | 98.0 | 0.00 | 316.33 |

AGB WT1/RM1 CBR - 28 days (562% increase)

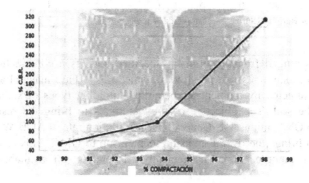

**Fig. 7.** Increase of CBR[24]

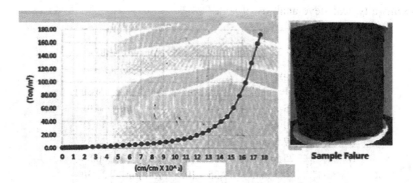

**Fig. 8.** Deformation characteristics of stabilized soil sample

It should be noticed that projects as in Table 1, Figs. 4, 5 and 6 have been in place for 2–7 years, and with AGB-WT surface seal, they provide a water resistant, hydrophobic surface.

Figure 9 presents the value proposition of AGB-WT, as a percentage of concrete. It can be observed that the use of AGB-WT with in-situ materials, along with top seal yields up to an 80% cost reduction, and at 50% when compared with traditional asphalt.

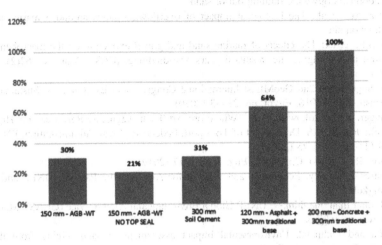

**Fig. 9.** Cost as a % of traditional concrete – installed sale price per m$^2$.

## 4 Conclusion

AGB-WT is a cost-effective patented environmentally friendly long-string cross-linking polymer-based soil stabilizer, can provide an alternative to standard asphalt and concrete roads, as they create stable road surfaces from in-situ soils. AGB-WT is designed to be used in accordance with standard road construction specifications, and when used in soil stabilization or in stabilizing base-layer, it increases load bearing capacity of in-situ soils by 400% to 600% and reduces the cost of road construction by 40% to 60%.

## References

AggreBind's Book of Civic Approvals and Independent Test Results, updated March 2021

AggreBind's Pre-Installation Evaluation (Process) Form, revised July 2020

Aggregate Use in Road Construction, Environmental Commissioner of Ontario, Annual Report, ECO Issues, February 2015

Al-Khanbashi, A., Abdalla, S.W.: Geotech. Geol. Eng. **24**, 1603–1625 (2006)

CIA, FactBook, Roadways (2013)

Dieppe (eastern English Channel), ICES J. Mar. Sci. **67**(2), 270–277 (2010)

Environmental Attestation & Independent Tests, updated March 2021

Kondolf, G.M.: Hungry water: effects of dams and gravel mining on river channels. Environ. Manag. **21**(533–551), 1997 (1997)

Green-tinted-building-products, The Magazine for Professional Engineers, NSPE, June 2012. https://www.nspe.org/resources/pe-magazine/june-2012/green-tinted-building-products

International Road Federation (IRF), World Road Statistics (WRS) (2018)

Krause, J.C., et al.: The physical and biological impact of sand extraction: a case study of the western Baltic Sea. J. Coastal Res. **51**, 215–226 (2010)

Van der Schalk, J.: "We Are Running Out of Sand", FreedomLab, 9 February 2018. https://freedomlab.org/we-are-running-out-of-sand

Desprez, M., et al.: The biological impact of overflowing sands around a marine aggregate extraction site

Boyd, S.E., et al.: The effects of marine sand and gravel extraction on the macrobenthos at a commercial dredging site: results 6 years post-dredging. ICES J. Mar. Sci. **62**(2), 145–162 (2005)

Proceedings of the 2nd GeoMEast International Congress and Exhibition on Sustainable Civil Infrastructures 2018, Egypt, pp. 21–33 (2019)

Meininger, R.C., Stokowski, S.J.: "Wherefore Art Thou Aggregate Resources for Highways", Public Roads, U.S. Department of Transport, Federal Highway Administration, PN, FHWA-HRT11-006, vol. 75 (2011)

Cabezas, R., Cataldo, C.: Cogent Eng. **6**(1), 1–17 (2019)

"Resource Efficiency: Economics and outlook for Asia and the Pacific", UNEP and CSIRO, Bangkok (2011)

"Sand, rarer than one thinks", UNEP Global Environmental Alert Service (GEAS), March 2014. www.unep.or/geas

Sreebha and Padmalal: Environmental impact assessment of sand mining from the small catchment rivers in the Southwestern Coast of India: a case study. Environ. Manag. **47**, 130–140 (2011)

United States Patent, US009260822B2, Friedman et al., February 2016

USAC-CBR-UCS-AOL_IN Final-09.2020

Amhadi, T.S., Assaf, G.J.: Overview of Soile Stabilization Methods in Road Construction, GeoMEast 2018

The Value of Aggregates, State of the Aggregate Resource in Ontario Study AECOM Canada Ltd., Paper 3 (2009)

Salam, Y.A., Netula, O.: Int. J. Enhanced Res. Sci. Technol. Eng. **5**(1), 3–11 (2016)

# Practical Applications of Big Data Science

Practical Applications of Drug
Science

# Road Roughness Estimation Using Acceleration Data from Smartphones

Arak Montha[1], Attaphon Huytook[1], Tatree Rakmark[1], and Ponlathep Lertworawanich[1,2(✉)]

[1] Department of Highway, Ministry of Transport, Bangkok, Thailand
ponla.le@doh.go.th
[2] Bureau of Road Research and Development, Department of Highways, Bangkok, Thailand

**Abstract.** The International Roughness Index (IRI) is a serviceability measure index for highway maintenance performance evaluation. At the Thailand Department of Highways (DOH), IRI is generally measured by Laser Profilers and Walking Profilers, which require experts to operate and process data. For the day-to-day road maintenance, real-time approximated IRI values are helpful to locate the damage area where roadwork is urgently needed. This study presents a simple method to estimate IRI from acceleration data collected from smartphones equipped with accelerometers. Acceleration data are segmented into 100-m sections and processed with the Fast Fourier Transform (FFT). For each 100-m road section, an average IRI is collected by using an accuracy Laser Profiler and compared with the processed acceleration data. The primary result shows that IRI has a linear relation with the sum of Magnitude of the FFT of the vertical acceleration obtained from the smartphones with the R-squared of 0.8745. In future, the DOH plans to extend this research findings and develop a smartphone application to collect real-time IRI for road maintenance management.

**Keywords:** Highways and roads · Road surface profile · Pavement condition evaluation · Pavement management · Mobile sensors · International roughness index (IRI) · Power spectral density (PSD)

## 1 Introduction

Road conditions play an important role in our everyday life. Abnormality of road surfaces greatly affect vehicles and accidents. Moreover, an increase of road roughness can adversely affect convenience for people who use these roads. Collecting road condition is significant for road maintenance. Collecting road roughness data in consecutive time intervals helps road operators to estimate the increasing of road roughness and to properly plan for road maintenance and monitoring. IRI is accepted as a road serviceability measure for maintenance and examination. IRI can be measured by several methods but most of them require expensive and complicated equipment. Experts are required to operate the equipment. Using a sensor in a smart phone is an easy way to collect road roughness data. To compare with other methods, mobile

© The Author(s), under exclusive license to Springer Nature Switzerland AG 2022
A. Akhnoukh et al. (Eds.): IRF 2021, SUCI, pp. 407–415, 2022.
https://doi.org/10.1007/978-3-030-79801-7_29

sensor is acceptable because of the result from smart phone and other equipment are nearly same. Reducing in cost can get more information to rearrange road maintenance properly.

## 2 Literature Review

Collecting road condition data by using mobile sensor has many types of data. It is essential to study how to convert information and select only useful data. Moreover, studying parameters for improving data from collecting road condition data by application from mobile sensor can be used to compare with IRI by existing method.

Sayers (1989) identified and discussed the differences and similarities between IRI and HRI. The similarities were the IRI and HRI indices are both obtained by using the quarter-car analysis. The distinction was that the IRI was obtained by applying the quarter-car to a single wheel track profile where the HRI was obtained by applying the quarter-car to a point-by-point average of two-wheel track profiles. Using power spectral density (PSD) function the relation between HRI and IRI was 0.71(IRIleft + IRIright)/2 < HRI < (IRIleft + IRIright)/2. The data collected in the Brazilian experiment (the IRRE) on asphalt, surface treatment, gravel, and dirt roads showed that relation between HRI and IRI was HRI $\approx$ 0.80(IRIleft + IRIright)/2. The differences between the left- and right-hand wheel tracks showed how one side of a lane has deteriorated more than the other.

Chen et al. (2016) developed a system recording acceleration data in three directions and spatial information such as time, location, and speed using an accelerometer and a GPS, respectively. The data were transferred to a central server after preprocessing to compute Power Spectral Density (PSD) and the IRI. This study approximated the IRI by and a linear equation using the standard deviation of roughness. The proposed system was capable of classifying roughness values into four categories according to the China standard. The total development cost was relatively low.

Douangphachanh and Oneyama (2013) conducted a study to estimate the IRI through smartphone accelerometer measurements. They used Android smartphones mounted on the dashboard of the test vehicle and two vehicles used in the experiment. The IRI was measured using Vehicle Intelligent Monitoring System (VIMS) for every 100 m. They found that the sum of the magnitudes of values obtained from the Fourier transform of accelerometer signals in the z-axis had a linear relationship with the IRI. The resulting relationship was statistically significant when the speed was less than 60 km/h.

Agostinacchio et al. (2013) developed numerical simulations on Matlab to identify the link among road surface characteristics, types of traffic vehicles, and consequent evolution of the vibrational phenomena. They used the ISO 8608 that classifies roads surface into eight classes from class A (minor degree of roughness) to class H (high degree of roughness) by comparing calculation of the PSD (Power Spectral Density) to generated road profiles, In their study, the ISO classes of considered profiles were A, B, C and D. During the simulations, a road section pavement of length equal to 250 m generated through the application of Fourier series. They used QCM (Quarter Car Model) that was the vehicle model that was effectively used to study the dynamic interaction between vehicle and road roughness profile by combining the masses values (m), the stiffness constant (k), and damping (c) to model three types of road vehicle (car, bus, and truck). The dynamic load, produced by the passage of these vehicles traveling at different steady speeds (20, 40, 60, 80, and 100 km/h). The results showed that the generation of vibration fundamentally depended on the longitudinal regularity of the road surface and only to a much lesser extent on the vehicles' increase in speed and the influence of the vehicle type.

Do and Le (2013) analyzed the performance of the Fast Fourier Transform (FFT) Filter which was applied to calculate the displacement from the signal measured from the accelerometer compare with Infinite Impulse Response (IIR) Filter and Finite Impulse Response (FIR) Filter. They used High-speed profilers to produce the true road surface profile and used an accelerometer to create an inertial reference that defines the height of the accelerometer located on the vehicle. The signals created in Matlab with different frequencies and multi harmonics are imported into the integrated processing. Standard error and peak error were used to estimate the calculated displacement data. The experiments showed that FFT filters gave results more accurate than did FIR and IIR not only for single frequency signal but also for multi-frequency signal. The standard error, the average peak error, and the maximum peak error of FFT Filter were less than those of FIR Filter and IIR Filter.

Douangphachanh and Oneyama (2014) studied the relation between IRI and the Magnitude of Vibration analyzed from acceleration data and added Gyroscope data, both collected from smartphone applications (Androsensor). The experiment used three types of the smartphones with different placements and four types of vehicles. The IRI was measured by using Vehicle Intelligent Monitoring System (VIMS) for every 100 m. The relationship between IRI and Magnitude of Vibration found that when added Gyroscope to analyze with Acceleration, the R2 values increased compared with using only one Acceleration data in every assumption.

Lima et al. (2016) developed an application called RoadScan on the Android operating system, to collect the acceleration data to analyze the standard deviation and classify the road surface condition. In addition, there also have crowdsourcing to collect information from users in the Google Maps API to develop the database and build the reliability of the application. After approximately three months of testing, the Google Maps API was possible to indicate road quality. The good quality happened when a smooth surface with low vibration was trafficked. Terrible quality usually occurred in a non-asphaltic road or road with too many deformities. As a result, this application presented an accurate overall outcome.

Mukherjee and Majhi (2016) observed the response of a smartphone in a vehicle. In the theoretical model: both quarter car and half car models. Then they compared the accelerations on a smooth road and a speed bump. They tested Android smartphone's accelerometers in the laboratory by comparing their records with laboratory accelerometers. They found a good correlation between the smartphone and the laboratory accelerations in the entire range of excitation. They recorded the accelerations on the main roads of Ahmedabad; India with four types of vehicles: 4-Wheeler MUV, 4-Wheeler Hatchback, 2-Wheeler, and 3-Wheeler running through two most common types of bumps: a single bump of width 3.5 m and height 0.12 m and multiple bumps comprise of a series of 6 bumps of width 0.35 m and height 0.04 m each and separated by of 0.5 m from center to center. The records had been filtered by Discrete Fourier Transform (DFT). The result presented that smartphone could record road bumps very accurately while riding all classes of vehicles.

Eljamassi and Naji (2017) used AHP (Analytical Hierarchy Process) method to compare several methods (Smartphone application, visual inspection survey, and aerial/Satellite images) used in the road network inventory and to determine the most efficient and less expensive for using in the Gaza Strip. Sony Xperia Z3 was used as data logger, video recording device. The test was conducted at average traveling speeds between 20 and 100 km/h. The road condition data were divided into four different levels of Road Quality (RQ): green for good, yellow for satisfactory, red for unsatisfactory, and black for poor. They combined collected data with GIS software to complete the method of using smartphone application to collect IRI. They used AHP as a multiple criteria decision-making tool to deal with complex decision making.

Sadjadi (2017) verified the accuracy of smartphone sensors as a measure of road roughness, compared with various reference data such as Road Quality Index rating (RQI) by road users' assessment. International Roughness Index (IRI) and Road Distress Index (RDI). Using Root Mean Square (RMS) calculated from Acceleration as road condition from smartphone compared with the IRI from Road Surface Profiler vehicle, also compared with RQI and RDI. The results showed that the relation between RMS and RQI was significantly correlated ($R2 = 0.8$). One could use a smartphone to express how road users felt. The relation between RMS and IRI ($R2 = 0.76$) showed that in the future, a smartphone could be able to replace a traditional method that has a limitation of operating cost.

Yang et al. (2020) applied a Road Impact Factor (RIF) transform to calibrate different smartphones for a consistent RIF-index by evaluating three calibration methodologies, 1) the reference mean method which calibrates smartphone signals toward the mean RIF, 2) the reference max method which calibrates the smartphone signals toward the maximum RIF, and 3) the reference road type method which uses the calibration coefficient determined for the paved road as the calibration coefficient for unpaved roads. Three different smartphones were selected to perform the field experiments. They were the iPhone8 which was reference measurement, iPhone 10, and Google Pixel (GP). They were mounted flat on the floor of the front passenger seat in 2015 Volkswagen Jetta. Data were collected by traversing the same road segment for 35 times. The two road segments, selected in Fargo, North Dakota, was a 400-m section of the paved road on 9th St NE, and a 300-m section of the unpaved road on 57th St N., The speed limit for the paved road is 25 mph, and for the unpaved road is 15 mph. Road tests indicated that the measured RIF-index varies significantly among smartphones.

This project shows application of smartphones in estimating road surface condition, especially to classify roughness condition of road sections by using original method. The ultimate goal from this project is our expectation to estimate road surface in tentative index not to get exact IRI from various data that we get from provincial officers by using mobile application, which made by us. That is used for running Road Management System and Road Maintenance Program.

## 3 Experimental

### 3.1 The experiment's Field

The experiment took place on Highway Number 357 (Concrete Surface) section length is 45 km in Suphanburi Thailand (Fig. 1).

**Fig. 1.** Map Highway Number 357 (Concrete Surface) section length is 45 km

### 3.2 Equipment and the Setting

This experiment's idea is to use the processed data from acceleration that collecting from smartphone application compare with IRI data from road surface profiler to find a trend and relation. This part will describe the equipment that use for this experiment and how to set them.

A. Smartphone and Application

We used Androsensor (Application on the Android operating system) with setting on 100 Hz frequency to collect acceleration 100 data per second then we used Huawei P9 that support the Android operating system to carry out this task and fixed it on a computer console case (Which operates RSP's IRI collecting system) behind a driver's seat (Figs. 2 and 3).

**Fig. 2.** Huawei P9

**Fig. 3.** Androsensor's user interface

B. Road Surface Profiler

Dynatest SRP MARK III is used for monitoring and collecting IRI with Graphical real-time display of the IRI, RN, laser elevations, inertial profile, macrotexture, and photo-logging. GPS and digital photo logging can be stored with profiler measurement data (Fig. 4).

**Fig. 4.** Dynatest SRP MARK III

### 3.3 Methodology

The used data in this experiment are acceleration and IRI which will be collected by smartphone and road surface profiler (SRP) respectively. So, we will transform acceleration into other factors such as Sum Magnitude, SQRT-PSD by Fast Fourier Transform Method. Then find a relation between those factors with IRI. We will divide methodology into three steps below.

Setup Step: We will setup the RSP to measure IRI every 5 m and setup the smartphone to collect acceleration 100 data every second (100 Hz). After that we will check the GPS time between the smartphone and RSP that should be the same to each other then we will know what the starting point and the finishing point is in every station.

Collecting Data Step: We started this experiment on 9 March 2021 take place on Highway Number 357 in Suphanburi, Thailand. The smartphone was fixed on a computer console case behind the RSP driver's seat. We pushed the recording button of the RSP measuring monitor and Androsensor application at the same time before the car moving. Then we drove the RSP on Highway Number 357 with speed 60–100 km/Hrs. both left and right lane.

Data Processing Step: After obtained data, we divided the length of both Highways into a 100 m length station. The GPS time from the RSP and the smartphone set the starting point and the finishing point, we also divided all data to suit each station. Then divided acceleration will transform to other factors by Fast Fourier Transform Method, divided IRI will average from every 5 m to 100 m to compare to each other. For more detail, please follow up on the data processing part below.

## 4 Analysis

In the experiment, we obtain acceleration data from a smartphone and IRI data with Road Surface Profiler (Dynatest RAP Mark III) then classify data with Microsoft Excel and process with Scilab Program (Fig. 5).

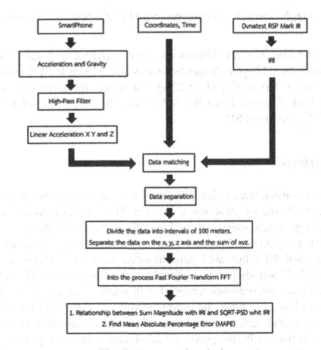

**Fig. 5.** Process of analysis

Acceleration data are vehicle's acceleration mix with gravity's acceleration, but we want to process only vehicle's acceleration, so gravity's acceleration will be filtered by High-Pass Filter. Data are segmented into 100-m intervals with time stamped. The acceleration data are processed with Fast Fourier Transform (FFT). The sum of FFT magnitude of acceleration data from the vertical axis is then plotted against the IRI measured from the specified equipment. Linear regression between the IRI and the sum of FFT magnitude of the vertical-axis acceleration is performed with the R-squared 0.8745. Additionally, the power spectral density (PSD) of the vertical-axis acceleration is also regressed against the IRI with the R-squared 0.7498 as shown in Fig. 6.

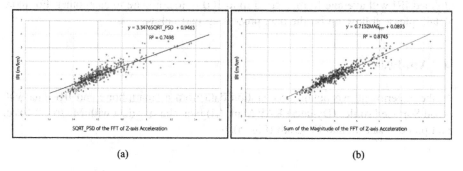

**Fig. 6.** Linear relationship between FFT of acceleration of z axis

In term of Mean Absolute Percentage Error (MAPE), find that MAPE from the relation between Sum Magnitude and IRI is less than MAPE from the relation between SQRT-PSD and IRI (6.65% and 35.22%). Therefore, the same conclusions can be drawn from both R-squared and the MAPE that the magnitude of FFT is a better predictor of IRI than the PSD.

## 5 Conclusions

The International Roughness Index (IRI) is a serviceability measure index for highway maintenance performance evaluation. At the Thailand Department of Highways (DOH), IRI is generally measured by Laser Profilers and Walking Profilers, which require experts to operate and process data. For the day-to-day road maintenance, real-time approximated IRI values are helpful to locate the damage area where roadwork is urgently needed. This study presents a simple method to estimate IRI from acceleration data collected from smartphones equipped with accelerometers. Acceleration data are segmented into 100-m sections and processed with the Fast Fourier Transform (FFT). For each 100-m road section, an average IRI is collected using an accuracy Laser Profiler and compared with the processed acceleration data. The primary result shows that IRI has a linear relation with the sum of Magnitude of the FFT of the vertical acceleration obtained from the smartphones with the R-squared of 0.8745. In future, the DOH plans to extends this research findings and develop a smartphone application to collect real-time IRI for road maintenance management.

# References

Agostinacchio, et al.: The vibrations induced by surface irregularities in road pavements - a matlab approach. Eur. Transp. Res. Rev. **6**, 267–275 (2013)

Android Application AndroSensor. Google Play Store. https://play.google.com/store/apps/details?id=com.fivasim.Androsensor&hl=en. Accessed 29 Jan 2021

Chen, K., Tan, G., Lu, M., Wu, J.: CRSM: a practical crowdsourcing-based road surface monitoring system. Wirel. Netw. **22**, 765–779 (2016)

Douangphachanh, V., Oneyama, H.: A study on use of smartphones for road roughness condition estimation. J. Eastern Asia Soc. Transp. Study **10**, 1551–1564 (2013)

Douangphachanh, V., Oneyama, H.: Exploring the use of smartphone accelerometer and gyroscope to study on the estimation of road surface roughness condition. In: The 11th International Conference on Informatics in Control, Automation and Robotics (ICINCO-2014), pp. 783–787 (2014)

Eljamassi, A.D., Naji, M.S.: Using smartphone and GIS to measure International Roughness Index (IRI) in Gaza strip-palestine road network. J. Eng. Res. Technol. **4**(3), 92–104 (2017)

Lima, A.P.R.O.: Using crowdsourcing techniques and mobile devices for asphaltic pavement quality recognition. In: 2016 VI Brazilian Symposium on Computing Systems Engineering (SBESC), Department of Computer and Systems (DECSI) Federal University of Ouro Preto (UFOP) - João Monlevade, Brazil (2016)

Mukherjee, A., Majhi, S.: Characterization of road bumps using smartphones. Eur. Transp. Res. Rev. **8**, 1–12 (2016)

Sadjadi: Investigating for road roughness using smartphone sensors. Int. J. Comput. Commun. **11**, 56–63 (2017)

Sayers, M.: Two Quarter-Car Models for Defining Road Roughness: IRI and HRI. Transportation Research Record 1215, University of Michigan, Transportation Research Institute, Ann Arbor, pp. 165–172 (1989)

Do, T.M., Le, T.C.: Performance analysis of FFT filter to measure displacement signal in road roughness profiler. Int. J. Comput. Electr. Eng. **5**(4), 356–361 (2013)

Yang, et al.: Calibration of smartphone sensors to evaluate the ride quality of paved and unpaved roads. Int. J. Pavement Eng. (2020)

# Improving Traffic Safety by Using Waze User Reports

Raitis Steinbergs and Maris Kligis[✉]

State Joint Stock Company Latvijas Autocelu Uzturetajs,
Krustpils iela 4, Riga, Latvia
{raitis.steinbergs,maris.kligis}@lau.lv

**Abstract.** Road inspection regularity and existing types made by road maintenance crew have not been good enough to be aware what is really happening on the roads. Road users' contribution in road traffic safety is very important to ensure fast reaction on different road hazards.

It is important to ensure not only the most common ways to report road hazards on state roads by phone, by email and on social media, but also expand data sources options in modern and user-friendly way.

Waze navigation application already had functionality to report road hazards – to warn other application users, but no one acted to solve these road hazards until someone reported them through existing communication channels supported by State Joint Stock Company Latvian Roads Maintainer and State Company Latvian State Roads.

To ensure better road traffic safety and faster reaction time on road hazards solving, Latvian Roads Maintainer gained access to Waze report feed and made a system for analyzing and processing Waze data. As the result - Latvian Roads Maintainer can improve road safety by faster reaction to road hazards reported by Waze users.

Today, up to 70% from total reports processed by Latvian Roads Maintainer are generated by Waze.

**Keywords:** Traffic safety · Road maintenance · Road hazards · Road user contribution · Waze

## 1 Introduction

To ensure better road traffic safety and more efficient response time, it is important to know what is happening on the roads. There used to be two ways to obtain information about hazards on state roads in Latvia– road users calling and road maintenance or road administration team regular road checks. The road maintenance on state roads in Latvia is performed by SJSC (State Joint Stock Company) Latvian Road Maintainer, but road maintenance supervision is carried out by State Company Latvian State Roads.

The current regulations determine the required frequency of road maintenance monitoring. The higher the maintenance category, the more frequently roads must be inspected. Road maintenance is carried out in priority order based on maintenance classes A (highest), B, C, D and E (lowest). They are determined after assessing the

---

© The Author(s), under exclusive license to Springer Nature Switzerland AG 2022
A. Akhnoukh et al. (Eds.): IRF 2021, SUCI, pp. 416–424, 2022.
https://doi.org/10.1007/978-3-030-79801-7_30

socio-economic importance of roads, traffic flow and road condition. This means that the roads of the most important and busiest routes are arranged first.

Road Supervisor Inspection Frequency is regulated by Regulations on the requirements for daily maintenance (Cabinet of Ministers 2021). For Latvian Roads Maintainer – Road specifications (Latvian State Roads 2020). On the other hand, both inspection frequencies are the same. More detailed inspection frequency and total length of state roads in Latvia for each maintenance category are displayed in Table 1 below.

**Table 1.** State road maintenance category length and inspection regularity (Latvian State Roads 2021)

| Maintenance category | Total length, km | Inspection frequency |
| --- | --- | --- |
| A | 1 946,417 | Once a week |
| B | 4 795,094 | Once every two weeks |
| C | 11 819,66 | Once a month |
| D and E | 1 570,765 | Once a quarter |
| Total | 20 131,936 | |

As per Table 1 above, more than half of roads are with C maintenance category and inspection frequency is only once a month.

When road maintenance crew performs road inspection, all dangerous obstacles to traffic safety are removed and information for planning other maintenance works are gathered, but due to many hazard types not being possible to predict, such as, obstacles (for example, tree) or dead animals, inspections scheduled once a week for A maintenance class roads are unlikely to identify any hazards on the roads efficiently. Of course, road inspection also has different objectives too. It is not only to discover hazards that need to be resolved within the set timeline.

Obstacles dangerous to traffic safety must be removed within 3 h on road sections with A, B and C category, within 6 h on road sections with D category and within 24 h on road sections with E category (Cabinet of Ministers 2021). On high maintenance category roads with the shortest hazard reaction time the traffic flow is the highest, Road users must contribute to ensure traffic safety. To ensure that, appropriate communication channel is required. For state road administration most used and most advertised is Traffic information center's support line. Typical road user can forget the number, or they have not heard it. As a result, there is still great possibility that hazard will not be reported or will be reported later. The lower the maintenance category, the lower possibility that hazard will be reported.

The main idea of developing and implementing a Waze data analysis and processing system was to use already existing data source, like Waze navigation application, that is the best strategy to improve awareness of what's happening on roads. The particular Waze data analyzing, and processing system was developed and implemented in the Winter season 2019/2020. Hazards reported by Waze users have been registered and solved as other hazards from other communication channels.

418    R. Steinbergs and M. Kligis

## 2 Waze Application

Waze is one of the most popular Navigation applications worldwide. It is the second most popular Navigation app for iOS (Apple Inc. 2020). Application have been downloaded from Android OS official application store "Google Play" more than 100 million times (Google Inc. 2020). Waze is not just a regular navigation app with route planning and giving turn-by-turn guidance. It also tells a lot about traffic, roadworks, police appearances, crashes and more in real-time because of the contributions from the road users. Waze is even used when travelling the well-known routes to understand the situation on the road ahead.

For a long time, Waze has been very popular in Latvia. There is no formal Waze statistics board, but 100 000 active users were the highest usage number few years ago. (All Media Latvia 2017; Leta 2015).

Roughly every 7th driver in Latvia uses Waze when taking the total number of registered drivers and vehicles with valid technical inspections (up to 3.5k tons) as per 1st May 2021 was almost 680,000 respectively (Celu satiksmes drosibas direkcija 2021).

In the Waze application it is possible to warn other drivers of different types of hazards, but only few of them can be solved by performing road maintenance work. Waze hazard types that meet the specifics of road maintenance work are:

- Pothole
- Roadkill (dead animal)
- Obstacle
- Missing road sign
- Floods
- Ice on road
- Snow on road

Latvian State Roads and Waze have been working closely since 2014. The information on roadworks, road condition in winter and road maintenance unit locations are shared with Waze to inform Waze users about possible road hazards ahead. Waze is sharing reports about potholes that are displayed in Latvian State roads public map as pothole concentration to improve pothole repair planning in spring (Latvian State Roads 2017). Reports about winter hazards reported by Waze users are also displayed on Latvian State roads public map to improve overall awareness about road conditions.

Waze reports are also available on the Waze public map at https://www.waze.com/livemap, but the display of reports is not suitable for effective prevention of dangerous traffic situations because:

- All notifications are displayed, regardless of their type and location
- Only partial hazard information is displayed
- Reports are not "stored" - the report may disappear from the map despite not being resolved;
- To filter reports based on the specific requirement is not available
- Unable to merge messages for a single event
- Unable to mark messages as processed

Based on the above points, it was impossible to work effectively with Waze reports as a stand-alone and equivalent to other communication channels. Also, it was necessary to develop a specific Waze report processing tool.

Waze always has been collaborative and offered access to their reports online through the Connected Citizens program. In 2019, as part of this program's framework, Latvian Roads Maintainer cooperation was approved, and an online dataflow on all events within Latvia validated.

## 3 Waze Report Processing and Analysis System

Access (see Fig. 1) to Waze data was the starting point to explore existing data and find out best solution for report analysis and processing.

**Fig. 1.** Waze report processing system

Waze uses the GeoRSS API to internally retrieve traffic data. Waze traffic data consists of the following information:

- General information: timestamp of the file, geographic area from which the data was retrieved, etc.
- Traffic alerts: traffic incidents reported by users.
- Traffic jams: traffic slowdown information generated by the service based on a user's location and speed.
- Unusual Traffic (Irregularities): alerts and traffic jams that affect an exceptionally large number of users.

Each alert gets a reliability score based on other user's reactions and the level of the reporter. Also, each alert gets a confidence score based on other user's reactions (Waze 2019).

The cooperation started with the Developer – Riga Technical University, to develop a specific web-based Waze data flow processing and analysis system. The data flow provided by Waze included not only reports about hazards and information on congestion, road construction and other public information. Each hazard report contains information about the time of publication, location (coordinates), hazard type, subtype, location (city) and street/road and several hazard reliability and confidence parameters.

The key task of the processing and analysis system was to filter only required information from the Waze data flow and display the reports in the desired way. The basic requirements of the solution were defined as follows:

- Report displays on the map the way it must update every 60 s and display current situation
- Report filtering by certain types of hazards, location, time and reliability, confidence score and other parameters;
- Ability to define which hazard types does not need to display or saved for later analysis
- Determine the belonging to the Latvian Roads Maintainer region
- Analysis of historical data by visually mapping their concentration
- Ability to combine reports and mark them as processed

Required data analysis functionalities were possible to implement as Waze already provided various additional details for the report. Drawing their polygons on the map for the Latvian Roads Maintainer region was the only detail added to the original requirements.

After collecting all filtered results from Waze database, Latvian Roads Maintainer Customer Service staff made a manual review and published reports on Customer Support System. Reports are prioritized based on hazard type. Following that, the set actions commence. The alert system of Latvian Roads Maintainer was already developed. This alert system creates tasks for respective region supervisor to check incoming alert. If there is a need for manual prevention of the task, the task is given to the work team to resolve the problem, and they would choose the best process to resolve it and provide feedback to the task.

Our goal was to add relevant Waze reports to this system and execute the set actions. As a result, Waze reports initiated action for respective Latvian Roads Maintainer region staff, cause manual prevention of the report. This action turned out to be several times faster than other communication channels, such as call, email etc. As a part of finding, it demonstrated that road safety improved and potentially life-threatening situations on the road are eliminated quicker.

Waze data flow processing and analysis system was developed in 2019. The analysis of Waze data flow as a full-fledged communication channel was started on 1st November 2019.

## 4 Impact of Waze Reports

Since 2018, incoming information about the situation on the roads in Latvian Roads Maintainer has been processed and managed centrally by the Customer Support Service. In public, Latvian State Roads Traffic Information Center remained as the main communication center. Inevitably road users called Latvian Roads Maintainer directly. This process had to be managed. As part of that, the Customer Support Service was established. Customer Support Service provides a more convenient communication between Latvian Roads Maintainer and Latvian State Roads. Moreover, determines the action status regarding the received information, and a digital journal was created for the registration and processing of incoming information.

Waze reports were recorded in the digital journal in the same way as the information received via any other communication channels (calls, SMS, e-mail, Facebook or Twitter entry) and passed on to the relevant region for execution.

Some hazard types like "Ice on-road" and "Snow on-road" relate to road maintenance in winter given that these conditions are monitored 24/7 during the winter season, the reports provide an additional source of information for responsible staff however, they are not recorded for now.

In the first 12 months since the start of Waze reports processing, the number of reports received by Latvian Roads Maintainer Customer Support Service more than doubled. As the result, the total share of Waze reports reached 64% (Latvijas autocelu uzturetajs 2021a). See Fig. 2 for a detailed info report distribution among different communication channels.

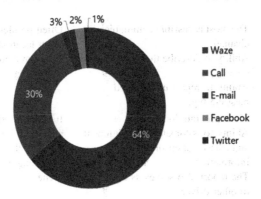

**Fig. 2.** Communication channels shares

Waze users most often reported hazards were Animals on the road - 70%, Obstacles on the road - 18% (Latvijas autocelu uzturetajs 2021b). See Fig. 3 for the most common types of hazards by Waze.

The main communication channel for Latvian State Roads Traffic Information Center is the support channel (calls). Including the reports received by Latvian State Roads, in 2020 they were a total of 9000 reports. 70% of these reports were received by phone, 26% from Waze, 4% from social networks or e-mail.

In essence, all reports (regardless of the communication channel) are the same, but there are some differences. The advantages and disadvantages of the information received via the support channel (calls) and Waze are gathered in Table 2.

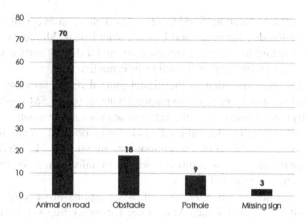

**Fig. 3.** Most common types of hazards

**Table 2.** Advantages and disadvantages of communication channels

| Communication channel | Advantages | Disadvantages |
|---|---|---|
| Support line (call) | The most accessible communication channel<br>Ability to describe the event in more detail<br>Ability to report unclassified message types | Often problems describe the exact location of the event<br>More time-consuming report processing |
| Waze app | More accurate location<br>Ability to automate the assignment and classification of additional information<br>The report also serves as a warning to other drivers | It is not possible to provide additional information to describe the event<br>Reporting is only possible at the scene |

## 5 Future Development Plans

After the implementation of the solution, the first improvements in the report processing automation were made. Two areas were identified for improvement:

- Hazard reports also often appeared on the part of roads that have been handed over to the Contractor during roadworks;
- Waze data flow also contained reports generated through the Latvian State Roads Winter Service System. Latvian Roads Maintainer Winter Service Team originally generated this.

To reduce the time required for report processing, based on the Road supervisor geospatial data about construction sites and the data flow submitted by the Road supervisor winter service system, the following improvements were made:

- Additional information was assigned to the report - belonging to the construction site. As a result, it is possible to filter the reports;
- Reports whose coordinates coincide with the coordinates of Latvian State Roads data stream reports were classified and can also be filtered.

Soon we plan to develop and implement the automation of Waze report flow processing. Firstly, the solution will be developed to add more accurate geospatial information (road addresses) to incoming hazard reports. By implementing a machine learning solution that combines reports that are about one event (based on historical report processing data), automatic publication can start. Also, functionality to warn staff about possible duplication of hazard will be developed, in case if there is similar incoming report from other communication channel, for example by phone.

# 6 Conclusions

The opening of the Waze communication channel has been a success to complement existing – traditional communication channels with a new way of digital reporting, without even developing the reporting platform itself but using an existing platform.

Using Waze reports as a similar report to the regular communication channel has been increased the number of hazards resolved and eliminated by more than a quarter. This makes it a good reason to believe that it is possible to increase road safety based on having quality information on the road state and hazards that need attention.

The future lies in connecting communication channels where information about what is happening on the road already exists, rather than creating new ones.

**Acknowledgements.** Waze data processing and analysis system was developed based on the agreement between Latvian Roads Maintainer and Riga Technical University. This manuscript was created to inform about the Waze project as a case study.

# References

All Media Latvia: Latvia is the seventh best country in the world to drive a car, according to a study by the Waze application (2017). Portal skaties.lv. https://skaties.lv/auto/latvija-ir-septita-labaka-valsts-pasaule-brauksanai-ar-auto-liecina-aplikacijas-waze-petijums/. Accessed 8 Dec 2020

Apple Inc.: Waze Navigation & Live Traffic on App Store (2020). https://apps.apple.com/us/app/waze-navigation-live-traffic/id323229106. Accessed 7 Dec 2020

Cabinet of Ministers: Regulations of the Cabinet of Ministers of April 16, 2021 No. 26, Regulations on the Daily Maintenance Requirements of State and Municipal Roads and Control of Their Execution, Paragraph 69, Riga, Latvia (2021)

Celu satiksmes drosibas direkcija: Report, Number of registered vehicles (2021). https://www.csdd.lv/transportlidzekli/registreto-transportlidzeklu-skait. Accessed 02 June 2021

Google Inc.: Waze - GPS, Maps, Traffic Alerts & Live Navigation (2020). https://play.google.com/store/apps/details?id=com.waze&hl=lv&gl=US. Accessed 7 Dec 2020

Latvijas autocelu uzturetajs: Report, Communication channel shares 2020, Riga, Latvia (2021a)

Latvijas autocelu uzturetajs: Report, Most common types of hazards 2020, Riga, Latvia (2021b)

Latvian State Roads: The winter maintenance season ends on state roads; please mark the holes in the Waze application (2017). https://lvceli.lv/aktualitates/uz-valsts-autoceliem-beidzas-ziemas-uzturesanas-sezona-aicinam-atzimet-bedrites-aplikacija-waze/. Accessed 8 Dec 2020

Latvian State Roads: Road specifications 2019, version 2, pp. 369–374, Riga, Latvia (2020)

Latvian State Roads: Road maintenance (2021). https://lvceli.lv/celu-tikls/buvnieciba-un-uzturesana/celu-uzturesana/. Accessed 12 Aug 2021

Leta: The Waze application has received more than 2,500 reports of pothole (2015). https://www.la.lv/aplikacija-waze-sanemti-vairak-neka-2500-zinojumi-par-bedrem. Accessed 8 Dec 2020

Waze: Waze Traffic-data Specification Document, Version 2.8, p. 6, CA, United States (2019)

# A Glimpse into the Near Future – Digital Twins and the Internet of Things

Howard Shotz[1](✉) and James Birdsall[2]

[1] Parsons, Centreville, USA
howard.shotz@parsons.com
[2] Parsons, Dubai, United Arab Emirates

**Abstract.** A glimpse into the near future, digital twins are an exciting and innovative technology designed to maximize existing road assets. Digital twins offer organizations the opportunity to deliver leading-edge asset management capabilities and integrated analytics by unlocking the wealth of information stored in existing pavement and bridge management systems. These capabilities offer immense benefit to transportation agencies by ensuring efficient, reliable, and available asset information critical to operations and planning.

This paper will offer a new perspective of capital investment strategy and maintenance operations. It will present a model for organizations to move beyond standard maintenance and operations and into a larger framework based on the Internet of Things (IOT). This will provide additional real-time intelligence about assets to enhance operations and maximize investments, while optimizing the total cost of ownership and state of good repair.

Topics addressed in this paper include:

- A discussion of how organizations are preparing to deploy digital twins within an overall asset lifecycle management framework.
- Use case examples of how organizations are planning to deploy digital twins by integrating source systems of record (SSOR) such as pavement and bridge management systems.
- An overview of the services, products, and technologies employed.
- The latest thinking on asset management, including human factors, cyber and physical security, big data analytics, artificial intelligence (AI), and machine learning (ML).

**Keywords:** Digital twins · Asset management · Bridge management · Paving management · Internet of Things

## 1 Introduction

A digital twin is a virtual representation of a physical asset, process, or system, as well as the engineering information that allows us to understand and model the asset, process, or system's performance both operationally and strategically. The digital twin may be represented as a highly detailed visual model in building information modelling (BIM) or geographic information system (GIS), as well as a process representation

---

© The Author(s), under exclusive license to Springer Nature Switzerland AG 2022
A. Akhnoukh et al. (Eds.): IRF 2021, SUCI, pp. 425–438, 2022.
https://doi.org/10.1007/978-3-030-79801-7_31

displayed via statistical data charts and reports. The distinction between a digital twin and a static model is that the digital twin is continuously updated with data from multiple source systems of record (SSOR) such as a paving management system (PMS), bridge management system (BMS), enterprise asset management system (EAM), safety management system (SMS), intelligent transportation system (ITS), and traffic accident reporting system (TARS).

Figure 1 illustrates how various activities and artifacts developed within engineering technology, operations technology, and information technology link to real-world infrastructure through activities such as immersive visualization via 3D modelling, augmented reality (AR), virtual reality (VR), timeline visualization (4D) and data analytics using machine learning (ML) and predictive modelling. Digital twin technology offers new opportunities for organizations to maximize the lifespan and reduce the total cost of ownership (TCO) of their existing roadway infrastructure. Digital twins can deliver leading-edge asset management capabilities and integrated analytics by unlocking the wealth of information stored in existing pavement and bridge management systems.

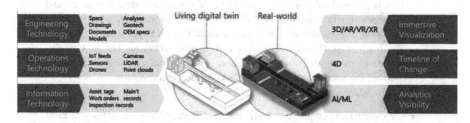

**Fig. 1.** Examples of digital twin technology converging with asset infrastructure to provide organizations with a complete picture of their infrastructure and processes

Digital twins can offer an immense benefit to transportation agencies by ensuring efficient, reliable, and available asset information critical to operations and planning. This information can help organizations evolve from reactive management (break/fix corrective maintenance) to proactive reliability centered maintenance (RCM). RCM is a robust maintenance methodology that ensures maintenance tasks are performed in an efficient, cost effective, reliable, and safe manner by following seven basic steps: identify system functions, identify failure modes, identify failure causes, identify effects of failure, identify consequences of failure, determine preventative tasks, and identify alternatives.

## 2 How Organizations are Preparing to Deploy Digital Twins Within an Overall Asset Lifecycle Management Framework

### 2.1 Asset Lifecycle Conceptual Framework

The framework illustrated in Fig. 2 proposes a conceptual model to connect SSOR at the operational level to the dashboards and digital twin systems at the presentation level. The four levels are described in detail below.

**Operational Level.** This level is comprised of several active distinct software applications used across the organization by various departments. These systems may be in-house legacy applications, on-premises commercial-off-the-shelf systems (COTS), or current generation cloud solutions that support many ongoing activities. In many organizations, they exist at different levels of asset maturity, data accuracy, and are comprised of distinct data ontologies and typologies that do not integrate with each other to provide an enterprise view.

**Integration Level.** The integration level is comprised of two primary components – the common data environment (CDE) and the asset registry/GIS spatial engine. Spatial and tabular information is linked to the SSORs at the operational level via point-to-point streaming API integration or middleware applications that extract/transform and load (ETL) the information in the CDE and asset registry.

**Strategic/Analytics Level.** To support analytics and strategy, the model incorporates Business Intelligence (BI) and analytic engines to further mine the CDE for trends and out of limit data. These will drive established metrics and key performance indicators (KPIs) based on use cases developed by the organization. Organizations may also send this packaged and analyzed information into one or more digital twins to support operations and planning.

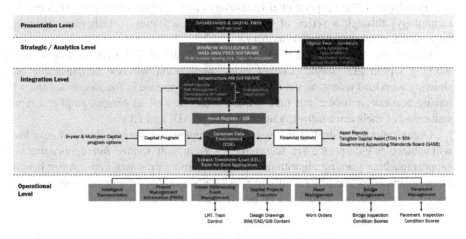

**Fig. 2.** Asset lifecycle conceptual model

**Presentation Level.** The presentation level synthesizes the information from the operational, integration, and strategic/analytics level into one user-friendly, bi-directional, integrated interface. The concept of unifying disparate information into a combined interface with a unified data ontology and data taxonomy is known as "the single pane of glass." See Fig. 3 for an example display showing how users will interact with the digital twin via the "single pane of glass" to analyze trends and access vast amounts of information from multiple systems with minimal training.

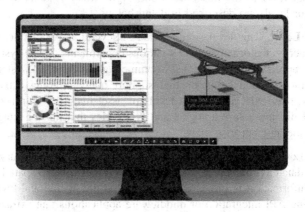

**Fig. 3.** "Single pane of glass" digital twin integrates BIM/GIS with multiple data SSORs

## 2.2 Applied Intelligence Conceptual Model

Figure 4 illustrates the powerful potential convergence of data and technology to create a digital twin with applied intelligence. Moving up from *Source Intelligence* (SSORs) in the lower left of the figure to the *Outcome* level in the upper right, we can see how each of the six layers provides a foundation for each subsequent layer moving upward in the figure. The digital twin (blue arrow) is operational at all levels and bridges between *Applied Data* (operational technology) and *Applied Technology* (information technology) through a series of five consecutive activities – *Authoring, Synching, Sequencing, Federating,* and *Visualizing*.

Within the *Visualizing* activity supported by the digital twin of Fig. 4, users working within the "single pane of glass" application will see demonstrable data showing workload reduction via *Automation* and an increase in *Prediction* accuracy by having accurate, reliable, and timely information as well as current graphical representations of their asset infrastructure via BIM, CAD, and GIS.

Organizations that deploy and maintain the implementation into the *Predictive* level will achieve an enterprise asset management system with *Applied Intelligence* that will support machine learning algorithms. Machine learning promises even greater benefits to an organization due to its ability to spot trends and support analysis with large and complex data sets that would not be possible with human operators.

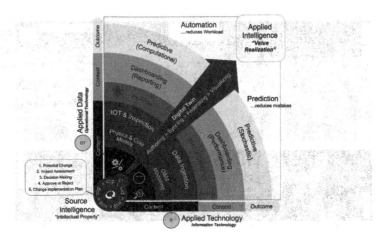

**Fig. 4.** Using digital twins to blend applied data and applied technology into applied intelligence

## 3 Use Case Examples of How Organizations are Planning to Deploy Digital Twins by Integrating Source Systems of Record Such as Pavement and Bridge Management Systems

### 3.1 Connecticut Department of Transportation (CTDOT)

Parsons is deploying Bentley's iTwin Services for CTDOT as a proof-of-concept design review, design validation and design insights on a new construction project shown in Fig. 5. These iTwin Services are hosted within Bentley's Cloud Service using Microsoft Azure. CTDOT will develop new digital process workflows for implementation during the preliminary design phase of a project. These digital process workflows will be tested and refined as part of a joint team between CTDOT and their vendors. The team will document and prepare a manual to summarize the use of iTwin Services for design and provide knowledge transfer and training in conjunction with Bentley's technical team. This manual will also develop technical submission requirements for the use of iTwin Services by the consulting community working under contract with CTDOT. The goal of the project is to understand the condition of the paving before failure using proactive analytics to compare the current problem state against a functional baseline.

### 3.2 Kuwait Public Authority for Roads and Transportation (PART)

PART oversees the road transport infrastructure including rail and metro systems on behalf of the government. They also plan, design, implement, and maintain road networks and land transportation systems. PART is using an AI supported iPhone application to conduct a drive-through asset inventory and road capture analysis. Figure 6 shows the output from the video capture application that can identify and

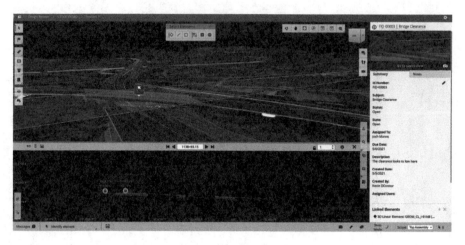

**Fig. 5.** CTDOT iTwin Services project

locate signs, traffic lights, guardrails, and other recognizable objects using post-processing AI analytics including road telemetry. This information is then available to multiple SSORs.

**Fig. 6.** Kuwait PART video capture and asset analytics application using artificial intelligence for survey

Figure 6 displays the application performing a pavement condition survey tied to an automated GIS "heat map" that displays favourable conditions in green and unfavourable conditions in red. The application will scan the video image and assign condition ratings based on AI/ML training. The Road AI system offers to collect survey data at a lower cost and can be done in a continuous fashion using non-specialist tools and personnel while still providing data that is consistent, accurate, and auditable. Using Road AI in this way enables:

- Early intervention with lower cost treatments based on asset condition to deliver 5–10% increase to the life of the asset
- Data driven forward programme planning – consistent, objective road condition data
- On demand reports – network deterioration modelling.

### 3.3 Dubai Roads and Transport Authority (RTA), Dubai, United Arab Emirates

The RTA is the major independent government roads and transportation authority in Dubai. RTA is the sole public transportation service provider and includes the Dubai Metro, Dubai Tram, Abras, Dubai Bus, Dubai Water Bus, Water Taxi, Dubai Ferry and Dubai Taxi. The RTA has started initial integration tasks to support a digital twin by linking several enterprise systems shown in Fig. 7. The enterprise ESRI GIS system of record is linked to the Maximo Enterprise Asset Management system. This integration provides several benefits. Metro Station deficiency information is collected in the field using mobile devices and sent to the EAM linked to GIS coordinates. The unified data set provides trend reporting and rule-based analytics on metro station asset health and inventory. It also integrates with the RTA's client side identity access management (IAM) for authentication, the RTA's centralized electronic document management system (EDMS), and with the RTA 's client side email service to send rule-based notifications.

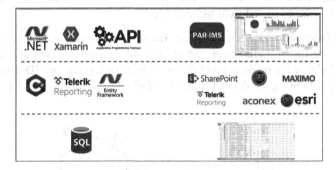

**Fig. 7.** System diagram of RTA's digital platform to support asset management and asset health monitoring

## 4 An Overview of the Services, Products, and Technologies Employed

The digital twin vendor marketplace is dynamic and rapidly spawning a broad range of technologies and conceptual ecosystems for road, rail and transit organizations. The marketplace can be divided into two categories: Data Contributors and Data Aggregators. Data Contributors are established technology vendors such as Microsoft, AutoDesk, ESRI, Bentley, Johnson Controls, IBM (Maximo), INFOR (EAM), that provide design and operations/maintenance applications, intermediate cloud services, and application programming interfaces (APIs). They help organizations develop and maintain asset information within SSORs and subsequently share that content both internally and externally to maximize interoperability, see Fig. 2 – Operational Level and Integration Level. Data Aggregators are a relatively new technology that collects content from Data Contributors, integrates with analytic services such as Snowflake or Copperleaf for computation, then presents the combined spatial and tabular data in a "single pane of glass" interface. Figure 2 – Strategic/Analytics Level and Presentation Level.

### 4.1 Data Contributors

Several leading vendors such Microsoft, AutoDesk, ESRI, Willow, and Bentley are part of a 200+ member, industry-driven standards organization known as the Digital Twin Consortium. The Digital Twin Consortium was chartered to unite industry, government, and academia to drive consistency in the vocabulary, architecture, security, and interoperability of digital twin technology. The Consortium advances the use of digital twin technology among all industries and asset types and is envisioned as a global ecosystem of users who are accelerating the digital twin market and demonstrating the value of digital twin technology.

**Bentley.** iTwin Services are available to Bentley's software ecosystem users, Fig. 8. Users "instantiate" a digital twin and the cloud-based service will create a comprehensive iModel of the project combining available context capture 3D data and project engineering data using the iModel Hub. The iTwin Service will continue to update both

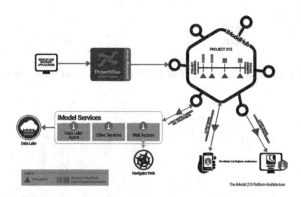

**Fig. 8.** Bentley iTwin Services application

the 3D data and engineering data as new changes come in, recording each change to a ledger. This means Bentley's iTwin Services offer a record of every change made throughout the project, enable project reviews for any stage of the project timeline, and offer visualization and analytics for any changes between different project states. The iTwin Services will also integrate with Bentley's SYNCHRO 4D construction modelling application.

**ESRI.** ESRI's digital ecosystem encompasses four types of content models: landscape information, building information, network information, and city information shown in Fig. 9. ESRI GIS helps organizations address three main problems: maintaining an accurate historical record, monitoring real-time operational performance, and testing or predicting future performance.

**Fig. 9.** ESRI ARC GIS digital twin services

**Johnson Controls.** OpenBlue Digital Twin is a suite of connected solutions for sustainability, occupant experience, safety and security. It features a suite of AI-powered service solutions such as remote diagnostics, predictive maintenance, compliance monitoring, and advanced risk assessments. The platform supports tailored services for HVAC, fire protection and security using data-driven insights for planning and decision-making, enhanced productivity and optimized performance, Fig. 10.

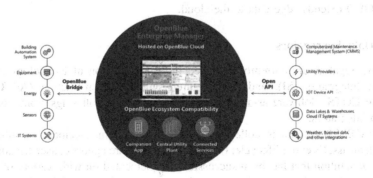

**Fig. 10.** OpenBlue Digital Twin Framework that provides connectivity from edge devices into the CDE

**Microsoft.** Azure Digital Twins is an Internet of Things (IoT) open-source platform that enables a digital representation of real-world objects, places, business processes, and people, Fig. 11. The platform provides an open modelling Digital Twins Design Language (DTDL) to create custom domain models of any connected environment. The platform also supports a live execution environment to bring the digital twins to life via live graph representation. Input from IoT and business systems connect assets, including IoT devices, using Azure IoT Hub, Logic Apps, and REST APIs. Output to time series insights, storage, and analytics use event routes to downstream services, including Azure Synapse Analytics.

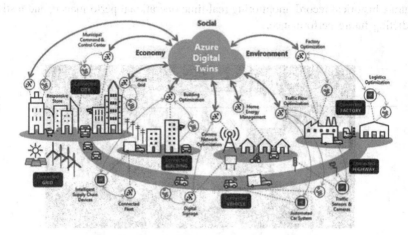

**Fig. 11.** Azure Digital Twin conceptual model

**OSIsoft.** OSIsoft has developed the PI system, a data management platform for industrial operations, Fig. 12. The platform consists of a series of three primary integrated sets of components: PI CORE, PI EDGE, and PI CLOUD. PI CORE provides industrial engineers, asset operators, and onsite analysts with real-time, high-fidelity operations data for informed decision-making. PI EDGE provides an integrated edge device, data storage, and adapters to collect sensor information not typically available. PI CLOUD extends edge data to the cloud.

## 4.2 Data Aggregators

Data Aggregators are currently marketed by a combination of technology start-ups, System Integrators (SI), and Architecture/Engineering/Construction (A/E/C) firms offering COTS, Software-as-a-Service (SaaS), as well as offerings from established system integrators.

ParADIM® automates collection, integration, and management of information through an asset's entire lifecycle. ParADIM® is software agnostic, user-friendly, and a universal solution that has been successfully implemented on wide variety of projects such as bridges, airports, railways, roads, treatment plants, and buildings. Primary features:

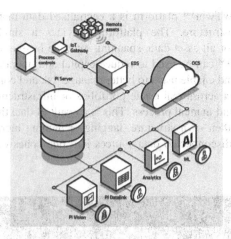

**Fig. 12.** The OSIsoft integration model

- Incorporates the appropriate tools, standards, and processes to develop parametric 3D design models in a common data environment promoting collaboration amongst various stakeholders
- Tools are deployed for model federation, design reviews, and coordination
- Aggregates the 3D model with project data through its lifecycle from design, through construction into operations and maintenance
- Leverage data analytics to provide unique design, construction, and operation insights for making informed decisions
- Provides a 3D visual for various potential scenarios, such as planned versus actual schedule, cost, means and methods, and heat maps for safety impact studies
- Provides mobile applications for accessing up-to-date data for productivity, quality, and safety (Fig. 13).

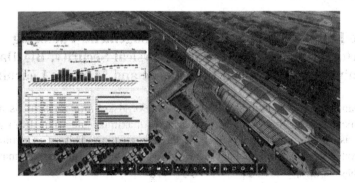

**Fig. 13.** ParADIM Digital Twin showing the "single pane of glass" interface of BIM and data integration

**Willow.** The WillowTwin™ platform is a centralised data management solution for buildings and infrastructure. The platform provides a single interface allowing actionable insights for all asset data, spanning multiple business functions and project stages. WillowTwin™ can be used to reduce capital and operating expenses, increase or protect revenue, and enable users to better manage risk and compliance. The analysis and actioning of data across real estate portfolios or infrastructure networks is a time consuming, costly, and manual process. This is a result of data that is siloed in discrete systems, each with their own structure, tagging or naming hierarchy. WillowTwin™ integrates formerly disconnected data sources into one cohesive, visual, and easy-to-use model (Fig. 14).

**Fig. 14.** Willow Digital Twin application

**EcoDomus.** EcoDomus is an information technology application focused on improving the way buildings are designed, built, managed, and retrofitted using BIM. EcoDomus brings BIM-based digital twin visualization and data integration to building owners and occupiers for improved design and construction data collection and handover, facility management, operations, and maintenance (Fig. 15).

## 5 The Latest Thinking on Asset Management, Factoring in Human Factors, Cyber and Physical Security, Big Data Analytics and Artificial Intelligence and Machine Learning

Asset management, as defined within *ISO 55000: Asset Management*, is an aligned management system that includes processes to maximize the value from assets and manage risk to the appropriate level for the organization. This is an evolution from earlier concepts that asset management was synonymous with facility management, maintenance and operations, or reliability.

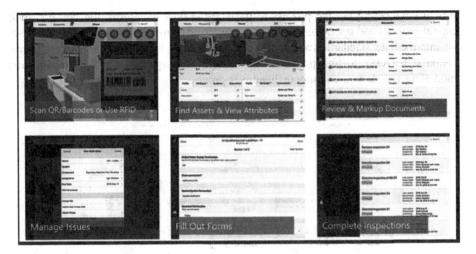

Fig. 15. EcoDomus mobile application

## 6 Conclusions

### 6.1 Findings

The recent evolution and convergence of cloud technology, the IOT, and mature ISO standards for asset management can support the implementation of digital twins. Customer-driven analytical requirements are encouraging leading technology vendors such as AutoDesk, ESRI, OSIsoft, and Bentley to provide open-source APIs for delivering their content.

### 6.2 Conclusions

Digital twins offer an immense benefit to transportation agencies by ensuring efficient, reliable, and available asset information critical to operations and planning. This information can help organizations evolve from reactive management (break/fix corrective maintenance) to proactive reliability centered maintenance (RCM). RCM is a robust maintenance methodology that ensures maintenance tasks are performed in an efficient, cost effective, reliable, and safe manner by following seven basic steps: identify system functions, identify failure modes, identify failure causes, identify effects of failure, identify consequences of failure, determine preventative tasks, and identify alternatives.

### 6.3 Recommendation

The authors recommend organizations prepare a digital governance plan as part of their strategic asset management plan (SAMP) and/or transit asset management plan (TAMP). The governance plan will identify opportunities for digital twins that are applicable to the user organization. Additionally, organizations may wish to develop a formal engineering change process to balance the flow of information between their capital projects and operational activities, Fig. 16. An engineering change process uses an intermediate standard, such as the capital facilities information handover specification (CIPHOS), to link project information modelling (PIM) to asset information modelling (AIM) in a continuous handover process.

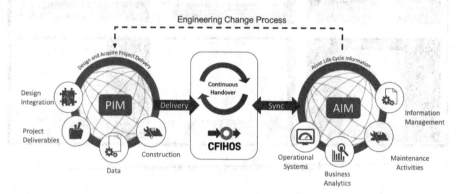

**Fig. 16.** EcoDomus mobile application

**Acknowledgements.** The authors acknowledge the foresight of Bruce Irwin at Parsons, for introducing us to the potential of digital twins and authoring our initial Asset Lifecycle Conceptual Model diagram that provides the conceptual framework for the advanced application of technology into asset lifecycle management.

A special recognition is also due to the deep thought contributions of David Armstrong at Bentley Systems for sharing his wonderful vision of the digital future, sweetened with just the right amount of wit and cynicism to be believable.

# References

Armstrong, D.: Asset Management, Machine Learning, and the Duality of Oreo Cookies. Unpublished (2021)
AutoDesk (2021). https://www.autodesk.com/solutions/digital-twin/architecture-engineering-construction
Bentley (2021). https://www.bentley.com/en/products/product-line/digital-twins
Cohesive Solutions. Asset Lifecycle Management: 6 Stages and Best Practices. https://blog.cohesivesolutions.com/asset-lifecycle-management-stages
ESRI (2012). https://www.esri.com/arcgis-blog/products/arcgis/aec/gis-foundation-for-digital-twins/
Johnson Controls (2021). https://www.johnsoncontrols.com/openblue/openblue-digital-twin
Microsoft (2021). https://azure.microsoft.com/en-ca/services/digital-twins/
O'Hanlon, T.: Understanding Asset Management and Lifecycle (2021). https://www.linkedin.com/pulse/understanding-asset-management-life-cycle-terrence-ohanlon/
Society of Automotive Engineers. SAE JA1011 Evaluation Criteria for Reliability-Centered Maintenance (RCM) Processes (1998)
Willow Inc. (2021). https://www.willowinc.com/company/about/

# Long Term Pavement and Asset Performance - 1

Long Term Pavement and User
Performance - I

# Field Monitoring of Road Pavement Responses and Their Performance in Thailand

Auckpath Sawangsuriya[✉]

The Department of Highways, Bangkok, Thailand
sawangsuriya@gmail.com

**Abstract.** From late 2017 to mid of 2018, the Bureau of Road Research and Development, Thailand Department of Highways (DOH) in collaboration with the Faculty of Engineering, Kasetsart University initiated three comprehensive field instrumented pavement sections in order to fully understand the load carrying behavior and characteristics of typical Thailand road conditions. The novel sections included a rigid pavement in Nakorn Chaisri district, Nakorn Phathom province and two flexible pavements in Potaram district, Rachaburi province and Sapphaya district, Chainat province. As part of national highway network under the responsibility of the DOH, each section adjacent to permanent weigh station and weigh-in-motion (WIM) system was uniquely researched and developed based on local Thailand conditions.

**Keywords:** Field instrumentation · Pavement response · Long-term performance · Rigid pavement · Flexible pavement

## 1 Introduction

The Department of Highways, Thailand (DOH) is responsible for the national highway network of approximately 60,000 km. The impact of increasing overweight truck traffic and climate on highways due to economic growth is a growing concern within the DOH. Current pavement design and rehabilitation procedures are based mainly on the Asphalt Institute (AI) and the Portland Cement Association (PCA) for flexible and rigid pavements, respectively. Both approaches utilized limited performance models developed from the AASHTO road test in the late 1950s (AASHTO 1993). In addition to the difference in material, traffic, and environment, the construction and rehabilitation techniques may not be relevant to local Thailand conditions. Characterization of pavement layer properties and responses under traffic loads, local materials, and weather conditions in Thailand is essential for accurate pavement performance prediction, pavement deterioration model as well as pavement management system.

In 2017, the DOH in cooperation with Kasetsart University constructed comprehensive field instrumented pavement sections: a rigid pavement on national highway No. 4, Nakorn Chaisri district, Nakorn Phathom province and two flexible pavements on national highway No. 4, Potaram district, Rachaburi province and national highway on Highway No. 32, Sapphaya district, Chainat province. For the sake of simplicity in the paper, a rigid pavement section was called Nakorn Chaisri test section and two flexible pavement sections were called Potaram and Sapphaya test sections. These

© The Author(s), under exclusive license to Springer Nature Switzerland AG 2022
A. Akhnoukh et al. (Eds.): IRF 2021, SUCI, pp. 441–449, 2022.
https://doi.org/10.1007/978-3-030-79801-7_32

sections were chosen in order to represent typical cross-sections for the flexible pavement structure (e.g. asphalt surface, base course, subbase course, and selected material above subgrade) and the rigid pavement structure (e.g. joint reinforced concrete pavement slab, sand cushion, subbase course, and selected material above subgrade) in Thailand. The joints were doweled and sealed. All test sections were situated near permanent weigh station and weigh-in-motion (WIM) system.

This paper summarized a comprehensive field instrumentation plan for three experimental sites in order to fully understand the load carrying behavior and characteristics of typical Thailand pavement structures as well as their performance. A discussion of instrumentation, type of sensors, installation procedures, and instrumentation layout is also presented herein.

## 2 Field Instrumented Pavement Sections

A comprehensive field instrumentation program was implemented to monitor responses and performance of pavement sections. The rigid pavement section was situated on national highway No. 4 at STA 41+750 (South Bound), approximately 100 m from WIM system and 550 m from Nakhon Chaisri stationery weight station, Nakhon Phatom province. The instrumented section had 3.5-m wide and 10-m long as shown in Fig. 1. The pavement structure consisted of 250-mm joint reinforced concrete pavement (JRCP) with doweled and sealed joints, 50-mm sand cushion, 100-mm soil-aggregate subbase, and 150-mm selected material over subgrade.

**Fig. 1.** Nakorn Chaisri test section.

The flexible pavement section was situated on national highway No.4 at STA 88+770 (South Bound), approximately 40 m from WIM system and 400 m from Potaram stationery weight stations, Ratchaburi province. The instrumented section had 3.5-m wide and 10-m long as shown in Fig. 2. The pavement structure consisted of 200-mm asphalt surface, 200-mm crushed rock base, 200-mm soil aggregate subbase, 250-mm selected material over subgrade. The other flexible pavement section was situated on national highway No.32 at STA 121+250 (South Bound), adjacent to WIM system and approximately 700 m prior to Supphaya stationery weight stations, Chainat province. The instrumented section had 3.5-m wide and 12-m long as shown in Fig. 3. The pavement structure consisted of 200-mm asphalt surface, 200-mm crushed rock base, and 200-mm soil aggregate subbase over subgrade.

**Fig. 2.** Potaram test section.

**Fig. 3.** Sapphaya test section.

To monitor its responses under actual traffic loads and local conditions, a series of sensors were installed in each experimental site. The field instrumented rigid pavement section consisted of sixteen strain gauges, sixteen pressure cells, sixteen moisture sensors, two observation wells, one data logger, one data processing unit, one AC to DC adaptor, and one UPS. The field instrumented flexible pavement section consisted of eight asphalt strain gauge, sixteen strain gauges, sixteen pressure cells, sixteen moisture sensors, eight thermocouples, two observation wells, one data logger, one data processing unit, one AC to DC adaptor, and one UPS.

The installation procedures for field instrumentations consisted of four major parts: (1) the excavation of existing pavement, (2) the placement and compaction of pavement layers, (3) the installation of field instrumentations, and (4) the quality control assessment. The existing pavement was excavated to the subgrade level. Each pavement material was filled and compacted in accordance with the original layer thickness.

The construction process also followed the DOH standard and specifications. Every instrument was carefully laid down to the specified level and location. The PE pipeline was utilized to protect and shield the cable. The conventional field moisture-density measurements were performed for compaction quality control assessment.

## 3  Field Testing

Falling weight deflectometer (FWD) was carried out to determine the surface deflection and the modulus of elasticity (E) of pavement structure on the basis of back-calculation procedure. In this study, the applied stress from the FWD was approximately 760 kPa, which was transmitted to the pavement sections through a 300-mm diameter circular plate as shown in Fig. 4. The FWD surface deflections of pavement at the center of a circular plate (0 mm) and at varied offsets from the center (e.g. 200, 300, 450, 600, 900, 1,200, 1,500, and 1,800 mm) are recorded. Note also that three replicated FWD measurements were made and the average values were reported. The pavement distress was also evaluated by visual inspection and surface deformation measurement.

**Fig. 4.** Field testing by the FWD.

In addition to the FWD and pavement distress evaluation, two types of tested Thai trucks, e.g. 6-wheel truck and 22-wheel semi-trailer, were utilized to generate local permitted axle and wheel loads on the pavement sections as shown in Fig. 5. The test data were classified into four axle groups including (1) single axle-single tire, (2) single axle-dual tire, (3) tandem axle-dual tire, and (4) tridem axle-dual tire. Table 1 summarizes the axle-wheel load characteristics. The pavement responses in terms of stress and strain were respectively measured using a series of pressure cells and strain gages installed under the asphalt surface and base course as well as above the subgrade. Note that these sensors were embedded within the pavement section both along and in between the wheel paths.

**Fig. 5.** Two types of tested Thai trucks: 6-wheel truck and 22-wheel semi-trailer.

**Table 1.** Summarizes the axle-wheel load characteristics

| Type of vehicle | Type of Axle/Tire | Location of measurement | Speed (km/hr) | Percentage of cargo weight |
|---|---|---|---|---|
| 6-wheel truck | Single Axle-Single Tire Single Axle-Dual Tire | Front Axle, Rear Axle | 0, 20, 60 | 0, 25, 50, 75, 100 |
| 22-wheel semi-trailer | Tandem Axle-Dual Tire Tridem Axle-Dual Tire | $2^{nd}$ Axle, Between $2^{nd}$ and $3^{rd}$ Axle, $5^{th}$ Axle | 0, 20, 60 | 0, 25, 50, 100 |

## 4 Pavement Responses

Typical pavement response data in terms of stress and strain are presented herein. A 6-wheel truck travelling at 20 km/hr and 60 km/hr generated stress and strain under the front and rear axles of different magnitudes. The influence of travelling speed on the pavement responses seemed to be more pronounce when the truck travelled at lower speed. The pulses of induced vertical stresses and tensile strains observed in the instrumented pavement sections are illustrated in Fig. 6. In addition, the induced vertical stress data collected from different locations of pressure cells during a passage of 18-wheel semi-trailer are shown in Fig. 7.

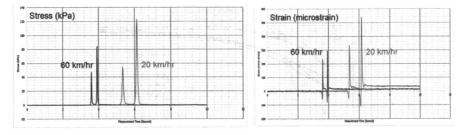

**Fig. 6.** Typical pavement response data in terms of stress and strain.

The finite-element analysis (FEA), the linear elastic analysis (LEA), and the elastic solution were used to examine the flexible pavement responses under the FWD. Every approach was calculated based on the assumption that the pavement layers exhibited homogeneous, isotropic, and linear-elastic. The surface deflections determined from the FEA, LEA, and elastic solution were compared with the FWD measured data. A plot of surface deflection vs. the offset distance is shown in Fig. 8. The backcalculated moduli (E-backcal) were used in the analysis. It was found that E-backcal tended to give reasonable surface deflections for both FEA and LEA, while the elastic solution gave the smallest surface deflections.

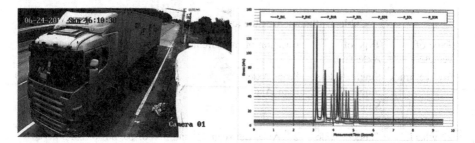

**Fig. 7.** Induced vertical stress during a passage of 18-wheel semi-trailer.

The strain distribution due to local permitted tandem axle-dual tire in the flexible pavement section is shown in Fig. 9 along with the theoretical values based on the closed form solutions and numerical analyses e.g. FEA, LEA etc. It should be noted that the backcalculated moduli from the FWD, assumed Poisson's ratio, and the applied wheel loads which were input in the closed form solutions were identical to those input in the FEA and LEA. The plots indicated that most measured responses were consistent with the theoretical values with some exceptions for the strains under the asphalt surface. The reason for this discrepancy might be due to the constitution of the pavement and the mechanical properties of layer materials, although the construction and quality control processes were in accordance with the local standard and specifications. As expected, both FEA and LEA provided similar results because the layer materials were modelled as homogeneous, isotropic, and linear-elastic. Furthermore,

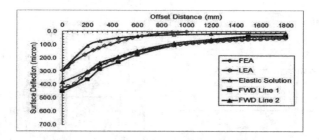

**Fig. 8.** Surface deflection data from the FWD

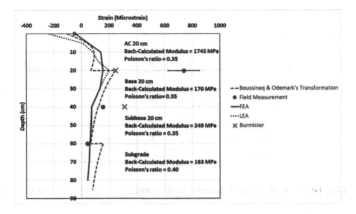

**Fig. 9.** Strain distribution due to local permitted tandem axle-dual tire.

both FEA and LEA tended to match reasonably well with the measured responses. Results suggested that caution must be exercised when using the typical modulus value in the current pavement design and rehabilitation practices.

It was also observed that both flexible pavement sections were subjected to surface deformation or rutting over time. Potaram test section seemed to exhibit higher rut depth than Sapphaya test section. This pavement rutting was one of the most common distresses found in the flexible pavement in Thailand.

## 5 Pavement Performance Monitoring

Long-term performance monitoring of one rigid pavement section and two flexible pavement sections were investigated and presented herein. The induced stresses at bottom of concrete slab and above subgrade of the rigid pavement section were monitored and their average values were reported over time as shown in Fig. 10. The average stress in the middle of base course, at bottom of base course, and above subgrade was determined to be approximately 100 kPa, 80 kPa, and 55 kPa, respectively. The induced strains at bottom of asphalt surface, at bottom of base course, and above subgrade of two flexible pavement sections were monitored and their average values were reported over time as shown in Figs. 11. The average strain at bottom of asphalt surface, at bottom of base course, and above subgrade was determined to be 280, 40, 25 microstrain, respectively. It was found that the average strain at bottom of asphalt surface tended to increase with time, while the average strain at bottom of base course and above subgrade tended to cease with time. Based on the performance monitoring, most pavement sections performed well according to the design criteria (AASHTO 1993) with some exceptions for the asphalt surface and base of flexible pavement sections. It was found that larger tensile strains than expected were measured under the asphalt surface and above the base of flexible pavement.

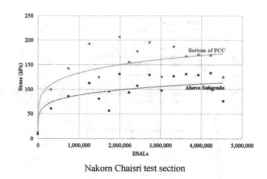

Nakorn Chaisri test section

**Fig. 10.** Long-term performance monitoring of a rigid pavement section.

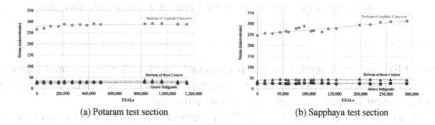

(a) Potaram test section          (b) Sapphaya test section

**Fig. 11.** Long-term performance monitoring of two flexible pavement sections.

## 6 Conclusions

From late 2017 to mid of 2018, the Bureau of Road Research and Development, Thailand Department of Highways (DOH) in collaboration with the Faculty of Engineering, Kasetsart University initiated three comprehensive field instrumented pavement sections in order to fully understand the load carrying behavior and characteristics of typical Thailand road conditions. The novel sections included a rigid pavement in Nakorn Chaisri district, Nakorn Phathom province and two flexible pavements in Potaram district, Rachaburi province and Sapphaya district, Chainat province. As part of national highway network under the responsibility of the DOH, each section adjacent to permanent weigh station and weigh-in-motion (WIM) system was uniquely researched and developed based on local Thailand conditions. This paper presents: (1) field installation of embedded instrumentation and data processing system for stress, strain, temperature, and moisture monitoring, (2) comparison of pavement responses from theoretical analysis, measured data, and numerical models, (3) field testing of the pavement layers using the falling weight deflectometer (FWD) and evaluation of Thai trucks under permitted axle loads including single axle-single tire, single axle-dual tire, tandem axle-dual tire, and tridem axle-dual tire, and (4) long-term pavement performance.

Three instrumented pavement sections were equipped with the most up-to-date technology and knowledge in road engineering. A series of embedded instrumentation including pressure cells, asphalt strain gauges, strain gauges, moisture sensors, and thermocouples were installed for monitoring the pavement response and long-term performance. Traffic data in ESALs from WIM system and pavement response data in stress, strain, moisture, and temperature were concurrently measured and automatically collected from construction in late 2017 through 2020. For the sake of field verification, Thai trucks with specified axle loads were tested in these pavement sections. A finite-element analysis (FEA) and a linear-elastic analysis (LEA) under three types of axle group load, e.g. single axle-dual tire, tandem axle-dual tire, and tridem axle-dual tire were performed and then compared with the field measurement data. The elastic moduli of pavement layers were determined from the FWD based on back-calculation procedure. Results suggested that the measured data were consistent with the closed-form solutions and numerical analyses (FEA and LEA) but with some exceptions at the bottom of the asphalt layer where larger measured tensile strains were observed. Such discrepancy was expected due to the constitution of the pavement and the mechanical properties of layer materials, although the construction quality control processes for these experimental sections complied with local standard and specifications. Finally, the study provided extensive data for design and analysis of Thailand road pavements as well as their long-term performance under local trafficking, material, construction, and seasonal variations due to moisture and temperature.

# Reference

AASHTO: Guide for design of pavement structures. American Association of State Highway and Transportation Officials, Washington, D.C. (1993)

# Supplementary Cementitious Materials in Concrete Industry – A New Horizon

Amin Akhnoukh[✉] and Tejan Ekhande

East Carolina University, Greenville, NC, USA
akhnoukhal7@ecu.edu

**Abstract.** Supplementary Cementitious Materials (SCMs) market share in ready mix concrete industry has been significantly increasing during the past 2 decades. SCMs as silica fume, fly ash, quartz flour, and blast furnace slag are currently used in partial replacement of conventional portland cement to improve mix properties including flowing ability, compressive strength, modulus of elasticity, and long-term performance. The main advantages of SCMs are attributed to the contribution of SCMs to the formation of additional concrete binder and the improved packing order of the cementitious material matrix.

Due to the afore-mentioned advantages, SCMs have been used in developing ultra-high-performance concrete, also known as reactive powder concrete. SCMs are successfully used in developing non-proprietary high-performance mixes used in high-rise construction and in fabricating high performance precast/prestressed bridge girders for new infrastructure projects. This paper displays the advantages of SCMs, and present different proprietary and non-proprietary mixes developed by incorporating SCMs, and their effect on different concrete properties. The wide spread of SCMs in concrete industry would potentially increase construction projects life span and minimize the need to routine maintenance.

**Keywords:** Silica fume · Fly ash · Slag · Reactive powder concrete · Supplementary cementitious materials

## 1 Introduction and Literature Review

Supplementary cementitious materials (SCMs) are increasingly incorporated in concrete mix designs. Different mixes are produced using one or two types of SCMs in precast facilities in the United States to attain the benefits of SCMs in concrete. Developed mixes using one type of SCMs (binary mixes) and two types (ternary mixes) are used in fabricating high performance precast/prestressed bridge girders (Akhnoukh 2008, 2018; Akhnoukh and Elia 2019). SCMS used in improving concrete properties include silica fume, also known as micro-silica, class C fly ash, class F fly ash, quartz flour, blast furnace slags, and carbon nanotubes.

SCMs are used in partial replacement of Portland cement. Current research findings prescribe a 10% to 40% cement replacement by weight dependent on the type of SCM used, the targeted change in fresh or hardened concrete properties, and the mix development condition including batching, mixing, and curing techniques, and/or

© The Author(s), under exclusive license to Springer Nature Switzerland AG 2022
A. Akhnoukh et al. (Eds.): IRF 2021, SUCI, pp. 450–457, 2022.
https://doi.org/10.1007/978-3-030-79801-7_33

mixing temperature. SCMs improvement to concrete properties are attributed to the following:

1. *SCMs granular size is finer than cement particles* which enable SCM particles to increase the packing order of the concrete mix, and minimize voids ratio, and results in reduced permeability, as shown in Fig. 1. This increased density improves hardened concrete long-term performance, mitigate alkali-silica reactivity (Akhnoukh et al. 2016), provides a higher resistance to chloride attacks, and reduce the rate of steel corrosion. In addition, specific types of SCMs (fly ash) has the tendency to increase the concrete mix slump and flowing ability. Thus, fly ash is extensively used in self-consolidating concrete mix development (Khayat 1999)

**Fig. 1.** Silica fume concrete improved packing order (SEM Microscope picture) (Panesar 2019)

2. *SCMs acts as a pozzolan* which tends to react with calcium hydroxide evolving from hydration process of Portland cement to form additional binder that increases concrete strength, modulus of rupture, and modulus of elasticity. The SCMs reaction is a two-step chemical reaction as follows:

**Step I: Portland cement hydration process:**

$$\text{Cement} + \text{Water} \longrightarrow \text{Calcium Silicate Hydrate} + \text{Calcium Hydroxide}$$
$$(H2O) \qquad\qquad (C\text{-}S\text{-}H) \qquad\qquad Ca(OH)_2$$

**Step II: Pozzolanic Micro-silica Reaction to Form Additional Binder:**

$$\text{Calcium Hydroxide} + \text{Microsilica} + \text{Water} \longrightarrow \text{Calcium Silicate Hydrate}$$
$$Ca(OH)_2 \qquad (SiO2) \qquad (H20) \qquad\qquad (C\text{-}S\text{-}H)$$

Due to their advantages, different research programs are currently considering the optimized used of SCMs in concrete mixes, including optimized batching, mixing, and pouring procedures. The optimized design of mix constituents considering different types of SCMs, chemicals, and fibers. Finally, the impact of the afore-mentioned parameters on fresh and hardened concrete properties.

## 2 SCM Concrete Mix Proprtioning, Batching, and Mixing Procedures

Mix proportioning of concrete mixes containing one or more SCMs targets the optimization of mix packing order to minimize voids, increase strength, and durability without altering mix slump and/or flowing ability. Different proprietary UHPC mixes are available in the construction market. Example of proprietary mixes are BSI "Beton Special Industrial," developed by Eiffage, Cemtec developed by LCPC, and Ductal developed by LaFarge. Different non-proprietary ultra-high performance and high-performance mixes are developed through research programs across the globe.

The primary difference between UHPC and HPC mixes is the reduced strength of HPC (UHPC mix compressive strength exceeds 21.7 ksi (Toutlemonde and Resplendino 2011; JSCE 2008), and contains a higher percentage of SCMs and incorporates random steel fibers to increase concrete ductility and post cracking stiffness.

Supplementary cementitious materials and different steel fibers increase the overall density of concrete mixes. Optimized mixing procedures for concrete with high percentages of SCMs are as follows:

I. Preblended the concrete mix granular ingredients for a 2–3 min. The granular ingredients include cement, SCMs, fine sand, and coarse aggregates (if present).
II. Mixing water is added to the preblended powder. Mixing water may include 50% of the high range water reducers (HRWR) included in the mix design. Wet mixing should continue for 8–10 min.
III. Remaining HRWR is added to the mix during the wet mixing process.
IV. Steel fibers included in UHPC are added before wet mixing is ended.

Due to the high packing order of granular materials and the possible inclusion of random steel fibers, high energy paddle mixers are required to produce mixes with sufficient rheology. Examples of high energy mixers are shown in Fig. 2.

**Fig. 2.** High energy paddle mixers – used for UHPC and HPC mixing

Concrete mixes with incorporated SCMs can be cured using regular moisture curing techniques. When very high early strength is required (as in precast/prestressed concrete industry), thermal curing could be applied. Thermal curing should be gradually applied to avoid developing hair-cracks within hardened concrete [14].

## 3 SCMs in Proprietary Ultra High Performance Concrete Mixes

Silica fume and quartz flour are used in proprietary UHPC mixes to increase mix strength and durability. Current proprietary mixes, commercially available in the US and EU markets have a high content of SCMs. In addition to their strength and durability advantages, SCMs partially reduces the cement consumption, hence reduce the carbon footprint (Elia et al. 2018; Akhnoukh 2013, 2018). Mix design and SCM content in major proprietary mixes are shown in Table 1 (Rossi et al. 2005; Aarup 2004; Howard et al. 2018; Akhnoukh and Buckhalter 2021).

**Table 1.** Proprietary UHPC mix designs in global construction markets

|  | Ductal | Cemtec | Cor-Tuf | CRC |
|---|---|---|---|---|
| Portland cement | 712 | 1,050 | 790 | 861 |
| Silica fume | 231 | 268 | 308 | 215 |
| Quartz flour | 211 | – | 216 | 215 |
| Total cementitious materials | **1,154** | **1,318** | **1,314** | **1,291** |
| Percentage of SCMs to C. materials | 38% | 20% | 40% | 33% |
| Sand | 1,020 | 514 | 765 | 792 |
| Water | 109 | 188 | 166 | 220 |
| HRWR | 30.7 | 44 | 14 | 9.45 |
| Accelerator | 30 | – | – | – |
| Fibers | 156 | 180 | 166 | 218 |
| W/CM | **0.21** | **0.17** | **0.14** | **0.18** |

The compressive strength of proprietary UHPC mixes ranges from 160 to 200 MPa. The strength is significantly increased due to the incorporation of SCMs and steel fibers. Steel fibers results in improved tensile capacity, and higher values for the modulus of elasticity (MOE) and modulus of rupture (MOR). Detailed mechanical properties of proprietary UHPC mixes are shown in Table 2.

454    A. Akhnoukh and T. Ekhande

**Table 2.** Average mechanical properties of proprietary UHPC mixes (Graybeal 2013)

| | UHPC value | Normal concrete Value | ASTM standard [22–28] |
|---|---|---|---|
| Design compressive strength | Greater than 150 MPa | 30–40 MPa | ASTM C39/C39 M |
| Flexural strength | Greater than 20 MPa | 3–5 MPa | ASTM C78/C78M-18 |
| First cracking strength | Greater than 4 MPa | 2 MPa | ASTM C1018-97 |
| Linear expansion coefficient | $12 \times 10^{-6}$ | $10^{-5}$ | ASTM C531-18 |
| Elastic modulus | 45 GPa | 20–25 GPa | ASTM C469/C469M-14 |
| Spread (Flowing Ability) | 55 to 75 cm | N/A | ASTM C1611M-18 |

## 4 SCMs in Non-proprietary High-Performance Concrete Mixes

HPC mixes incorporates different SCMs with variable ratios according to the mix design purpose. Silica fume is used to increase the binder content, and improve mix mechanical properties. Whereas fly ash is primarily used to increase flowing ability. HPC mixes are produced using similar batching, mixing, and curing procedures as compared to UHPC. Steel fibers are eliminated due to its high cost, while chemicals for increased flowing ability are used to maintain high flowing ability using low water-to-powder ration. Different HPC mixes are shown in Table 3. Optimized SCMs are to be attained through a trial-and-error procedure to attain the targeted mix properties. The incorporation of SCMs may substantially affect the final cost of the mix, considering that one ton of micro-silica cost $900 compared to one ton of cement that cost $110 to $135.

**Table 3.** Non-proprietary HPC mixes using local materials in US market

| | Mix #1 | Mix #2 | Mix #3 | Mix #4 | Mix #5 |
|---|---|---|---|---|---|
| Portland cement | 630 | 625 | 630 | 670 | 630 |
| Silica fume | 90 | 80 | 90 | 145 | 90 |
| Class C fly ash | 180 | 80 | 180 | 145 | 180 |
| Total cementitious materials | **900** | **785** | **900** | **960** | **900** |
| Percentage of SCMs to C. materials | 30% | 20% | 30% | 32% | 30% |
| Sand | 1350 | 1450 | 950 | 1350 | 950 |
| Water | 135 | 155 | 145 | 145 | 140 |
| HRWR | 37 | 21 | 37 | 43 | 43 |
| Fibers | – | – | – | – | – |
| W/CM | **0.18** | **0.22** | **0.20** | **0.19** | **0.19** |

The afore-mentioned non-proprietary mixes had an average 24-h compressive strength of 80 MPa, and a final 28-day compressive strength of 110 MPa. Detailed compressive strength testing results are shown in Fig. 3.

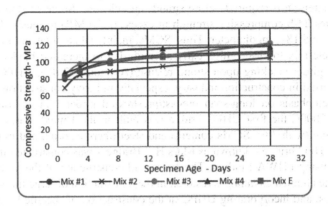

**Fig. 3.** Compressive strength test results for non-proprietary HPC mixes

## 5 Applications of SCM Incorporated Concrete in Construction Industry

Concrete mixes, with different SCMs, are currently used in different applications that requires high early strength, superior mechanical properties, and increased durability. The following represents main applications in residential and heavy construction applications:

I. High rise residential construction

   HPC, including SCMs, mixes are currently used in the construction of structural members in high rise buildings. The use of HPC in high rise construction started in the 1970s in the metropolitan areas within the United States including New York, Los Angeles, and Chicago. Recently HPC mixes with high flowing ability (SCC characteristics) are used in pumping concrete floors in the City of Jeddah's Kingdom Tower, and the Iconic Tower in Egypt's New Administrative Capital (shown in Fig. 4).

**Fig. 4.** Iconic Tower – Egypt's New Administrative Capital

II. Precast/prestressed bridge construction

Prefabricated prestressed girders with spans up to 60 m are increasingly used in bridge construction, including accelerated bridge construction projects (Akhnoukh 2020). In order to increase the productivity of prestressing facilities, very high early strength is required before strands are released. Some prestressing facilities requires 24-h compressive strength in excess of 70 MPa to release larger strands of 15- and 18-mm. diameter. Thus, SCMs, mainly silica fume, are incorporated in concrete mix to ensure high early strength due to increased binder content, and avoid girder cracking upon strand release (Akhnoukh 2010). UHPC are increasingly used in construction industry, especially in heavy construction applications, with emphasis on long-span precast/prestressed girder bridges construction. In early 2000s, the first UHPC bridge was built in the United States using Ductal concrete (with 38% SCMs content) and steel fibers incorporated in the mix. The first UHPC bridge – known as Mars Hill Bridge – was constructed in the State of Iowa using FHWA funding. The successful construction of the Mars Hill bridge and the attained advantages resulted in the construction of large number of UHPC bridges, and incorporating UHPC in the construction of specific bridge members, as shown in Fig. 5 (Sritharan et al. 2018).

**Fig. 5.** UHPC overlay on Mud Creek Bridge Deck: (a) grooving of the surface; (b) finishing surface

## References

Aarup, B.: CRC – a special fibre reinforced high performance concrete. In: Proceedings of the First International RILEM Symposium on Advances through Science and Engineering. Publication Pro 048 (2004)

Akhnoukh, A.K.: Development of high performance precast/prestressed bridge girders. Dissertation. University of Nebraska-Lincoln, Nebraska, USA (2008)

Akhnoukh, A.K.: The effect of confinement on transfer and development length of 0.7-inch prestressing strands. In: Proceedings of the 2010 Concrete Bridge Conference: Achieving Safe, Smart & Sustainable Bridges, Arizona, USA (2010)

Akhnoukh, A.K.: Overview of nanotechnology applications in construction industry in the United States. Micro Nanosyst. **5**(2), 147–153 (2013)

Akhnoukh, A.K.: Implementation of nanotechnology in improving the environmental compliance of construction projects in the United States. Particulate Sci. Technol. **36**(3), 357–361 (2018). https://doi.org/10.1080/02726351.2016.1256359

Akhnoukh, A.K.: Accelerated bridge construction projects using high performance concrete. Case Stud. Constr. Mater. **12**, e00313 (2020), https://doi.org/10.1016/j.cscm.2019.e00313

Akhnoukh, A.K., Elia, H.N.: Developing high performance concrete for precast/prestressed concrete industry. Case Stud. Constr. Mater. **11**, e00290 (2019). https://doi.org/10.1016/j.cscm.2019.e00290

Akhnoukh, A.K., Soares, R.: Reactive powder concrete application in the construction industry in the United States. In: Proceedings of the 10th Conference of Construction in the 21st Century, Srilanka (2018)

Akhnoukh, A.K., Kamel, L.Z., Barsoum, M.M.: Alkali-silica reaction mitigation and prevention measures for Arkansas local aggregates. World Acad. Sci. Eng. Technol. Int. J. Civil Environ. Eng. **10**(2) (2016). https://doi.org/10.5281/zenodo.1338860

Akhnoukh, A.K., Buckhalter, C.: Ultra-high-performance concrete: constituents, mechanical properties, applications, and current challenges. J. Case Stud. Constr. Mater. **15**, e00559 (2021). https://doi.org/10.1016/j.cscm.2021.e00559

ASTM C1018 – 97. Standard test method for flexural toughness and first-crack strength of fiber-reinforced concrete (using beam with third-point loading). ASTM International (1997)

ASTM C1611M-18. Standard test method for slump flow of self-consolidating concrete. ASTM International (2018)

ASTM C39/C39M-01. Standard test method for compressive strength of cylindrical concrete specimens. ASTM International (2001)

ASTM C469/C469M-14. Standard test method for static modulus of elasticity and poisson's ratio of concrete in compression. ASTM International (2014)

ASTM C531-18. Standard test method for linear shrinkage and coefficient of thermal expansion of chemical-resistant mortars, grouts, monolithic surfacings, and polymer concrete. ASTM International (2018)

ASTM C78/C78M-21. Standard test method for flexural strength of concrete (using simple beam with three-point loading). ASTM International (2021)

Elia, H., Ghosh, A., Akhnoukh, A.K., Nima, Z.A.: Using nano- and micro-titanium dioxide (TiO2) in concrete to reduce air pollution. J. Nanomed. Nanotechnol. **9**(3) (2018). https://doi.org/10.4172/2157-7439.1000505

Graybeal, B.: Ultra-high-performance concrete: a state-of-the-art report for the bridge community. Federal Highway Administration Report FHWA-HRT-13–060. 2013 (2013)

Howard, L.L., et al.: Mechanical behavior of Cortuf ultra-high-performance concrete (2018)

JSCE. Recommendations for design and construction of high-performance fiber reinforced cement composites with multiple fine cracks (HPFRCC). Japan Society of Civil Engineers (2008)

Khayat, K.H.: Workability, testing, and performance of self-consolidating concrete. Am. Concr. Inst. Mater. J. **96**(3), 346–353 (1999)

Panesar, D.K.: Supplementary cementing materials. In: Developments in the Formulation and Reinforcement of Concrete. Textbook, 2nd edn. (2019)

Rossi, P., Arca, A., Parant, E., Fakhri, P.: Bending and compressive behaviors of a new cement composite. Cem. Concr. Res. J. 35 (2005). https://doi.org/10.1016/j.cemconres.2004.05.043

Sritharan, S., Doiron, G., Bierwagen, D.: First application of UHPC bridge deck overlay in North America (2018)

Toutlemonde, F., Resplendino, J.: Designing and building with UHPFRC (2011). https://doi.org/10.1002/9781118557839

# Long Term Anti Corrosion Measures for Magosaki Viaduct by Cover Plate Method

Hiromasa Kobayashi[✉], Yukio Usuda, Yuki Kishi, and Yukio Nagao

Honshu-Shikoku Bridge Expressway Company Limited, Kobe, Hyogo, Japan
hiromasa-kobayashi@jb-honshi.co.jp

**Abstract.** Magosaki Viaduct is a three-span continuous steel plate girder bridge with a total length of 134 m and has served for 35 years since opening in 1985. The viaduct is an approach to the Ohnaruto Bridge crossing the Naruto Strait where corrosion environment is harsh. Because of the wide carriageway, the viaduct has 14 main girders and thus has a large number of members and massive area to be repainted. Repainting works had been implemented twice so far in 1999 and 2008.

By the comparison of long-term anti-corrosion measures, a cover plate method, in which the whole girders are covered with panels, was selected. This method is advantageous in maintenance work because maintenance personnel can walk on the cover plate as the permanent scaffold and approach the main girders and underside of the road deck for proximity inspection. Besides it was found that this method would provide better anti-corrosion performance and need less life cycle cost than repeating full repainting. Execution of the method started in 2017 and completed in 2018.

In order to evaluate the anti-corrosion performance of the method, relative humidity in the cover plate and the amount of adhered salt on the main girders were measured in the next year of the installation. As a result, it was confirmed that the relative humidity was comparatively stable and the amount of salt did not increase.

This paper reports a study on long-term anti-corrosion measures, and its design, installation work and evaluation of anti-corrosion performance of the cover plate method.

**Keywords:** Steel girder bridge · Life cycle cost · Long-term anti-corrosion measures · Cover plate method · Titanium plate

## 1 Introduction

Magosaki viaduct (Fig. 1, Photo 1(Left)) was opened to traffic as an approach to the Ohnaruto Bridge in 1985. Though the steel viaduct is exposed to a severe corrosion environment, a lead-based paint was applied to the base layer at the construction phase by a painting specification for bridges in a mild corrosion environment. So, after 14 years of the opening, in 1999, the full repainting was conducted by using an epoxy resin paint and a fluoropolymer coating as an intermediate coat and a top coat,

© The Author(s), under exclusive license to Springer Nature Switzerland AG 2022
A. Akhnoukh et al. (Eds.): IRF 2021, SUCI, pp. 458–473, 2022.
https://doi.org/10.1007/978-3-030-79801-7_34

respectively. However, as shown in Photo 1(Right), another full repaint was required again in 2008, and one of reasons was that the edge of the lower flange had not been chamfered. Therefore, anti-corrosion measures other than full repainting were studied to solve these problems below.

1) Severe corrosion environment
   Magosaki viaduct easily gets floating sea salt because the viaduct faces the Naruto Strait which faces the Pacific Ocean, and strong winds often blow there.
2) Large paint surface
   Magosaki viaduct has 14 main steel girders. The bridge length and width is 134 (m) and 31 (m), respectively. Therefore, there are a large paint area of approximately 16,300 (m$^2$).
3) Including lead-based paint
   As shown in Table 1, the original coat of Magosaki viaduct includes toxic components such as the lead and the hexavalent chromium; therefore, appropriate protective equipment, processing and disposal as a special controlled industrial waste are required in accordance with the related laws in removing the original coat.

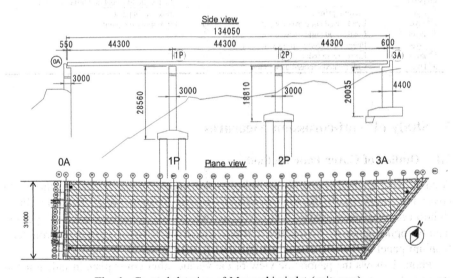

**Fig. 1.** General drawing of Magosaki viadct (unit: mm).

**Photo 1.** (Left) Panoramic view of Magosaki viaduct before cover plate installation. (Right) The corrosion situation of the lower frange of Magosaki viaduct before repair work.

**Table 1.** History and constitution of painting in Magosaki viaduct.

| Event | Construction(1985) | Full repainting(1999) |
|---|---|---|
| Surface Preparation | - | ·Manual scraping and wire brushing<br>·Leave the live coat and remove the rust. |
| 1$^{st}$ layer | Etching primer | Epoxy resin paint |
| 2$^{nd}$ layer | Lead based anticorrosive paint | Fluoropolymer paint |
| 3$^{rd}$ layer | Lead based anticorrosive paint | - |
| 4$^{th}$ layer | Phenol micaceous iron oxide paint | - |
| 5$^{th}$ layer | Epoxy resin paint | - |
| 6$^{th}$ layer | Chlorinated rubber-based paint | - |

## 2 Study of Anti-corossion Measures

### 2.1 Outline of Cover Plate Method

A cover plate method is one of the anti-corrosion countermeasures. This method is to cover the girders with the high corrosion resistant material (e.g. Titanium, Fibre Reinforced Plastics: hereafter referred to as "FRP") as a cladding for a bridge and mainly applied to a newly constructed bridge. Due to the severe corrosion environment, titanium panel was studied as a high durability material in the viaduct.

Photo 2 shows the panoramic view of the viaduct after cover plate installation and Photo 3 shows the structure of the titanium cover plate. The cover plate is made with 3 layers (inner surface, content, external surface) and its materials are galvanized steel, urethane foam and pure Titanium, respectively. Airtightness is ensured by the panel connection structure such as the fitting joint and the butt joint. Following effects were expected by the method.

1) Anti-corrosion function
   Inside of the cover plate is insulated from degradation factors such as ultraviolet ray, rainwater and floating salt, and it contributes to preventing corrosion of the steel.

2) Permanent scaffolding function
   Maintenance personnel can walk on the cover plate as the permanent scaffold and approach main girders, underside of the deck and bridge accessories for proximity inspection and repair work.
3) Falling object protection
   In case of fragment falling from the concrete deck, it could be caught in the cover plate.
4) Aesthetics
   The cover plate performs as decorative boards to improve landscape.

 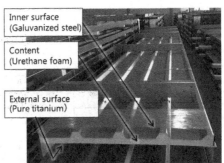

**Photo 2.** Panoramic view of Magosaki viaduct after cover plate installation.

**Photo 3.** Structure of Titanium covere plate. (Total thickness; 73 mm, Unit weight; 10.0 kg/m$^2$)

## 2.2 Functionality Assessment

The cover plate method can be expected better corrosion protection function than full repainting due to the above mentioned anti-corrosion function. Besides, the method is advantageous in maintenance work because scaffolding function enables proximity inspection without bridge inspection facilities. In addition, there are a lot of tourists who visit a sightseeing facility under 0A of the viaduct, so the cover plate could prevent the falling object from hurting them, and its aesthetics improves the landscape. Hence, it was found that cover plate method was more profitable than full repainting.

## 2.3 Cost Assessment

In cost assessment, 4 anti-corrosion measures were studied; full repainting (Plan A, Plan B), cover plate method (Plan C, Plan D). Firstly, compared preliminarily between full repainting (Plan A, Plan B), and also cover plate method (Plan C, Plan D), respectively. Next, based on the result of preliminary comparison, Life Cycle Cost (hereinafter called LCC) of each beneficial plan were analysed.

462     H. Kobayashi et al.

As shown in Table 2, Plan A is repaint from base to top coat with Organic zinc rich paint and Fluoropolymer paint, as full repainting. Plan B is repaint from middle to top coat with Epoxy resin paint and Fluoropolymer paint. As cover plate method, Plan C is covering the whole girder with titanium panels, and Plan D is covering the girder from upper flange to web with FRP panels.

### 2.3.1   Preliminary Comparison

In case of full repainting, the initial cost of Plan A is higher than Plan B, however, the cumulative cost is lower due to the high durability. Besides, Plan A takes advantages in repaint with Organic zinc rich paint, blast cleaning and chamfering the edge. Therefore Plan A was considered to be superior to Plan B.

**Table 2.** Comparison of full repainting and cover plate method.

| Measures method | Full repainting | | Cover plate method | |
|---|---|---|---|---|
| | Plan A | Plan B | Plan C | Plan D |
| | Repainting from base to top layers. | Repainting from middle to top layers. | Covering the whole girders with titanium panel. | Covering the girder webs with FRP panel. |
| Schematic drawing | | | | |

In case of cover plate method, the initial cost of Plan C is higher than Plan D for the cover plate installation, however, Plan C has advantageous in long-term durability and unnecessity to replace the cover plate compared with Plan D. Furthermore, whole girders, including lower flange, could be inspected and repaired from inside of cover plate. It was inferred that Plan C was more beneficial than Plan D.

### 2.3.2   LCC Analysis

Based on the result of preliminary comparison, LCC of Full repainting (Plan A) and Cover plate method (Plan C) over 100 years were compared. As initial cost of Full repainting, (a) chamfering work of steel member edges, (b) blast cleaning including disposal cost of lead-based paint, (c) full repainting from base to top coat (Organic zinc rich paint, Epoxy resin paint and Fluoropolymer paint), (d) scaffolding, (e) repair for the underside of the deck, were taken into consideration. On the other hand, in case of Cover plate method, (a) installation of titanium cover plate, (b) repainting for the lower flange of the main girders, were assumed as initial cost.

Table 3 shows conditions for LCC analysis. In case of full repainting, it was assumed that middle and top coat would be repainted after light treatment every 20 years. On the other hand, in Cover plate method, it was assumed that local repair only for intensively corroded parts would be conducted 5 years after cover plate installation and every 60 years since then. Owing to its anti-corrosion function and permanent scaffolding function, replacement of inspection passages and bearing repair were not included in maintenance cost. Repair for joint structure between substructure and cover plate was assumed every 30 years because rubber gasket was used in the joint structure and susceptible to deterioration by abrasion, rainwater and ultraviolet ray.

Figure 2 shows the result of LCC analysis between full repainting (Plan A) and cover plate method (Plan C). Initial cost of full repainting was lower than cover plate method, however, total cost was reversed after 25 years. LCC for 100 years of cover plate method was about 60% of that of full repainting.

As a result, it was inferred that cover plate method with titanium panel was more beneficial than full repainting in Magosaki viaduct from the aspects of functionality and cost.

**Table 3.** Conditions of maintenance cost for LCC analysis.

| Case | | Full repainting (Plan A) | Cover plate method (Plan C) |
|---|---|---|---|
| Repaint | Specification | Middle and top coat | Local repair |
| | Period | 20 years | 60 years |
| Repair for underside of the road deck | | 30 years | 50 years |
| Replacement of inspection passage | | 50 years | - |
| Bearing repair | | 30 years | - |
| Repair for joint structure between substructure and cover plate | | - | 30 years |

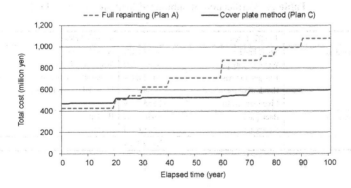

**Fig. 2.** The result of LCC analysis between full repainting and cover plate method.

# 3 Design of Cover Plate Method

In this chapter, verification against increased dead weight as well as wind load, clearance in the box girder, and panel connection structure with high airtightness and maintainability are described as design of cover plate method.

## 3.1 Verification of Weight

Though the viaduct was designed to accommodate 6 lanes in original design, it tentatively accommodates 4 lanes. In other words, it has some loading capacity margins against live load. Therefore, increased dead loads by the cover plate method was compared to original design load as a verification of weight. Table 4 shows the item of original design load and loading condition. In original design, noise barrier was planned to install, however, it has not been installed and planned. Water supply pipe and 1 out of 2 inspection passages, and handrail of main girder were removed in cover plate installation work because they were not necessary.

Table 5 shows verification of subtracted load and increase load for 4 traffic lanes and 6 traffic lanes. Both cases were below original design load. Therefore, it was confirmed that the increase load of cover plates did not affect the loading capacity.

**Table 4.** Item of original design load and loading condition.

| Item of load | Loading condition |
|---|---|
| Dead load | |
| Pavement | Loaded |
| Road deck | Loaded |
| Concrete haunch | Loaded |
| Median strip | Loaded |
| Wheel guard | Loaded |
| Balustrade | Loaded |
| Steel girder | Loaded |
| Noise barrier | Not loaded |
| Water supply pipe | Removed |
| Inspection passage | Removed (1 out of 2) |
| Handrail of main girder | Removed |
| Live load | |
| TT-43 | Loaded |
| L-20 | Loaded (4 out of 6) |

**Table 5.** Verification of subtracted load and increase load.

| Case | 4 traffic lanes | | 6 traffic lanes | |
|---|---|---|---|---|
| Subtracted load | Noise barrier | -943.6 kN | Noise barrier | -943.6 kN |
| | Water supply pipe | -268.0 kN | Water supply pipe | -268.0 kN |
| | 1 inspection passage | -134.0 kN | 1 inspection passage | -134.0 kN |
| | Handrail of main girder | -59.0 kN | Handrail of main girder | -59.0 kN |
| | Live load(L-20) | -2,555.0 kN | | |
| Increase load | Back panel | +934.7 kN | Back panel | +934.7 kN |
| | Side panel | +315.6 kN | Side panel | +315.6 kN |
| Total | | -2,709.3 kN | | -154.3 kN |

## 3.2 Clearance in the Box Girder

Figure 3 shows the schematic drawing of sectional view in Magosaki viaduct. Considering the maintainability for multiple main girder structure, enough clearance between lower flange and back panel (Photo 4) is necessary to pass across transverse direction at any place. So the clearance was verified with mock-up model. As a result, the clearance was decided 600 (mm) because maintenance personnel could pass under the girders with getting down on their hands and knees.

In addition, enough clearance between main girder and side panel (Photo 5) is necessary to set a ladder for proximity inspection for the underside of road deck. Therefore, the distance between side panel from the edge of lower flange was decided 780 (mm) so that general ladder could be set in the clearance. Besides, it was confirmed that rainwater was hard to enter because the existing creasing on the underside of overhang deck which was located outside of box girder could be used.

Therefore, the clearance between lower flange and back panel was decided 600 (mm). And the distance between side panel and the edge of lower flange was decided 780 (mm).

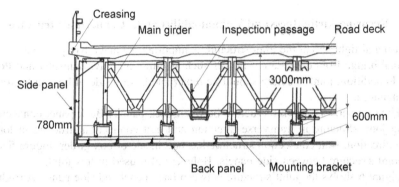

**Fig. 3.** Schematic drawing of sectional view in Magosaki viaduct.

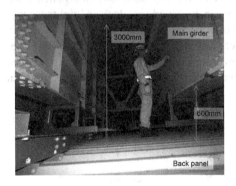 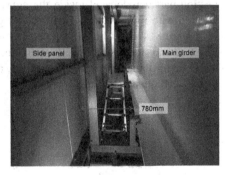

**Photo 4.** Clearance between lower flange and back panel.

**Photo 5.** Clearance between main girder and side panel.

### 3.3 Verification of Wind Load

The panel connection structure against wind load was verified assuming the wind load from the wind pressure distribution got from the wind tunnel test which was conducted for Haneda airport taxiway bridge (Fujikawa et al. 2007). At first, applying the building standard had been considered in setting external pressure coefficient to the basic structure computed from permanent load (dead load + live load), applying the building standard was judged unsuitable for this verification because the shape of the structure is very different from roof and wall in the building standard. In this case of cover plate method, outer shape was similar to Haneda airport taxiway bridge. Besides, dimension of cover panel and mounting bracket were almost the same.

As a result of verification, it was confirmed that the panel connection structure safely resist against wind load.

### 3.4 Ensuring Airtightness and Maintainability at the Connection Structure

Ensuring airtightness at the joint structure is important so that cover plate insulates the original bridge from degradation factors such as ultraviolet ray, rainwater and floating salt. In addition, joint structure should be repairable from inside of box girder for easy maintenance.

Figure 4 shows the joint structure between back panels. Back panel connection is fitting joint structure in transverse direction and butt connection structure in longitudinal direction. Butt connection structure has bolt inside of box girder. Figure 5 shows the joint structure between side panels. Bolts are also used in this joint.

Figure 6 shows the joint structure between back panel and side panel. Airtightness was ensured with rubber packing on the titanium plate to make distance at the corner between back panel and side panel.

Figure 7 shows the joint structure between overhang deck and side panel. Connection part was located inside the creasing on the overhang deck. Airtightness was ensured with rubber packing and rubber gasket on the titanium joint plate.

Figure 8 shows the joint structure between substructure and back panel, substructure and side panel as well. Airtightness was ensured with rubber packing between the titanium joint plate and substructure, and rubber gasket between the titanium joint plates on both substructure and panel.

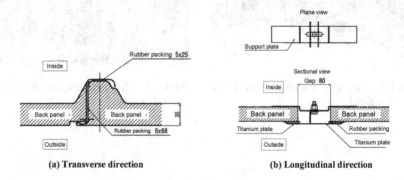

**Fig. 4.** Joint structure between back panels.

# Long Term Anti Corrosion Measures for Magosaki Viaduct     467

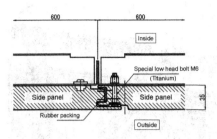

**Fig. 5.** Joint structure between side panels.

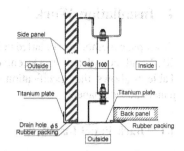

**Fig. 6.** Joint structure between back panel and side panel.

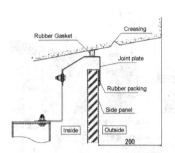

**Fig. 7.** Joint structure between overhang deck and side panel.

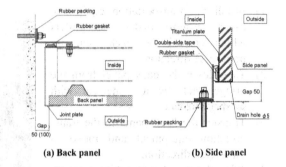

**Fig. 8.** (a) Joint structure between substructure and back panel. (b) Joint structure between substructure and side panel.

**Table 6.** Specification of back panel and side panel.

| Panel | Back panel | Side panel |
|---|---|---|
| Unit weight (N/m$^2$) | 98 | 98 |
| Size (mm) | 1000× 7600 | 600× 2950 |
| Total quantity | 528 | 452 |

## 4 Installation Work

Cover plate method work started in January-2018 and completed in December-2018. Table 6 shows the specification of back panel and side panel.

### 4.1 Back Panel Installation

Photo 6 shows how to install the back panel. 8–16 panels were installed in a day. Following 6 procedures were repeated.

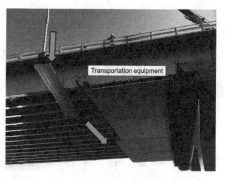

**Photo 6.** Installatin of back panel.

1) Install the transportation equipment (hoist system) under the girders which carried the back panel in transverse direction.
2) Take down the back panel from the bridge deck by a crane car with closing left lane.
3) Transfer the back panel to the transportation equipment and move to transverse direction.
4) After the panel was set to a designated position, suspend the panel temporarily from the main girder with lever blocks.
5) Attach the back panels to the lower flange with the mounting bracket. Here, put a rubber plate between the lower flange and the mounting bracket as shown in Photo 7.
6) After installation of 4 panels in transverse direction, replace the transportation equipment to next position.

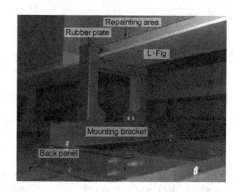

**Photo 7.** Connection between back panel and the lower flange.

### 4.2 Side Panel Installation

Photo 8 shows how to install the side panel. 40–50 panels were installed in a day. Following 3 procedures were repeated.

**Photo 8.** Installation of side panel.

1) Take down the side panel from the bridge deck by the crane car with closing left lane.
2) Fix side panel temporarily with the mounting bracket and carry the panel in longitudinal direction with human power.
3) Attach the side panel to the designated position.

### 4.3 Other Work

The lower flange and the hanging piece were repainted because it was difficult to repair after the cover plate installation. Likewise, underside of road deck where is in proximity to the mounting bracket of side panel was inspected by hammering test, and damaged part was repaired such as chipping, rust prevention and section repair. After installation of cover plate, all girders were wiped to remove adhered salt with a damp cloth.

## 5 Evaluation of Anti-corrosion Performance

Covered with titanium panels, the steel and the coat of paint in the box girder are insulated from degradation factors such as ultraviolet ray, rainwater and floating salt. In order to identify the improvement of corrosive environments inside the panel, adhered salt, temperature and relative humidity were measured at inside and outside of the box girder after 1.5 years of cover plate installation. In this chapter, the measurement results are reported to evaluate quantitatively the anti-corrosion performance.

### 5.1 Measurement of Adhered Salt

#### 5.1.1 Method and Material

The adhered salt was measured at inside and outside of box girder by using the Chloride ion detector tube in December-2018 (soon after the cover plate installation and wiping the whole girders), June-2020 (after 1.5 years of the cover plate installation). Figure 9 shows the measurement points of adhered salt. As shown in Photo 9, the measurement points at outside of box girder were on side panel and back panel. And the measurement points at inside of box girder were on the main girder web as shown in Photo 10. The number of measurements was 15. The instruments and procedure of adhered salt measurement were shown below.

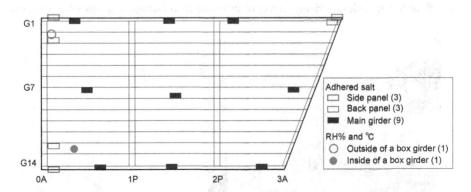

**Fig. 9.** Measurement points of adhered salt, temperature and relative humidity. The number of () shows the quantity of measurement points.

**Photo 9.** Measurement points of 0A Magosaki viaduct.

**Photo 10.** Measurement point of adhered salt on main girder.

*Instruments of adhered salt measurement*

1) Gauze (300 mm * 300 mm)
2) Wide-mouthed cup (200 ml)
3) Rubber glove
4) Purified water
5) Chlorine ion detector tube

*Procedure of adhered salt measurement*

1) Decide measurement section (0.25 m$^2$).
2) Wear rubber glove and pour 100 ml purified water into the wide-mouthed cup.
3) Soak gauze into the wide-mouthed cup and wring out the gauze.
4) Wipe measurement surface laterally and vertically (Photo 11, a).
5) Repeat above 3)–4) 3 times.
6) Wash rubber glove with 50 ml purified water on the cup to pour the water into the cup.
7) Cut both ends of Chlorine ion detector tube and immerse the one end of the tube into the 150 ml solution (Photo 11, b).
8) Wait for the solution rising to top of tube and read boundary of discoloured layer.

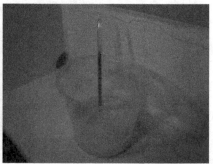

(a) Wiping measurement surface.

(b) Immersing the end of chlorine ion detector tube into the solution.

**Photo 11.** Procedure of adhered salt measurement.

## 5.1.2 Results and Discussion

Table 7 shows the result of adhered salt measurement. For outside of box girder, adhered salt on side panel was 1–4 (mg/m$^2$) because the panel was easy to be washed by rainwater. And adhered salt on back panel was 150–350 (mg/m$^2$) because the panel was not easy to be washed by rainwater. On the other hand, adhered salt on main girder was 3–10 (mg/m$^2$) in November-2018 (soon after cover plate installation and wiping the whole girders), and 1–14 (mg/m$^2$) in June-2020 (after 1.5 years of cover plate installation). So apparent increase of adhered salt on the main girder was not identified after 1.5 years of the cover plate installation. Therefore, it can be said that main girders are insulated from floating salt and anti-corrosion performance is improved.

**Table 7.** Result of adhered salt measurement.

| Measurement date | | Adhered salt (mg/m$^2$) | |
| --- | --- | --- | --- |
| | | November-2018 | June-2020 |
| Outside of box girder | Side panel | - | 1-4 |
| | Back panel | - | 150-350 |
| Inside of box girder | Main girder | 3-10 | 1-14 |

## 5.2 Measurement of Temperature and Relative Humidity

### 5.2.1 Method and Material

Temperature and relative humidity has been measured at inside and outside of box girder since November-2019 as shown in Photo 12. Measurement device can measure and log temperature and relative humidity with a high accuracy and a wide measurement range. As shown in Photo 13, this device consists of data logger and sensor. Measurement range is −25–75 °C and 0–99% RH, and the deviation is ±0.3 °C and ±2.5% RH. Temperature and relative humidity were measured once a day at 1 point each of inside and outside.

### 5.2.2 Results and Discussion

Figure 10 shows the measurement result of temperature and relative humidity at inside and outside of box girder at 6 a.m. This graph shows every 6 a.m. data from November-2019 to June-2020. In case of temperature, inside is approximately 8–25 °C and outside is about 5–18 °C. Though the temperature of inside and outside is almost the same, inside is 4 °C higher than outside on average as shown in Table 8.

In terms of relative humidity, inside is approximately 45–80% RH and outside is about 20–90% RH level. Although fluctuation of inside relative humidity seems to follow one of outside with time differences, the average of relative humidity is 10% RH lower than outside. Nevertheless, the average of inside relative humidity is about 62% RH so it is not low enough for rust prevention.

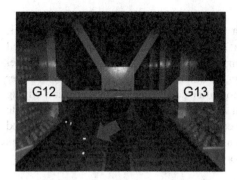
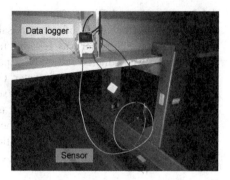

**Photo 12.** Measurement point of RH% and °C inside of box girder.

**Photo 13.** Measurement device consists of data logger and sensor.

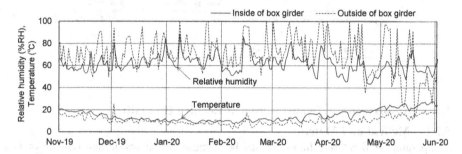

**Fig. 10.** Measurement result of temperature and relative humidity at inside and outside of box girder.

**Table 8.** Average of measurement result of temperature and relative humidity.

| Item | Average of measurement result | |
|---|---|---|
| | Temperature | Relative humidity |
| Outside of box girder | 10°C | 72%RH |
| Inside of box girder | 14°C | 62%RH |

## 6  Summary and Conclusion

Magosaki viaduct was repainted after 14 years of opening and full repaint was necessary again. For full repaint, there were some problems such as severe corrosion environment, large paint face and including lead-based paint. Due to this, cover plate method, which is mainly applied for a newly constructed bridge as an anti-corrosion measure, was evaluated for Magosaki viaduct from the aspects of functionality and cost. As a result, it was found that cover plate method was more beneficial than full repainting.

In the design of cover plate method, verification of weight, clearance in the box girder, verification of wind load, ensuring airtightness and maintainability at the connection structure were reported. In the installation work, install procedures of cover plate and other work were reported.

In the evaluation of anti-corrosion performance, corrosion environment was examined quantitatively by measurements of adhered salt, temperature and relative humidity at inside and outside of box girder after 1.5 years of cover plate installation. As a result of adhered salt measurement, inside of box girder was insulated from floating salt. So it was found that anti-corrosion performance was improved. On the other hand, inside temperature was 4 °C higher than outside on average. Although fluctuation of inside relative humidity seems to follow one of outside with time differences, the average of relative humidity is 10% RH lower than outside. Nevertheless, the average of inside relative humidity was about 62% RH so it was not low enough for rust prevention.

In closing, Honshu-Shikoku Bridge Expressway Company Limited (HSBE) continue to measure adhered salt, temperature and relative humidity to evaluate anti-corrosion performance of cover plate method. In addition, HSBE intend to execute proximity inspection and repair work for main girder, underside of deck and accessories using permanent scaffolding function of this cover plates.

# Reference

Fujikawa, N., Okamoto, Y., Ishihara, T., Noguchi, T.: Study on calculation of wind load of Haneda airport taxiway bridge cover plate [Translated from Japanese]. In: Proceedings of the 62nd Japan Society of Civil Engineers Annual Meeting, Japan, pp. 311–312 (2007)

# Degradation of Friction Performance Indicator Over the Time in Highways Using Linear Mixed Models

Adriana Santos[1]($\boxtimes$), Elisabete Freitas[2], Susana Faria[3], Joel Oliveira[2], and Ana Maria A. C. Rocha[4]

[1] Ascendi IGI S.A., Porto, Portugal
asantos@ascendi.pt
[2] ISISE, University of Minho, Guimarães, Portugal
{efreitas, joliveira}@civil.uminho.pt
[3] CBMA, University of Minho, Guimarães, Portugal
sfaria@math.uminho.pt
[4] ALGORITMI Center, University of Minho, Braga, Portugal
arocha@dps.uminho.pt

**Abstract.** Pavement Management Systems use models describing performance indicators and can be used to plan future maintenance and rehabilitation activities taking into account the safety and economic requirements. The COST Action 354 (Long Term Performance of Road Pavements), with the participation of experts from 15 European countries, has established a list of seven performance indicators. Skid resistance and texture are interrelated in the sense that texture can be considered to be one of the explanatory variables of skid resistance and both have a major bearing on road safety.

Using the database provided by a Portuguese motorway concessionary, this study aimed to develop and validate linear mixed-effects models to describe the degradation of friction performance indicator, throughout the time, based on the formula established by COST, and identify whether the traffic, climate conditions, pavement structure, and geometric characteristics of the highway influence that behavior. To accomplish these main objectives, linear-mixed models with two random effects (one in the intercept and another associated with the slope in the Time variable) were considered for a 2-level data set (level 2 distinguishes the road sections and level 1 represents the repeated measures made over time). Two approaches were made: a model that includes only the variables inherent to traffic and climate conditions and another including the factors inherent to the highway characteristics.

Results confirm that the traffic, climate conditions, pavement structure, and geometric characteristics of the highway influence the degradation of friction performance indicator.

**Keywords:** Friction · Skid resistance · Performance · Degradation · Linear mixed models (LMM) · Pavement management system

---

© The Author(s), under exclusive license to Springer Nature Switzerland AG 2022
A. Akhnoukh et al. (Eds.): IRF 2021, SUCI, pp. 474–489, 2022.
https://doi.org/10.1007/978-3-030-79801-7_35

# 1 Introduction

The success of pavement management systems (PMS) depends greatly on the performance prediction models used, as the accuracy of the predictions determines the reasonableness of the decisions. Many pavement performance prediction models have been proposed over the years. Most of them were developed for a particular region or country, under specific traffic and climatic conditions. Therefore, they cannot be directly applied to other countries or conditions (Erlingsson et al. 2008).

The age of pavement, traffic, and climatic actions, such as rain, sun, or ice, are the major contributors to asphalt pavement degradation (Li et al. 2017). The effect of these climatic agents for very long periods can accelerate the process of asphalt mixture deterioration (Sol-Sánchez et al. 2015). The climatic regions are usually determined based on the maximum and minimum temperature and rainfall (Dong et al. 2021). Other factors can influence the performance of friction over time, as the ones recently identified: geometry (Viner et al. 2004), type of layer, and pavement surface texture characteristics (Kogbara et al. 2016).

There are many methods that can be used to develop pavement degradation models. Haas (2003) grouped the performance prediction models into classes: Empirical, Mechanistic-empirical, both usually developed through regression analysis, and subjective, which use, for example, Markovian transition process models, and Bayesian models. Markov chains have been, indeed, extensively applied for modelling the deterioration of infrastructures over time, such as bridges, pavements, or waste-water systems (Moreira et al. 2018). Other technics were used, for example, Yu et al. (2007) applied linear mixed-effects models to predict future conditions of a specific pavement section, and Lorino et al. (2012) developed nonlinear mixed-effects models to identify and quantify the impact of structural and climatic factors on cracking evolution. Recently, Artificial Intelligence (AI) techniques were also used to model pavement degradation, friction precisely (Marcelino et al. 2017).

This study proposes using linear mixed effects models (LMEM) to describe the degradation of friction performance indicator throughout time and identify whether the traffic, climate conditions, pavement structure, and geometric characteristics of the highway influence that behavior. The friction performance indicator adopted was the one established by COST Action 354. The main objective of this Action was the definition of uniform European performance indicators and indexes for road pavements taking the needs of road users and road operators into account (Litzka et al. 2008).

To achieve the objectives of this study, linear-mixed models with two random effects (one in the intercept and another associated with the slope in the Time variable) were considered for a 2-level data set (level 2 represents the road sections and level 1 represents the repeated measures made on each road section over time). For both levels, two approaches were made:

- Model I - considers only the variables inherent to traffic and climate conditions. This model intends to provide the network manager with a tool that enables to predict and/or assess the degradation of friction performance indicator through time, when there are no data on vertical or horizontal alignment, or the type of surface course cannot be differentiated;

- Model II - in addition to the variables considered in model I, it includes the factors inherent to the highway characteristics;
- Model III - this model includes the variable texture in addition to model II variables.

All analyses were performed using the R statistical software and the Statistical Package for the Social Sciences.

## 2 Data Set

The friction data were collected on different roads of the highway network of the Portuguese Concessionaire Ascendi, as shown in Fig. 1. Although they are inserted in different climate environments and topographic conditions, they were constructed mainly with granite aggregates as these are abundant in those regions.

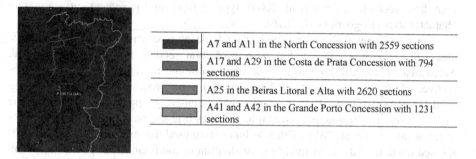

Fig. 1. Number of sections by Concession, Highway, and Location.

For this study, a sample with 7204 pavement sections with 100 m length was considered. Each section was identified by the corresponding kilometric point (PKi). Three monitoring campaigns were included, time 0, 1, and 2. The first measurements (time 0) were carried out 6 to 8 months after the road opening to traffic and, after that, every 4 years. In this period (8 years), there were no maintenance operations on the pavement. This procedure copes with this concessionary quality control plan, establishing the section length and friction monitoring campaigns frequency.

The details to calculate the skid resistance performance indicator proposed by the COST Action 354, including proposed technical parameters and transfer functions, are described in Table 1. As in this study, the wet skid resistance was measured with a Grip Tester at 50 km/h (Longitudinal Friction Coefficient (LFC)), according to the technical standard CEN/TS 15901-7:2009 (CEN/TS 15901-7, 2009), the transfer function adopted was formula [2]. This indicator represents the dependent variable ($PI\_F$) of the models. For more details see the COST Action 354.

The variables characterizing each 100 m section were grouped in a database as follows: pavement structure; traffic; climate conditions; geometric characteristics of the vertical and horizontal alignments; lane; and hypsometry. Some of these variables were identified in the state-of-the-art review; others were defined according to the objectives of the work.

## Degradation of Friction Performance Indicator    477

**Table 1.** Details for skid resistance indicator (COST Action 354, 2008).

| 3. Proposed Transfer function(s), usage and Limitation | | |
|---|---|---|
| Proposed Transfer function(s): | $PI\_F = Max(0; Min(5; (-17.600*SFC+11.205)))$ | [1] |
| | $PI\_F = Max(0; Min(5; (-13.875*LFC+9.338)))$ | [2] |
| | Transformation [1] should only be used for SFC devices running at 60Km/h | |
| | Transformation [2] should only be used for LFC devices running at 50Km/h | |
| Limitations of Proposed Transfer function(s): | Transformation [1] and [2] are both suitable for all pavement types (flexible, semi-rigid and rigid). | |
| | Transformation [1] and [2] are both suitable for motorway and primary roads. | |

| 4. Range and Sensitivity of Transfer function(s) | | | | | |
|---|---|---|---|---|---|
| | Very Good | | $\rightarrow$ | | Very Poor |
| Skid Resistance, PI_F | 0 to 1 | 1 to 2 | 2 to 3 | 3 to 4 | 4 to 5 |
| SFC (60Km/h) – Transformation [1] | 0.64 – 0.58 | 0.58 – 0.52 | 0.52 – 0.47 | 0.47 – 0.41 | 0.41 – 0.35 |
| LFC (50Km/h) – Transformation [2] | 0.67 – 0.60 | 0.60 – 0.53 | 0.53 – 0.46 | 0.46 – 0.38 | 0.38 – 0.31 |

The traffic data were obtained using automatic vehicle counter systems that are installed on highways and toll plazas. The selected traffic variables were:

- Annual Average Daily Traffic (*AADT*);
- Accumulated Annual Daily Traffic (*Ac.ADT*);
- Accumulated Annual Daily Traffic per direction of circulation (*Ac.AADT/direction*), it is the *Ac.ADT* divided by 2;
- Accumulated Annual Daily Traffic per direction and per lane (*Ac.AADT/direction/lane*) – for sections with 2 lanes: RL = 95% and LL = 5%, and 3 lanes: SL = 80%, RL = 15% and LL = 5%;
- Accumulated Annual Daily Heavy Traffic (*Ac.AADT heavy*);
- Accumulated Annual Daily Heavy Traffic that circulated in the daytime period (between 07:00 and 19:00 h) (*Ac.AADT heavy/day*) and also the night-time period (between 19:00 and 07:00 h), characterized by *Ac.AADT heavy/night*.

The unity for all those variables is vehicles per million, calculated up to the time of the observation (t = 0, t = 1, t = 2).

Climatic factors include temperature, humidity, precipitation, snowfall, etc. It was considered adequate to use the data available on the site of the Portuguese Institute of the Sea and Atmosphere (IPMA) to characterize it. Therefore, the Climate Normals (average values) concerning 1971–2000 period (IPMA 2019) associated to the zones involved in the study were used to set the following temperature factors:

- Higher value of the maximum temperature, °C (*Higher max temp*);
- Average maximum temperature, °C (*Average max temp*);
- Average maximum temperature, °C (*Average max temp*);
- Average mean temperature, °C (*Average med temp*);
- Average minimum temperature, °C (*Average min temp*);
- Lower value of the minimum temperature, °C (*Lower min temp*);
- Average number of days with maximum temperature $\geq 30$ °C (*Average n.° days temp max30*);

478 A. Santos et al.

- Average number of days with maximum temperature $\geq 25$ °C (*Average n.° days temp max25*);
- Average number of days with minimum temperature $\geq 20$ °C (*Average n.° days temp min20*) and
- Average number of days with minimum temperature $\leq 0$ °C (*Average n.° days temp min0*).

Other climate variables were considered:

- Average of the total quantity of precipitation, mm (P);
- Daily maximum quantity of precipitation, mm (PM) and
- Relative humidity, % (RH).

The data for the first two variables are available in the Climatological Atlas of the Iberian Peninsula or Iberian Climate Atlas (ACI) (IPMA 2019), and the data for the third variable is available in Portal do Clima (2015).

When a highway is designed, many factors determine plan and profile characteristics, as the speed, safety, comfort of the users, and cost of construction and operation. Therefore, it becomes relevant to characterize each section according to the geometrical features to establish how design decisions influence the degradation of friction performance indicator. In this study, the variables considered were:

- *Lane*: left (*LL*), right (*RL*), and slow lane (*SL*);
- *Plan*: straight alignment (*SA*), clothoid (*C*), and circular curve (*CC*);
- *Profile*: Slope (*S*), concave concordance curve (*Ccc*), and convex concordance curve (*Ccv*);
- *Hypsometry*: *Low Altitude* (the hypsometric curves are inserted in the interval [0, 200 m]), and *Medium Altitudes* (the hypsometric curves are inserted in the interval] 200, 2000 m]).

Because the type of surface course can influence the degradation of friction performance indicator in road pavements, this variable also was considered. Thus, the sections in this study have the following types of surface course:

- Porous asphalt (*PA*); Gap-graded asphalt concrete with Bitumen Modified with a high percentage of Rubber Modified Binder (*GGA.BMR*) and Gap graded asphalt concrete surface course (*GGAC*).

For more details see Santos et al. (2021).

## 3 Data Analysis

Table 2 presents the descriptive analysis of some quantitative variables. From the analysis of the values obtained, the minimum value obtained for friction performance indicator is 0 (classification very good), the maximum is 5 (classification very bad), and the mean value is 1.41 (classification good). The variables for temperature that present the largest difference between the minimum and the maximum value are: the higher value of the maximum temperature (Higher max temp) and the average number of days with maximum temperature $\geq 25$ °C (Average n.° days temp max 25).

Table 2. Statistical description for some quantitative variables.

| Variables | Definition | Min. | Max. | Mean | Std. deviation |
|---|---|---|---|---|---|
| *PI F* | Friction performance indicator | 0.00 | 5.00 | 1.41 | 1.33 |
| *AADT* | Annual average daily traffic | 0.0016 | 0.3290 | 0.0546 | 0.0574 |
| *Ac.ADT* | Accumulated annual daily traffic | 0.0722 | 120.0850 | 18.9549 | 20.9676 |
| *Higher max temp* | Higher value of the maximum temperature | 22.00 | 40.50 | 32.43 | 6.23 |
| *Average med temp* | Average mean temperature | 7.70 | 21.40 | 15.64 | 4.33 |
| *Lower min temp* | Lower value of the minimum temperature | −7.30 | 11.40 | 1.76 | 4.65 |
| *Average n.° days temp max30* | Average number of days with maximum temperature $\geq 30$ °C | 0.00 | 14.70 | 2.92 | 4.22 |
| *Average n.° days temp max25* | Average number of days with maximum temperature $\geq 25$ °C | 0.00 | 25.80 | 7.74 | 8.47 |
| *Average n.° days temp min20* | Average number of days with minimum temperature $\geq 20$ °C | 0.00 | 1.50 | 0.29 | 0.42 |
| *Average n.° days temp min0* | Average number of days with minimum temperature $\leq 0$ °C | 0.00 | 4.90 | 0.53 | 1.04 |
| *P* | Averages of the total quantity of precipitation | 11.80 | 231.15 | 83.38 | 57.27 |

Figure 2 shows box plots of PI F in the time: 0, 1 and 2. The figure shows that PI F increases with time.

To confirm that, the Friedman test revealed statistically significant differences between the friction performance indicator values at the various times periods. Applying the Wilcoxon signed rank test, it is observed that the values of the friction performance indicator (PI F) on time 2 are significantly higher than the values of the indicator (PI F) on time 0 and time 1. The relationship between the response variable (PI F) and continuous explanatory variables was analyzed using a Spearman correlation coefficient. In this case, the variable response has a moderate positive correlation with the explanatory variable Ac.ADT (rs = 0.627, p-value < 0.01), and for the explanatory variables related to climate conditions, the variables that have the highest negative correlation are: the Higher max temp (rs = −0.21, p-value < 0.01), Average n.° days temp max25 (rs = −0.219, p-value < 0.01) and RH (rs = −0.353, p-value < 0.01).

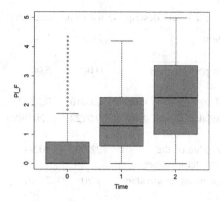

**Fig. 2.** Friction indicator performance for each time period under study.

## 4 Linear Mixed-Effects Model (Concept and Formulation)

The LMEM is also called the longitudinal model since the data are usually taken in a temporal order (Yu et al. 2007). The term mixed is used in this context to denote that the model contains both fixed and random effects.

The distinctive feature of a LMEM is that the mean response is modeled as a combination of population characteristics that are assumed to be shared by all individuals and subject-specific effects that are unique to a particular individual. The former are referred to as fixed effects, while the latter are referred to as random effects (Yu et al. 2007).

The linear mixed-effects model expresses for the $i$-th kilometric point is given by Equation:

$$Y_i = X_i\beta + b_i + \varepsilon_i, \qquad i = 1, \ldots, n \tag{1}$$

where:
- $Y_i$ $(Y_{i1} \ldots Y_{iT_i})^T$ represents a vector (of size $T_i \times 1$) of responses (friction performance indicator) for the i-th kilometric point;
- $X_i$ is the known fixed-effects covariates matrix (of size $T_i \times p$);
- $\beta$ is a vector (of size $p \times 1$) unknown regression coefficients (or fixed-effects parameters);
- $b_i$ is a vector (of size $q \times 1$) of random-effects associated with the $i$-th kilometric point;
- $\varepsilon_i$ is the $T_i \times 1$ vector of errors for observations in $i$-th kilometric point.

Moreover
- $b_i$ ~$N_q$ (0, $D$)
- $\varepsilon_i$ ~$N_{Ti}$ (0, $R_i$)
- $b_i$ e $\varepsilon_i$ are independent for the same $i$-th kilometric point and of each other,

where $D$ is the q x q covariance matrix for the random effects, and $R_i$ is the $T_i$ x $T_i$ covariance matrix of the errors in kilometric point $i$.

In the case of study, as age of the pavement is a temporal variable, two types of models were developed, first considering only the random effect on the intercept (kilometer point PKi) and then models with two random effects; one on the intercept and the other on the slope of the Time variable. The approaches to the models were made considering: i) the null model; ii) the model considers only the variables inherent to traffic and climate conditions; iii) in this model, in addition to the previous variables, were included the factors intrinsic to highway characteristics; iv) in the last model it was included the variable texture.

Thus, based on Eq. (1), the linear mixed models for two levels data set with a random effect on the intercept (Models PKi) are described by:

| | | |
|---|---|---|
| Null Model PKi | $PI\ F\ COST, PKi_{ti} = \beta_0 + b_{0i} + \varepsilon_{ti}$ | (2) |
| Model PKi I | $PI\ F\ COST, PKi_{ti} = \beta_0 + \beta_1\ Time_{ti} + \beta_2\ Ac.ADT_{ti} +$ <br> $\beta_3\ Higher\ max\ temp_{ti} + \beta_4\ Average\ no.days\ temp\ max25_{ti} +$ <br> $\beta_5\ P_{ti} + \beta_6\ RH_{ti} + b_{0i} + \varepsilon_{ti}$ | (3) |
| Model PKi II | $PI\ F\ COST, PKi_{ti} = \beta_0 + \beta_1\ Time_{ti} + \beta_2\ Ac.ADT_{ti} +$ <br> $\beta_3\ Higher\ max\ temp_{ti} + \beta_4\ Average\ no.days\ temp\ max25_{ti} + \beta_5\ P_{ti} +$ <br> $\beta_6\ Lane_{ti} + \beta_7\ Plan_{ti} + \beta_8\ Profile_{ti} + \beta_9\ Surface\ Layer_{ti} +$ <br> $\beta_{10}\ Landform_{ti} + b_{0i} + \varepsilon_{ti}$ | (4) |
| Model PKi III | $PI\ F\ COST, PKi_{ti} = \beta_0 + \beta_1\ Time_{ti} + \beta_2\ Ac.ADT_{ti} +$ <br> $\beta_3\ Higher\ max\ temp_{ti} + \beta_4\ Average\ no.days\ temp\ max25_{ti} + \beta_5\ P_{ti} +$ <br> $\beta_6\ Lane_{ti} + \beta_7\ Plan_{ti} + \beta_8\ Profile_{ti} + \beta_9\ Surface\ Layer_{ti} +$ <br> $\beta_{10}\ Landform_{ti} + \beta_{11}\ Texture_{ti} + b_{0i} + \varepsilon_{ti}$ | (5) |

where $b_{0i} \sim N(0, \sigma_P^2), \varepsilon_{it} \sim N(0, \sigma_\varepsilon^2)$, $i = 1, \ldots, 7204$ and $t = 0, 1, 2$.

The linear mixed models for two levels data set with two random effects, one on the intercept and the other on the slope of the time variable (Models T|PKi), are described by:

| | | |
|---|---|---|
| Null Model T|PKi | $PI\ F\ COST, PKi_{ti} = \beta_0 + b_{0i} + b_{1i}Time_{ti} + \varepsilon_{ti}$ | (6) |
| Model T|PKi I | $PI\ F\ COST, PKi_{ti} = \beta_0 + \beta_1\ Time_{ti} + \beta_2\ Ac.ADT_{ti} +$ <br> $\beta_3\ Higher\ max\ temp_{ti} + \beta_4\ Average\ no.days\ temp\ max25_{ti} +$ <br> $\beta_5\ P_{ti} + \beta_6\ RH_{ti} + b_{0i} + b_{1i}Time_{ti} + \varepsilon_{ti}$ | (7) |
| Model T|PKi II | $PI\ F\ COST, PKi_{ti} = \beta_0 + \beta_1\ Time_{ti} + \beta_2\ Ac.ADT_{ti} +$ <br> $\beta_3\ Higher\ max\ temp_{ti} + \beta_4\ Average\ no.days\ temp\ max25_{ti} + \beta_5\ P_{ti} +$ <br> $\beta_6\ Lane_{ti} + \beta_7\ Plan_{ti} + \beta_8\ Profile_{ti} + \beta_9\ Surface\ Layer_{ti} +$ <br> $\beta_{10}\ Landform_{ti} + b_{0i} + b_{1i}Time_{ti} + \varepsilon_{ti}$ | (8) |
| Model T|PKi III | $PI\ F\ COST, PKi_{ti} = \beta_0 + \beta_1\ Time_{ti} + \beta_2\ Ac.ADT_{ti} +$ <br> $\beta_3\ Higher\ max\ temp_{ti} + \beta_4\ Average\ no.days\ temp\ max25_{ti} + \beta_5\ P_{ti} +$ <br> $\beta_6\ Lane_{ti} + \beta_7\ Plan_{ti} + \beta_8\ Profile_{ti} + \beta_9\ Surface\ Layer_{ti} +$ <br> $\beta_{10}\ Landform_{ti} + \beta_{11}\ Texture_{ti} + b_{0i} + b_{1i}Time_{ti} + \varepsilon_{ti}$ | (9) |

where $b_i = \begin{bmatrix} b_{0i} \\ b_{1i} \end{bmatrix} \sim N(0, D), \varepsilon_{it} \sim N(0, \sigma_\varepsilon^2)$, $i = 1, \ldots, 7204$ and $t = 0, 1, 2$.

482 A. Santos et al.

The estimation of fixed effects and random effects was performed based on the maximum likelihood and restricted likelihood methods and applying the lme function of Program R. For the variation-covariance matrix of the random errors two types of matrices: the identity matrix and the first order autoregressive matrix AR (1) were considered.

The comparison of the models for the different $R_i$ matrix structures was performed based on the AIC and BIC information criteria, as well as on the Likelihood Ratio Test (LRT). Table 3 shows the comparison of the Models PKi, and Table 4 show the results for the Models T|PKi, for the different $R_i$ matrix structures.

**Table 3.** Comparison of the Models PKi for the different $R_i$ matrix structures.

| Structure ($R_i$) | Models PKi | AIC | BIC | -2LL | Test | LRT | p-value |
|---|---|---|---|---|---|---|---|
| Identity | I | 55 353,04 | 55 416,88 | 55 337,04 | | | |
| AR(1) | I | 51 738,11 | 51 809,94 | 51 720,12 | I vs. I | 3 616,92 | <0,0001 |
| Identity | II | 52 261,14 | 52 388,83 | 52 229,14 | | | |
| AR(1) | II | 48 217,50 | 48 337,21 | 48 187,50 | II vs. II | 4 041,64 | <0,0001 |
| Identity | III | 52 260,47 | 52 396,14 | 52 226,48 | | | |
| AR(1) | III | 48 203,84 | 48 331,53 | 48 171,84 | III vs. III | 4 054,64 | <0,0001 |

**Table 4.** Comparison of the Models T|PKi for the different $R_i$ matrix structures.

| Structure ($R_i$) | Models T|PKi | AIC | BIC | -2LL | Test | LRT | p-value |
|---|---|---|---|---|---|---|---|
| Identity | I | 54 276,73 | 54 348,56 | 54 258,74 | | | |
| AR(1) | I | 51 636,81 | 51 716,62 | 51 616,82 | I vs. I | 2 641,92 | <0,0001 |
| Identity | II | 50 590,42 | 50 718,11 | 50 558,42 | | | |
| AR(1) | II | 47 918,20 | 48 053,87 | 47 884,20 | II vs. II | 2 674,22 | <0,0001 |
| Identity | III | 50 592,17 | 50 727,84 | 50 558,18 | | | |
| AR(1) | III | 47 895,69 | 48 039,34 | 47 859,68 | III vs. III | 2 698,50 | <0,0001 |

As can be seen, when using the Identity (I) structure for the variance-covariance matrix of random errors, the information criteria values are greater than obtained when applying the first order autoregressive matrix structure AR (1). In addition, from the analysis of the Likelihood Ratio Test (LRT) results, it appears that the choice of the AR (1) correlation structure is adequate. The estimates of the fixed parameters of the models, the respective standard errors and p-values of the t-test, and the estimates of random parameters and confidence intervals, considering the matrix structure of the variance-covariance of the errors first-order autoregressive matrix AR (1) are shown in Table 5 for the Models PKi and Table 6 for the Models T|PKi.

## Table 5. Results of the friction performance indicator for the Models PKi.

| Models PKi | Model I PKi | | | Model II PKi | | | Model III PKi | | |
|---|---|---|---|---|---|---|---|---|---|
| | Estimate ($\beta$) | Stand. Error | p-value | Estimate ($\beta$) | Stand. Error | p-value | Estimate ($\beta$) | Stand. Error | p-value |
| **Fixed-Effect parameter** | | | | | | | | | |
| *Intercept* | 1.818 | 0.133 | <0.0001 | 1.815 | 0.084 | <0.0001 | 1.717 | 0.088 | <0.0001 |
| *Time* | 0.209 | 0.016 | <0.0001 | 0.245 | 0.016 | <0.0001 | 0.241 | 0.016 | <0.0001 |
| *Ac.ADT* | 0.033 | 0.001 | <0.0001 | 0.026 | 0.001 | <0.0001 | 0.026 | 0.001 | <0.0001 |
| *Higher max temp* | -0.057 | 0.004 | <0.0001 | -0.075 | 0.003 | <0.0001 | -0.075 | 0.003 | <0.0001 |
| *Average no. days temp max25* | 0.056 | 0.002 | <0.0001 | 0.059 | 0.002 | <0.0001 | 0.060 | 0.002 | <0.0001 |
| *P* | 0.002 | 0.000 | <0.0001 | - | - | - | - | - | - |
| *RH* | - | - | - | - | - | - | - | - | - |
| Lane (ref: *LL*) | | | | - | - | - | - | - | - |
| [Lane=*RL*] | | | | 0.456 | 0.010 | <0.0001 | 0.453 | 0.010 | <0.0001 |
| [Lane=*SL*] | | | | 0.171 | 0.018 | <0.0001 | 0.169 | 0.018 | <0.0001 |
| Plan (ref: *SA*) | | | | - | - | - | - | - | - |
| [Plant=*C*] | | | | -0.010 | 0.030 | 0.7452 | -0.010 | 0.030 | 0.7452 |
| [Plant=*CC*] | | | | 0.085 | 0.029 | 0.0031 | 0.085 | 0.029 | 0.0031 |
| Profile (ref: *Ccc*) | | | | - | - | - | - | - | - |
| [Profile =*Ccv*] | | | | - | - | - | - | - | - |
| [Profile=*S*] | | | | - | - | - | - | - | - |
| Layer (ref: *GGAC*) | | | | - | - | - | - | - | - |
| [Layer =*PA*] | | | | 0.559 | 0.027 | <0.0001 | 0.492 | 0.032 | <0.0001 |
| [Layer =*GGA.RMB*] | | | | -0.501 | 0.067 | <0.0001 | -0.501 | 0.067 | <0.0001 |
| Hypsometry (ref: *Medium Altitudes*) | | | | - | - | - | - | - | - |
| [Hypsometry = *Low Altitudes*] | | | | 1.103 | 0.029 | <0.0001 | 1.100 | 0.029 | <0.0001 |
| *Texture* | | | | | | | 0.111 | 0.028 | 0.0237 |
| **Random parameters** | | IC 95% | | | IC 95% | | | IC 95% | |
| $\sigma_P^2$ | 0.451 | [0.408; 0.497] | | 0.113 | [0.094; 0.132] | | 0.111 | [0.094; 0.132] | |
| $\sigma_\varepsilon^2$ | 0.711 | [0.691; 0.731] | | 0.665 | [0.645; 0.685] | | 0.665 | [0.646; 0.685] | |
| $\rho$ | 0.459 | [0.444; 0.476] | | 0.486 | [0.470; 0.500] | | 0.486 | [0.471; 0.501] | |
| **Comparison** | | | | | | | | | |
| LRT | I vs. Null | | 7363.52** | II vs. I | | 3 532.62 | III vs. II | | 15.66 |

**Table 6.** Results of the friction performance indicator for the Models T|PKi.

| Models T|PKi | Model T|PKi I | | | Model T|PKi II | | | Model T|PKi III | | |
|---|---|---|---|---|---|---|---|---|---|
| | Estimate ($\beta$) | Stand. Error | p-value | Estimate ($\beta$) | Stand. Error | p-value | Estimate ($\beta$) | Stand. Error | p-value |
| Fixed-Effect parameter | | | | | | | | | |
| Intercept | 2.205 | 0.925 | <0.0001 | 1.874 | 0.085 | <0.0001 | 1.778 | 0.088 | <0.0001 |
| Time | 0.208 | 0.017 | <0.0001 | 0.267 | 0.017 | <0.0001 | 0.248 | 0.017 | <0.0001 |
| Ac.ADT | 0.033 | 0.001 | <0.0001 | 0.025 | 0.001 | <0.0001 | 0.024 | 0.001 | <0.0001 |
| Higher max temp | -0.062 | 0.003 | <0.0001 | -0.078 | 0.003 | <0.0001 | -0.080 | 0.003 | <0.0001 |
| Average no. days temp max25 | 0.048 | 0.002 | <0.0001 | 0.060 | 0.002 | <0.0001 | 0.061 | 0.002 | <0.0001 |
| P | 2.205 | 0.925 | <0.0001 | - | - | - | - | - | - |
| RH | - | - | - | - | - | - | - | - | - |
| Lane (ref: LL) | | | | - | - | - | - | - | - |
| [Lane=RL] | | | | 0.459 | 0.010 | <0.0001 | 0.453 | 0.010 | 0,0095 |
| [Lane=SL] | | | | 0.213 | 0.018 | <0.0001 | 0.169 | 0.018 | 0,0181 |
| Plan (ref: SA) | | | | - | - | - | - | - | - |
| [Plant=C] | | | | 0.035 | 0.030 | 0.7452 | 0.039 | 0.029 | 0,0289 |
| [Plant=CC] | | | | 0.137 | 0.029 | 0.0031 | 0.140 | 0.028 | 0,0276 |
| Profile (ref: Ccc) | | | | - | - | - | - | - | - |
| [Profile =Ccv] | | | | - | - | - | - | - | - |
| [Profile=S] | | | | - | - | - | - | - | - |
| Layer (ref: GGAC) | | | | - | - | - | - | - | - |
| [Layer =PA] | | | | 0.500 | 0.027 | <0.0001 | 0.407 | 0.032 | <0.0001 |
| [Layer =GGA.RMB] | | | | -0.482 | 0.067 | <0.0001 | -0.489 | 0.071 | <0.0001 |
| Hypsometry (ref: Medium Altitudes) | | | | - | - | - | - | - | - |
| [Hypsometry = Low Altitudes] | | | | 1.181 | 0.029 | <0.0001 | 1.183 | 0.029 | <0.0001 |
| Texture | | | | | | | 0.143 | 0.028 | <0.0001 |
| Random parameters | | IC 95% | | | IC 95% | | | IC 95% | |
| $\sigma_P^2$ | 0.576 | [0.510; 0.651] | | 0.332 | [0.293; 0.376] | | 0.339 | [0.300 ; 0.382] | |
| $\sigma_T^2$ | 0.061 | [0.048; 0.078] | | 0.082 | [0.069; 0.0978] | | 0.080 | [0.067 ; 0.096] | |
| $\sigma_{P,T}^2$ | -0.106 | [-0.100; -0.108] | | -0.137 | [-0.123; -0.151] | | -0,140 | [-0.126 ;-0.155] | |
| $\sigma_\varepsilon^2$ | 0.661 | [0.640; 0.682] | | 0.585 | [0.567; 0.603] | | 0.587 | [0.416 ; 0.449] | |
| $\rho$ | 0.426 | [0.408; 0.443] | | 0.431 | [0.414; 0.447] | | 0.433 | [0.569 ; 0.606] | |
| Comparison | | | | | | | | | |
| LRT | I vs.Null | 3 453.12** | | II vs. I | 3 732.62** | | III vs.II | 24.52** | |

** p-value <0,0001

For the Models PKi, all variables are statistically significant (p-value < 0.05), except for the *Precipitation* variable (in the case of models II and III) and the variable inherent to the characteristics in *Profile*. For the Models T|PKi, all variables are statistically significant (p-value < 0.05), except for the *Precipitation* and *Profile* variables. In both models, the relative humidity of the air (*RH*) was not significant.

Regarding the null model, the Portuguese National Road Administration establishes a minimum friction standard value of 0.60 GN when t = 0. So, at this time, the friction performance indicator should get a very good classification. The intercept parameter estimates $\left(\widehat{\beta_0} = 1.37 \text{ and } \widehat{\beta_0} = 1.02\right)$ for the Model PKi and the Model T|PKi, respectively, reflect it. From the results obtained for the Model PKi III (Table 5), it is observed (when the other characteristics remain constant) that, on average, the friction performance indicator increases with: *Time* ($\widehat{\beta} = 0.241$), Accumulated Annual Average Daily Traffic (*Ac.ADT*) ($\widehat{\beta} = 0.026$), which is expected considering the state of the art, and Average number of days with maximum temperature of 25 °C (*N med days temp max25*) ($\widehat{\beta} = 0.060$). Also, on average, the friction performance indicator decreases with the Higher maximum temperature ($\widehat{\beta} = -0.075$).

For the factors inherent to the highway characteristics, it can be observed that, on average, the friction performance indicator increases in the right lane sections ($\widehat{\beta} = 0.453$) and in the slow lane sections ($\widehat{\beta} = 0.169$) when compared with the sections in the left lane. Likewise, it appears that the average value of the friction performance indicator increases in the circular curve sections ($\widehat{\beta} = 0.085$) when compared with the sections in straight alignment. Regarding the type of surface course, on average, the friction performance indicator increases in the Porous Asphalt sections (*BBd*) ($\widehat{\beta} = 0.492$) and decrease in the Gap-Graded Asphalt concrete with Bitumen Modified with a high percentage of Rubber (GGA.BMR) ($\widehat{\beta} = -0.501$) when compared with Gap Graded Asphalt Concrete (*GGAC*) sections. Also, on average, the friction performance indicator increases in the *Low Altitude* sections ($\widehat{\beta} = 1.100$) when compared with *Medium Altitude* sections and increases with the increase in the *Texture* value ($\widehat{\beta} = 0.111$).

Analyzing the results obtained for the Models T|PKi is observed that the values of the estimated parameters were the same order of magnitude of the values obtained for the models with one random effect on the intercept. Because Models I, II, and III are nested, the Likelihood Ratio Test (LRT) was applied to assess the quality of the adjustment of the models. The results show a significant decrease in the log-likelihood associated with the inclusion of the variables inherent to the highway characteristics, indicating the best performance of Model II. Also, the criteria of information AIC and BIC in Model II and Model III are lower than in Model I, indicating that these models are the preferred to fit the data. Then, the comparisons between Model I and Model II confirm that the variables inherent to the highway characteristics have a significant effect on the degradation of the friction performance indicator.

The comparison of the models with a random effect on the intercept (Models PKi) with the models with two random effects (Models T|PKi) is presented in Table 7. As can be seen, Model I T|PKi is the model chosen for the evaluation of the degradation of the friction performance indicator in case the network manager does not have the data inherent to the road characteristics. Otherwise, the network manager will be able to assess the degradation of the friction performance indicator based on the Model II T| PKi and Model III T|PKi (if available with Texture data).

**Table 7.** Comparison of the models T|PKi with matrix structure AR(1).

| Models | | AIC | BIC | -2LL | Test | LRT | p-value |
|---|---|---|---|---|---|---|---|
| Model Null | PKi | 59 091.63 | 59 123.55 | 59 083.64 | | | |
| | T|PKi | 55 081.93 | 55 129.82 | 55 069.94 | Null vs. Null | 4 013.70 | |
| Model I | PKi | 51 738.11 | 51 809.94 | 51 720.12 | | | |
| | T|PKi | 51 636.81 | 51 716.62 | 51 616.82 | I vs. I | 103.30 | <0,0001 |
| Model II | PKi | 48 217.50 | 48 337.21 | 48 187.50 | | | |
| | T|PKi | 47 918.20 | 48 053.87 | 47 884.20 | II vs. II | 303.30 | <0,0001 |
| Model III | PKi | 48 203.84 | 48 331.53 | 48 171.84 | | | |
| | T|PKi | 47 895.69 | 48 039.34 | 47 859.68 | III vs. III | 312.16 | <0,0001 |

## 5 Validation

The purpose of model validation is to check the accuracy and performance of the model. A scatter plot of observed vs. fitted values was drawn to assess the predictive performance of all models considered in this study and the coefficient of determination ($R^2$) was calculated (Fig. 3). The determination coefficient ($R^2$) for Model III suggests a good goodness-of-fit between observed and fitted values, confirming that the traffic, climate conditions, pavement structure, geometric characteristics of the highway, and texture influence the degradation of friction performance indicator.

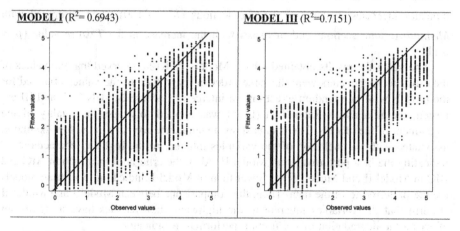

**Fig. 3.** Observed versus fitted values of friction for both models. The line is a 1:1 line.

Cross validation is a technique used to evaluate predictive models by partitioning the original sample into a training set to train the model and a test set to evaluate it. Two cross-validation methods were taken into account: the holdout method, k-fold method, and leave-one-out method (Kohavi 1995). The k-fold method is considered one of the most robust methods for estimating model accuracy, so it was adopted. Thus,

Degradation of Friction Performance Indicator     487

the original sample was randomly partitioned into k equal sized subsamples (k = 10), one sample was used to train the model, and the remaining ones were used to test it later to quantify the prediction error, as the mean squared difference between the observed, and the predicted outcome values.

To assess models' predictive performance, the following error measures were used: the Root Mean Squared Error (RMSE), the Mean Absolute Error (MAE), and the Mean Bias Error (MBE). A smaller RMSE, MAE, or MBE value indicates a more accurate prediction. For each model, Table 8 shows the values obtained for these errors and the respective standard deviation (SD). As can be seen, the values obtained for prediction errors are high. One possible explanation for these results is that the ranges of values defined for the friction parameter in the quality control plan for the highways integrating these concessions do not fully fit the ranges of values defined by the COST Action.

**Table 8.** RMSE, MAE, MBE and respective standard deviation for each model.

| | RMSE | (SD) | MAE | (SD) | MBE | (SD) |
|---|---|---|---|---|---|---|
| Model I T\|PKi | 0,8990 | (0,0017) | 0,7017 | (0,0016) | −0,0008 | (0,0006) |
| Model III T\|PKi | 0,7972 | (0,0016) | 0,6178 | (0,0017) | −0,0074 | (0,0006) |

## 6  Conclusions

This study presented linear mixed-effects models to describe the degradation of the friction parameter proposed in COST Action 354, and identified several factors affecting it.The models presenting the best fit are those developed for a two-level dataset with two random effects: one in the intercept and another in the intercept associated with the slope in the Time variable designated by PI F T\|PKi Models, when the first-order autoregressive matrix structure (AR(1)) is used, for the variance-covariance matrix of random errors.

Also, it was demonstrated that the traffic, climate conditions, pavement structure, geometric characteristics of the highway, and texture influence the degradation of friction performance indicator. The signal of the parameters estimates showed that the friction performance indicator increases with the age of pavement, Accumulated Annual Average Daily Traffic (*Ac.ADT*), Average number of days with maximum temperature of 25 °C (*N med days temp max25*), and decrease with the Higher maximum temperature (*Higher max temp*). When considering the factors inherent to the highway characteristics, on average, the friction performance indicator increases in the right lane (RL) sections and the slow lane (SL) sections when compared with the sections in the left lane (LL), and in the circular curve sections (CC) when compared with the sections in straight alignment (SL). Concerning the type of surface course, on average, the friction performance indicator increases in the Porous Asphalt sections (*BBd*) and decrease in the Gap-graded asphalt concrete with Bitumen Modified with a high percentage of Rubber Modified Binder sections (*GGA.BMR*) when compared with Gap Graded Asphalt Concrete (*GGAC*) sections. Also, on average, the friction performance indicator increases in the Low Altitude sections when compared with Medium Altitude sections and increases with the increase in the Texture value.

The methodology followed offers the network manager simple models that require only traffic and climate conditions variables and more accurate models if the network database includes factors inherent to the highway characteristics and texture. In the future, aiming at improving the accuracy of the models, a transfer function should be developed considering what is established by the quality control plan of these concessions and the flowchart of the COST Action 354, and new models applying the presented methodology should be developed.

**Acknowledgements.** This article was developed with the support of FCT – Fundação para a Ciência e Tecnologia (Foundation for Science and Technology) and withheld information belonging to the project PEst-OE/ECI/UI4047/2020. To ASCENDI that provided the resource to its highways network, for the analysis and collection of data, without which it would not have been possible to reflect and build these models.

# References

Dong, Q., Chen, X., Dong, S., Zhang, J.: Classification of pavement climatic regions through unsupervised and supervised machine learnings. J. Infrastruct. Preserv. Resilience **2**(1), 1–15 (2021). https://doi.org/10.1186/s43065-021-00020-7

Erlingsson, S., Saba, R., Hoff, I., Huvstig, A.: NordFoU – pavement performance models. Part 2 (2008). https://doi.org/10.1201/9780203885949.ch22

Haas, R.: Good technical foundations are essential for successful pavement management, key note paper. In: Proceedings of MAIREPAV 2003, Guimaraes, Portugal (2003)

IPMA. IPMA - Instituto Português do Mar e da Atmosfera (2019). Accessed Jan 2019

Kogbara, R.B., Masad, E.A., Kassem, E., Scarpas, A.T., Anupam, K.: A state-of-art review of parameters influencing measurement and modelling of skid resistance of asphalt pavements. Constr. Build. Mater. **114**, 602–617 (2016)

Kohavi, R.: A study of cross-validation and bootstrap for accuracy estimation and model selection. IJCAI **14**, 1137–1145 (1995)

Li, Q.J., Zhan, Y., Yang, G., Wang, K.C.P., Wang, C.: Panel data analysis of surface skid resistance for various pavement preventive maintenance treatments using long term pavement performance (LTPP) data. Can. J. Civil Eng. **44**(5), 358–366 (2017). https://doi.org/10.1139/cjce-2016-0540

Litzka, J., Leben, B., La Torre, F., Weninger-Vycudil, A., et al.: The Way Forward for Pavement Performance Indicators across Europe 2008 | Action 354. FSV (2008)

Lorino, T., Lepert, P., Marion, J.-M., Khraibani, H.: Modeling the road degradation process: non-linear mixed effects models for correlation and heteroscedasticity of pavement longitudinal data. Procedia – Soc. Behav. Sci. **48**, 21–29 (2012). https://doi.org/10.1016/j.sbspro.2012.06.984

Marcelino, P., Antunes, M.L., Fortunato, E., Gomes, M.C.: Machine learning for pavement friction prediction using scikit-learn. In: Oliveira, E., Gama, J., Vale, Z., Cardoso, H.L. (eds.) EPIA 2017. LNCS (LNAI), vol. 10423, pp. 331–342. Springer, Cham (2017). https://doi.org/10.1007/978-3-319-65340-2_28

Moreira, A.V., Tinoco, J., Oliveira, J.R.M., Santos, A.: An application of Markov chains to predict the evolution of performance indicators based on pavement historical data. Int. J. Pavement Eng. **19**(10), 937–948 (2018). https://doi.org/10.1080/10298436.2016.1224412

Portal do Clima (2015). Accessed Jan 2019. http://portaldoclima.pt/en/

Santos, A., Freitas, E.F., Faria, S., Oliveira, J.R.M., Rocha, A.M.A.C.: Prediction of friction degradation in highways with linear mixed models. Coatings **11**, 187 (2021)

Sol-Sánchez, M., Moreno-Navarro, F., García-Travé, G., Rubio-Gámez, M.C.: Laboratory study of the long-term climatic deterioration of asphalt mixtures. Constr. Build. Mater. **88**, 32–40 (2015)

Viner, H., Sinhal, R., Parry, T.: Review of UK skid resistance policy. Preprint SURF 2004 (2004)

Yu, J., Chou, E.Y., Luo, Z.: Development of linear mixed effects models for predicting individual pavement conditions. J. Transp. Eng. **133**(6), 347–354 (2007). https://doi.org/10.1061/(ASCE)0733-947X(2007)133:6(347)

# Road Safety Risk Diagnosis

# Application of an Innovative Network Wide Road Safety Assessment Procedure Based on Human Factors

Andrea Paliotto[(✉)], Monica Meocci, and Valentina Branzi

Department of Civil and Environmental Engineering, University of Florence, Florence, Italy
{andrea.paliotto,monica.meocci,
valentina.branzi}@unifi.it

**Abstract.** Often the identification of critical road segments relies on accidents as primary data. Nevertheless, accidents data are sometimes prone to errors. Moreover, relying on accidents, means waiting for them to occur before intervening. A modern road administration must intervene before accidents occur to save lives. Furthermore, it is nowadays proven that drivers' behaviour is the leading accident influencing factor. Thus, it is crucial to deepen the analysis of the road-driver interaction by means of application of human factors' principles to road safety. This means to judge how much the road is perceived by the driver and influences their behaviour. This helps to identify which are the triggering factors of accidents. This paper proposes a new analysis procedure which allows to make a network safety assessment and screening considering the human factors aspects. The process relies on road visual surveys and evaluations which allows to make a more detailed and proactive analysis of the road. Furthermore, the data required to implement the analysis are few and this makes it suitable to use also when data are hardly available (e.g., low- and middle-income countries). The paper presents the application of the procedure to three different road stretches of a total length of about 38 kms: one in Italy, one in Germany and one in Slovenia. The application of the procedure demonstrates that it can be easily applied by Human Factors trained inspector and that it may provide a ranking of the network suitable to identify the most critical section of the network.

**Keywords:** Road safety · Human factors · Network safety assessment · Proactive approach · Low- and middle-income countries · Road safety inspection

## 1 Introduction

The process to analyse the safety of the road network, is the so-called Network Safety Assessment (NSA). This process assumes different names within different of road administrations (RAs), however the objective of the process is always the same: it is "the process of identifying sites for further investigation and potential treatment" (FHWA 2016), or "to identify sections of the network that should be targeted by more detailed road safety inspections and to prioritise investment according to its potential to

---

© The Author(s), under exclusive license to Springer Nature Switzerland AG 2022
A. Akhnoukh et al. (Eds.): IRF 2021, SUCI, pp. 493–510, 2022.
https://doi.org/10.1007/978-3-030-79801-7_36

deliver network-wide safety improvements" (European Parliament; Council of European Union 2019). The identification of the riskiest sections of a road and the classification of all the sections belonging to the road network, allows to prioritize the interventions, choosing the site which has the highest impact on road safety. Nowadays, NSA procedures mainly rely on accidents data, both considering "standards" indices such as the accidents frequency, the accidents density, the accidents rate (PIARC 2013), and the safety potential (SAPO) (Kathmann et al. 2016), and also considering more advanced procedures, like the use of Accident Prediction Models (APMs) and Empirical-Bayes procedures. APMs generally account for the influence of road physical characteristics on accident occurrence and are developed analysing the historical accident trends occurring on similar sites using statistical procedures (Hauer 2015). Trought the application of the Empirical-Bayes procedures, the use of APMs reduces the possible errors related to accident data, such as the regression to the mean and the non-linearity with traffic. The use of APMs has proven to be a consistent method for making reliable crash frequency predictions, and thus improve a reliable NSA, but the implementation of an APM requires a large amount of data. Those data are not always available, mainly in low- and middle-income countries. Moreover, sometimes it is necessary to adapt existent AMPs to local conditions (La Torre et al. 2016; La Torre et al. 2019), and very often APMs do not account for all the possible factors influencing an accident occurrence, therefore they still rely to a great extent on accident data (Empirical-Bayes procedures).

Accident data are an important source of information to improve road safety analysis, but unfortunately, they are sometimes prone to errors or even just missing. These conditions are present in high-income countries, but especially in low- and middle-income countries. Besides, as highlighted by the WHO accidents' statistics (WHO 2019), low- and middle-income countries are the countries where most efforts should be taken. Those limitations on accidents data and databases highlight the need of some new approaches, which may allow to analyse the safety level of a network, without the need of a wide accident database. An answer to this demand comes from the development of approaches based on road inspections, which must evaluate the road safety without considering the number of accidents. This type of approach is called "proactive approach" because it is based on the road in-built safety analysis. Proactive approaches should be the focus of RAs to avoid as much as possible any risk of accidents before they occurred.

One of the most used proactive approach procedures, is provided by the Road Safety Inspections (RSIs). At first, RSIs were mainly used as an instrument to carry out specific analysis on sites identified as "high risk sections" from an NSA. However, nowadays RSIs are widely used in a systematic and periodic process to screening the network. The results of an inspection provide a risk analysis, which can be used to classify different road sections. One RSI procedure, which has widely spread all over the world during the last years, is the iRAP Star Rating procedure (iRAP 2021). The procedure relies on visual inspections that can be carried out both manually by inspectors or automatically by equipped vehicles. The procedure requires that the main road characteristics are considered, both geometrical and functional. The influence of these aspects is considered and a Star Rating is obtained for the analysed segment. This allows to implement a network safety ranking. Another interesting proactive approach to a NSA procedure, was proposed by Cafiso et al. (2007). This approach relies on RSI and provide a safety index suitable for the ranking of the analysed sections. Moreover, another research from Cafiso et al. (2006), along with the iRAP procedure, demonstrated that RSI which are carried out following a systematic and structured procedure, produce the same results, even if carried out by different inspectors. These considerations suggested that RSIs can be chosen as an instrument to carry out NSAs, even more in those countries where expensive equipment and instruments, road data, and accidents data, are not easily available.

This paper illustrates the application of an innovative RSI procedure, whose scope is the implementation of an NSA and thus, the definition of a network safety ranking. The proposed procedure, namely Human Factors Evaluation (HFE), relies on the Human Factors principles. Human Factors are defined by PIARC as "Those psychological and physiological threshold limit values which are verified as contributing to operational mistakes in machine and vehicle handling. In the case of road safety, the Human Factors concept considers road characteristics that influence a driver's right or wrong driving actions" (PIARC 2016). Analysing road safety by means of Human Factors, means to identify the accidents triggering factors. These factors are related to the driver's comprehension of the road, considering his human capabilities and limitations (Čičković 2016; Birth et al. 2015).

Due to its nature and composition, this HFE procedure can be carried out both as a stand-alone procedure or along with other RSI procedures, which consider other road safety aspects. Human Factors aspects are only partially considered in "standard" RSI and without a systematic approach. Comparing to such RSI procedures, the HFE procedure provides a structured and systematic approach focused on Human Factors. Together with the description of the procedure, this paper describes the application of the procedure to three case studies. The case studies analysed are three rural highway stretches, one from Italy (SR2), one from Germany (B38), and one from Slovenia (106), with a total length of about 38 km.

## 2 Methodology

The proposed RSI procedure relies on road surveys and on the application of the Human Factors Evaluation Tool proposed by PIARC (PIARC 2019). The procedure is a three-steps procedure that follows a top-down process which aids at identify specific issues, and a bottom-up process required to provide suitable results for a network analysis.

Figure 1 shows the conceptual scheme of the procedure, identifying the three steps to follow.

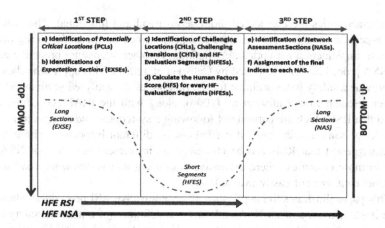

Fig. 1. Conceptual scheme of the Human Factors Evaluation procedure

The first step of the procedure represents a first global analysis of the road. In this step the Potentially Critical Locations (PCLs) must be identified. PCLs are those locations that require a change in the driving programme. Curves, at-grade intersections, crossings (cyclist, pedestrian, or railway crossings), driveways and accesses, points where the cross section significantly change (adding or removing a lane), points where the road function change (e.g., from rural environment to urban environment), stopping areas such as bus stops and lay-by, are all defined as PCLs. Even if RAs have no database about these locations, they can be identified by driving along the road. Table 1 shows the PCLs considered in the procedure. During the first step the Expectations Sections (EXSEs) must also be defined. EXSEs are road sections characterized by the same expectations for the driver, which means that the driver expects to find some specific road elements in the section and to behave in a specific manner (according to self-explaining roads concept) (Theeuwes 2021). More precisely, driving performances are related to driver expectations based on what drivers perceive from the road and its environment, what they have assimilated during their driving, and what they face in front of them. EXSEs are defined based on the section main characteristics, which are the road category, the road winding, and the road perception of possible interaction (PPI).

Application of an Innovative Network Wide Road Safety Assessment Procedure    497

**Table 1.** Summary of the considered Potentially Critical Locations (PCLs)

| PCL type | PCL | Characteristics |
|---|---|---|
| Curve | Curve0 | Curve requiring a small change or no change in the speed |
| | Curve10 | Curve requiring a medium change in the speed (around 10 km/h) |
| | Curve20 | Curve requiring a high change in the speed (around 20 km/h) |
| At-grade intersection | Roundabout | Junction with rotatory circulation, where drivers from each direction must yield to the car in the junction |
| | Signalised | Junction organized with traffic lights |
| | With priority | Junction where the drivers travelling on the analysed road do not have to stop or even yield at the junction |
| | Without priority | Junction where the drivers travelling on the analysed road must stop or yield at the junction (excluding roundabouts) |
| Crossing | Pedestrian Crossing | Road area where pedestrians cross the road |
| | Cyclist Crossing | Road area where cyclists cross the road |
| Driveway | Minor residential | Access to one or few houses |
| | Major residential | Access to a large group of houses |
| | Minor commercial | Access to little or few commercial, industrial, or agricultural activities |
| | Major commercial | Access to big or many commercial, industrial, or agricultural activities |
| Stopping area | Lay-by | Areas along the road that allow for a car to stop (for emergency or not) |
| | Bus stop | Reserved lay-by or simply vertical signs and marking along the road |
| | Parking lot | Parking lots along the road |
| Railway level crossing (LC) | With mobile bar | Railway level crossing intersection with vertical signs, markings, and bars |
| | Without mobile bar | Railway level crossing intersection with only vertical signs and markings |
| Lane change | Added/removed lane | Added/removed lane that changes the cross-section configuration (it includes speed change lanes) |
| | Diverging lane | Lane splits into two (or more) lanes, which follow different directions |

The road category includes motorway, rural highway, rural local, urban arterial, urban connectors, and urban local. This paper focus on the application of the procedure only to rural highways.

The road winding is defined based on the value of the curvature change rate (CCR), which is defined by Eq. (1). The CCR calculation requires road geometry data, however with the current GIS software and satellite images, the geometry of the road can be often easily defined, even if geometrical data are not available. The possible error made is negligible.

$$CCR = \frac{\sum_{i=1}^{n} |\gamma_i|}{\sum_{i=1}^{n} L_i} \qquad (1)$$

where:

CCR = curvature change rate [gon/km];
$\gamma_i$ = angular change of the geometric element "i" [gon];
$L_i$ = length of the geometric element "i" [km];
n = number of geometric elements of the road section (tangents, circular curves, spirals) [-];

Three levels of road windings have been identified based on the value of the CCR. These values can be applied to rural highways.

- High winding: CCR > 350 gon/km
- Medium winding: 350 gon/km > CCR > 160 gon/km
- Low winding: CCR < 160 gon/km

Concerning the definition of the PPI, an evaluation should be made by the inspectors. Two levels of PPI are defined for rural highways: low and medium. The "high" PPI level can only be present in urban area.

- Low level: only rural area can be considered at low level. Low level means very few or no perceived possible interactions. The surrounding environment is almost totally natural, without any trace of anthropization, if not the road itself.
- Medium level: this level represents the upper level of rural area. Such PPI is often representative of suburban area, where the density of houses and commercial activities is reduced. Medium level can also address rural road stretches which pass through small village or groups of houses along the road, or also which pass through an area with many driveways and at-grade intersections, due to the presence of many activities and factories.

Table 2 presents some examples of medium and low level of PPI.

**Table 2.** Examples of Low PPI and Medium PPI, photos by Google Street View

Application of an Innovative Network Wide Road Safety Assessment Procedure 499

A road survey is not essential to complete Step 1 of the procedure, nevertheless, is highly recommended. It consists of driving along the analysed road, providing a clearer view of the road features and its environment. This will allow the inspector to easily and reliably define the EXSE. A video should be taken during the process and the inspector could explicit some comments while driving that are recorded by the camera. This is a quick survey so is it possible to make this kind of survey on an extended part of the network within the same day. Once the EXSEs have been defined by the road category, the road winding, and the road PPI, it is possible to move to the second step.

The second step is the core of the evaluation. To calculate a numerical risk value, the PIARC Human Factors Evaluation Tool (HFET) has been used (PIARC 2019). The HFET structure allows to evaluate a single potentially critical locations (PCL) within a road segment, thus it should be applied to all the PCLs. To speed up the procedure without a waste of time also judging PCLs that clearly do not present any issue, a criterion to identify the riskiest PCLs has been defined. That is the reason for the introduction of the first step. Comparing the PCLs present along the EXSE to the characteristics of the EXSE itself, it is possible to define the Challenging Locations (CHLs). CHLs are unexpected PCLs. For example, a curve with a low radius that requires a medium speed (e.g., 60 km/h) to be travelled under safe conditions, it is unexpected if the EXSE is characterized by low winding and low PPI (and thus a high speed). Instead, the same curve will be expected if the EXSE is characterized by high winding and low PPI. The relationship between different type of PCLs, the level of expectancy, and the operating speed, has been investigated considering a review of design standards of several countries (Italy, Germany, England, Slovenia, Portugal, Australia, Canada, Austria, Switzerland). Following the results of the process and their own experience, the authors propose the criteria shown in Table 3 to determine if a PCL is not expected, based on the road characteristics. In the same table the expected operating speed range is also provided (it derives from the above-mentioned analysis of different design standards). The results presented in Table 3, can be used as a reference, but the inspector can judge independently about the risk of a specific PCL, considering the characteristics of the EXSE and the specific characteristic of the PCL itself. The identification of EXSEs is a crucial step because it clarifies to the inspector the main characteristics of the road stretch. Furthermore, inspectors must consider not only the expectation about a specific PCL (derived from the EXSE), but also its visibility and clarity. This means that the inspector can include in the group of CHLs, those PCLs which have some problems concerning the visibility or the field of view (first and second rule of Human Factors). The visibility can be primarily evaluated considering a minimum available distance equal to 6 s (which corresponds to the sum of the time required for the response and the anticipation sections of the first rule of Human Factors) (PIARC 2016). This value accounts for the presence of a Decision Sight Distance (DSD). If such distance is not present, then some visibility issues might occur. If the operating speed data are available, it is possible to use the data to calculate such distance, otherwise the speed value can be defined based on the EXSE characteristics (see Table 3). The field of view analysis requires to analyse how the road appear to the driver. Thus, a road survey is required or, at least, the analysis of digital images (e.g., Street View) and photos of the PCL. To account for the configuration of the field of view and the visibility, a PCL must be judged considering both the directions, and

500 A. Paliotto et al.

therefore it can be a CHL considering one direction, but not considering the other. This is a first quick qualitative screening the inspector must carry out before applying the HFET. To clearly identify CHLs, inspectors should be aware of the Human Factors principles (PIARC 2016).

**Table 3.** PCL expected (-) or not (x), and expected operating speed range, based on the rural highway winding and PPI

| PCLs | Combination of winding / PPI level (H = high, M = medium, L = low) | | | | | |
|---|---|---|---|---|---|---|
| | L/L | L/M | M/L | M/M | H/L | H/M |
| Curve0 | - | - | - | - | - | - |
| Curve10 | x | x | - | - | - | - |
| Curve20 | x | x | x | x | x | - |
| Roundabout | x | - | x | - | x | - |
| Signalized intersection | x | x | x | x | x | x |
| Intersection with priority | - | - | - | - | - | - |
| Intersection without priority | x | x | x | x | x | x |
| Minor residential | - | - | - | - | - | - |
| Major residential | x | - | x | x | x | x |
| Minor commercial | - | - | - | - | - | - |
| Major commercial | x | x | x | x | x | x |
| Added lane | x | x | x | x | x | x |
| Removed lane | x | x | x | x | x | x |
| Diverging lane | x | x | x | x | x | x |
| Parking lot | x | x | x | x | x | x |
| Lay-by | - | - | - | - | x | x |
| Bus stop | - | - | - | - | - | - |
| LC with mobile bar | x | x | x | x | x | x |
| LC without mobile bar | x | x | x | x | x | x |
| Pedestrian crossing | x | - | x | - | x | x |
| Cyclist crossing | x | x | x | x | x | x |
| Expected operating speed range [km/h] | 70–100 | 70–100 | 70–100 | 50–80 | 50–80 | 50–80 |

Once the CHLs have been identified, the area to evaluate with the HFET must be chosen. This area is the road stretch preceding and including the CHL and it is called Challenging Transition (CHT). CHT started typically 6–10 s before the CHL (PIARC 2019) and can include other elements of the road that are not CHLs, or even other CHLs. If the latter is the case, thus the two overlapping CHTs (one for each CHL) must be merged, creating a single CHT. This final CHT is then called Human Factors Evaluation Segment (HFES) because this is the segment that will be evaluated with the

HFET. HFESs can be different considering the two directions of travel. The HFET is then applied to each HFES.

The HFET proposed by PIARC (2019) has been adjusted to be suitable for the application within the procedure. The checklists are divided by human factors rules (first level), investigation topic (second level), subsections (third level) and requirements (fourth level). Table 4 shows an extract of the investigation topic 2.1 "Density and shape of the field of view" of the second rule. This investigation topic comprises two subsections (2.1.1 and 2.1.2). The subsections are each composed by three requirements (a, b, and c).

Both for each CHL and for each HFES the checklists must be fulfilled and the Human Factors Score (HFS) must be calculated, following the criteria proposed by PIARC (2019). The CHLs with the worst HFS for each subsection is then selected to calculate the result for each Human Factors rule and considering all the rules together (Total). As presented in PIARC (2019), the HFS is a percentage value which identifies three level of risk: $HFS < 0.40$ means high risk of accidents, $0.40 \leq HFS \leq 0.60$ means medium risk, and $HFS > 0.60$ means low risk. An example of calculation of the HFS is provided in Table 4. The two CHL analysed are a pedestrian crossing and a curve ("curve10", see Table 1). Concerning the first subsection (2.1.1), the worst result of the HFS belongs to the curve, while in the second subsection (2.1.2) the worst result of the HFS belongs to the crossing. The rating of the worst CHL for each subsection will be select and used to calculate the final value of the HFS for each rule and for the Total (e.g., the "worst" column of the example will be used).

At the end of step 2, each HFES for each direction have been judged and four HFSs have been assigned to each HFES: one for each of three rules, and one considering all the rules together (Total).

The third step concerns the aggregation of the results into longer sections, which must be suitable for an NSA. The criteria to define those sections must be chosen by RAs. Too short sections (HFES) provide a detailed evaluation of a limited road stretch. The sectioning can be made according to the criteria of a fixed length, or based on the section characteristics (e.g., traffic volumes). This paper proposes to consider sections of a fixed length of about 1 km. Those sections are called Network Assessment Sections (NASs). A NAS can comprise a variable number of HFESs. HFESs, must be wholly included within the same NAS, thus the length of NAS can slightly varies around 1 km. First, for each NAS the worst HFS result must be identified for each rule and for the Total, considering all the HFES belonging to the NAS (considering both the direction together). Table 5 shows the results of four HFES all belonging to the same NAS (one considering the increasing km posts direction, three considering the decreasing km posts direction). The colours red, yellow, and green represent respectively the high, medium, and low risk level. The HFS to be assigned to this NAS (NAS7) will be 0.17 for the First Rule, 0.55 for the Second Rule, 0.62 for the Third Rule, and 0.52 for the Total, which are the lowest results of each column.

**Table 4.** Example of calculation of the HFS

| 2.1 | **Density and shape of the field of view** | **Crossing** | **Curve** | **Worst** |
|---|---|---|---|---|
| *2.1.1* | *Monotony of road section and surroundings* | | | |
| a | Diversified planting or buildings (variation in height, distance, …) | 1 | 1 | 1 |
| b | Good light contrast and good colour contrast in the environment | 1 | 0 | 0 |
| c | Road environment well structured | 0 | 0 | 0 |
| | *Total 2.1.1* | *67%* | *33%* | *33%* |
| *2.1.2* | *Long/far visible approaching sections before CHL* | **Crossing** | **Curve** | |
| a | Road environment well structured | 0 | 1 | 0 |
| b | Road environment composed by a sinuous "rhythmic" alignment | 1 | 1 | 1 |
| c | Stretches are avoided where a "wider or longer" space is perceived | 1 | 1 | 1 |
| | *Total 2.1.2* | *67%* | *100%* | *67%* |
| | **Total 2.1** | *67%* | *67%* | *50%* |

**Table 5.** Example of a series of HFES belonging to the same NAS from SR2.

| HFES | Starts [km] | Ends [km] | Length [km] | I Rule | II Rule | III Rule | Total |
|---|---|---|---|---|---|---|---|
| INC 14 | 288.550 | 288.800 | 0.250 | 0.46 | 0.55 | 0.64 | 0.55 |
| DEC 11 | 288.400 | 288.600 | 0.200 | 0.17 | 0.60 | 0.62 | 0.52 |
| DEC 12 | 288.750 | 288.950 | 0.200 | 0.33 | 0.69 | 0.67 | 0.61 |
| NAS 7 | 288.300 | 289.200 | 0.900 | 0.17 | 0.55 | 0.62 | 0.52 |

Finally, a code will be assigned to each of the NAS, based on the obtained HFS. With this code, the ranking will be made. The code is composed by five terms: AB – C – D/E. The meanings of those terms are:

- A = the worst level of NAS's HFS, including all the Rules and the Total (R = red, at least one score < 0.40 within the rules, Y = yellow, no red scores and at least one score < 0.60 within the rules, G = green, all other results);
- B = the number of results of level "A" (see before) between the NAS's HFS, considering the worst results of all the Rules and the Total (min = 0, max = 4);
- C = the NAS's HFS Total result;
- D = weighted average expressed in percentage of the Total HFSs of all the HFESs belonging to the NAS, calculated following Eq. (2);
- E = standard deviations of Total results expressed in percentage of the HFES Total within the NAS, calculated following Eq. (4);

$$\mu_j = \sum_{i=1}^{N_j} p_i x_i \qquad (2)$$

# Application of an Innovative Network Wide Road Safety Assessment Procedure    503

where:

$\mu_j$ = weighted average of the Total HFSs of all the HFEs belonging to the NAS "j" [-];
$N_j$ = total number of HFES belonging to NAS "j" [-];
$p_{ij}$ = frequency of the element "i", which is calculated following Eq. (3) [-];
$x_i$ = Total HFS of the HFES "i" [-].

$$p_{ij} = \frac{l_i}{\sum_{i=1}^{N_j} l_i} \tag{3}$$

where:

$l_{ij}$ = the length of the HFES "i", belonging to the NAS "j" [km].

$$\sigma_j = \sqrt{\sum_{i=1}^{N_j} p_{ij}(x_i - \mu_j)^2} \tag{4}$$

where:

$\sigma_j$ = standard deviation of NAS "j" [-].

For the calculation of the weighted average value and the standard deviation, all the road segments belonging to the NAS must be considered, even if they are not HFES. For this reason, a total HFS of 100% was assigned to these segments, considering both the directions. This value represents a stretch with a low risk of accidents. Considering the example in Table 5, NAS 7 is long 0.9 km. If both the directions are considered, the length can be assumed as 1.800 km (doubled). Within this, 0.250 + 0.200 + 0.200 + 0.250 = 0.900 km are HFES. The remaining 0.900 will be classified with a score of 100%. The weighted average value and the standard deviation will be then calculated. The code assigned to the NAS 7 will be R1-52-79/21.

The ranking of the NASs will follow a two-level ranking. The first ranking identifies the risk level of the NAS, and thus it is based into four groups:

- very high risk: "AB" part of the code is equal to "R4";
- high risk: "AB" part of the code is equal to "R2" or "R3";
- medium risk: "AB" part of the code is equal to "R1", or "Y4";
- low risk: all the remaining cases.

These four risk levels are consistent with the number of different risk level required from the new European Directive (European Parliament; Council of European Union 2019).

The second level ranking considers instead a ranking within the same risk level, that is a classification of the NAS belonging to the same group (very high risk, high risk, medium risk and low risk). This second ranking is made following the criteria presented in Table 6. It must be considered each parameter composing the code, following the order they are presented (AB-C-D/E). This means that considering the

504     A. Paliotto et al.

same results of part "A" of the code, priority will be given to the NAS which has a higher part "B". If equal, then the one with lowest value of part "C" will have the priority, then the one with the lowest part "D", and if all the previous parameters are the same, the priority will be given to the NAS with the higher part "E" value. Generally, part D and E of the code are not necessary for the purpose of the ranking, however they provide important information about the composition of the NAS, which can help the RA in the following stage of intervention. At the end of step 3, the ranking is obtained of all the NASs belonging to the road network.

**Table 6.** Ranking criteria within the same risk level group.

| Priority | A | B | C | D | E |
|---|---|---|---|---|---|
| Higher | R | High value | Low value | Low value | High dispersion |
| Lower | G | Low value | High value | High value | Low dispersion |

## 3   Inspectors Training and Site Inspections

Inspectors should be qualified through participation in training courses about the HF principles. It is also suggested that a trained but inexperienced inspector, cooperates with at least one experienced inspector in the first two/three inspections it will carry out. A site inspection procedure must be carried out during the Step 2 (and is recommended also during step 1). During the inspection, a camera mounted on the windshield, should be used to record the road while driving, and to record the comments from the inspection team, then the video will be post-processed in office and the HFET applied (PIARC 2019). The inspection will follow the standard procedure for road safety inspections, with both a daylight and night survey. This allows to easily include the use of the HFET during a standard road safety inspection procedure (PIARC 2012). The procedure presented in this paper can also be carried out by an inspector, nevertheless it is suggested that the inspection is carried out by a team of three inspectors.

## 4   Applications to Case Studies and Results

The procedure was applied to three rural highway stretches. It has been decided to include in the analysis also short urban area segments located along the rural roads (with an extension less than 500 m). The three roads selected are: SR2, in Italy, B38 in Germany, and 106 in Slovenia. Although the three roads are all the same type, they present some peculiarities. The SR2 stretch develops through a hilly terrain, where it is characterized by many curves of small radius. Then it runs along a river valley, maintaining a low curvature ratio. In the last part, it crosses some urban areas. Many driveways are present. The stretch ranges between km post 280.6 and 292.4 (11.8 km). The B38 stretch develops in an almost plain terrain with few mayor grades. It presents many curves with high radius, many signalised intersections, and few driveways. It is about 10.8 km long and extends from km post 0.6 starting from 6118–001 to 6118–036. The 106 stretch develops entirely in a hilly terrain. It presents many curves of different radius, nevertheless it is a fast stretch. It passes through some minor villages, and it presents many curves and some at-grade intersections. It is approximately 16 km long and it coincides with the section 261. Figure 2 shows the calculated CCR of the analysed roads. The SR2 demonstrates to be the road with higher CCR. B38 and 106 have a similar CCR in their first half, but in the second half 106 has a medium CCR, while B38 continues to have a low CCR.

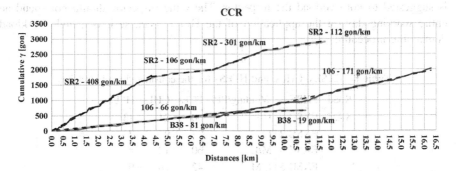

**Fig. 2.** CCR section of the analysed roads

The PPI has been then evaluated and EXSEs for each road stretch have been identified. Each PCL has been identified too. Long urban segments (more than 500 m) have been excluded from the analysis. Table 7 shows the statistics about the PCLs belonging to the analysed stretches of each road. SR2 has the higher density of each type of PCLs.

**Table 7.** Statistics of PCLs for each roads (n = number of PCLs, n/km = density of PCLs).

| Road | Length | Curves | | At-grade intersections | | Crossings | | Driveways | | Stopping areas | | Lane changes | | Total PCLs | |
|---|---|---|---|---|---|---|---|---|---|---|---|---|---|---|---|
| | km | n | n/km | n | n/km | n | n/km | n | n/km | n | n/km | n | n/km | n | n/km |
| SR2 | 10.85 | 60 | 5.5 | 20 | 1.8 | 15 | 1.4 | 54 | 5.0 | 17 | 1.6 | 3 | 0.3 | 169 | 15.6 |
| B38 | 9.9 | 15 | 1.5 | 15 | 1.5 | 0 | 0.0 | 13 | 1.3 | 0 | 0.0 | 0 | 0.0 | 43 | 4.3 |
| 106 | 16 | 60 | 3.8 | 14 | 0.9 | 2 | 0.1 | 39 | 2.4 | 12 | 0.8 | 2 | 0.1 | 129 | 8.1 |

The PCLs characterized by the major deficiencies in terms of expectations, quality of the field of view and visibility, have been promoted to CHLs. CHTs have been defined for all the CHLs and the ones which overlapped, have been joined together. Consequently, the HFESs have been identified. Table 8 provides a summary of the results obtained from the previous steps. The results highlight the reduction of the number of PCLs that must be analysed (CHLs) compared to the number of PCLs present along the roads. The first screening process which led from the total number of PCLs to a reduced number of CHLs required some hours for each road stretch, but it is an extremely reduced time compared to the time required to apply the HFET to all those PCLs. Concerning the results on these three road stretches, the percentages of PCLs that must be analysed (i.e., CHLs) are 30% for SR2, 53% for B38, and 37% for 106.

The average number of HFES that must be evaluated for each road km is 3.4 for the SR2, 1.9 for B38, and 2.1 for 106. The average time for an experienced inspector to apply the HFET is about 20/25 min for each HFES. It can be estimated that about 8 km can be analysed by a single inspector in one working day and thus about 40 km of the road can be analysed in a week by one inspector. If the inspector has no experience in the application of the HFET, then the estimated time should be doubled. Furthermore, it is suggested to not overload the inspector. The same inspector should not spend an entire working day on the application of the HFET, because it is a high workload demanding procedure.

**Table 8.** EXSEs, CHLs, and HFESs for the considered road stretches.

| Road | EXSE | Winding/PPI | PCLs | CHLs | HFESs |
|---|---|---|---|---|---|
| SR2* | EXSE 1 | H/L | 60 | 21 | 13 |
| | EXSE 2 | L/L | 27 | 9 | 7 |
| | EXSE 3 | M/L | 60 | 12 | 8 |
| | EXSE 5 | L/M | 42 | 10 | 9 |
| | **Total** | | **169** | **50** | **37** |
| B38* | EXSE 2 | L/L | 43 | 23 | 19 |
| | **Total** | | **43** | **23** | **19** |
| 106 | EXSE 1 | L/L | 42 | 13 | 11 |
| | EXSE 2 | L/M | 14 | 5 | 3 |
| | EXSE 3 | M/L | 55 | 22 | 15 |
| | EXSE 4 | M/M | 18 | 8 | 5 |
| | **Total** | | **129** | **48** | **34** |

*SR2 EXSE 4 and B38 EXSE 1 have been excluded because they were urban sections

The results of the application of the HFET to all the HFESs, are the HFSs for each rule and for the Total, as presented in the previous example of Table 5. Once all the HFESs have been evaluated for each of the analysed road, the NASs have been identified and each code calculated. The NAS length has been fixed to 1 km. The length may slightly vary because an entire HFES must be included in the same NAS. The results of the application of the procedure are shown in Table 9. The last two columns of the table show respectively the level of risk assigned to that NAS and the ranking of the NAS within the same road.

# Application of an Innovative Network Wide Road Safety Assessment Procedure   507

**Table 9.** Outputs of the NAS evaluation and relative rankings

| Road | NAS ID | Length [km] | Code | Risk Level | Ranking |
|---|---|---|---|---|---|
| SR2* | NAS 1 | 1.200 | R1-63-95/13 | Medium | 9 |
| | NAS 2 | 1.950 | R4-34-54/24 | Very High | 1 |
| | NAS 3 | 0.950 | R1-44-66/24 | Medium | 5 |
| | NAS 4 | 1.150 | R1-40-72/26 | Medium | 3 |
| | NAS 5 | 1.000 | R1-42-74/29 | Medium | 4 |
| | NAS 6 | 1.250 | R1-46-72/22 | Medium | 7 |
| | NAS 7 | 0.900 | R1-52-79/21 | Medium | 8 |
| | NAS 9 | 1.000 | R1-45-80/21 | Medium | 6 |
| | NAS 10 | 1.250 | R3-41-75/28 | High | 2 |
| B38* | NAS 2 | 1.300 | R1-45-80/24 | Medium | 5 |
| | NAS 3 | 1.200 | Y3-58-75/19 | Low | 9 |
| | NAS 4 | 1.000 | R1-64-92/14 | Medium | 7 |
| | NAS 5 | 1.000 | R1-45-74/26 | Medium | 4 |
| | NAS 6 | 1.200 | R1-40-68/25 | Medium | 2 |
| | NAS 7 | 1.200 | R3-38-71/28 | High | 1 |
| | NAS 8 | 1.100 | R1-42-77/26 | Medium | 3 |
| | NAS 9 | 0.700 | G4-100-100/00 | Low | 10 |
| | NAS 10 | 1.190 | R1-51-77/24 | Medium | 6 |
| 106 | NAS 1 | 1.000 | R1-47-82/22 | Medium | 10 |
| | NAS 2 | 1.300 | R2-44-60/23 | High | 5 |
| | NAS 3 | 1.300 | Y4-45-74/26 | Medium | 12 |
| | NAS 4 | 1.200 | Y3-53-93/17 | Low | 14 |
| | NAS 5 | 0.700 | Y1-61-94/14 | Low | 15 |
| | NAS 6 | 1.100 | R2-43-54/20 | High | 4 |
| | NAS 7 | 1.000 | R2-41-80/26 | High | 2 |
| | NAS 8 | 1.100 | R1-47-87/22 | Medium | 11 |
| | NAS 9 | 1.000 | R1-43-70/28 | Medium | 7 |
| | NAS 10 | 1.100 | R2-42-72/28 | High | 3 |
| | NAS 11 | 0.800 | Y3-52-73/22 | Low | 13 |
| | NAS 12 | 1.100 | R1-43-71/27 | Medium | 8 |
| | NAS 13 | 1.200 | R1-41-73/27 | Medium | 6 |
| | NAS 14 | 1.200 | R2-35-72/30 | High | 1 |
| | NAS 15 | 0.900 | R1-43-74/28 | Medium | 9 |

*SR2 NAS 8 and B38 NAS 1 have been excluded because they are urban sections

As expected, the results provide a high number of medium-risks HFES, with some high-risk section, and very few low-risk sections. This is because the low-risk PCLs have been discarded in the first screening process. A total of 35 NASs have been obtained for a total road length of about 38 km. Within the 35 NASs, one from the SR2 and one from B38 have been excluded because they develop entirely within an urban area longer than 500 m. Within the analysed 33 NASs, 5 have been classified as low-risk level, 20 as medium-risk level, 7 as high-risk level, and 1 as very high-risk level.

The SR2 stretch shows overall a higher risk. No low-risk NASs are present, and some of the medium-risk NASs provide a very low result for the Total value of the HFS (central part of the code). In this stretch a very high-risk section has been identified. This section includes a series CHLs such as sharp curves, driveways and two at-grade intersections. The visibility of most of those CHLs is poor, and the field of view presents many deficiencies, specifically along the curves. Poor framing of the outer curve is present, and the curvature and development of those curve are hard to be judged. Considering the NAS 10, it is judged as high-risk mainly because of a misleading road environment. The road passes through an urbanized area, with many pedestrian crossings, driveways, and at-grade intersections, however the change in the road environment it is not clearly perceived. A speed limit of 30 km/h is set along this section, however, during the inspection it was possible to evaluate that most of the cars travelled at speeds above 50 km/h.

The B38 stretch shows less risky sections. The road is easier to comprehend, curves have great radii, and the visibility of PCLs is generally good. Nevertheless, some NASs present some problems, as NAS 7, which was judged as at high-risk. In this NAS an at-grade intersection is located immediately after a curve on a crest. Furthermore, the driver's attention is diverted from the intersection by the presence of a single tree across the road. The curve has a poor framing of both the outer and inner curve, with a change in the direction of the elements parallel to the road. The result is higher speeds with the underestimation of risk, poor visibility, and a hardly perceptible radius of curvature. A photo of the CHLs is provided in Fig. 3 (left side). Considering the other NASs, some medium-risk NAS present a low result for the Total value of the HFS.

The 106 stretch include 3 low-risk sections, 7 medium-risk sections, and 5 high-risk sections. The two most critical sections are NAS 14 and NAS 7. NAS 14 includes a series of curves and a lay-by. The curves present problems concerning the field of view and are not consistent with the operating speed. The lay-by is not visible. Both problems are present when driving north. NAS 7 comprehends an at-grade intersection in a curve. The road sections develop into the wood, except for the location of the curve and the intersection. The sudden disappearance of the trees removes an important reference for the driver, who may feel disoriented. The intersection is also not visible, and it is difficult to estimate the curvature of the curve, due to the missing references. A photo of the CHLs is provided in Fig. 3 (right side).

**Fig. 3.** Photos of CHLs from B38 NAS 7 southbound (left) and from 106 NAS 7 southbound (right)

The results are suitable for a prioritization of the intervention. The main priority list is provided by the risk level. Moreover, once a NAS has been selected to intervene, it is possible to analyse the detailed results obtained from the Step 2 to find out specific deficiencies and decide how to intervene. A first qualitative analysis with the RAs managing the analysed roads, suggests that the results are consistent with the recorded accidents and the impressions of the RAs.

## 5 Conclusions

A new procedure for implementing an NSA based on Human Factors principles has been proposed. The objective of the procedure is to analyse the accidents triggering factors related to the interaction between the driver and the road (and its environment), providing a classification of the road network based on different risk levels (European Parliament; Council of European Union 2019). One of the objectives of the procedure is to be easily applicable even in the presence of a lack of data, which is a recurring condition in low- and middle-income countries, where the major number of road fatalities occurs. Therefore, the proposed procedure relies on road safety inspections, which are based on surveys of existing roads. The procedure doesn't require specific data, nor expensive instruments. The procedure is based on the use of the HFET from PIARC, which has been improved to be suitable for its use in the procedure. The procedure is divided into three main steps: a first step which give a general analysis of the road characteristics, a second step where the evaluations are made, and a last third step where the results are processed to provide a safety ranking suitable for an NSA.

The procedure has been applied to three case studies with interesting results. The positive outcome is that the final ranking based on safety levels, allow to prioritize RA's interventions, and that the procedure seems to be easy to implement. The main limitation of the procedure is that the inspectors must be trained about Human Factors and that it requires additional time compared to standard accidents-based analysis. The results of the procedure highlight the problems related to the driver-road interaction, and they do not provide any information about other road factors which may influence safety (such as obstacles in the road margins, and road surface conditions). Nevertheless, the procedure can be easily implemented together with other road safety inspection procedure, which account for different aspects. Moreover, it can be implemented without the necessity of a large amount of data; thus, it can be easily applied also in low- and middle-income countries. Following research on this procedure should focus on its application to other roads and also to compare the results with those obtained from other NSA procedures to test its effectiveness.

## References

Birth, S., Demgensky, B., Sieber, G.: Relationship between Human Factors and the likelihood of single-vehicle crashes on Dutch motorways. Report for Rijkswaterstaat Dienst Verkeer en Scheepvaart, Delft (2015)

Cafiso, S., La Cava, G., Montella, A.: Safety index for evaluation of two-lane rural highways. Transp. Res. Rec. **2019**, 136–145 (2007). https://doi.org/10.3141/2019-17

Čičković, M.: Influence of human behaviour on geometric road design. Transp. Res. Procedia **14**, 4364–4373 (2016)

European Parliament; Council of European Union. Directive (EU) 2019/1936 of the European Parliament and of the Council of 23 October 2019. L 305, 26.11.2019 (2019)

FHWA: Reliability of Safety Management Methods - Network Screening. FHWA-SA-16-037: Federal Highway Administration Office of Safety, Washington, DC (2016)

Hauer, E.: The Art of Regression Modeling in Road Safety. Springer, Cham (2015). https://doi. org/10.1007/978-3-319-12529-9. ISBN 978-3-319-35446-0

iRAP: iRAP Star Rating and Investment Plan. International Road Assessment Programme (iRAP) 2021 (2021). http://www.irap.org

Kathmann, T., Ziegler, H., Pozybill, M.: Road safety screening on the move. Transp. Res. Procedia 14, 3322–3331 (2016). https://doi.org/10.1016/j.trpro.2016.05.281. ISSN 2352-1465

La Torre, F., et al.: Development of a transnational accident prediction model. In: Proceedings of 6th Transport Research Arena, 18–21 April, Warsaw, Poland (2016)

La Torre, F., Meocci, M., Domenichini, L., Branzi, V., Tanzi, N., Paliotto, A.: Development of an accident prediction model for Italian freeways. Accid. Anal. Prev. **124**, 1–11 (2019). https://doi.org/10.1016/j.aap.2018.12.023

PIARC: Road Safety Inspections Guidelines for Safety Checks of Existing Roads. PIARC, publ. 2012R27EN (2012)

PIARC: Accidents Investigations Guidelines for Road Engineers. 2013R07EN (2013). ISBN 978-2-84060-321-4

PIARC: Human Factors Guidelines for a Safer Man-Road Interface. PIARC, publ. 2016R20EN (2016)

PIARC: Road safety evaluation based on Human Factors method. World Road Association (PIARC). 2018R27EN (2019)

Theeuwes, J.: Self-explaining roads: What does visual cognition tell us about designing safer roads? Cognitive Research: Principles and Implications, SpringerOpen (2021). https://doi.org/10.1186/s41235-021-00281-6

WHO: Global Status Report on Road Safety 2018. WHO (2019)

# A Review of the Spatial Analysis Techniques for the Identification of Road Accident Black Spots and It's Application in Context to India

Shawon Aziz[✉] and Sewa Ram

Department of Transport Planning, School of Planning and Architecture,
New Delhi, India
shawonaziz786@yahoo.in, s.ram@spa.ac.in

**Abstract.** A Road accident is an unexpected negative outcome of growth in the transportation infrastructure sector. Identification of road stretches where accidents occur frequently (also known as Accident Black spots) is the first step towards development of a safer road network. Over the past decades, several analysis tools and techniques have been adopted to identify accident black spots. Among these tools, Geographic Information System (GIS) stands out for its ability of not only storing and presenting data spatially but also for performing complex spatial analysis using statistical tools that rely on geographically referenced data.

The paper presents GIS Approach to examine spatial patterns of road accidents and determine if they are spatially clustered, dispersed or random. The paper discusses the various spatial analysis techniques that have been applied in the past by Researchers worldwide in a chronological order for the identification of accident black spots. Spatial Auto-correlation Techniques such as Moran's I and Getis-Ord (Gi) statistics have been discussed with an example from Indiana USA to examine spatial patterns and cluster mapping of accidents. Examples of application of Kernel Density Estimation (KDE) have been illustrated using previous research conducted on Des Moines city of Iowa State in USA, Melbourne in Australia.

The paper also gives a comparative analysis of the three techniques i.e., Moran 'I, Getis-Ord and KDE. At last the Paper discusses the road accident scenario in India and the scope of application of GIS techniques in Indian Cities for identification of road accident black spots.

**Keywords:** Accident · Hot spot · Spatial analysis · Kernel · Getis-Ord · Moran's I

## 1 Introduction

As per the Global status report on road safety released by the World Health Organization, about 1.35 million people die each year because of road traffic accidents and nearly 93% of the world's fatalities on the roads occur in low and middle-income countries which possess only about 60% of the world's vehicles (WHO 2018). Moreover, Road traffic injuries are the leading cause of death for children and young adults aged 5–29 years (WHO 2018). In India, as per the Ministry of Road Transport

© The Author(s), under exclusive license to Springer Nature Switzerland AG 2022
A. Akhnoukh et al. (Eds.): IRF 2021, SUCI, pp. 511–524, 2022.
https://doi.org/10.1007/978-3-030-79801-7_37

and Highways, in the year 2019 a total of 4,49,002 road accidents were reported by States and Union Territories (UTs) killing 1,51,113 people and causing injury to 4,51,361 persons. The number of accidents and deaths in 2019 translates into an average of 1,230 accidents and 414 deaths every day and nearly 51 accidents and 17 deaths every hour. (MoRTH, Road Accidents in India, 2019). These figures have very marginally decreased from the statistics of year 2018 when 4,67,044 accidents took place killing 1,51,417 people and injuring 4,69,418 road users in India (MoRTH 2018). Thus, in various developing countries like India, road accidents have become an important concern as it has emerged as a negative and unexpected outcome of growth in the transportation infrastructure. Identifying high risk locations for road accidents can serve as a useful tool to Government agencies to allocate their limited resources more efficiently and find effective countermeasures.

In road safety studies, when several accidents frequently occur in close proximity to a location on a road, it is described as a hazardous location, more commonly known as an accident black spot (Elvik 2007). Such areas of the concentrated accidents have also been referred as crash hotspots (Anderson 2007), sites with promise (Hauer 1996), hot pieces (Gundogdu 2010) and road traffic crashes vulnerability area (Benedek et al. 2016).

There are various methods to identify the accident black spots. This paper illustrates the use of Geographic Information System (GIS) for identification of road accident black spots. The GIS approach helps to examine the spatial patterns of road accidents and evaluates if accidents are spatially clustered, dispersed, or random. This paper discusses spatial autocorrelation techniques such as Moran's I, Getis-Ord (Gi) statistic along with case study examples to examine how these techniques can be used to study spatial patterns and cluster mapping of accidents. Furthermore, use of Kernel Density Estimation (KDE) has been discussed with examples to illustrate how KDE can be used to generate crash concentration maps that show the spatial density as well as network density of accidents. The paper at end discusses the scope of possible application of such methods on roads in India to assess the spatial pattern of road accidents.

## 2 Literature Review

The Tobler's first law of Geography states that "everything is usually related to all else but those which are near to each other are more related when compared to those that are further away" (Tobler 1970). Thus, there may exist a spatial inter-relationship between road accidents (also known as spatial autocorrelation) which is advocated by the Tobler's first law of Geography as stated above. The existing statistical techniques and models for accident black spot identification are based on an assumption that the accidents in a sample are independent of one another. The Assumption that samples are "independent" is violated if accidents are spatially related. Moreover, since standard error in a sample of observation is inversely proportional to square root of sample size as shown in Eq. (1):

$$Standard\ Error = \frac{Standard\ Deviation}{\sqrt{Sample\ Size}} \tag{1}$$

Thus, if in a sample of road accidents, a few accidents are spatially related, then the actual number of independent accident samples decrease and thus the standard error increases. So, in any classical statistical techniques, the regression coefficient can be biased and its precision is exaggerated incase the accidents hold a spatial relationship and are not independent samples.

Thus, the primary reason behind employing spatial techniques for detection of accident black spots is that the classical statistical techniques neglect the geographical relationship between the different locations (Ouni and Belloumi 2018). The Spatial methods which are employed for identification of clustering patterns of road traffic crashes, produce two kinds of results. The first one is identifying global clustering tendency of accidents within road section. These second result is identifying local clustering tendency of accidents within the road section (Ouni and Belloumi 2019).

There are a number of indices that attempt to measure spatial autocorrelation for such as Moran's I and Getis-Ord Gi statistic. These indices can be used as Global or Local measures depending on the scope of the analysis. The global autocorrelation is the extent to which points that are close together in space have similar values, and the local autocorrelation is the extent to which points that are close to a given point or area have similar values (Abdulhafedh 2017). Other methods for studying spatial patterns of crash data which is most widely used is the Kernel Density Estimation (KDE). The clustering tendency of road accidents can be evaluated using (KDE). The brief description of Moran's I, Getis Ord (Gi) statistic and KDE are given in the sub sections.

## 2.1  Moran's I

Moran's I is used to measure spatial autocorrelation (feature similarity) based on both accident locations & accident values simultaneously. It evaluates whether the pattern expressed is clustered, dispersed, or random. For n number of observations, Moran's I is calculated by Eq. (2) as follows:

$$I = \frac{n}{S_0} \frac{\sum_i^n \sum_j^n W_{ij}(x_i - \bar{x})(x_{ij} - \bar{x})}{\sum_i^n (x_i - \bar{x})^2} \qquad (2)$$

Where,

$\bar{x}$ is the mean of the variable x;
$W_{ij}$ are the elements of the weight matrix;
$S_0$ is the sum of the elements of the weight matrix: $S_0 = \sum_i^n \sum_j^n W_{ij}$

Values for this index typically, range from $-1.0$ to $+1.0$, where a value of $-1.0$ indicates negative spatial autocorrelation, and a value of $+1.0$ indicates positive spatial autocorrelation. When nearby points or segments have similar values, their cross product is high. Conversely, when nearby points or segments have dissimilar values, their cross-product is low (Abdulhafedh 2017). The expectation of Moran's I is given by Eq. (3):

$$E(I) = \left(\frac{-1}{n-1}\right) \tag{3}$$

with a Moran's I value larger than E(I), indicates positive spatial autocorrelation, and a Moran's I less than E(I), indicates negative spatial autocorrelation. In Moran's formulation, the weight variable, $w_{ij}$, is a contiguity matrix. If zone j is adjacent to zone i, the product receives a weight of 1.0. Otherwise, the product receives a weight of 0 (Abdulhafedh 2017). The z-scores of Moran's I can be computed by Eq. (4):

$$Z_1 = \frac{I - E(I)}{\sqrt{V(I)}} \tag{4}$$

Were E(I) is the expected value of I, and V(I) is the variance of I, as shown in Eq. (5):

$$(I) = E(I^2) - E^2(I) \tag{5}$$

Interpretation

For a null hypothesis, it can be assumed that "there is no spatial clustering of the accidents with the geographic features in the study area". When the p-value is small & the absolute value of the Z score is large enough that it falls outside of the desired confidence level, the null hypothesis can be rejected. A statistically significant positive z-score indicates that the distribution of the observations is spatially autocorrelated producing a hot spot with high values clustering together, whereas a negative z-score indicates that the observations tend to be more dissimilar producing a cold spot with low values clustering together. A z-score close to zero indicates that observations are randomly and independently distributed in space.

## 2.2 Getis-Ord Gi Statistic

The Getis-Ord Gi statistic is another index of spatial autocorrelation that can distinguish between positive spatial autocorrelation with high values from positive spatial autocorrelation with low values. The General (Global) Gi statistic computes a single statistic for the entire study area, while the local Gi statistic is an indicator for local autocorrelation for each data point. There are two types of Gi statistics, although almost the two types produce identical results.

The first one, Gi, does not include the autocorrelation of a zone with itself, whereas the i G∗ includes the interaction of a zone with itself (i.e., the Gi statistic does not include the value of Xi itself, but only the neighborhood values, but i G∗ includes Xi as well as the neighborhood values), and formally both can be computed by the given Eqs. (6) and (7):

$$G_i(d) = \frac{\sum_{j \neq i}^{n} W_{ij}(d) x_j}{\sum_{j \neq i}^{n} x_j} \tag{6}$$

$$G_i^*(d) = \frac{\sum_{j=1}^{n} W_{ij}(d)x_j}{\sum_{j=1}^{n} x_j} \qquad (7)$$

where, d is the neighborhood (threshold) distance, and $w_{ij}$ is the weight matrix that has only 1.0 or 0.0 values, 1.0 if j is within d distance of i, and 0.0 if its beyond that distance. These formulae indicate that the cross-product of the value of X at location i and at another location j is weighted by a distance weight, $w_{ij}$ which is defined by either a 1.0 if the two locations are equal to or closer than a threshold distance, d, or a 0.0 otherwise. The G statistic can vary between 0.0 and 1.0. The statistical significance of the local autocorrelation between each point and its neighbors is assessed by the z-score test and the p-value. The expected G value for a threshold distance, d, is given by Eq. (8):

$$E[G(d)] = \frac{W}{n(n-1)} \qquad (8)$$

Were, W is the sum of weights for all pairs of location (w = $\sum_i^n \sum_j^n W_{ij}$), and n is the number of observations. Assuming normal distribution, the variance of G(d) is given by Eq. (9):

$$Var[G(d)] = (G^2) - E^2(G) \qquad (9)$$

The standard error of G(d) is the square root of the variance of G as shown in Eq. (10). Therefore, a z-test can be computed by Eq. (11):

$$S.E[G(d)] = \sqrt{Var[G(d)]} \qquad (10)$$

$$Z[G(d)] = \frac{G(d) - E[G(d)]}{S.E[G(d)]} \qquad (11)$$

Interpretation

For a null hypothesis, it can be assumed that "there is no spatial clustering of the values". When the absolute value of the Z score is large & the p-value is very small, the null hypothesis can be rejected. If the null hypothesis is rejected, then the sign of the Z score becomes crucial to differentiate a Hot spot from a cold spot. If the Z score value is positive, it means that high values cluster together in the study area (hot spot formation). If the Z Score value is negative, it means that low values cluster together (cold spot formation).

## 2.3 Planar Kernel Density Estimation

Planer Kernel density estimation is a non-parametric representation of the Probability Density Function (PDF) for a random variable. Kernel distribution is used when a parametric distribution cannot properly describe the data i.e., when samples in a dataset are not independent of each other. Kernel Density Estimation produces a smooth density surface of point events over a 2-D geographic space (planar space) as shown in Fig. 1. The goal of KDE is to develop a continuous surface of density estimates of discrete events such as road accidents by summing the number of events within a search bandwidth.

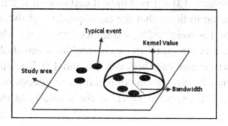

**Fig. 1.** Illustration of PKDE for spatial point pattern analysis

PKDE method estimates the accident density by counting the number of accidents in an area. The area is known as kernel. The total study area is divided by predetermined number of cells. PKDE is evaluated by fitting a smooth function called kernel over every accident point and then computing the distance from that point to the reference location based on a mathematical function and adding the value of all the surfaces for that reference location (Lakshmi et al. 2019).

The general form of a KDE in a 2-D space is given by Eq. (12):

$$\lambda(s) = \sum_{1}^{n} \frac{1}{\pi r^2} k \frac{d_{is}}{r} \qquad (12)$$

Where $\lambda(s)$ is the density at location s, r is the search radius (bandwidth) of the KDE, k is the weight of a point i at distance $d_{is}$ to location s. The kernel function k is usually considered as a function of the ratio between $d_{is}$ and r. As a result, the longer the distance between a point and location s, the less that point is weighted for calculating the overall density. All points within the bandwidth r of location s are summed for calculating the density at 's'. Different kernel functions exist like Gaussian, Quartic, Triangular functions which are used in accident analysis (Abdulhafedh 2017). Quartic function is used in ArcGIS, which is popular in accident analysis (Lakshmi et al. 2019) and it is given by Eq. (13):

$$k \frac{d_{is}}{r} = K \left( 1 - \frac{d_{is}^2}{r^2} \right) \qquad (13)$$

Where K is a scaling factor to ensure the total volume under Quartic curve is 1.0, and usually used as ¾ (Abdulhafedh 2017). To find the KDE value, two key parameters must be chosen: the kernel function 'k'; and the search radius (bandwidth) 'r'. Literature review suggests that the type of the distribution of the kernel function k has a very little effect on the results compared to the choice of search bandwidth 'r' (O'Sullivan and Wong 2007). The value of search bandwidth 'r' is very important because it usually determines the smoothness of the estimated density and can affect the outcome. If bandwidth is narrow, it will not produce a continuous smooth surface, whereas a wider bandwidth will suppress spatial variation of events. Hence, an optimal value of 'r' must be chosen that minimizes the sum of the squared errors of the kernel estimation (Scott 1992). ESRI ArcGIS 10.2 uses the following formula as a default optimal search radius as shown in Eq. (14):-

$$r = 0.9 \times mm \left( SD, \sqrt{\frac{1}{ln(2)}} \times D_m \right) \times n^{-0.2} \tag{14}$$

Where SD is the standard distance, $D_m$ is the median distance and n: the number of points if no population field is used, or if a population field is supplied, n is the sum of the population field values, and min: means that whichever of the two options that results in a smaller value will be used (Abdulhafedh 2017).

### 2.4 Network Kernel Density Estimation

Events such as Road accidents, gas pipe leakages, street crimes occur along a road network i.e., they are events constrained on a road network. Thus, in analyzing the hot spots of network-constraint events such as road accidents, the assumption of homogeneity of 2-D space does not hold and the relevant PKDE methods may produce biased results (Okabe et al. 2009). Thus, Network Kernel Density Estimation (NKDE) can be applied for a network analysis of road accident data analysis. Flahaut et al. (2003) was the pioneer in developing NKDE for traffic accident analysis, but his attempt was limited to a single stretch of road, not a road network (Flahaut et al. 2003). The network KDE is a 1-D measurement while the planar KDE is a 2-D measurement (Okabe et al. 2009). Thus, the major difference between network KDE and planar KDE is that the density is measured per linear unit instead of area unit. The network KDE uses the following equation for the linear density estimation of network constrained events as shown in Eq. (15):

$$\lambda(s) = \sum_{i=1}^{n} \frac{1}{r} k \left( \frac{d_{is}}{r} \right) \tag{15}$$

## 3 Chronology of Research Developments

Various Researchers in the recent past have applied spatial analysis techniques to evaluate the spatial relationship between discrete events such as road accidents. Some of such significant researches have been enlisted in Table 1.

## Table 1. Chronology of research developments for spatial analysis of road accidents

| S. no. | Author and publication year | Research theme |
|---|---|---|
| 1 | (Kim and Nitz 1995) | Study of Spatial clustering of road accidents. Case study: Honolulu |
| 2 | (Black and Thomas 1998) | Network Autocorrelation Analysis on Belgium's Motorways |
| 3 | (Flahaut 2004) | Impact of Infrastructure and Local Environment on Road Unsafety by Logistic Modeling with Spatial Autocorrelation |
| 4 | (Schweitzer 2006) | Spatial analysis of vehicle accidents in Southern California |
| 5 | (Yamada and Thill 2007) | Spatial clustering of vehicle accidents in Buffalo, New York |
| 6 | (Erdogan et al. 2008) | Traffic Accident Analysis using GIS Case Study: City of Afyonkarahisar, Turkey |
| 7 | (Erdogan 2009) | Study of Spatial clustering of road accidents. Case study: Turkey |
| 8 | (Gundogdu 2010) | Spatial analysis using GIS techniques. Case study of Konya |
| 9 | (Turong and Somenahalli 2011) | Pedestrian-vehicle accident hot spot analysis |
| 10 | (Budhiharto and Saido 2012) | Accident blackspot identification and ambulance route mobilization. Case study: City of Surakarta, Indonesia |
| 11 | (Cela et al. 2013) | Integrating GIS and spatial analysis techniques in an analysis of road traffic accidents in Serbia |
| 12 | (Effati et al. 2014) | Geospatial neuro-fuzzy approach for identification of hazardous zones in regional transportation corridors |
| 13 | (Tortum and Atalay 2015) | Spatial analysis of road mortality rates in Turkey using Moran's I and Getis Ord statistic |
| 14 | (Abdulhafedh 2017) | Spatial Autocorrelation Indices and Kernel Density Estimation Case Study: Indiana Road Network USA |
| 15 | (Lakshmi et al. 2019) | Identification of Traffic Accident Hotspots using GIS Case Study Des Moines city of Iowa state USA |

## 4 Case Studies

The following case studies illustrate the application of spatial analysis tools by various researchers in order to determine spatial autocorrelation and spatial clustering.

### 4.1 Indiana Network USA (Abdulhafedh 2017)

The study was carried on Indiana Road Network in USA by Azad Abdulhafedh of the Department of Civil and Environmental Engineering, University of Missouri-Columbia. As shown in Table 2, for the Global Moran's I statistic, the null hypothesis states that the attributes (i.e. accidents) being analyzed are randomly distributed among the features in the study area (i.e. no global spatial autocorrelation exists for the entire network). However, since the p-value being generated is less than 0.01 (using a confidence level of 99%), then this indicates that the Moran's I spatial autocorrelation is significant, and hence, we can reject the null hypothesis. Similarly, the Global (General) Getis-Ord Gi, the p-value being generated is less than 0.01 (using a confidence level of 99%), then this indicates that the General Gi statistic spatial autocorrelation is significant, and hence, we can reject the null hypothesis.

**Table 2.** Moran's I and Getis-Ord Gi Summary for Indiana road network.

| Spatial autocorrelation indices | Index | Expected index | Variance | z-score | p-value | Decision |
|---|---|---|---|---|---|---|
| Moran's I | 0.135847 | −0.000335 | 0.000006 | 53.817107 | 0.000000 | Significant |
| Global Getis-Ord $G_i$ | 0.128449 | 0.106109 | 0.000001 | 19.233837 | 0.000000 | Significant |

Figure 2 shows the HH, LL, HL, LH identified by Moran's I. Figure 3 shows the HH, LL, HL, LH of Moran's I with rendering that clearly illustrates the range of the z-scores of the identified clusters between the range LL < −2.0, and the HH > 2.0. It can be seen that the Gi∗ statistic has identified a larger number of significant hot spots (157 HHs) and significant cold spots (307 LLs) than the Moran's I (102 HHs, 287 LLs). However, Moran's I has identified a larger number of significant outliers (79 HLs, 82 LHs) than the Gi∗ (48 HLs, 0.0 LHs). Figure 4 shows the HH, LL, HL, LH identified by the Gi∗ statistic. Figure 5 shows the HH, LL, HL, LH of Gi∗ with rendering that clearly illustrates the range of the z-scores of the identified clusters between the range of LL < −2.0, and the HH > 2.0.

**Fig. 2.** Hot spots Moran's I statistic

**Fig. 3.** Hot Spots by Moran's I with z-scores rendering

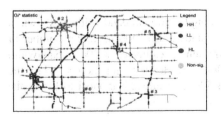

**Fig. 4.** Hot Spots by Gi* statistic

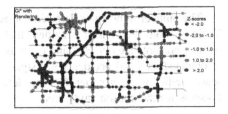

**Fig. 5.** Hot Spots by Gi* statistic with z-scores rendering

## 4.2 Des Moines City of Iowa State USA (Lakshmi et al. 2019)

The authors of this paper carried-out a case study on accidents occurred in Des Moines city of Iowa state to show the effect of varying bandwidths in creating density maps based on Planar Kernel Density Estimation (PKDE). Crash data for the years 2008 to 2012. Figures 6, 7, 8 and 9 illustrate the effect of influence of different bandwidths on density estimation for detecting hotspots. Based on published literatures (Table 1), cell size was set as 100 m and density estimate was calculated for varying bandwidths 250 m, 500 m, 750 m and 1000 m. It is quite evident that 500 m bandwidth is appropriate in this case as clusters are distinct. Hotspots are hardly detected when the bandwidth is 250 m and clusters get merged in higher bandwidths.

**Fig. 6.** Bandwidth 250 m

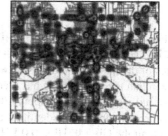

**Fig. 7.** Bandwidth 500 m

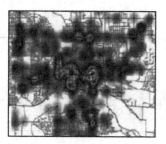

**Fig. 8.** Bandwidth 750 m

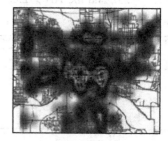

**Fig. 9.** Bandwidth 1000 m

The number of hotspots detected corresponding to different bandwidths is shown in Table 3. From the table it is evident that at a bandwidth is 500 m, larger number of distinct hotspots are detected and the maximum density estimate for 500 m bandwidth is obtained as 1340 accidents per square kilometer. Hotspots get merged when the search radius is greater than 500 m. Hence a 500 m bandwidth is ideal to locate the hotspots for the chosen study area.

Table 3. Results from PKDE analysis - Des Moines city of Iowa state USA

| S. No | Bandwidth (in m) | No. of hotspots detected | Density estimate (No. of accident counts per square km) |
|---|---|---|---|
| 1 | 250 | 9 | 8573 |
| 2 | 500 | 12 | 1340 |
| 3 | 750 | 5 | 2297 |
| 4 | 1000 | 4 | 1710 |

## 5  Application in Context of India

As per the Ministry of Road transport and Highways (Circular Number- RW/NH-29011/2/2015/P&M (RSCE) dated 07.12.2015) in India, "A Road Accident Blackspot is a stretch of National Highway of about 500 m in length in which either 5 road accidents (in all three years put together involving fatalities/grievous injuries) took place during the last 3 calendar years or 10 fatalities (in all three years put together) took place during the last 3 calendar years".

The current ministry protocol is not applicable to State Highways and other roads in India. As shown in Figs. 10 and 11, the National Highways which comprises of 2.03% of total road network accounted for 30.6% of total road accidents and 35.7% of deaths in 2019. State Highways which account for 3.01% of the road length accounted for 24.3% and 25.5% of accidents and deaths respectively. Other Roads which constitute about 95% of the total roads were responsible for the balance 45%of accidents and 38.8% deaths respectively. Thus, the Highways (both National and State) which accounted for about 5% of total road network witnessed a disproportionately large share of accidents of 55% and accident-related killings of 61% during the year 2019. (MoRTH, Road Accidents in India, 2019)The local risk factors on all such roads are different which need a different approach and methodology of identification. Also, it is important to explore the possibility of spatial auto-correlation and spatial clustering of accidents on various road sections along National highways, state highways and other roads in India.

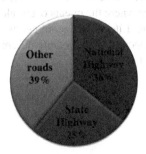

Fig. 10.  Fatalities due to Road Accidents in 2019. Source: (MoRTH 2019)

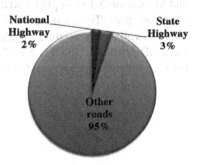

Fig. 11.  Length of Road in India in the Year 2019. Source: (MoRTH 2019)

With the augmentation in technology several developing countries like India are adopting for a road database management system where at each police station, a predefined MoRTH accident data recording format is being filled along with First Information Report (FIR) by the Policemen for each road accident. The MoRTH Road accident data recording format collects in detail the pertaining to the location and surrounding, GPS Co-ordinates, geometric design attributes, details of the vehicles involved and types of accident victims. Such data collected for each accident is being stored to develop a road accident database management system. This practice is being introduced in various States in India in accordance to the guidelines of the Supreme Court Committee of Road Safety which was constituted by the Honorable Supreme Court of India vide its order dated 22nd April 2014 in Writ Petition (C) Number 295 of 2012 (MoSPI 2019). Thus, in near future, a micro level as well as a macro level spatial analysis of road accidents on various types of roads in India would be possible.

## 6 Conclusion

Hot spot analysis is used to highlight areas which have higher than average concentration of accidents, and it is a useful technique for visualizing the concentration of discrete events on networks such as road accidents. This paper reviews two methods: Moran's I and Getis-Ord (Gi) statistic based on spatial autocorrelation and another third method i.e. kernel density estimation (KDE) to examine the spatial patterns of vehicle accidents and determines if accidents are spatially clustered, dispersed, or random. The Global values of both Moran's I and Gi statistically determine the possibility of spatial clustering of the overall accidents on a road network while the local Moran's I and Gi* identify the pattern of clusters on the road network. Kernel density estimation is a way to estimate the Probability Density Function (PDF) of a random variable in a non-parametric way.

Review of published research shows that the choice of kernel function k is less important than the impact of the bandwidth in KDE results. Also, literature review indicates that there is no an exact approach to find optimum bandwidth research in KDE. In fact, different studies are used and recommended different bandwidth (Pour and Moridpour 2015). In planar KDE, the space is characterized as a 2-D homogeneous Euclidian space. Thus, it does not hold effective in cases where the events occur along a road network such as road accidents. Therefore, the Literature recommends Network KDE method, which is a 1-Dimentional measurement. A few notable differences between different hot spot analysis methods are shown in Table 4.

## A Review of the Spatial Analysis Techniques  523

Table 4. Comparison of hot spot analysis methods

| S. no. | Moran's I | Getis-Ord | Kernel density |
|---|---|---|---|
| 1 | Measures spatial correlation, identifies hot spots with high-high values, and cold spots with low-low values | Measures spatial correlation, identifies hot spots with high-high values, and cold spots with low-low values | Measures probability density function, identifies only hot spots in term of density per unit area or per unit linear distance |
| 2 | Identifies outliers (dispersed incidents) with high-low values and low-high values | Does not identify outliers | Does not identify outliers |
| 3 | Does not include the interaction of a zone with itself but only with its neighborhoods in measuring spatial correlation | Includes the interaction of a zone with itself in addition to its neighborhoods in measuring spatial correlation | Does not include the interaction of a zone with itself but only with its neighborhoods in measuring kernel density |
| 4 | Reports an index-value, and a z-score | Reports a combined index-value and a z-score | Reports a linear density value, and a z-score |
| 5 | Reports a p-value | Reports a p-value | Does not report a p-value |
| 6 | Presents the statistical significance of clustering | Presents the statistical significance of clustering | Does not present the statistical significance of clustering |
| 7 | Ranges from $-1.0$ to $+1.0$ | Ranges from $-1.0$ to $+1.0$ | Any positive value |

Thus, it is evident that, each method identifies different clustering patterns, therefore the literature recommends using a combination of these methods in hot spot analysis. Comparable results from these methods can produce more reliable and effective interpretations among the spatial clustering patterns.

# References

Abdulhafedh, A.: Identifying vehicular crash high risk locations along highways via spatial autocorrelation indices and kernel density estimation. World J. Eng. Technol. **5**, 198–215 (2017)

Anderson, T.: Comparison of spatial methods for measuring road accident hot spots. A case study of London. J. Maps **3**(1), 55–63 (2007)

Benedek, J., Ciobanu, S.M., Man, T.C.: Hot spots and social background of urban traffic crashes: a case study in Cluj-Napoca (Romania). Accid. Anal. Prev. **87**, 117–126 (2016)

Black, W., Thomas, I.: Accidents on Belgium's motorway: a network autocorrelation analysis. J. Transp. Geogr. **6**, 23–31 (1998)

Budhiharto, U., Saido, A.: Traffic accident black spot identification and ambulance fastest route mobilization process for the city of Surakarta. Jurnal Transportasi **12**(3) (2012)

Cela, L., Shiode, S., Lipovac, K.: Integrating GIS and spatial analysis techniques in an analysis of road traffic accidents in Serbia. Int. J. Traffic Transp. Eng. **3**(1), 1–15 (2013)

Effati, M., Rajabi, M., Shabani, S., Samadzadegan, F.: A geospatial neuro-fuzzy approach for identification of hazardous zones in regional transportation corridors. Int. J. Civil Eng. 12(3), 289–303 (2014)

Erdogan, S.: Explorative spatial analysis of traffic accidents and road mortality among the provinces of Turkey. J. Safety Res. 40(5), 341–351 (2009)

Erdogan, S., Yilmaz, I., Baybura, T., Gullu, M.: Geographical information systems aided traffic accident analysis system case study: city of Afyonkarahisar. Accid. Anal. Prev. 40, 174–181 (2008)

Flahaut, B.: Impact of infrastructure and local environment on road unsafety: logistic modeling with spatial autocorrelation. Accid. Anal. Prev. 36, 1055–1066 (2004)

Flahaut, B., Mouchart, M., Martin, E., Thomas, I.: The local spatial autocorrelation and the kernel method for identifying black zones: a comparative approach. Accid. Anal. Prev. 35(6), 991–1004 (2003)

Gundogdu, I.B.: Applying linear analysis methods to GIS-supported procedures for preventing trafficaccidents: case studyof Konya. Saf. Sci. 48(6), 763–769 (2010)

Hauer, E.: Identification of sites with promise. Transp. Res. Rec. J. Transp. Res. Board 1542, 54–60 (1996)

Kim, K., Nitz, L.: Spatial analysis of honolulu motorvehicle crashes. Accid. Anal. Prev. 27(5), 663–674 (1995)

Lakshmi, S., Srikanth, I., Arockiasamy, M.: Identification of traffic accident hotspots using geographical information system (GIS). Int. J. Eng. Adv. Technol. (IJEAT) 9(2) (2019)

MoRTH: Road Accidents in India. Ministry of Road Transport and Highways, Government of India, New Delhi (2019)

MoSPI: The Ministry of Statistics and Programme Implementation (2019). http://mospi.nic.in/sites/default/files/iss-circular/SCCoRS.pdf

Okabe, A., Toshiaki, S., Kokichi, S.: A kernel density estimation method for networks, its computational method and a GIS-based tool. Int. J. Geogr. Inf. Sci. 23, 7–32 (2009)

O'Sullivan, D., Wong, D.W.: Surface-based approach to measuring spatial segregation. Geogr. Anal. 39(2), 147–168 (2007). https://onlinelibrary.wiley.com/doi/abs/https://doi.org/10.1111/j.1538-4632.2007.00699.x

Ouni, F., Belloumi, M.: Spatio-temporal pattern of vulnerable road user's collisions hot spots and related risk factors for injury severity in Tunisia. Transport. Res. F: Traffic Psychol. Behav. 56, 477–495 (2018)

Ouni, F., Belloumi, M.: Pattern of road traffic crash hot zones versus probable hot zones inTunisia: a geospatial analysis. Accid. Anal. Prev. 128, 185–196 (2019)

Schweitzer, L.: Environmental justice and hazmat transport: a spatial analysis in southern California. Transp. Res. Part D: Transp. Environ. 11, 408–421 (2006)

Scott, D.W.: Multivariate Density Estimation: Theory, Practice and Visualization Print (1992). ISBN 9780471547709. Online ISBN 9780470316849. Wiley Series in Probability and Statistics. https://onlinelibrary.wiley.com/doi/book/10.1002/9780470316849

Tobler, W.: A computer movie simulating urban growth in the Detroit region. Econ. Geogr. 46, 234–240 (1970). Supplement: Proceedings. International Geographical Union. Commission on Quantitative Methods. https://www.jstor.org/stable/143141?origin=crossref&seq=1

Tortum, A., Atalay, A.: Spatial analysis of road mortality rates in Turkey. Proc. Inst. Civil Eng.-Transp. 168(6), 532–542 (2015)

Turong, L., Somenahalli, S.: Using GIS to identify pedestrian-vehicle crash hot spots and unsafe bus stops. J. Public Transp. 14, 6 (2011)

WHO: Global Status Report on Road Safety. World Health Organization (2018)

Yamada, I., Thill, J.: Local indicators of network-constrained clusters in spatial point patterns. Geogr. Anal. 39, 268–292 (2007)

# Road Network Safety Screening of County Wide Road Network. The Case of the Province of Brescia (Northern Italy)

Michela Bonera[(✉)], Benedetto Barabino, and Giulio Maternini

Department of Civil, Architectural, Environmental Engineering
and of Mathematics (DICATAM), University of Brescia,
Via Branze 43, 25123 Brescia, Italy
m.bonera010@unibs.it

**Abstract.** Although EU roads are the safest in the world, the target of halving the road deaths by 2020 was not achieved. Road Infrastructure Safety Management procedures are key to improve road safety performances, and their implementation is required for primary road networks. Specifically, Road Network Screening enables to apply a wide-level analysis to identify the most critical segments of the network, and direct in-depth investigations more efficiently. Crash prediction models (CPMs) are extremely useful tools for quantitative road safety analysis, and road network screening can greatly benefit from their application. The Highway Safety Manual (HSM) is one of the main references worldwide, but the reported models are subjected to transferability issues due to their site-specific formulation. Most of previous studies on CPMs focused on HSM calibration models or investigated the effect of several factors over the crash frequency on specific road type. However, to our knowledge, Europewide few attempts were performed to develop a road network screening by mean of CPM. This paper covers these gaps by developing a specific CPMs to screen county-road network and identify most critical segment. The model was applied to the main road network of the Province of Brescia (Northern Italy). Few, but significant variables were identified in the model and maps were produced to rank the road network based on the crash frequency values. This model can serve as a relevant decision support tool for all bodies responsible in the definition of road safety interventions and related resources allocation, prior than crashes occur.

**Keywords:** Road network screening · Crash frequency · Crash count · Road safety · Crash modelling · Road safety ranking

## 1 Introduction

European roads got safer over the last decades, but still the burden of road crashes remains too high. According to the latest available statistics, Europewide, 23400+ road fatalities and no fewer than 1,23 million injuries were registered in 2018 (Eurostat 2020). Moreover, the considerable reduction in road crashes and deaths achieved in 2020 (compared to 2010) cannot be considered a full positive result, given that this was

© The Author(s), under exclusive license to Springer Nature Switzerland AG 2022
A. Akhnoukh et al. (Eds.): IRF 2021, SUCI, pp. 525–541, 2022.
https://doi.org/10.1007/978-3-030-79801-7_38

due to the great reduction in transfers and mobility imposed by the Covid-19 restrictions and not to an overall improvement in road safety (ETSC 2020). However, the target of halving the number of road deaths by 2020 compared to 2010 was not achieved by most of EU Member States. Such huge societal and health issue has clear economical externalities, given that road crashes costs range from 0,4 to 4% of the national Gross Domestic Product (GDP) across European Union (EU) Member States (Wijnen et al. 2019). Conversely, such resources could have been employed in a more effective way, for instance by implementing pro-active strategies to reduce such silent pandemic in advance, as expected by the new EU road safety policy for 2030 (European Commission 2020).

It is well known that, besides human driving behavior and vehicle safety standards, the infrastructures play a core role in the road safety challenge, being the fifth pillar of road safety. On the one hand, greater attention should be paid to this element to guarantee higher safety standards to all road users. On the other hand, road infrastructure interventions typically require time and huge amounts of resources, so that Road Authorities and Public Administrations must optimize the always little ones available. Hence, it is relevant for them to have effective decision support tools that can guide the prioritization of interventions and, therefore, the allocation of resources.

Heading to this goal, the EU delivered specific Directives that require Member States to implement specific Road Infrastructure Safety Management (RISM) procedures on their TEN-T network roads and strongly recommend implementing such procedure also on non-TEN-T roads (Elvik 2010; European Union 2008; European Union 2019). RISM procedures include all tasks to test, monitor and improve the safety performance of a road network, by covering all the possible stages of a road infrastructure life (i.e., from the project stage to the operating stage) and at different investigation levels (e.g., at a network-wide or site-specific level) (Elvik 2010; Hauer et al. 2002; Persia et al 2016). Specifically, Road Network Screening (RNS) is conceived as the first step of the whole RISM process and it is applied to assess the safety performance of the road network and identify the most critical segments (Elvik 2010; Park and Sahaji 2013; Stipancic et al. 2019). Therefore, RNS is relevant for who manages wide road networks and needs to know how and where to intervene.

Numerous methods are available for the implementation of RNS in the literature, and they mainly refer to (i) step-by-step procedures that lead to the calculation of simple (e.g., Gupta and Bansal 2018; European Road Assessment Program 2020, Borghetti et al. 2021) and composite indicators (e.g., Yannis et al. 2013; Viera Gomes et al. 2018), or (ii) the application of Crash Prediction Models (CPMs) (e.g., Ambros et al. 2018; Lord and Mannering 2010; Mannering and Bhat 2014; Yannis et al. 2016).

As for (i), such procedures are easy to implement in terms of mathematical computation and generally do not require too specific information as input (e.g., number of crash and/or people involved, severity level, road, and context features), so that they are usually preferred by road safety practitioners. However, they refer to already occurred events and reflect a passive approach, which is not the recommended strategy.

As for (ii), such methods can return an estimate prediction of the expected outcome (e.g., crash frequency over a specific time unit), so that such procedures are preferable for the implementation of RNS, as they enable the identification of safety drawbacks before the occurrence of crashes. However, such models are not immediate to develop and implement, being more demanding in terms of computational and interpretability skills by the user side. To help overcome such issues, much research has been developed to test the transferability of the Highway Safety Manual (HSM) prediction procedures (AASHTO 2010) to other contexts, being the HSM an international reference for CMP (e.g., Dragomanovits et al. 2016; Farid et al. 2016; Farid et al. 2018; La Torre et al. 2016; La Torre et al. 2019). However, dependability issues arose, due to inconsistencies among road crash definitions and differences in road infrastructural design and operational features associated to context other than the American one.

Hence, some studies tried to find more performing and efficient calibration procedures, by acting on the parameters and coefficients included in the HSM model or by developing new ones (e.g., La Torre et al. 2016; La Torre et al. 2018; Bonera and Maternini 2020). Some others opted for the development of site-specific or road-specific CPM by mean of Generalized Linear Models or Bayesian Modelling techniques (e.g., Lord and Mannering 2010; Jiang et al. 2014; Stipancic et al. 2019).

All the previous studies provided valuable insights about road crash prediction modelling. However, some gaps persist. First, most of the studies focused on calibration methods for the HSM model, to be able to transfer this model also to other contexts, such as the European one. However, as shown in the literature, the proper transferability of such criteria on different contexts may be questionable or not fully satisfying. Next, most studies focused on specific road categories (e.g., highways, rural road), so that it might be the case that the same model cannot fit over different roads. Indeed, calibration procedures returned quite satisfactory results only for some road classes (Bonera and Maternini 2020). As for studies that focused on the development of site-specific CPM, for most of them, the objective was to evaluate the effect of the different infrastructural and operational factors (e.g., AADT levels, road design elements) over crash occurrence. In doing so, detailed data were collected to assess such influences. Conversely, no study considered the wide road network, to evaluate the related safety performance of the whole network at a wide-level investigation (i.e., road network screening). Moreover, to our knowledge, no study associated such methods to a road network ranking process and provided spatial visualizations of the results (e.g., maps), which are very useful instead, especially to road managers and public administration to immediately identify which section requires further attention.

This paper aims to cover the previous gaps by exploring the application of a context-specific CPM as a road network screening tool. More precisely, the objective is the definition of a road network ranking based on the predicted crash frequency on each road sections over a time span (5 years). Furthermore, being the model oriented to a wide-level analysis of the road network, few and easy-to-collect variables were included, to foster the interpretability, updatability, and replicability of the method. Next, based on the value of the predicted crash frequency, maps were created as a decision support tool for road screening. Indeed, such maps provide a clear and immediate visualization of the safety performance of the road network to be used by road managers and administrators.

A total of 6000+ crash data were collected, related to the main road network of the Province of Brescia (northern Italy), which is one of the most populated Italian area, with a wide road network and the second highest number of road crashes in the Lombardy Region.

528    M. Bonera et al.

This paper attempts to contribute to both theory and practice. On the theoretical side, this paper sheds light on a research area which is not completely addressed so far, to the best of our knowledge. From a practical perspective, this paper helps implement a very useful RISM procedure for those responsible for road safety, who need effective decision support tools to coordinate and supervise road safety actions at a wider level to prioritize interventions.

The remaining paper is structured as follows: Sect. 2 briefly describes the research context, the development of the county-specific CPM, and the data collection and processing; Sect. 3 provides a brief discussion of the results; Sect. 4 concludes the paper and provides some glimpses on the future research perspectives.

## 2 Materials and Methods

### 2.1 Research Context

The Province of Brescia was selected as case study, being the largest province in the Lombardy Region (northern Italy), with a population of about 1.250 million people (the fifth in the Italy per inhabitants). On the one hand, Brescia represents one of the most important industrial, economic, and social areas in Italy and a crucial hub of incoming/outgoing daily traffic. On the other hand, a relevant number of road crash occurred on the province's road network over the last years, with an average of 3.000+ crashes, 4.000+ injuries and around 80 deaths each year (Polis-Lombardia 2020). Most of road crashes occurred on urban roads (almost 74%) but the mortality rate on non-urban roads is four time higher than that on urban roads. Given the geographical position of the Province, this experiment provides a good case study of road network screening, as it is representatives of different road infrastructure environment (i.e., mountain, rolling and flat terrain), and therefore of many similar areas, especially in northern Italy.

As required by the RISM European Directive, this study considered just the main roads of the network (i.e., motorways, State roads, and County roads), for a total extent of 2.500+ km. A total of 6.039 road crashes occurred on the main road network of the Province of Brescia over the five-years period 2014–2018 were included in the study.

### 2.2 Modelling Crash Frequency for the Road Network Screening

To perform road network screening, a crash frequency prediction model was used as a driver of the probability of crash occurrence over a specific road section over a given unit of time. According to previous research, different techniques have been applied to model road crash frequency. Generalized Linear Models (GLM) with a Negative Binomial (NB) distribution are the most employed, being suitable for modelling non-negative discrete data and able to account for overdispersion (Lord and Mannering 2010; Mannering and Bhat 2014). Crash frequency is generally computed as a simple function of exposure variables only (i.e., traffic volumes) or, in more complex models, including a set of additional variables that might contribute to crash occurrence (e.g., road design features, operational characteristics, context factors, etc.). In addition, a non-linear relationship exists between the crash frequency and exposure variable so that, if the exposure variable is zero, the crash frequency is equal to zero too (Barabino et al. 2021). Indeed, an exponential formulation is generally used to model such relationship. Let:

- $F_i$ be the estimated crash frequency for each segment i of the network, with i = 1, ..., m.
- $x_j$ be a generic variable included in the model associated to factor j, with j = 1, ..., n
- $\alpha$ be the model coefficient to be estimated.
- $\beta$ be the coefficient related to the exposure variable to be estimated.
- $\gamma_j$ be the coefficient related to the j-factor to be estimated, with j = 1, ..., n.

The functional form of the crash frequency prediction model is:

$$F_i = \alpha E_i^{\beta} \cdot \exp\left(\sum_{j=1}^{n} \gamma_j x_j\right), \quad \forall i = 1, \ldots, m \tag{1}$$

The ratio between the regression deviance and the degree of freedom (i.e., the deviance ratio - d.r.) was referred in this study as a measure of the goodness of fit of the models, along with its statistical significance shown by the $\chi 2$. Next, some further checks of the prediction capabilities were performed, by computing the percentage of predicted crashes over observed crashes and the related Root Mean Squared Error (RMSE). Let $O_i$ be the observed crash frequency for each segment i of the network over the study period, with i = 1, ..., m. The RMSE was computed as follows:

$$RMSE = \sqrt{\frac{\sum_{i=1}^{m} (O_i - F_i)^2}{m}} \tag{2}$$

The closer the RMSE to zero the better the result of the model. Then, based on the values of the predicted crash frequency of Eq. (1), maps were created to rank all the road segments. Among the several ranking techniques available in the literature, in this work the network screening scale was set based on the quartile's distribution of the estimated crash frequency values. Five ranges of crash frequency were defined, and ranges' thresholds were set in correspondence to the lower (Q1 = 25th percentile), the middle (Q2 = 50th percentile), and the upper quartiles (Q3 = 75th percentile). Also, the interquartile range (IQR, i.e., the difference between Q3 and Q1) of the distributions of the ordered crash frequency values was used to enable the identification of the most critical paths. Table 1 shows the five-level ranking scale, in which the lower the crash frequency, the safer the segment.

**Table 1.** Definition of the ranking scale for the estimated crash frequency.

| Level | Ranges values | | Crash frequency | Colour |
|---|---|---|---|---|
| | Lower limit | Higher limit | | |
| 1 | 0 | Q1 | Low | |
| 2 | Q1 | Q2 | Medium – low | |
| 3 | Q2 | Q3 | Medium | |
| 4 | Q3 | (Q3 + 1,5 IQR) | Medium – high | |
| 5 | (Q3 + 1,5 IQR) | MAX | High | |

## 2.3 Data Collection and Preparation

As mentioned before, crash frequency can be affected by several variables that might contribute to crash occurrence (e.g., road design features, operational characteristics, context factors, etc.). Depending on data availability and according to the scope of the analysis, different data at several detail extent should be collected. However, the more the factors the more complex the model will result.

Road network screening reflects a wide-level analysis of the safety performance of a road network. Hence, the data used in such procedure must be suitable for the wide-level investigation, thus consistent to the scale of the analysis and specifically to the segmentation of the road network itself. If the segments considered are kilometers long in order of size, then too detailed or location-specific information may not be appropriate for such analysis, while variables that represent more general and long-wide characteristics of the segments should be preferred (e.g., road class, number of lanes per directions, presence of median, etc.).

For these reasons, few and easy-to-collect data were used in this study and specifically gathered from official sources, to ensure greater data attainability and standardization. Variables to be included in the crash frequency prediction modelling were selected among those resulted significant from the literature and that available for the specific context. This allowed to make the model development easier and to enable simpler interpretability, replicability, and updateability of the process itself over time and over other context. For this work, data were gathered from three different sources for crash-related information, traffic volumes and road infrastructure, respectively. Then, data were pre-processed by mean of spreadsheet and a geographical information system (GIS) software (i.e., QGis) (Fig. 1).

Disaggregated road crash data were provided by Polis-Lombardia, which is the official body for statistics in the Lombardy Region. According to the national template for road crash records, the road crash data contained all the main information about the event (e.g., crash outcome, location, date), but also road infrastructure related information (e.g. road site, type of carriageway, pavement conditions), context-related information (e.g., time, day).

Traffic data for the whole road network considered were provided by Regione Lombardia. Such data were retrieved from the road traffic simulation that the Region developed for the whole regional network over the past years. Specifically, road traffic data were provided in terms of Average Annual Daily Traffic (AADT), both for light and heavy vehicles.

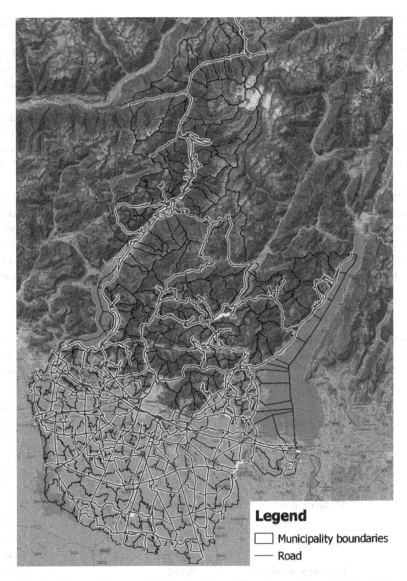

**Fig. 1.** Main road network of the Province of Brescia considered in the study

Few additional data were collected to enrich road infrastructure information from basemap data or other sources (e.g., geographical maps, Province of Brescia officers, etc.). More precisely, due to data availability and collection feasibility, data about the territorial units for statistics (i.e., NUTS), type of terrain, and road hierarchy were just included.

532     M. Bonera et al.

Once data were collected, a segmentation of the road network was performed, and each segment was identified as the section of a given road (defined by its name or code) within the boundaries of each municipality crossed by the road itself (defined by the NUTS). Then, road crash data were aggregated based on crash location information (i.e., road name or code and NUTS), to be associated to each segment of the network. Next, each segment was also associated with its specific traffic volumes and road infrastructure characteristics, retrieved from both the road crash database and the additional road infrastructure information (i.e., road hierarchy, type of terrain, number of lanes per direction). A total of 657 segments was obtained, with a length ranging from 1 to 30 km, but most of them (97%) with a length up to 7 km, as obtained also in previous research (Hosseinpour et al. 2014).

For modelling purposes, all variables were coded as binary (e.g., urban/non-urban, day/night) or in some cases categorical (i.e., more than two categories within the same variable). Table 2 provides self-explanatory descriptive statistics of the variables included in the database, divided into three categories (i.e., infrastructural-related, operational-related, and context-related variables). In the "description" column, the variables reported in italic are the reference variables for the model computation.

**Table 2.** Descriptive statistics of the road crash database

| Category | Independent variable | Description | % Value |
|---|---|---|---|
| Infrastructure | Road type | *Urban* | *36.50%* |
| | | Non-urban | 63.50% |
| | Road class | *Primary* | *31.64%* |
| | | Non primary | 68.14% |
| | Type of terrain | *Flat* | *64.75%* |
| | | Rolling | 21.58% |
| | | Mountain | 13.45% |
| | Number of lanes per direction | *One* | *67.68%* |
| | | More than one (e.g., two or three) | 32.09% |
| | Median presence | *Divided carriageway* | *26.33%* |
| | | Undivided carriageway | 73.67% |
| | Site type | Segment | 65.54% |
| | | *Intersection* | *34.46%* |
| | Pavement condition | Paved | 97.63% |
| | | *Ruined* | *2.37%* |
| | Surface conditions | *Dry* | *64.75%* |
| | | Not dry | 35.25% |
| | Road signs type | Horizontal | 19.89% |
| | | Vertical | 13.67% |
| | | Horizontal and Vertical | 57.06% |
| | | *Absent* | *9.38%* |

(*continued*)

## Table 2. (*continued*)

| Category | Independent variable | Description | % Value |
|---|---|---|---|
| Operational | % Heavy good vehicle | <25% | 55.46% |
| | | 25%–50% | 31.64% |
| | | *>50%* | *12.90%* |
| Context | Season | Spring - Summer | 49.94% |
| | | *Winter-Fall* | *50.06%* |
| | Day type | Weekday | 57.85% |
| | | *Festive* | *42.15%* |
| | Daytime | *Day* | *59.55%* |
| | | Night | 40.45% |

# 3 Results and Discussion

In this study, two models were developed for the crash frequency prediction. First, a base model (BM) was built, by including the exposure variable only (i.e., traffic volumes). Hence, in Eq. (1) no further independent variables were considered beside the AADT. Next, further models were developed, by including more independent variables to account for infrastructural, operational, and context factors, to obtain a better estimation of the crash frequency. Backward selection was also applied to improve the model performance. Indeed, backward selection (or elimination) is an iterative procedure that, starting from a full model (i.e., with all the variables included), excludes step-by-step the least significant variables to improve the model results. Hence, the model at the best fit obtained with the backward selection was considered as the final model (FM). Crash frequency models were developed by using the statistical software Genstat. Tables 3 and 4 report the model results for the BM and FM, respectively. Also, the summary statistics are reported, which are used to assess the model fit.

**Table 3.** Results of the base model (BM) for crash frequency prediction

| Predictor category | Predictor specific factors | | Coefficient estimate | P-value |
|---|---|---|---|---|
| Exposure | Model constant (i.e., $\alpha$) | | 0.2493 | 0.004 |
| | Natural log of AADT (i.e., $\beta$) | | 0.1642 | **<0.001** |

*Summary statistics*

| Source | Degree of freedom | Deviance | Mean deviance | Deviance Ratio (d. r.) | $\chi2$ |
|---|---|---|---|---|---|
| Regression | 1 | 24 | 23.8168 | 23.82 | **<0.001** |
| Residual | 1941 | 1469 | 0.7567 | | |
| Total | 1942 | 1493 | 0.7685 | | |

534     M. Bonera et al.

**Table 4.** Results of final model (FM) for crash frequency prediction

| Predictor category | Level 1 | Level 2 | Coefficient estimate | P-value |
|---|---|---|---|---|
| Exposure | | Model constant | −2.161 | 0.005 |
| | | Natural log of AADT | 0.105 | **<0.001** |
| Infrastructure | Road type | Non-urban road | 0.1309 | 0.018 |
| | Type of terrain | Rolling | −0,169 | 0.008 |
| | | Mountain | −0.667 | **<0.001** |
| | Number of lanes per direction | More than one | −0.6079 | **<0.001** |
| | Site type | Segment | 0.2041 | **<0.001** |
| | Pavement condition | Unpaved | −0.663 | 0.026 |
| | Surface condition | Wet or icy | −0.5649 | **<0.001** |
| | Road signs | Horizontal and Vertical | 0.9120 | **<0.001** |
| Operational | % Heavy Good Vehicle (HGV) traffic | <25% | −0.1245 | 0.145 |
| | | 25%–50% | −0.4048 | **<0.001** |
| Context | Season | Spring - Summer | 0.1284 | 0.013 |
| | Daytipe | Weekday | 0.3083 | **<0.001** |
| | Daytime | Night | −3.018 | **<0.001** |

*Summary statistics*

| Source | Degree of freedom | Deviance | Mean deviance | Deviance ratio (d.r.) | $\chi 2$ |
|---|---|---|---|---|---|
| Regression | 14 | 567.0 | 405.029 | 40.50 | **<0.001** |
| Residual | 2412 | 821.2 | 0.3405 | | |
| Total | 2426 | 1388.2 | 0.5722 | | |

Overall, the two models properly fitted the data, given that a large d.r. combined with a small p-value (<0.001) represent a satisfactory goodness-of-fit. More precisely, the FM fits the data better than the BM, with a d.r. equal to 40.50. With reference to Table 4, it is also possible to appreciate that most of the variables included in the model were found to be strongly significant (in bold) or significant up to 0.05 (underlined). Specifically, as for infrastructural characteristics, segments of non-urban roads and the presence of horizontal and vertical road signs were found to positively affect crash occurrence (i.e., increase crash frequency). Conversely, rolling, or mountainous terrain, segment with at least two lanes per direction, unpaved roads, or roads with wet or icy surface seem to contribute to the decrease of crash frequency.

In terms of operational characteristics, the greater the traffic volumes the higher the crash frequency. Conversely, roads with a quota of HGV ranging 25–50% of the total traffic seem to be less prone to crashes.

In terms of context characteristics, weekdays and spring-summer days seems to be critical for road crashes, as they may be more likely to happen. Conversely, night-time reduces the number of crashes on roads.

Although most of these results are confirmed by the previous literature, some considerations can be made related to the most counterintuitive results, such as the ones related to the presence of road signs, both vertical and horizontal, and the presence of two or more lanes per direction. As for road signs, on the one hand they should help the drivers understanding and respecting the driving rules as well as the road environment itself. On the other hand, too many signs or bad-maintained ones could lead to drivers neglecting or not seeing them. Indeed, similar results – even though only related to horizontal road signs – were found in previous study (Park and Lord 2007).

The presence of two or more lanes per direction should help decreasing the number of crashes over a road segments, as also found in previous studies (Ma and Kockelman 2006). Indeed, two or more lanes per directions are generally associated to divided carriageways (e.g., presence of median). If the two directions of the traffic flows are kept apart, a great quota of head-on or side-swipe crashes can be avoided, as no interference exists between the opposed flows. Highways and principal roads usually have more than two lanes per direction. However, in this case, the greater influence of the other variables (especially AADT) may reduce the beneficial effect of having two or more lanes per directions.

Then, the two models were compared, by computing the related ratio of predicted over observed crashes and the RMSE by applying Eq. (2) both over (i) the 5 years period and (ii) the five levels of the ranking scale. The results are shown in Tables 5 and 6, respectively. As for (i), the BM produced an underestimation of crashes, while the FM produced an overestimation of crashes. More precisely, an average $-11\%$ and a $+12\%$ of road crashes was predicted with the BM and FM, respectively. Looking at the RMSE, although results are quite similar between the two models, the FM showed a total value for the RMSE over the 5 years lower than the ones of the BM (Table 5).

**Table 5.** Total Observed vs Predicted crashes for the base (BM) and final (FM) models over the 5 years period

| Year | Observed | Predicted_BM | Pred_BM/Obs | Predicted_FM | Pred_FM/Obs | RMSE_BM | RMSE_FM |
|---|---|---|---|---|---|---|---|
| 2014 | 1,214 | 1,119 | $-8\%$ | 1,382 | 14% | 4.87 | 8.09 |
| 2015 | 1,220 | 1,098 | $-10\%$ | 1,375 | 13% | 6.04 | 7.48 |
| 2016 | 1,159 | 1,026 | $-11\%$ | 1,290 | 11% | 6.57 | 6.42 |
| 2017 | 1,204 | 1,050 | $-13\%$ | 1,350 | 12% | 7.47 | 7.08 |
| 2018 | 1,242 | 1,101 | $-11\%$ | 1,385 | 12% | 6.09 | 7.03 |
| Total | 6,039 | 5,395 | $-11\%$ | 6,783 | 12% | 9.88 | 8.93 |

As for (ii) and focusing on the "High" level of the ranking scale (which is the most critical in terms of road safety), although both models predicted fewer crashes than the occurred ones, FM performed better than BM as it was able to predict the 81% of the observed crashes for all the segment in "High" level (against the 66% of the BM), and it returned a RMSE lower than the one of BM (24.59 and 32.67, respectively).

Conversely, focusing on the remaining ranking levels, the BM seemed to perform better than the FM, due to lower percentage of predicted over observed crashes and lower values of RMSE. However, being "High" crash frequency segments the ones with the greatest priority for road authorities, the FM should be considered preferable in term of prediction capabilities with respect to the reference model, because it can return more accurate results for such roads. Hence, the FM was used for the road network screening and the ranking of the road segments.

**Table 6.** Observed vs Predicted crashes per ranking level (over the 5 years period)

| Year | Observed | Predicted_BM | Pred_BM/Obs | Predicted_FM | Pred_FM/Obs | RMSE_BM | RMSE_FM |
|---|---|---|---|---|---|---|---|
| High | 2,474 | 1,525 | 66% | 1,905 | 81% | 32.67 | 24.59 |
| Medium – High | 1,724 | 1,716 | 101% | 2,172 | 129% | 3.56 | 8.78 |
| Medium | 997 | 1,099 | 111% | 1,371 | 138% | 2.15 | 6.02 |
| Medium – Low | 615 | 770 | 125% | 976 | 158% | 1.82 | 4.51 |
| Low | 229 | 285 | 124% | 359 | 154% | 0.62 | 1.71 |

Therefore, by applying the FM equation, estimated crash frequency was obtained for all the segments of the network considered. Specifically, for such computation all the significant variables up to the 0.05 were included. Depending on the relative estimated road crash frequency computed over a 5-year period all the road segments were associated with a color according to Table 1. More critical roads are the ones in red and dark-red, while less critical roads are the one in green and yellow. Figure 2 and 3 report the maps of the highway's road network and the County and State road network, respectively. Indeed, such roads are managed by different bodies so that it is

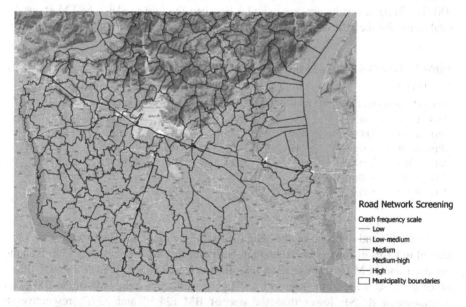

**Fig. 2.** Road network screening and Crash Frequency ranking (Highways)

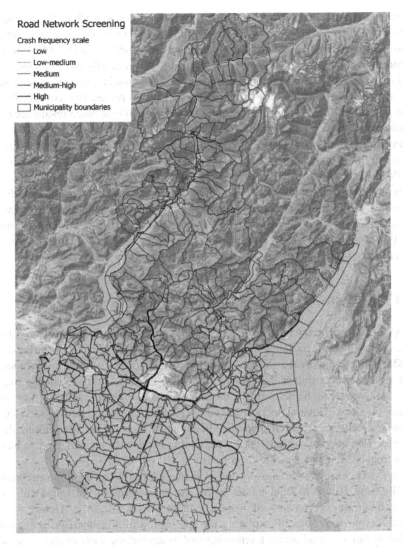

**Fig. 3.** Road network screening and Crash Frequency ranking (Major roads)

useful to rank them separately. Thanks to these representations, for a given road, the most critical segments can be easily identified, and more accurate investigations can be planned to enhance the road safety performance of the network.

Figure 2 shows how most of the highways register a medium-high value of crash frequency. Indeed, as highways, such roads register a very high traffic volume, and they also belong to the TEN-T network. Moreover, they are in operation since a quite long time so that improvements might be necessary. Just only one highway presented satisfactory crash frequency values (i.e., low, and medium-low) being this one quite new infrastructure and having lower traffic volumes compared to the others. However, highways are usually subjected to specific maintenance and control procedure and they

538    M. Bonera et al.

are managed separately, as they represent the main road network at the national and international level. Hence, it is of greater importance to focus on county and state roads, which are instead managed by the public body and local administrators, so that resources to implement road safety improvements are generally more limited.

Figure 3 shows how most of the critical road segments in terms of crash frequency (i.e., red, and dark red colour) mainly belong to major routes, that connect the main areas of the Province from south to north or from west to east. Such results are quickly explained: all those are non-urban roads with high traffic volumes and are interested by a non-negligible quota of HGV, as they directly connect economic and industrial hubs. Moreover, they are mainly built on flat terrain or on flat viaducts in the northern part of the Province. In addition, they are mainly one-lane-two-way undivided roads. These roads are paved and usually covered by draining surface (so that it can be kept dry, as it happens for highways) and are equipped with massive horizontal and vertical road signs. Hence, such roads and specifically their most critical segments require particular attention and further in-depth investigation (also by mean of on field inspections) might be necessary to identify safety issues.

# 4   Conclusions

European roads got safer over the last decades, but still the burden of road crashes remains too high. According to the European Road Infrastructure Safety Management (RISM) Directive, Member States are required to implement specific procedures on their road networks (either TEN-T and non-TEN-T roads), and specifically road network screening can be relevant in the identification of road segments that require greater attention. Several techniques are available for the implementation of road network screening such as crash prediction models, which can help monitoring the road safety performance of the network by computing in advance an estimated value of the crash frequency, given specific conditions.

However, Europewide, most of the available studies focused on the calibration of the Highway Safety Manual model or on the assessment of the impact of specific factors on crash occurrence, with respect to defined road categories. Hence, this study covers the previous gaps by exploring the application of a context-specific crash prediction model as a road network screening tool to support the work of those responsible for road safety.

By applying generalised linear model theory with a Negative Binomial distribution, two models were developed and validated to predict road crash frequency on the main road network of the Province of Brescia (northern Italy). Based on the results of the best-fitted model, a ranking of the road segments was provided, and maps were produced as a decision support tool. Furthermore, being the model oriented to a wide-level analysis of the road network, few and easy-to-collect variables were included, to foster the interpretability, updatability, and replicability of the method. Indeed, the variables helped in explaining the screening results of the network, especially for the most critical segments.

The study provided a good experiment as a total 2.500+ kms of roads were analysed with a total of 6.039 road crashes occurred on the main road network of the

Province of Brescia over the five-years period 2014–2018. An exposure-only model (Base Model) and multivariable model (Final Model) were developed, and their prediction capabilities compared. The Final Model resulted to be preferable to the Base Model and it was used to rank the segments of the network, based on their predicted crash frequency level. Maps were produced accordingly, which helped visualizing the most critical segments of the network.

Such models and procedures represent a valuable decision support tool for road authorities, and they may be easily applicable also on other Italian areas (Counties or Regions), being the data source (e.g., crash reports) the same and presenting quite similar road network characteristics.

Nonetheless, the study can be further improved. First, more variables (e.g., road features) may be included to increase the prediction capabilities and test their effect on crash frequency (Elvik et al. 2009). Then, models should be tested also on other road networks of the surrounding Provinces to confirm the results obtained in this study. Second, as suggested by previous studies, along with crash frequency also crash severity should be considered to assess the road safety performance of the network and rank the road segments. A viable solution would be considering a bivariate (frequency and severity) model that account for both the road safety parameters as applied by recent research in public transport (Porcu et al. 2020; Barabino et al. 2021; Porcu et al. 2021).

**Acknowledgments.** This research was carried out thanks to the contribution of Polis-Lombardia, within the Agreement relative to the support for study activities of the Regional Road Safety Monitoring Centre of the Lombardy Region.

# References

AASHTO: Highway Safety Manual, First Ediction. American Association of State Highway and Transportation Officials (2010)

Ambros, J., Jurewicz, C., Turner, S., Kieć, M.: An international review of challenges and opportunities in development and use of crash prediction models. Eur. Transp. Res. Rev. **10** (2), 1 (2018). https://doi.org/10.1186/s12544-018-0307-7

Barabino, B., Bonera, M., Maternini, G., Olivo, A., Porcu, F.: Bus crash risk evaluation: an adjusted framework and its application in a real network. Accid. Anal. Prev. **159**, 106258 (2021). https://doi.org/10.1016/j.aap.2021.106258

Bonera, M., Maternini, G.: Methodology for the application of predictive method for road safety analysis in urban areas. The case study of Brescia. Transp. Res. Procedia **45**, 659–667 (2020)

Borghetti, F., Marchionni, G., De Bianchi, M., Barabino, B., Bonera, M., Caballini, C.: A new methodology for accidents analysis: the case of the State Road 36 in Italy. Int. J. Transp. Dev. Integr. **5**(3), 278–290 (2021)

Dragomanovits, A., et al.: Use of accident prediction models in road safety management – an international inquiry. Transp. Res. Procedia **14**, 4257–4266 (2016)

Elvik, R.: Assessment and applicability of road safety management evaluation tools: current practice and state-of-the-art in Europe (Issue 0349) (2010)

Elvik, R., Vaa, T., Hoye, A., Sorensen, M. (eds.): The Handbook of Road Safety Measures. Emerald Group Publishing ETSC (2020). Pin briefing. The impact of covid-19 lockdowns on road deaths in April 2020 (2009)

European Commission: EU road safety policy framework 2021-2030-Next steps towards "Vision Zero". Publications Office of the European Union, Luxembourg (2020)

European Road Assessment Program: RAP Crash Risk Mapping: Technical Specification (Issue January) (2020)

European Union: Directive 2008/96/EC of the European Parliament and of the Council on Road Infrastructure Safety Management (2008)

European Union: Directive 2019/1936 of the European Parliament and of the Council amending Directive 2008/96/EC on Road Infrastructure Safety Management (vol. 2019, Issue 1315) (2019)

Eurostat: Eurostat regional yearbook - 2020 edition (2020). https://ec.europa.eu/eurostat/web/products-statistical-books/-/ks-ha-20-001

Farid, A., Abdel-Aty, M., Lee, J.: Transferring and calibrating safety performance functions among multiple States. Accid. Anal. Prev. **117**(April), 276–287 (2018)

Farid, A., Abdel-Aty, M., Lee, J., Eluru, N., Wang, J.H.: Exploring the transferability of safety performance functions. Accid. Anal. Prev. **94**, 143–152 (2016)

Gupta, A., Bansal, A.: Integrating traffic datasets for evaluating road networks. In: Proceedings - 12th IEEE International Conference on Semantic Computing, ICSC 2018, 2018-January, pp. 411–416 (2018)

Hauer, E., Kononov, J., Allery, B., Griffith, M.S.: Screening the road network for sites with promise. Transp. Res. Rec. **1784**, 27–32 (2002)

Hosseinpour, M., Yahaya, A.S., Sadullah, A.F.: Exploring the effects of roadway characteristics on the frequency and severity of head-on crashes: Case studies from Malaysian Federal Roads. Accid. Anal. Prev. **62**, 209–222 (2014)

Jiang, X., Abdel-Aty, M., Alamili, S.: Application of poisson random effect models for highway network screening. Accid. Anal. Prev. **63**, 74–82 (2014)

La Torre, F., et al.: Development of a transnational accident prediction model. Transp. Res. Procedia **14**, 1772–1781 (2016)

La Torre, F., Meocci, M., Domenichini, L., Branzi, V., Tanzi, N., Paliotto, A.: Development of an accident prediction model for Italian freeways. Accid. Anal. Prev. **124**, 1–11 (2019)

La Torre, F., et al.: Accident prediction in European countries – development of a practical evaluation tool. In: Proceedings of 7th Transport Research Arena TRA (2018)

Lord, D., Mannering, F.: The statistical analysis of crash-frequency data: a review and assessment of methodological alternatives. Transp. Res. Part A: Policy Pract. **44**(5), 291–305 (2010)

Ma, J., Kockelman, K.M.: Poisson regression for models of injury count, by severity. Transp. Res. Rec. J. Transp. Res. Board **1950**, 24–34 (2006)

Mannering, F.L., Bhat, C.R.: Analytic methods in accident research: methodological frontier and future directions. Anal. Methods Accid. Res. **1**, 1–22 (2014)

Park, E.S., Lord, D.: Multivariate poisson-lognormal models for jointly modeling crash frequency by severity. Transp. Res. Rec. **2019**, 1–6 (2007)

Park, P.Y., Sahaji, R.: Safety network screening for municipalities with incomplete traffic volume data. Accid. Anal. Prev. **50**, 1062–1072 (2013)

Persia, L., et al.: Management of Road Infrastructure Safety. Transp. Res. Procedia **14**, 3436–3445 (2016)

Polis-Lombardia: L'incidentalità sulle strade della Lombardia nel 2019 (2020). www.polis.lombardia.it

Porcu, F., Olivo, A., Maternini, G., Barabino, B.: Evaluating bus accident risks in public transport. Transp. Res. Procedia **45**, 443–450 (2020)

Porcu, F., Olivo, A., Maternini, G., Coni, M., Bonera, M., Barabino, B.: Assessing the risk of bus crashes in transit systems. Eur. Transp./Trasporti Europei **81** (2021)

Stipancic, J., Miranda-Moreno, L., Saunier, N., Labbe, A.: Network screening for large urban road networks: using GPS data and surrogate measures to model crash frequency and severity. Accid. Anal. Prev. **125**, 290–301 (2019)

Viera Gomes, S., Cardoso, J.L., Azevedo, C.L.: Portuguese mainland road network safety performance indicator. Case Stud. Transp. Policy **6**(3), 416–422 (2018)

Wijnen, W., et al.: An analysis of official road crash cost estimates in European countries. Saf. Sci. **113**, 318–327 (2019)

Yannis, G., et al.: Use of accident prediction models in road safety management - an international inquiry. Transp. Res. Procedia **14**, 4257–4266 (2016)

Yannis, G., et al.: Road safety performance indicators for the interurban road network. Accid. Anal. Prev. **60**, 384–395 (2013)

# An iRAP Based Risk Impact Analysis at National Highway-1 for a Proposed Route Connecting Coastal Areas of Bangladesh

Armana Sabiha Huq[✉]

Bangladesh University of Engineering and Technology, Dhaka, Bangladesh
ahuq002@fiu.edu, ashuq@ari.buet.ac.bd

**Abstract.** The coastal areas of Bangladesh are poorly connected with the mainland only by waterways halting their economic growth and ability to exacerbate the potential tourism sector. According to the World Travel and Tourism Council (WTTC) and United States Agency for International Development (USAID), the tourism and fisheries sector contributes only 2% and 4.4% in the national GDP of Bangladesh despite having potentiality in these sectors in the coastal region of Bangladesh. According to the department of statistics in the University of Chittagong, about 15% of the road accidents in 2017 occurred in the Barisal and Khulna divisions. This research aims to assess the risk associated with the adjacent points of the national highways for a proposed road alignment to be built in the country's coastal region connecting with the mainland. This study conducts the Corridor Selection Matrix to select the best roadway alignment and a semiquantitative approach to analyze the effect of different factors for road accidents in the National Highway-1 (N1) at the Chittagong district. Analyzing the existing road traffic condition by field visit, International Road Assessment Program (iRAP) and Global Information System (GIS) most risky areas found for different exposure conditions, locality and pedestrian characteristics will be determined. This research will help to understand the road accident severity and its pattern in Chittagong with the growing number of development i.e. tourism in particular and make a proper safety framework to prevent road accidents in the country by better understanding the related factors.

**Keywords:** Risk assessment · Road accident · Transportation risk factors · Road safety · Bangladesh transportation risk · iRAP

## 1 Introduction

Road accidents are being a major issue in developing countries now a day. It is hampering both the economic growth and infrastructural development of a country. According to the World Health Organization (WHO), 80% of the total deaths caused by road accidents occurred in middle-income countries, which has surprisingly increased by 60% from 2013. About 20.7 people out of 1,00,000 die each year from road accidents in Southeast Asia which is the second-highest in the world after Africa. (Global status report on road safety 2018). 4,000 casualties occurred in the whole world because of road accidents which cause about $518 billion per year, highlighted a major

---

© The Author(s), under exclusive license to Springer Nature Switzerland AG 2022
A. Akhnoukh et al. (Eds.): IRF 2021, SUCI, pp. 542–555, 2022.
https://doi.org/10.1007/978-3-030-79801-7_39

concern about the safety and reduction of accidents all over the world (Islam Bin and Kanitpong 2008).

From Bangladesh's perspective, road accidents largely affect the country's social life and economic development. About 3,000 people die every year and 2,650 people are injured in road accidents in Bangladesh (Ahsan et al. 2021). More than 43% of deaths related to road accidents occur on the national highways, covering around 5% of the total roads in the country (Hoque and Mahmud 2009). Dhaka-Chittagong Highway (N1) is a backbone for the economy of Bangladesh. Every year traffic growth on this highway is increasing by 21.03%, which is much higher than normal growth (Ullah et al. n.d.). Also, vehicles need to use this highway to travel the south-east side of the country from the south-west part, which takes a long time and kilometers to travel. For this reason, an alternative roadway is badly needed to divert the traffic from N1 connecting the southern coastal parts of Bangladesh. This road will also join the country's coastal region to the mainland, which has no sufficient road network at the time. In this research, we tried to objectify:

- Finding the best road alignment to connect the coastal region of the country with the mainland through Chittagong.
- Accidental pattern and factors for the portion of N1 to connect with our proposed route alignment.
- Detailed risk impact analysis of the portion of N1 where our proposed route will intersect.

## 2 Literature Review

The existing road traffic risk and safety analysis literature mainly focus on multi-criteria decision-making modeling, simulation modeling, Hidden Markov Modeling, mathematical modeling, statistical data analysis, and questionnaire survey. The subject is briefly presented below. (Tam2 1998) developed a simulation model based on the Monte Carlo technique, used Zhuhai Neilingding Crossing in South China as a case study for risk analysis. This study evaluated the individual probability of traffic and revenue forecasts every year. Also, the authors conducted a Sensitivity analysis to examine the effect of some parameters related to probability values and the toll charges on traffic demand. (Bubbico et al. 2004) Transportation risk analysis instruments like geographic information systems (GIS) can be helpful to gather more local reports about any accident location. (Ozbas and Altıok 2009) conducted a study in the Strait of Istanbul to analyzed safety-related risk for transit vessel maritime traffic and the analysis was performed by including a probabilistic accident risk model. Also, they developed a mathematical risk model. Accidental risk behavior related to change in geographical, weather and traffic conditions of the area obtained by Scenario analysis. The authors found two main factors- local traffic density and pilotage turned out to affect the risks at the Strait of Istanbul.

(Nordfjærn et al. 2011) presented differences in road traffic risk between Norway, Russia, India, Ghana, Tanzania, and Uganda. Also picked up difference's attitudes and behavior of the driver. Analysis was completed based on a questionnaire survey. The authors concluded that road traffic risks were higher in low-income countries in Sub-Saharan Africa than in Norway, Russia and India.The model used factors such as risk, population characteristics, and road traffic attitudes to predict driver behavior that were appropriate for Norway, Russia, and India.On the other hand, it was not appropriate for Sub-Saharan African countries. Reseaech shows lack of this human factor consideration which would be very effective in reducing accidents rate. (Banik et al. 2011) carried out a study to highlights the road traffic accidents and road safety situation in Sylhet. Authors conducted a questionnaire survey on transportation related groups where was drivers, pedestrians, traffic police and collected accident data from local newspaper. They pointed out some reasons that were responsible for the accident such as vehicle head, loss of control of the vehicle, overtaking, sudden breaking, carrying the extra load by vehicle, the speed of the vehicle, collision, while taking turns, peoples'disability, rail crossing etc. Found that the most dangerous highway is Dhaka-Sylhet highway and the most vulnerable thana is Sylhet Sadar thana.

(Bergel-Hayat et al. 2013) conducted a study to highlight the link between an overall weather situation and the risk of road accidents. They considered three weather factors such as rainfall, temperature, and frost to see the effect on road accidents in France, Netherlands, and Athens. They conduct data analysis on the 20 years accident with the weather conditions. They found that in the Netherlands and France, risk and fatalities of accidents are directly connected to rainfall and with the square of temperature wherein Athens accident decreases with rainfall and temperature affect the accidents only in winter. Both France and Netherlands have shown a decrease in accidents in the time of frost (Bhavsar et al. 2017) presented the causes of autonomous vehicle failure in mixed traffic streams. The authors used the fault tree–based risk analysis method and the outcome of this analysis can be used to develop risk minimization strategies. Reviewing the published literature and publicly available data sources used to estimate the failure probabilities. The result showed that continuous innovations in computing, communication technologies, and installing backup sensors significantly affect reducing failure probability. Mechanical failure did not consider in this study. (Cioca and Ivascu 2017) evaluated the causes of accidents, risk and performance of road safety indicators. They used a qualitative and semiquantitative method based on different variables like collision mode, road configuration, occurrence condition, category of road, type of vehicle involved, personal factors and driving license in Romania from 2012–2016. The authors concluded that the factors that affected traffic and road accidents were Vehicle and personal adjustment.

(Ul Baset et al. 2017) focused on the current prevalence and risk factors of road traffic injury (RTI) for different age groups based on a structured questionnaire survey in rural Bangladesh. Analysis showed that RTI deaths occurred in more than one-third of pedestrians and men had higher mortality and morbidity rates than women. The risk factors for fatal and non-fatal RTI were carried out Pedestrian or student, lower socioeconomic condition and no education. (Zheng et al. 2018) studied the improvement of intelligent vehicles and assessed how to prevent road traffic accidents of intelligent vehicles. This study mentions a model for quantitative traffic risk analysis based on Hidden Markov Model (HMM). Authors mentioned some parameter related to road traffic environments like pedestrians, cyclists, vehicles, and this study only works on vehicle motion. (Hossain and Faruque 2019) conducted a study to evaluate the road accidents reasons and review the situation in Bangladesh. They proposed the road traffic accident assessment based on recorded data from 1971 to 2017. The analysis found that the Dhaka division has the most road traffic accidents (34%) and deaths due to road traffic accidents (32%). Most of the injuries occurred with the 21 to 30 age group and death occurred with the 11 to 30 age group, the cause of accidents and deaths is head-to-head collisions and run over by vehicles. Most road accidents occur during major religious festivals because there is more traffic than usual. (Zafri et al. 2020) showed the road crossing characteristic of pedestrians and the risks associated with a different pedestrian group. The study has completed considering five risk factors: nearest vehicle speed, nearest vehicle type, vehicle flow, interrupted by vehicle, and minimum gap, using the combined AHP-TOPSIS method. Obtained most at risk were a male and young pedestrian group, where groups of women and the older pedestrian group were much less at risk. Also evaluated the behaviors that cause accidents and increase the risk of the pedestrians, besides evaluated the safest behaviors which will control the problems. (Stanković et al. 2020) developed a fuzzy MARCOS algorithm to support multi-criteria decision-making model for traffic risk assessment. To determine the degree of risk Authors analyzed a road network 7.4 km length which divided into 38 short section (200 m each) and found that $23^{rd}$ section is the most hazardous section.

## 3 Methodology

We tried to focus on the risk study for a roadway project in the coastal areas of Bangladesh in a qualitative and semiquantitative way. First, the alignment of the best route for industrial, infrastructure, and tourism development has been selected based on different factors by the corridor selection matrix. The route chosen connects to the existing national highway N1 in Kumira union of Sitakunda Upazila under the Chittagong districts, where we analyzed the risk for transportation. Data for the survey and finding the accident pattern is collected from the Accident Research Institute (ARI), Bangladesh University of Engineering and Technology (BUET). The selected portion of the highway was determined both by field exploration through google street view and by the International road assessment program (iRAP) (Fig. 1).

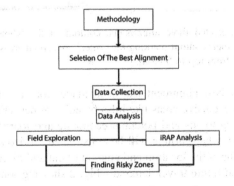

**Fig. 1.** Methodology of our work

## 4 Selection of Best Alignment

Despite having great potential, the people of the coastal region of Bangladesh faces difficulties contributing to the country's economy by not having proper and faster connectivity with the mainland. The second big city of the country is located near the project area, but to travel to Chittagong, the people have to rely on boats and water transport which is risky and slow at the same time. For this reason, three notable routes were selected as a preliminary selection to connect the coastal areas with the Chittagong district. They were:

- Alignment 1: Chittagong-Sandwip-Urirchar-Noakhali-Hatiya-Monpura-Bhola
- Alignment 2: Chittagong-Sandwip-Noakhali-Hatiya-Monpura-Bhola
- Alignment 3: Chittagong-Sandwip-Urirchar-Noakhali-Bhola

The purpose of the proposed highway is to improve the living conditions of the rural people living in this area. As well as make faster transportation of goods from this area. This highway will also enhance the tourism industry to flourish in the country's coastal region, potentially becoming a tourist hotspot. Moreover, the site is also suitable for building economic zones under the Bangladesh Economic Zone Authority (BEZA).

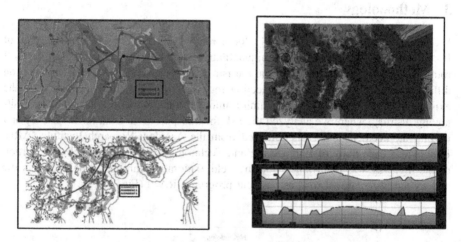

**Fig. 2.** (Clockwise) Selected three alignments, contour of the whole project area, selected alignments along with their contour marking the critical concerned areas and elevation profile of alignments 1, 2 and 3 from top to bottom.

The selection of best alignment has been made from a comparison of different parameters. Such as exclusive areas like hilsa breeding zones, critical concerned areas, control points like big towns and potential economic and tourist zones, length of the route, seawater depth where bridge will be constructed, horizontal and vertical alignments like the contour of the location and amount of cut and fill required, cost analysis, cost-benefit ratio, and future travel demand—Fig. 2 showing some of these features of our project.

Table 1 shows the internal rate of return (IRR) of the 3 alignments. Alignment 2 gives the best IRR and alignment 1 gives the least. According to the other factors for all three alignments, the best one is selected by using a corridor selection matrix where each route is rated using a scale of 0 − 5, with 0 representing the best alternative and 5 being the worst for each one of the critical site elements (Table 2).

**Table 1.** Internal rate of return of the selected three alignment

Cost/Benefit in the form of IRR

| Alignment | Bridge (km) | Road (km) | Cost of Bridge In Billions BDT | Cost of Ro In Billions BDT ad | Total Cost in Billions BDT | Toll BDT/Veh | AADT- 2030 | Toll Collected per day in Billions BDT | Years To Repay | IRR |
|---|---|---|---|---|---|---|---|---|---|---|
| 1 | 64.25 | 112 | 12.65 | 28.4 | 41.1 | 1,500 | 17,000 | 0.00612 | 18.4 | 5.44 |
| 2 | 54.72 | 60.3 | 70.78 | 15.4 | 26.2 | 1,500 | 15,000 | 0.0054 | 13.28 | 7.52 |
| 3 | 69.5 | 37.5 | 13.69 | 9.6 | 23.3 | 1,500 | 7,000 | 0.0025 | 14.88 | 6.72 |

**Table 2.** Corridor selection matrix for the selection of best alignment

| Criteria | No build | Alignment 1 | Alignment 2 | Alignment 3 |
|---|---|---|---|---|
| Exclusive Area | 0 | 2 | 3 | 4 |
| Concerned Area | 0 | 1 | 2 | 1 |
| Control Point | 4 | 1 | 3 | 4 |
| Length of Road | 5 | 3 | 2 | 1 |
| Length of Bridge | 0 | 2 | 1 | 3 |
| Travel Time | 5 | 3 | 2 | 1 |
| Vertical-Horizontal Alignment | 5 | 2 | 1 | 3 |
| Cost and IRR | 0 | 2 | 3 | 1 |
| Travel Demand | 4 | 1 | 2 | 3 |
| Seawater Depth | 0 | 2 | 2 | 3 |
| Total | 23 | 20 | 21 | 24 |

So, alignment 1 is chosen as the best alignment among the three alignments from the corridor matrix.

## 5 Road Accident Situation in the Selected Coridor

Accident data for the year 2010–2015 was collected from the Accident Research Institute (ARI) of Bangladesh University of Engineering and Technology (BUET). Bangladesh Police collects every accidental data all over the country and registered in the Accident Research Form (ARF). With the help of the Modular Accident Analysis Program (MAAP), we collect the data for our survey locations in Mirarsarai and Sitakunda Upazila. Data shows a detailed overview of the random pattern in the Chittagong district, Mirarsarai and Sitakunda Upazila. We analyzed the data based on different Junction Types, Road Geometry, Road Features, Vehicle Type, Casualties by Sex and Age, Collision Type.

548     A. S. Huq

**Table 3.** Percentage of the accident occurred in Sitakunda and Mirarsarai Upazila with the Chittagong District from 2010 to 2015

| Type | Total Collision | | | | | | | | | |
|---|---|---|---|---|---|---|---|---|---|---|
| | Fatal | | | | | Non-Fatal | | | | |
| | Chitta gong | Sitak unda | % of Chitta gong | Mirar sarai | % of Chitta gong | Chitta gong | Sitak unda | % of Chitta gong | Mirar sarai | % of Chitta gong |
| Juncti on | 224.0 | 81.0 | 36.2 | 20.0 | 8.9 | 50.0 | 10.0 | 20.0 | 5.0 | 10.0 |
| Road Geom etry | 225 | 81 | 36.0 | 20 | 8.9 | 52 | 11 | 21.2 | 5 | 9.6 |
| Road Featu re | 222 | 80 | 36.0 | 20 | 9.0 | 50 | 11 | 22.0 | 5 | 10.0 |
| Vehic le Type | 306 | 101 | 33.0 | 29 | 9.5 | 77 | 19 | 24.7 | 9 | 11.7 |
| By Sex | 357 | 112 | 31.4 | 48 | 13.4 | 87 | 21 | 24.1 | 2 | 2.3 |
| By Age | 240 | 70 | 29.2 | 38 | 15.8 | 49 | 9 | 18.4 | 2 | 4.1 |

From Table 3, we can see that Sikaunda shares 29.2–36.2% of total accidents in different types for the whole Chittagong district. In non-fatal cases, the values range from 18–20%, which is higher than the other Upazilas of Chittagong. From the subdivision, we found that around 38% of accidents in the Chittagong district have occurred in Sitakunda at the straight roads and 25% occurred in the same Upazila at curves and sloppy regions. Also, from all 9 accidents in the district at T-junctions, 6 were in the Sitakunda (66%) and 224 accidents that occurred for not having proper road features in the district 79 (37%) were in Sitakunda and 19 (9%) were in Mirarsarai Upazila.

**Table 4.** Collision based on vehicle Type in the Sitakunda, Mirarsaraai Upazilas and their % concerning whole Chittagong.

| Vehicle type wise collision | | | | | | | | | | |
|---|---|---|---|---|---|---|---|---|---|---|
| Type | Fatal | | | | | Non-fatal | | | | |
| | Ctg | Sitakunda | % of Ctg | Mirarsarai | % of Ctg | Ctg | Sitakunda | % of Ctg | Mirarsarai | % of Ctg |
| Cycle | 2 | 1 | 50 | 0 | 0 | 2 | 0 | 0 | 0 | 0 |
| Motorbike | 16 | 4 | 25 | 2 | 12.5 | 4 | 0 | 0 | 0 | 0 |
| Bus | 60 | 17 | 28.3 | 5 | 8.3 | 19 | 4 | 21.1 | 1 | 5.3 |
| Car | 4 | 2 | 50 | 0 | 0 | 2 | 0 | 0 | 1 | 50 |
| Heavy Truck | 89 | 44 | 49.4 | 13 | 14.6 | 9 | 3 | 33.3 | 5 | 55.6 |
| Baby Taxi | 33 | 7 | 21.2 | 4 | 12.1 | 10 | 1 | 10 | 0 | 0 |
| Others | 20 | 9 | 45 | 0 | 0 | 5 | 0 | 0 | 1 | 20 |

Table 4 shows the specific vehicle type and their contribution in the two Upazilas of our survey area. Finally, Fig. 3 shows the specific type of collisions in the Sitakunda Upazila where pedestrians vulnerability and head-on collision are the significant types responsible for more casualties. From 2010 to 2015 heavy truck was the most dangerous vehicle that took 43% of lives than other types of vehicles. Bus was in the second position, responsible for 16.8% of total fatalities. Among the people who lost their lives in the Sitakunda-Mirarsarai portion of N1 during 2010–2015, most aged between 21–40 years. From the casualties in the whole district based on sex, 34% were males and 16% were females form Sitakunda, as shown in Fig. 3.

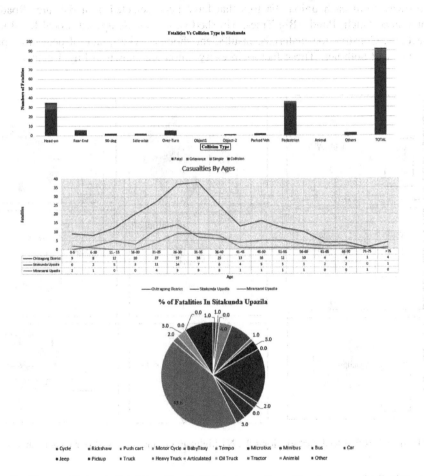

**Fig. 3.** (Top to Bottom) Fatalities Vs. Collision type, Casualties by age distribution and percentage vehicle type to cause fatalities in Sitakunda Upazila from 2010 to 2015

## 6 Current Scenario of Study Area

For the current pandemic, we couldn't move outside and visit the area physically, so we took help from google street view to study the risk factors in the location of our study area. Our study started at the Kamaldaha of Mirarsarai Upazila and ended at the Bhatiyari union of Sitakunda Upazila. The corresponding latitude and longitudes of the start and end position of the survey area are 22.65419408973514; 91.63783802876509 and 22.452025544121238; 91.73296263187791. We thoroughly explored the whole area and tried to find out the potential risks and vulnerable structures alongside the roadway from each union. Factors that have been selected as a risk are: Roadside buildings, Ditch, Ponds, Big Trees, Tin shed structures, Scrapyard, Local Road intersections, Construction materials, insufficient shoulder width, lack of pedestrian pavements, and potholes. These factors in every union are shown in Fig. 4 below:

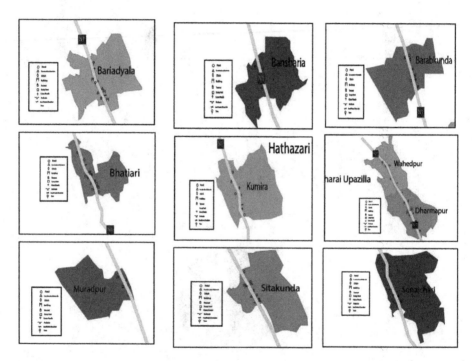

**Fig. 4.** Classified Risk factors along the N1 highway in nine unions of our located survey area.

The concentration of the point of risk factors in the whole survey area is placed on a single map that signifies the region's riskiest areas. The location of the risky areas was documented with their Latitude and Longitudes while exploring the locations and from these data, a GIS-based map having the locations of all classified risky zones was made. Figure 5 shows all the points of risky factors in the whole region. Some deviation has been seen while plotting the location point in the map because of the coordinate system WGS84.

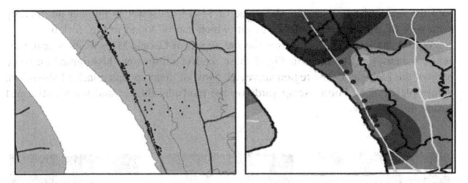

**Fig. 5.** (From Left to right) Locations of the risky zones in the survey area and Risk severity of the locations according to their risk factors and ratings

To better understand the risk impacts, we determined the locations where the concentration of the risk factors is the most. This was done both from the visual exploration and from the map locations by selecting the clusters of location points. Then, giving each of these locations a rating of risk factors according to the classified risk associated with them, we tried to find the distribution of risky areas. Figure 5 shows the severity of the risks in the specific locations, with their corresponding ratings extended from 1 (mildest) to 14 (Severe).

These areas are finally plotted in the whole map, including all the region's unions, to visualize the locations simply. For example, Fig. 6 shows places in the proximity of the N1 highway going through the Shitakunda and Mirarsarai Upazilla.

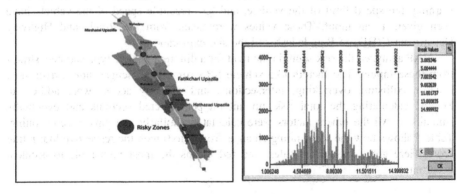

**Fig. 6.** (From Left to right) Risky zones highlighted in red in the whole survey area and Distribution of the risk severity factors

Figure 6 (right) shows that risk severity in the whole area is slightly left-skewed normally distributed, signifies that most of the risk severity areas are moderately risky, having the values of 5–7, which are situated in the northern parts of our surveyed area.

And the major risky areas with the value of 11–14 are lower in number and mainly located in the southern part of the region, which can be found in Fig. 6 (left).

We also took some pictures from the street view of Google Maps to showcasing the actual scenarios of the area. In Fig. 7 these road conditions are highlighted on some parts of the highway of the region surveyed. Most of them contain a lack of shoulders, vulnerable tin shed house, scrap yards on the roadside, ditches and waterbodies just beside the highway.

**Fig. 7.** Actual Road conditions in the research area

## 7 iRAP Analysis

To find the risky zones in this portion of the highway, we have analyzed our data in iRAP. Altogether 33 points have been selected in the highway area from where different data such as road width, shoulder width, the position of trees, lanes, quality of curve, grade, skid resistance, delineation, street lighting, pedestrian crossing facilities, vehicle parking, service roads, sight distance, AADT, differential speed, school zone warning, the speed limit of the vehicle, and 85 percentile speed of the vehicle have been given as an input. These values were taken from the Roads and Highway Department (RHD) of Bangladesh and the site exploration.

iRAP analyzed the risk score in the road for a different vehicle type and pedestrians from these inputs. The factors like vehicle SRS run-off, passenger side, driver side, head-on collision, overtaking, intersections, and property access were addressed. Finally, calculating the total risk amount on per selected sections and eventually smooths it. All the others factors were calculated similarly and gave a score rating. Table 5 shows how the score rating changes from 5 to 0 with the respective Star rating score. Here green means the safest and red means the most vulnerable to accident occurrence.

**Table 5.** SRS Rating Bands and colors. source: https://irap.org/methodology

| Star Rating Score |||||
|---|---|---|---|---|
| Star Rating | Vehicle Ocupencies and Motorcyclist | Byclists | Pedestrians |||
| ^ | ^ | ^ | Total | Along | Crossing |
| 5 | 0 to <2.5 | 0 to <5 | 0 to <5 | 0 to <0.2 | 0 to <4.8 |
| 4 | 2.5 to <5 | 5 to <10 | 5 to <15 | 0.2 to <1 | 4.8 to <14 |
| 3 | 5 to <12.5 | 10 to <30 | 15 to <40 | 1 to <7.5 | 14 to <32.5 |
| 2 | 12.5 to <22.5 | 30 to <60 | 40 to <90 | 7.5 to <15 | 32.5 to <75 |
| 1 | 22.5+ | 60+ | 90+ | 15+ | 75+ |

Figure 8 shows the star rating score and corresponding star rating for the bicyclist, vehicle occupancy, motorcyclist and pedestrians SRS values. We can see all of the four graphs indicate a score rating 1 on most of the highways. Which means the portion is risky for all kind of movements. Moreover, some areas are showing higher values than others. From Fig. 9, we can get an idea about the riskiest areas where all the graph shows similar high impact for the accident probability. However, these portions of sites were also detected as dangerous zones in the field exploration as well. Therefore, both analytical and visual data converged in the same part of the N1 in our survey area, reflecting the correctness of the results.

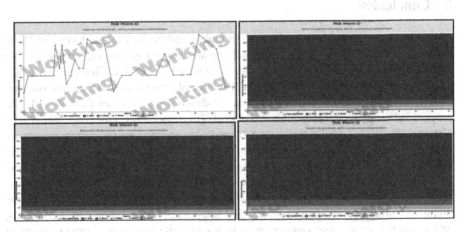

**Fig. 8.** SRS values for (clockwise) pedestrians, vehicle occupancy, motorcyclists, and bicyclists

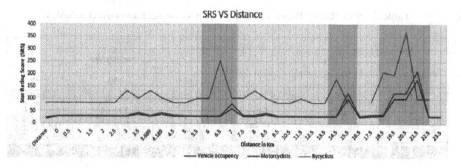

**Fig. 9.** Areas in the surveyed portion of N1 selected as a red zone in the graph of SRS values Vs. Distance of vehicle occupancy, motorcyclists, and bicyclists.

The analysis says that from the starting point, which was Kamaldaha in the Mirarshrai Upazilla, risky zones are at 6–7 km, 14.5–16.5 km and 18.5–22.5 km range indicating Sitakunda, Barabkunda and Sonaichari unions which were already marked as risky zones in this report by field exploration. These results are also justified by the research of the Accident Research Institute (ARI), from Bangladesh University of Engineering and Technology (BUET) says most of the portion of the N1 is rated 2 stars out of 5 for the mentioned sectors (Saleh and Islam, n.d.).

## 8 Conclusion

The research has a practical significance to understand the risk in the portion of N1 at Sitakunda Upazila. It shows the accidental rate and percentage for different factors and specific types of vehicle, sex and ages for the connection point at Chittagong with the alignment that we selected for the coastal region of Bangladesh. Alignment one can actively contribute to the country's economic development with an IRR of 5.14. But proper road safety is a must to achieve that goal. This paper reflects the risky zones of the National Highway 1 in the region of Sitakunda-Mirarsarai Upazila, which contributes about 30% in all the accidents of Chittagong District. Filed exploration with google street view and iRAP analysis combined shows the accident-prone areas where the Star Rating Score was more than safe. However, the whole road was more or less unsafe, with a 2-star score in most parts. Insufficient shoulder width, Big trees, ditches and Ponds alongside the highway, vulnerable tin-shed structures and scrap yards, intersection with local roads, and accessibility for the non-motorized vehicles were the fundamental reasons for most accidental events. Finally, the riskiest areas were selected as Sitakunda, Barabkunda, and Snaichari Unions based on iRAP analysis and filed exploration.

**Acknowledgement.** We want to express our gratefulness to the Accident Research Institute, BUET, Bangladesh Police and Roads and Highway Department of Bangladesh for helping us providing the necessary data and information. We also want to show our gratitude to Mr. Atiq Azad, Graduate Research Assistant, BUET for accident data extraction from the MAAP5 software.

# References

Ahsan, J., Roy, S., Huq, A.S.: An in-depth estimation of road traffic accident cost in bangladesh. **2021**(2012), 12–13 (2021)

Banik, B.K., Chowdhury, M.A.I., Hossain, E., Mojumdar, B.: Road accident and safety study in sylhet region of Bangladesh. J. Eng. Sci. Technol. **6**(4), 503–515 (2011)

Bergel-Hayat, R., Debbarh, M., Antoniou, C., Yannis, G.: Explaining the road accident risk: weather effects. Accid. Anal. Prev. **60**, 456–465 (2013)

Bhavsar, P., Das, P., Paugh, M., Dey, K., Chowdhury, M.: Risk analysis of autonomous vehicles in mixed traffic streams. Transp. Res. Rec. **2625**, 51–61 (2017)

Bubbico, R., Cave, S.D., Mazzarotta, B.: Risk analysis for road and rail transport of hazardous materials: a GIS approach. J. Loss Prev. Process Ind. **17**(6), 483–488 (2004)

Cioca, L.I., Ivascu, L.: Risk indicators and road accident analysis for the period 2012–2016. Sustainability **9**(9), 1–15 (2017)

Hoque, M., Mahmud, S.M.S.: Accident Hazards on National Highways in Bangladesh (2009)

Hossain, M.S., Faruque, M.O.: Road traffic accident scenario, pattern and forecasting in Bangladesh. J. Data Anal. Inf. Process. **07**(02), 29–45 (2019). https://doi.org/10.4236/jdaip.2019.72003

Islam Bin, M., Kanitpong, K.: Identification of factors in road accidents through in-depth accident analysis. IATSS Res. **32**(2), 58–67 (2008)

Nordfjærn, T., Jrgensen, S., Rundmo, T.: A cross-cultural comparison of road traffic risk perceptions, attitudes towards traffic safety and driver behaviour. J. Risk Res. **14**(6), 657–684 (2011)

Ozbas, B., Altıok, T.: Risk analysis of the vessel traffic in the strait of Istanbul. Risk Anal. Int. J. **29**(10), 1454–1472 (2009)

Saleh, S., Bin Al Islam, S.M.A.: Safety investigation and assessment of Dhaka-chittagong national highway in Bangladesh, pp. 1–13 (n.d.)

Stanković, M., Stević, Ž, Das, D.K., Subotić, M., Pamučar, D.: A new fuzzy marcos method for road traffic risk analysis. Mathematics **8**(3), 1–17 (2020)

Lam, W.H.K., Tam, M.L.: Risk Analysis of Traffic and Revenue Forecasts. Manager, March, 19–27 (1998)

Ul Baset, M.K., Rahman, A., Alonge, O., Agrawal, P., Wadhwaniya, S., Rahman, F.: Pattern of road traffic injuries in rural bangladesh: burden estimates and risk factors. Int. J. Environ. Res. Public Health **14**(11), 1354 (2017)

Ullah, M.A., Hoque, S., Nikraz, H.: Traffic Growth rate and Composition of Dhaka-Chittagong Highway (N-1) of Bangladesh: The Actual situation 2 Objectives and Scope of the Study, pp. 201–206 (n.d.)

Zafri, N.M., Rony, A.I., Rahman, M.H., Adri, N.: Comparative risk assessment of pedestrian groups and their road-crossing behaviours at intersections in Dhaka, Bangladesh. Int. J. Crashworthiness (2020)

Zheng, X., Zhang, D., Gao, H., Zhao, Z., Huang, H., Wang, J.: A novel framework for road traffic risk assessment with HMM-based prediction model. Sensors **18**(12), 4313 (2018)

# Maintenance Strategies - 1

# Improving Pavement Condition at an Accelerated Pace: The City of Phoenix Accelerated Pavement Maintenance Program

Ryan Stevens[✉]

The City of Phoenix, Phoenix, AZ, USA
ryan.stevens@phoenix.gov

**Abstract.** In 2018, the Phoenix City Council took action to increase resources available to triple the amount of major street rehabilitation across the City. This two hundred-million-dollar advance sought to quickly and dramatically improve the pavement condition of the City's major streets through a five-year resurfacing program. With more than 4,850 miles of publicly maintained streets in the City, the Accelerated Pavement Maintenance Program (APMP) project has significant positive impact on Phoenix residents, businesses, and visitors as people travel throughout the City. The Street Transportation Department's APMP team worked hard to deliver outstanding results for the community in a short timeframe through hard work, rigorous analysis, collaboration, and innovative thinking. To accelerate the existing five-year plan into a three year timeframe and add additional projects, the team increased communication and coordination with stakeholders, both internal and external, to avoid pavement cuts, engaged in a major public outreach and involvement campaign using innovative approaches, partnered with paving contractors and industry, and used an analysis of pavement condition to identify and prioritize projects using the principles of asset management. The APMP progressed ahead of schedule, delivering approximately eighty percent of the programmed miles within the first three years. In the last three years, the APMP has completed the rehabilitation of nearly eleven percent of the entire street network, more than ever before. After completing several years of increased maintenance activity, the City is taking steps to protect this investment by improving the ongoing maintenance and preservation of City streets.

**Keywords:** Maintenance · Funding · Public engagement · Technology · Asset management · Pavement management · Project management

## 1 Introduction

The City of Phoenix, Arizona is the fifth largest city in the United States. Approximately 1.7 million people reside in Phoenix, occupying an area of 1,340 km². As the capital of the State of Arizona, Phoenix is the centre of the Phoenix-Mesa-Scottsdale Metropolitan Statistical Area, which has a population of approximately 4 million residents. As such, Phoenix's street network is heavily utilized, with residents, commuters, and visitors traveling within, to, and through Phoenix as part of their daily activities.

© The Author(s), under exclusive license to Springer Nature Switzerland AG 2022
A. Akhnoukh et al. (Eds.): IRF 2021, SUCI, pp. 559–574, 2022.
https://doi.org/10.1007/978-3-030-79801-7_40

560 R. Stevens

As a large city, Phoenix's 7,800 km street network is critical infrastructure facilitating the safe transport of people, goods, and services using many modes of transportation including vehicles, bicycles, public transit, and micro-mobility solutions. To maintain pavement infrastructure, Phoenix utilizes a Pavement Management System to prioritize preservation and rehabilitation treatments using available funding.

In response to increasing calls from the public to improve the pavement quality citywide, the Phoenix City Council, the elected governing body composed of the Mayor and eight Council Members, unanimously approved a financing and implementation plan to establish the Accelerated Pavement Maintenance Program (APMP) in 2018 with the goal of tripling the amount of streets receiving major rehabilitation over five years. To deliver this program, Phoenix's Street Transportation Department undertook the largest maintenance program in its history using data-driven decision-making processes, gained input from the greater community, and effectively engaged with local industry to deliver the APMP ahead of schedule. This paper will discuss the steps used to develop the APMP, challenges and opportunities presented during implementation, and report current progress of the program.

## 2 Phoenix Street Network

Phoenix's 7,800 km street network includes arterial/collector streets and local streets. Table 1 details the average number of vehicular travel lanes and the Average Annual Daily Traffic (AADT) for the various street classifications throughout Phoenix.

**Table 1.** City of phoenix street classification parameters

| Street classification | Number of vehicular lanes | Width (m) | AADT |
|---|---|---|---|
| Arterial | 5–7 | 22.5–31.7 | 20,000–60,000 |
| Major collector | 3–4 | 15.2–19.5 | 8,000–20,000 |
| Minor collector | 2–3 | 12.2–15.2 | 1,500–8,000 |
| Local | 2 | 9.8–8.5 | 300–1,500 |

In the city's pavement management system, the street network is split into two sub-networks:

- Arterial Street Network: includes Arterial and Major Collector streets
- Local Street Network: includes Minor Collector and Local streets

Figure 1 shows how the overall street network is split into its sub-networks.

By using sub-networks, the pavement management system analysis can use separate cost/benefit models. Resurfacing specifications and funding resources can also vary based on the street classification.

Phoenix's street layout mainly follows a consistent grid layout, with the Arterial Street Network primarily aligned South to North and West to East at 1.6 km (one mile) increments. However, Phoenix's location in the Salt River Valley creates natural disruptions to the grid pattern in some areas. Each member of the City Council represents

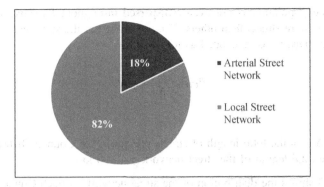

**Fig. 1.** Phoenix sub-network split

a specific region of Phoenix. Thus, Phoenix can be viewed as eight Council Districts of equal population, each with their own priorities, concerns, demographics, and character. Figure 2 shows the City of Phoenix street network and the eight Council Districts.

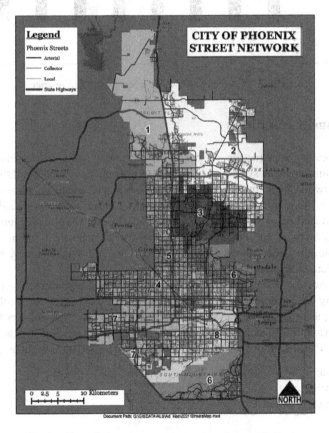

**Fig. 2.** City of phoenix street network and council districts

Since the population is not evenly dispersed throughout Phoenix, some Council District have more streets than others. The percentage of the street network located in each Council District is calculated as follows:

$$P_{CD} = \frac{S_{CD}}{S_N} * 100\%  \tag{1}$$

Where: $S_{CD}$ = the total length of streets present in the Council District; $S_N$ = the total length of the street network = 7,800 km

Figure 3 shows the distribution of the street network in Each Council District.

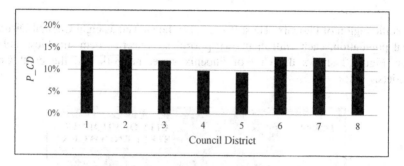

**Fig. 3.** Street network distribution by council district

## 3 Pavement Condition

Prior to creating construction programs for pavement maintenance, Phoenix uses a pavement management system methodology to assess pavement conditions and perform a network-level analysis using various budget scenarios. Phoenix has used automated systems to measure pavement distresses since 2008 and in 2018, upgraded to an Automated Road Analyzer 9000 (ARAN). The ARAN primarily utilizes a Laser Crack Measurement System (LCMS) and lasers to measure roughness. Pavement distress data are collected on a continual basis, taking approximately two years to complete data collection of both sub-networks.

As pavement condition data are collected, the data are processed to rate, classify, and quantify the pavement distresses used for further analysis. Data are then exported to the pavement management system for further processing. Phoenix uses the Deighton Total Infrastructure Management System (dTIMS) for its pavement management system. Using dTIMS, Phoenix turns the pavement distress data into three indices:

- Structural
- Environmental
- Roughness

The pavement management system uses trigger values for the three indices to determine what preservation and rehabilitation treatment can be selected. Pavement deterioration models, customized for Phoenix, are used to forecast the values of the indices in the future. By supplying the costs of maintenance treatments and funding levels including inflation, the pavement management system selects roads for various treatments for yearly construction programs. An overall Pavement Condition Index (PCI) is also calculated to report general road conditions for comparison throughout Phoenix.

The PCI is used as a composite index to easily communicate information about pavement quality to the public and decision makers. Using 2018 data as a baseline, the PCI scale and percent of the network in each condition state is shown in Table 2.

**Table 2.** Pavement condition distribution

| PCI value | Condition | Amount of street network |
|-----------|-----------|--------------------------|
| 90–100 | Excellent | 0.58% |
| 70–89 | Good | 39.78% |
| 45–69 | Fair | 54.75% |
| 20–44 | Poor | 4.88% |
| 0–19 | Very Poor | 0.00% |

Prior to finalizing yearly construction programs, field visits are made by Phoenix's engineers to check for any pavement distress that may have been missed during collection by the ARAN or other conditions that may invalidate the system's recommendation. This step to confirm or negate the recommendation allows for feedback to the system to make better recommendations in the future. Finally, a review of all the recommended projects is conducted by other city departments and outside utilities to ensure preservation and rehabilitation projects are sequenced appropriately with other work in the street.

## 4 Preservation and Rehabilitation Treatments

Phoenix utilizes many preservation and rehabilitation treatments to maintain the pavement. The treatments 'toolbox' includes treatments to address streets from good to poor condition, to keep good streets in good condition, and restore streets in poor condition. The street classifications for which the various treatments can be used are demonstrated in Table 3.

564    R. Stevens

**Table 3.** Pavement preservation and rehabilitation treatments used in phoenix

| Treatment name | Rehabilitation or preservation | Applicable street classifications | | | |
|---|---|---|---|---|---|
| | | Arterial | Major collector | Minor collector | Local |
| Mill & Overlay | Rehabilitation | ✓ | ✓ | ✓ | ✓ |
| Microsurfacing | Preservation | ✓ | ✓ | | |
| Slurry Seal | Preservation | | | ✓ | ✓ |
| Chip Seal | Preservation | | | ✓ | ✓ |
| Sealcoat | Preservation | | | ✓ | ✓ |
| Fog Seal | Preservation | ✓ | ✓ | ✓ | ✓ |

*Mill & Overlay*

Mill & Overlay is a street rehabilitation technique that requires the removal of the top layer of asphalt pavement using the grinding action of a large milling machine. After the top layer is removed, a new layer of pavement, approximately 2.5 cm thick, is placed on top of the milled pavement. Phoenix typically uses SBS polymer-modified and tire-rubber modified asphalt binders in the mill & overlay, with an effective asphalt binder grade of PG76-22. Mill & Overlay is used when preventative treatments no longer benefit the street or address the pavement distresses. This treatment is able to be used on all street classifications. The asphalt content varies when utilized in the Arterial Street Network and the Local Street Network. As a part of mill & overlay projects, curb ramps are evaluated and upgraded to meet American with Disabilities Act (ADA) standards where required.

*Microsurfacing*

Microsurfacing is a mix of a polymer-modified asphalt emulsion, water, other chemicals, and small aggregate to replace fine aggregates lost and provide a sealed wearing surface to the existing street. Two mixes are primarily used: Type II with 4.75 mm maximum nominal aggregate size and Type III with 9.5 mm maximum nominal aggregate size. Microsurfacing is used for the Arterial Street Network. As a part of microsurfacing projects, curb ramps are evaluated and upgraded to meet current ADA standards where required.

*Slurry Seal*

Like microsurfacing, slurry seal is a mixture of asphalt emulsion, water, chemicals and aggregate to provide a new wearing surface. Unlike microsurfacing, the asphalt emulsion breaks, or hardens, primarily by evaporation, not chemical processes. Two mixes are primarily used: Type II with 4.75 mm maximum nominal aggregate size and Type III with 9.5 mm maximum nominal aggregate size. Slurry seal is used on the Local Street Network.

*Chip Seal*

Chip seal is a treatment where nearly uniform-sized aggregate is applied on top of a polymer-modified asphalt membrane. Phoenix uses a chip seal where the aggregate is pre-coated with asphalt binder. This is done to promote better aggregate retention. Chip seal is used in a high-volume and low-volume traffic formulation depending on pavement distresses and is used on the Local Street Network in industrial and residential settings. Two mixes are primarily used; Low Volume with 12.5 mm maximum aggregate size and High Volume with 19.0 mm maximum aggregate size.

*Sealcoat*

Various asphalt sealcoat products are used to treat streets in good condition. The products address multiple distresses ranging from hairline cracking, early loss of fines, and initial oxidation. Sealcoat products are used on the Local Street Network, as the friction characteristics are not appropriate for the higher speed Arterial Street Network.

*Fog Seal*

Fog seal is an asphalt emulsion used to prevent early oxidation. Traditional and polymer-modified fog seals with or without rejuvenator are used depending on the circumstance. As a liquid product, without fillers or aggregate, fog seal can be used on all street classifications.

In advance of all resurfacing treatments, cracks in the pavement surface with a width of at least 6.35 mm are sealed with a hot-applied asphalt sealant. Cracks greater than 7.62 cm are sealed with a mastic product. Areas of severe deterioration are addressed by patching or other types of repairs.

## 5 Funding for Pavement Maintenance

Funding for pavement maintenance in Phoenix comes from a combination of state taxes and local sales taxes. Administered by the State of Arizona, the Arizona Highway User Revenue Fund (HURF) is comprised mainly of:

- Gasoline taxes
- Use-fuel taxes
- Motor-carrier taxes
- Motor vehicle registration fees
- Other fees

Revenues from the above taxes and fees are collected by the State of Arizona which are then distributed to the cities, towns, and counties; they are also used for Arizona Department of Transportation (ADOT) projects. In the fiscal year ending June 30, 2019, a total of $1.7 Billion US dollars were collected with Phoenix receiving a distribution of approximately $136 Million.

In 2015, Phoenix voters approved Transportation 2050 (T2050) expanding investment in Phoenix for bus service, light rail construction and street improvements through a sales tax. The previous transit plan, known as T2000, was a voter-approved tax that primarily funded transit service in Phoenix. T2050 places additional emphasis on street needs including street maintenance, new pavement, bike lanes, sidewalks and ADA accessibility. The 35-year T2050 plan was developed by a citizen-led committee of transportation experts and community. It became effective on January 1st, 2016.

The funding for T2050 is generated by a 0.7% sales tax and is estimated to generate approximately $16.7 B over the 35-year life of the plan. Figure 4 shows how the T2050

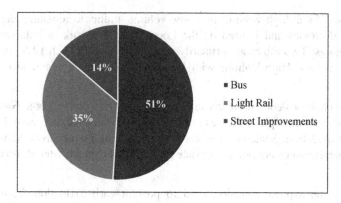

**Fig. 4.** T2050 distribution breakdown

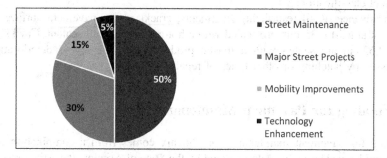

**Fig. 5.** Street improvement program allocation

funding is apportioned across the street improvement, bus transit, and light rail transit plan areas. Fourteen percent of the revenues are dedicated to street improvements, estimated to generate $2.3B through the life of the 35-year program. Figure 5 shows the breakdown of street improvements funded by T2050.

Coming out of The Great Recession of 2007 and 2008, funding for pavement maintenance improved. With a focus on improving the pavement maintenance program, expanding the use of preservation treatments, and additional funding provided by T2050, the total budget for the Pavement Maintenance Program approached $50M per year. This allowed the City of Phoenix to consistently perform maintenance on over 600 km per year. The change in funding and amount of pavement treated is shown in Fig. 6.

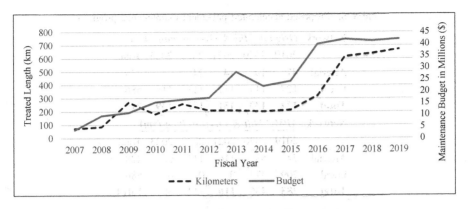

**Fig. 6.** Pavement maintenance funding over time

## 6 Developing the APMP

By Fiscal Year 2019 (beginning in July 2018), the Phoenix City Council approached the Street Transportation Department to examine options to dramatically improve pavement conditions. Although funding had increased over the years, the proportion of streets in fair and poor condition was increasing. Upon receiving feedback from residents and the Council, Phoenix performed an analysis using data from the pavement management system to provide the Council with the information needed to make decisions regarding how to address the pavement conditions citywide.

To provide City Council the needed context, Phoenix conducted an analysis to determine how many kilometres of streets would fall below a PCI of 70, the threshold of the 'Good' condition category. The pavement management system uses deterioration models to forecast future conditions. By conducting planning-level funding analyses, Phoenix was able to estimate 6,750 km would fall below the PCI 70 threshold. The outcomes from this analysis is summarized in Table 4.

**Table 4.** Funding needed to reach PCI 70 over five years

| Street network | Kilometres | $/km | Funds needed over 5 years |
| --- | --- | --- | --- |
| Arterial street network | 877 | 625,000 | $545M |
| Local street network | 5,874 | 206,250 | $1.1B |
| **Total street network** | **6,751** | – | **$1.6B** |

On October 3, 2018, the Phoenix City Council directed the Street Transportation Department to implement a plan to advance $200M of T2050 funding to improve the condition of the Arterial Street Network over 5 years. A consequence of this action was that a future light rail line extension was delayed open in 2050, instead of earlier in the T2050 program. Phoenix implemented a strategy to accelerate the existing Arterial mill & overlay projects in the 5-year plan. Phoenix also identified $50M dollars available

568     R. Stevens

**Table 5.** Proposed accelerated pavement maintenance program

| Network | Original Mill & Overlay Plan (km) | | | | | |
|---|---|---|---|---|---|---|
| | 2019 | 2020 | 2021 | 2022 | 2023 | Total |
| Arterial | 47 | 19 | 14 | 24 | 24 | 129 |
| Local | 127 | 114 | 117 | 113 | 115 | 586 |
| **Total** | **174** | **133** | **131** | **137** | **139** | **715** |
| Network | APMP Mill & Overlay Plan (km) | | | | | |
| | 2019 | 2020 | 2021 | 2022 | 2023 | Total |
| Arterial | 74 | 58 | 116 | 113 | 68 | 428 |
| Local | 209 | 377 | 0 | 0 | 0 | 586 |
| **Total** | **283** | **435** | **116** | **113** | **68** | **1,014** |

within the Street Transportation Department 5-year budget without cancelling other street improvement projects. This would allow local streets to be accelerated with the arterial streets.

Phoenix had a 5-year plan for pavement maintenance projects. In examining the existing plan for Fiscal Years 2019 through Fiscal Years 2023, it was determined the accelerated program would provide for 402 km of arterial streets to receive a mill & overlay treatment. In addition to accelerating existing projects, new locations were to be identified using both pavement condition data and community input. The total proposed changes of mill & overlay projects in the 5-year plan is presented in Table 5. Maintenance projects for non-mill & overlay treatments were not accelerated by the Council's action.

## 7  Public Outreach and Engagement

Since the APMP provided the opportunity to add additional mill & overlay projects, Phoenix undertook a community outreach and engagement process to identify the new mill & overlay projects. As part of the criteria for selection, Phoenix first analysed the streets with a PCI below 70 as the candidates to be considered for mill & overlay. With those streets identified, Phoenix engaged the public to help determine where pavement condition and resident priorities aligned. Although not possible in all cases, streets were prioritized in the APMP where these two factors were in alignment.

To engage with residents and obtain their valuable input, Phoenix embarked on a major public outreach effort. The primary focus of the outreach process was to get in front of the public wherever possible. Phoenix worked with the various City Council offices to get on meeting agendas for all kinds of public meetings including:

- Council-hosted meetings and events
- Community Association meetings
- Homeowners Association meetings
- Neighbourhood Block Watch meetings
- Major City-sponsored events

Presentation materials were developed in both English and Spanish. Phoenix estimates that over 10,000 residents were reached across over 80 individual meeting events. Occasionally, staff were presenting at multiple meetings per day where residents were able to ask questions and provide feedback through traditional comment cards, phone calls and email; they were also able to provide feedback on an interactive mapping platform.

*Resident Input Application*
In addition to traditional public input, Phoenix knew it was important to get as much resident involvement as possible. One of the most effective ways to get input is to engage people at a time and place convenient to them. Phoenix decided to add an avenue of digital engagement by creating a web-based application. The digital platform had to fulfill three purposes:

1. Display streets that would be advanced in the APMP
2. Display candidate streets with pavement condition below PCI 70
3. Collect and display crowd-sourced input on residents' priorities

Using a web-based, GIS-driven platform, residents were able to place virtual pins on a map of the Phoenix street network showing the relevant information. Three different pins could be selected: rough pavement, cracking, and potholes. Pins that identified potholes were used to create work requests for immediate repair by maintenance crews. The other pins were used to see where clustering pointed to major concerns from residents. The streets where these pins aligned with the appropriate pavement condition were selected to be added into the APMP program as long as there were no other factors that precluded its addition to the plan, such as construction conflicts or other planned improvements through the Capital Improvement Program. Over 6,700 discrete pin drops were collected over a four-month public input period from February 2019 to June 2019. Phoenix promoted and explained the application in public meetings and on Twitter to further expand the use of the application. Figure 7 shows a screenshot of the interface residents used to provide their input on the APMP.

*Pavement Maintenance Dashboard*
It is important for Phoenix to be transparent with the public on how the APMP is progressing. Using a GIS-based Story Map, Phoenix created a maintenance dashboard to allow the public to monitor progress on the APMP and other pavement treatments. By connecting to a project management database, the dashboard displays projects completed and not completed in the pavement maintenance program. Residents can select any project to get more information on the treatment, such as when it was done, when it is expected to be done, and overall program. Tabs in the Story Map also provide information on the APMP and include directions for users on how to use the map. Figure 8 shows the dashboard interface and is available to the public in English and Spanish.

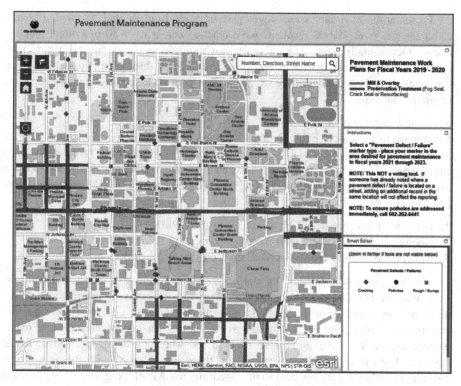

**Fig. 7.** Resident input application dashboard

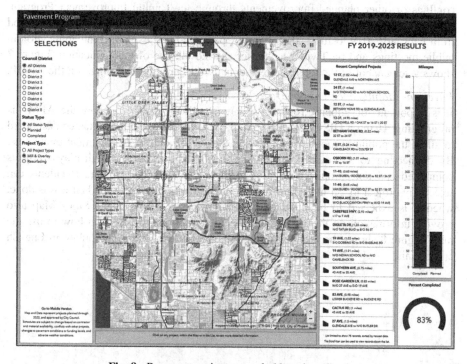

**Fig. 8.** Pavement maintenance dashboard application

## 8 Construction and Progress

Phoenix began many pre-construction activities ahead of the City Council's action. This included preparing the roadway surfaces by sealing cracks, patching heavily deteriorated areas, upgrading ADA curb ramps where required, and conducting a thorough review of construction conflicts with other projects in the right of way. One of the keys to successfully beginning the prep work was early communication with stakeholders. In Summer 2018, Phoenix began letting its contractors know about the potential advancement of projects and worked with them to begin preliminary work. Phoenix relayed the plan to the contracting community and local industry throughout the process to gear up for triple the amount of work normally seen in a year. This allowed material suppliers to order the correct amount of material and the contractors to procure any equipment and manpower needed to take on the extra work.

A key group of stakeholders were private and public utility companies. In Phoenix, there are public and private utility companies with infrastructure in and beneath the City streets. These utility companies include water, wastewater, natural gas, electricity, environmental monitors, and telecommunications. Each company has construction programs to install and maintain their infrastructure in the public right-of-way. Since these activities usually involve cutting the asphalt pavement to get to underground infrastructure, it was extremely important for Phoenix to communicate and coordinate with all utilities so work was done in a way that the mill & overlay was placed last, avoiding cuts to the overlay as much as possible. Early and frequent communication with the utility companies resulted in changes among all parities' plans to avoid pavement cuts.

Phoenix experiences warm weather for paving most of the year, but ambient temperatures drop during the winter months to a level which creates conditions not ideal for paving, resulting in low production. Although the City Council authorized the APMP in October 2018, cooling temperatures did not allow for the increased construction to begin immediately. Additionally, the Thanksgiving and Christmas holidays are celebrated in November and December; a time where Phoenix and the Arizona Department of Transportation does not want to restrict roadways. Work was able to be begin, though not at an accelerated pace. During this time, the major focus was on crack sealing and completing ADA curb ramp upgrades for the 2019 paving season. By March 2019, the prep work had been completed for many project locations. This, in addition to consistent rising temperatures, allowed the APMP to begin construction. Mill & overlay projects proceeded to be constructed at a much greater pace, tripling the amount of paving usually done in a single year. The amount or paving done through the APMP from October 2018 to May 2021 is shown in Fig. 9. In 2019 alone, more mill & overlay projects were constructed than in the preceding three years combined, demonstrating to the public Phoenix's commitment to delivering the APMP on time or quicker than promised. Figure 10 shows mill & overlay construction in calendar years 2016 through 2021.

Projects scheduled in calendar year 2021 are still in progress as of this writing.

Progress on the APMP will continue through June 2023. The program is front-loaded with projects being delivered earlier than initially planned. As of this writing, 77% percent of the APMP has been completed, while only 60% of the timeframe for

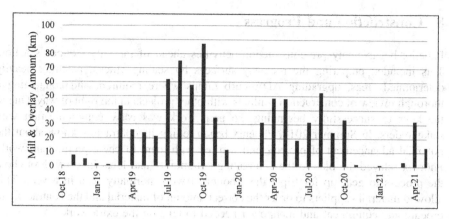

**Fig. 9.** APMP mill & overlay monthly progress

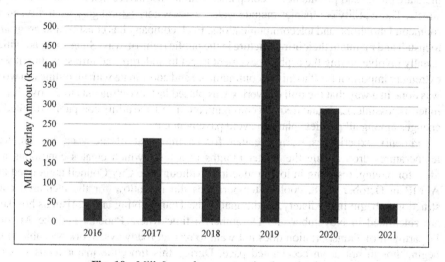

**Fig. 10.** Mill & overlay construction by calendar year

the program has elapsed. By the end of the 2021 paving season in October 2021, 880 km are projected to be completed. This represents 87% of the total APMP. Additionally, the remainder of the pavement maintenance program (micro-surfacing, slurry seals, etc.) continues to be constructed on the rest of the street network as planned.

The mill & overlay projects provide Phoenix the opportunity to examine the roadway striping configurations. Since a new layer of asphalt pavement is being placed, the pavement surface is a blank slate to re-stripe, allowing for changes to the lane dimensions, the number of vehicular travel lanes, and the addition of dedicated bicycle lanes. Through a thorough engineering review and public engagement process, Phoenix has been able to implement several striping changes that allowed for new bicycle lanes and improvements to existing bike lanes, which is summarized in Table 6.

## Table 6. Bicycle facilities implemented in the APMP

| Striping change | Kilometres |
|---|---|
| New Bicycle Lanes | 43 |
| New Bicycle Lanes with Buffers | 20 |
| New Bicycle Lane Buffers on Existing Bicycle Lanes | 14 |
| **Total Improved Bicycle Infrastructure** | **77** |

# 9  Conclusions

For three years, Phoenix has undertaken the largest paving effort in its history to address the public's concerns about its roadway condition. Phoenix has more than tripled the typical yearly amount of pavement rehabilitation without forgoing other planned preservative pavement surface treatments. Phoenix is poised to deliver the APMP in 2023 on time and on budget with hard work, rigorous analysis, collaboration, and innovative thinking.

To quickly execute the desires of the City Council, Phoenix greatly benefitted from having a pavement management system with a multi-year rehabilitation and maintenance plan. By having an existing plan developed using pavement management principles, Phoenix was able to demonstrate to the City Council exactly which roads would be included and accelerate the existing plan. Using pavement condition data, engineers were able to show the public which additional roads would be candidates for rehabilitation with a mill & overlay, allowing for residents to provide input on streets that would be their priority for treatment. Using resident input driven by pavement condition, the public was able to have a real impact on the program while selecting appropriate roads from an engineering perspective. The pavement management system analysis was also a key demonstration to the City Council of the current and future resource needs for pavement maintenance.

Effective communication was critical to the successful delivery of the APMP, thus far. Phoenix brought in its contractors and material suppliers early in the process to ensure construction equipment, materials, and personnel would be in place to deliver this ambitious program. Contracts were examined and amended to prevent any delays from contract procurement or administration. New and innovative communication tools were used to keep the public involved in the development of the APMP and provide up to date progress reports as paving projects are completed. Phoenix also communicated with other stakeholders working in the right-of-way to minimize cuts to new pavement by effectively communicating construction schedules and expectations.

The APMP, when complete in 2023, will improve the roadway condition of Phoenix's street network. In a short time, Phoenix has rehabilitated over 10% of the street network. In particular, the acceleration and addition of major street mill & overlays on the Arterial Street Network will have a positive impact for residents, commuters, and visitors as they travel through Phoenix. However, Phoenix has identified over $1B in street rehabilitation needs to address all streets below PCI 70; the $200M APMP only addresses the Arterial Street Network. Though local street rehabilitation was advanced, no new projects were added on local streets, leaving a gap in the outer years of the plan. Funding is planned to return to pre-APMP levels, which may cause issues in the future now that a large portion of the street network will be in the same condition all at the same time. This can result in a wave of future maintenance Phoenix is not currently able to fund. Phoenix estimates that maintenance funding should be doubled, at a minimum, to protect the APMP investment and further improve roadway conditions for the long term. There is much work to be done to identify a sustainable, long-term funding solution for pavement maintenance.

# Smart Infrastructure Asset Management System on Metropolitan Expressway in Japan

Hirotaka Nakashima[1]([⊠]), Taishi Nakamura[1], Yusuke Hosoi[1],
Koji Konno[2], and Hinari Kawamura[2]

[1] Metropolitan Expressway Co. Ltd., Tokyo, Japan
h.nakashima1110@shutoko.jp
[2] Shutoko Engineering Co. Ltd., Tokyo, Japan

**Abstract.** The enormous aging road infrastructure is serious present issue in Japan. This critical problem is caused by the rapid increase of deterioration and lack of engineers due to the population decline. In order to deal with this problem, we need an innovative approach for the improvement of productivity in road asset management as soon as possible. This innovative system should accurately predict the diagnosis and deterioration of the road structure with the improving the efficiency and rationalization of inspection technique, which then enables timely and appropriate repair and reinforcement.

Metropolitan Express Company Limited, which is an expressway administrator in Tokyo Japan, utilizes smart infrastructure management system that supports to optimize life cycle cost and realize sustainable infrastructure. This innovative system makes use of ICT (Information and Communication Technology), and connects various information efficiently by using IoT (Internet of Things). Furthermore, it can allow to visualize prospective tasks and to help judgement of engineers by applying AI (Artificial Intelligence) technology.

In this paper, the outline of asset management system applied in Metropolitan Expressway Company for our maintenance work is introduced. As core technologies of this asset management system, data platform based on GIS (Geographical Information System), the utilization of digital twin by means of 3 dimensional point cloud data, and patrolling and inspection system that makes use of the patrol vehicle with high performance drive recorder are mainly described.

**Keywords:** ICT (Information and Communication Technology) · 3 dimensional point cloud data · GIS (Geographical Information System) platform · Full high-vision video image · CIM (Construction Information Modeling/Management)

## 1 Introduction

Japan experienced the rapid economic growth from around 1960. Metropolitan expressway, which currently extends approximately 330 km in the greater Tokyo area, is one of the infrastructures that has been intensively developed since that time. As shown in the Fig. 1, about 40% of its network is over 40 years old, and nearly 60% of it is more than 30 years old. The structures of metropolitan expressway have been aging.

At the same time, there are concerns that the number of inspection and maintenance engineers will decrease due to a reduction of working-age population by low birth rate.

© The Author(s), under exclusive license to Springer Nature Switzerland AG 2022
A. Akhnoukh et al. (Eds.): IRF 2021, SUCI, pp. 575–586, 2022.
https://doi.org/10.1007/978-3-030-79801-7_41

Therefore, there is an urgent need to bridge the gap between the reliable maintenance and management of the enormous aging infrastructures and the shortage of engineers by improving productivity through technological development. It is necessary to build a maintenance and management system that enables accurate and timely repairs and reinforcements by precisely diagnosing structures and predicting the deterioration with efficient and rationalized inspections based on advanced technologies.

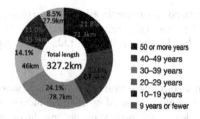

**Fig. 1.** Percentage of years elapsed since opening

## 2 Introduction of Maintenance and Management System for Infrastructure

Metropolitan Expressway Co. Ltd. has developed and currently utilizes smart infrastructure management system (i-DREAMs intelligence-Dynamic Revolution for Asset Management system) that supports to optimize life cycle cost and realize sustainable infrastructure. This innovative system makes use of ICT (Information and Communication Technology), and organically links various information by using IoT (Internet of Things). Furthermore, it can allow to visualize prospective tasks and to help judgement of engineers by applying AI (Artificial Intelligence) technology. Figure 2 shows an overview diagram.

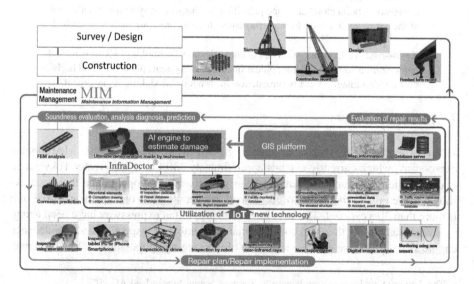

**Fig. 2.** Overview diagram of i-DREAMs

i-DREAMs is an integrated system that provides total support for maintenance and management work, such as prompt confirmation of data for infrastructure administrator to make optimal decisions in the work processes of survey/design, construction and maintenance and management. In addition, AI-based machine learning of accumulated big data such as inspection data, sensing data, and traffic volume can be used to more efficiently estimate the occurrence and cause of damage. This allows for timely and appropriate infrastructure maintenance and management. Furthermore, by accumulating the information obtained through a series of processes again, it is possible to spiral up the maintenance cycle.

## 2.1 Maintenance and Management by Infradoctor

InfraDoctor, which is the core technology of i-DREAMs, is a system that supports the maintenance and management of infrastructure using GIS and 3D point cloud data. Conventionally, the maintenance system was added a separate search function with digitalized paper-based data. However, as shown in Fig. 3, by integrating those data on the GIS platform in this new system enables to search related data of structures such as drawings, inspection and repair records a simultaneously or instantly. In addition to selecting from the map, searching by keyword with the intelligent search function makes it easier to check the target object and helps to reduce the work time.

Fig. 3. Usage of GIS platform

Furthermore, the GIS platform is linked with the 3D point cloud data of the entire Metropolitan Expressway network acquired by the MMS (Mobile Mapping System), which is shown in Fig. 4. As a result, the 3D structure of the actual structure can be confirmed on the system the same as by drawings. In other words, the digital twin using the 3D point cloud data is formed. As for MMS, it can collect data efficiently by using the equipped laser scanner and all around camera while traveling on highway at the same speed as general vehicles without traffic regulation.

**Fig. 4.** MMS vehicle

In the following, the functions of InfraDoctor that make it possible to improve the efficiency of maintenance work by using a digital twin that utilizes 3D point cloud data are described.

### 2.1.1 Site Condition Survey and Measurement on the System

It is required to accurately grasp the road width and various dimensions of structures in maintenance work such as design of road structures for repair and reinforcement and drafting of inspections and construction plans. The coordinates of each point cloud have an equivalent accuracy with that obtained from a conventional on-site surveying.

By using this function, the dimensions can be accurately measured on the system as shown in Fig. 5. Moreover, the clearance limit of road structure can be also verified directly from the system by applying these dimension measurement functions. As the result, comparing to conventional method, it enables to reduce the time for traffic control and its consultation of high traffic sections and large-scale intersections, and to easily measure at high locations where on-site measurement is difficult. Accordingly, efficiency of condition survey and measurement can be significantly improved.

**Fig. 5.** Measurement on the system

### 2.1.2 Creation of 2D and 3D CAD Drawing

Usually, there are drawings of individual works, but not integrated drawings including other accessories at the location where repairs and reinforcement work were conducted several times. To deal with this situation, as shown in Fig. 6, we implemented a function of extracting the contour line of a structure and creating a 2D CAD drawing automatically by using point cloud data. Through this function, it is possible to easily create a general drawing for maintaining and managing a structure that does not have a drawing itself, or a drawing that includes a structure with a different administrator.

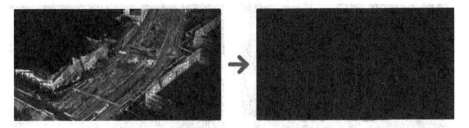

Fig. 6. 2D CAD drawing

In addition, as shown in Fig. 7, FEM (Finite Element Method) model of the current shape can be easily created by using the function that automatically extracts planes and curved surfaces from point cloud data and creates a 3D CAD drawing of the structure. This makes it possible to accurately and efficiently perform deterioration diagnosis and predictive analysis. This 3D model is also used as the data for CIM (Construction Information Modeling/Management) described later.

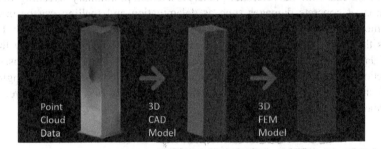

Fig. 7. 3D CAD drawing and 3D FEM model creation

### 2.1.3 Simulation in 3D Work Space

By semi-automatically arranging traffic control equipment such as safety cones based on stipulated rules on 3D point cloud data, a function was developed to easily simulate traffic control plan of work zone. Therefore, it is possible to confirm in advance whether the traffic control plan of work zone is appropriate or not from the driver's point of view.

A function that enables a dynamic simulation of any construction machine in the work space of 3D point cloud data by using the tool of 3D model of construction machine prepared beforehand was developed. As a result, it is possible to confirm in advance the selection of heavy construction / inspection machinery that is suitable for the actual site condition, the arrangement position, and the interference with surrounding structures on the system, and feed it back to the actual operation. Hence, it is possible to draft the design of repair work and construction plan, which are suitable for site condition efficiently. Figure 8 shows examples of these simulation.

**Fig. 8.** Simulation in 3D work space

### 2.1.4 Damage Detection of Structure

A function to detect the structural damage was developed by creating the reference plane of the structure from the point cloud data and calculating the differences of each point cloud data with that reference plane. As a result, preliminary screening and timely detection of concrete damages such as delamination and spalling can be practically implemented, thus improving the accuracy and efficiency of inspection work. Figure 9 shows the results of verification of the possibility of damage detection by the point cloud data for the damage locations of concrete spalling detected by the closed visual inspection. The similar damage of concrete spalling from the wall surface ranging from 3 to 7 mm could be detected from the point cloud data, hence its effectiveness was confirmed.

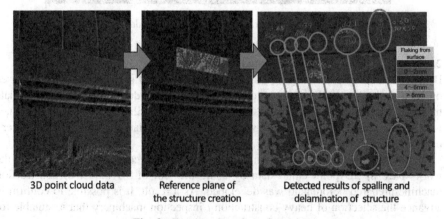

**Fig. 9.** Damage detection of structure

Further, the road surface property value can be grasped by quantitatively evaluating the unevenness state of the road surface with spatial frequency processing since the elevation value at each point of the 3D point cloud data can be regarded as a signal waveform. By using this method, the required information for pavement inspection, which are local damages such as ruts, flatness, and potholes, can be detected from 3D point cloud data of InfraDoctor. On the other hand, it is possible to detect cracks automatically in a shorter time than conventional method by acquiring a high-definition image through the line sensor camera mounted on the MMS and using AI for the acquired image. Figure 10 shows the image of automatic detection by using each device.

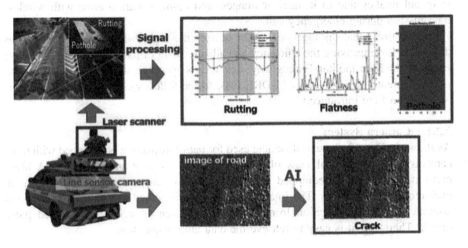

Fig. 10. Application to pavement inspection

In addition, as shown in Fig. 11, the inspection results can be visualized by recording the survey results on GIS platform and displaying it in different color for each value.

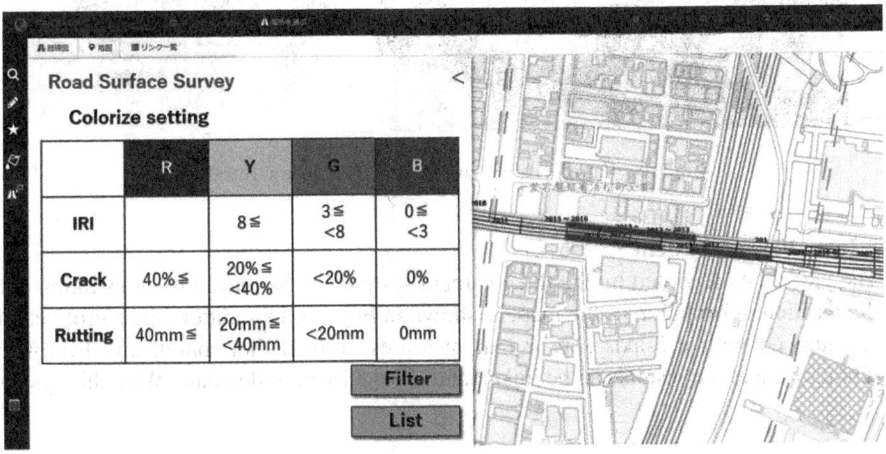

Fig. 11. Display for inspection result on GIS platform

## 2.2 Maintenance and Management by Infrapatrol

The patrol inspection is conducted for finding out the abnormal events on the expressway. Specifically, inspection engineers confirm the presence of damages, abnormalities, or falling objects on the structures of the road, the road surface, or road facilities. In the conventional patrol inspection, inspectors ride on the patrol vehicle and conduct visual inspection from inside the car. In case they find the damage, they take a record of it by using the hand-holding digital camera. However, it is not easy to stop patrol vehicle on the expressway to confirm the damage of structure. Therefore, this conventional method has potential problems such as overlooking of damage, shooting miss and retakes due to it, unclear images, and communication error with wireless transmission during emergency call.

InfraPatrol, which is the system that utilizes the high-vision drive recorder, was developed for improving the efficiency and advancement of patrol inspection to ensure safety on the Metropolitan Expressway. Besides, its operation started in FY2017 as one of the elemental technologies of i-DREAMs. In the following, five advanced systems of InfraPatrol are introduced.

### 2.2.1 Camera System

As shown in Fig. 12, the patrol vehicle used for patrol inspection is equipped with three cameras with an angle of view of 60°, to ensure a viewing angle of 180°. A high-performance camera is equipped with the vehicle to capture clear images even in backlight or dark tunnels. The image data captured is linked to time of shooting and accurate location information from the GPS, gyro sensor, and vehicle speed pulse sensor. Therefore, it is easy to retrieve the data after acquisition.

**Fig. 12.** Patrol vehicle and camera system for InfraPatrol

### 2.2.2 Viewer System

The video data recorded by the three-camera system can be played in cooperation with the location information of the map, as shown in Fig. 13. In addition, the patrol vehicle is equipped with a 3-axis sensor that measures vertical, horizontal, and front-back vibrations, and the results can be linked and checked with video data. With this system, it is possible to check the details of video data as well as event image from any specified time and location on the map. As the result, there is no need to re-run and re-photograph due to shooting miss.

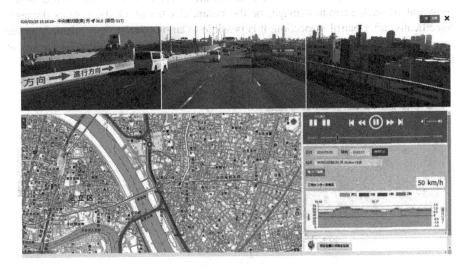

**Fig. 13.** Display of 3 viewer system

### 2.2.3 Emergency Report System

Emergency report system was built in case of serious damage such as a pothole on the road surface is found during patrol inspection. In this system, when inspectors find serious damage, they press the emergency button on the remote control switch to send the image of the damaged part to the management office in real time by LTE (Long Term Evolution) communication. At the same time, the video footage of approximately 10 s before and after the damage is recorded.

Furthermore, as shown in Fig. 14, at the management office, once emergency button is pressed, employees at the management office are notified about it by notification lamp. In addition, the exact location of the damage and the damage situation can be immediately confirmed with a video for about 10 s before and after on a large monitor. This system enables swift response to emergency work.

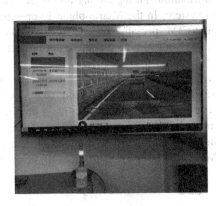

**Fig. 14.** Notification lamp of emergency report system

### 2.2.4 Automatic Damage Detection System

If any serious damage is found during the patrol inspection, emergency response will be taken by the above-mentioned emergency report system. In addition, automatic damage detection system with movie data was built in order to confirm whether or not oversight of such serious damage, or the potential damage of the serious damage. Specifically, after patrol inspection, it is possible to automatically detect damage to the road surface potholes or joints by the singularity extraction technique as shown in Fig. 15, using the video data recorded on the portable hard disk.

**Fig. 15.** Automatic damage detection system

### 2.2.5 Movie Sharing System by Cloud Server

The video of all the steps taken by the 3 camera system is stored in the cloud server, and it is possible to search from location information, time, sensor information, and play back the video on a PC in the management office to check. By accumulating these video data, various analyses can be conducted, and it can be linked to the deterioration prediction of the damaged part.

## 2.3 Introduction of Cim for Advanced Maintenance

In order to improve productivity in a series of project stages from construction to maintenance, it is planned to introduce CIM (Construction Information Modeling/Management) linked with i-DREAMs. Generally, CIM means to create 3D model from the upstream stage of the project, such as the planning or the survey stage, and to develop the 3D model in the post-process such as the construction stage. It is intended to facilitate information sharing among all project members and improve the construction production system. In the Metropolitan Expressway Company, it is considered the introduction of CIM specialized in maintenance as one of the forms of CIM, since the maintenance of existing structure is the main project rather than construction or renewal one.

As described above, 3D point cloud data of almost all the structure in the Metropolitan Expressway has been already acquired, so this point cloud data is used for CIM. It is effective to use a 3D data to manage a huge number of damages in several structures. Therefore, in the CIM, which is specialized for maintenance, the tags for each structure are created by marking the information of damage position on 3D point cloud data of i-DREAMs. Further, the information of the structure damage unit, such as inspection, repair information and so on, is associated with the tag and managed thoroughly. Using i-DREAMs with CIM function, makes it possible to efficiently obtain the damage position, damage summary, and past repair information in the office. There is no need to go to the field. The CIM image, which is created by trial basis during tunnel repair project, is Fig. 16.

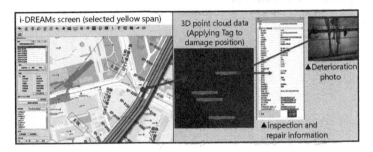

**Fig. 16.** Image of utilization of CIM in i-DREAMs

In the current inspection and repair, it is often necessary to conduct a field survey based on the damage record of the paper medium, so it takes a lot of time to identify the damage position due to the huge number of damages. By using i-DREAMs, damage information can be obtained efficiently even in the office, and it is expected that the repair project's efficiency will improve greatly, such as reduction of traffic regulations during the field survey. In addition, it is possible to verify the repair effect and select an appropriate repair method based on the past repair effect. It is expected to improve the productivity in the maintenance and management.

## 3 Conclusions

The improvement of efficiency through technology development is required in preparation for the expected shortage of engineer due to decline of working-age population for the maintenance and management of enormous aging infrastructure. In this situation, smart infrastructure asset management system, which has been developed and currently is utilized by Metropolitan Expressway Co., Ltd, is introduced.

We will keep developing this system for further advancement as the introduction of CIM. It is expected that this system can be utilized not only for highway structures in the Metropolitan Expressway but also for other infrastructure and can contribute to the development of safe and secure social infrastructure.

## References

Osada, T., et al.: Improvement of patrol inspection on metropolitan expressway. J. Civil Eng. 62–65 (2018)

Witchukreangkrai, E., et al.: Smart infrastructure asset management system of metropolitan expressway in Japan. In: Proceeding of 26th World Road Congress, PIARC, TC D1. Abu Dhabi, United Arab Emirates (2019)

# Decarbonizing Road Transport

# Evaluation of the $CO_2$ Reduction Effect of Low Rolling Resistance Asphalt Pavement Using the Fuel Consumption Simulation Method

Yu Shirai[1]([⊠]), Atsushi Kawakami[2], Masaru Terada[3], and Kenji Himeno[4]

[1] Nippo Corporation, Research Institute, Saitama, Japan
shirai_yuu@nippo-c.jp
[2] Public Works Research Institute, Pavement Research Team, Ibaraki, Japan
[3] Public Works Research Center, Road Research Department, Ibaraki, Japan
[4] CHUO University, Tokyo, Japan

**Abstract.** This study verified the $CO_2$ emission reduction effect of the Low Rolling Resistance Asphalt Pavement (LRRAP) using the fuel consumption simulation method for evaluating the automobile consumption rate. In this verification, the relationship between the rolling resistance coefficient reduction ratio and $CO_2$ emission reduction ratio was first studied for two types of vehicles: heavy duty vehicles and light duty vehicles. Subsequently, the $CO_2$ reduction was evaluated when applying LRRAP to expressways and major roads across Japan. The $CO_2$ emission reduction effect of the difference in traffic volume was also studied. The study indicated that the reduction effect of applying LRRAP corresponds to approximately 1.1% for the entire transportation sector in Japan. Also, since the routes with a larger traffic volume indicated a larger application effect, it clarified that $CO_2$ can be efficiently reduced by prioritizing route maintenance with a larger traffic volume.

**Keywords:** Low rolling resistance asphalt pavement · $CO_2$ emissions · Fuel economy simulation method · Porous pavement · WLTC mode

## 1 Introduction

Currently, global decarbonization efforts are actively underway, and the Japanese Government has set the target of net zero emissions by 2050. (Prime Minister's Office of Japan 2020). The transport sector accounts for about 20% of Japan's carbon dioxide emissions ($CO_2$), about 90% of which comes from vehicles. (Ministry of Land, Infrastructure, Transport and Tourism 2020). Pavement technology for reducing the tire/road rolling resistance is expected to improve the energy consumption while driving and reduce $CO_2$ emissions from internal combustion engine vehicles fueled by gasoline or other liquid fuels. It is also expected to extend the driving range of zero-emission vehicles such as electric vehicles (EVs).

We developed a Low Rolling Resistance Asphalt Pavement (LRRAP) as an asphalt pavement technology with low rolling resistance and conducted studies to investigate mechanisms for reducing the rolling resistance (Ishigaki et al. 2013; Shirai et al. 2016; Kawakami et al. 2017). In previous study regarding the verification of the $CO_2$ emission reduction effect of LRRAP, we have measured the $CO_2$ emissions during

© The Author(s), under exclusive license to Springer Nature Switzerland AG 2022
A. Akhnoukh et al. (Eds.): IRF 2021, SUCI, pp. 589–601, 2022.
https://doi.org/10.1007/978-3-030-79801-7_42

590 Y. Shirai et al.

constant speed driving by a fuel consumption measuring vehicle on LRRAP. And we have verified the $CO_2$ emission reduction effect by trial calculation using the provisional rolling resistance coefficient of LRRAP. However, since these examinations are limited conditions, we thought that more accurate verification was necessary.

In the verification, the actual rolling resistance coefficient reduction rate of LRRAP and the $CO_2$ emission reduction effect in the actual driving mode were verified. The rolling resistance coefficient used was the value obtained from the experiment. The fuel consumption simulation method for automobiles was used to calculate $CO_2$ emissions. This simulation method is a highly accurate method that is also used in the official fuel efficiency evaluation of Japanese automobiles.

First study, a) the change rate of the rolling resistance coefficient and the $CO_2$ reduction rate were studied, and subsequently, b) the reduction effect when applying it to major roads across Japan was considered. Finally, c) the difference in the reduction effect of the traffic volume difference was verified.

## 2 Study Method

### (1) Study conditions

The conditions in this evaluation are listed in Table 1. As shown, expressways, national roads, and prefectural roads were evaluated as major roads across Japan. Also, the verification of the $CO_2$ emission reduction effect of the difference in traffic volume, the route between Ebina JCT and Atsugi IC on the Tomei Expressway was targeted as a route with heavy traffic volume. The types of pavement compared were LRRAP (5) and porous (13). Porous (13) is standardly used for major roads in Japan. Also, the verification was conducted for two types of vehicles: heavy duty vehicles and light duty vehicles. For road extension and traffic volume, we used traffic census data from the Ministry of Land, Infrastructure, Transport and Tourism.

Concerning heavy duty vehicles for simulation, since vehicles exceeding GVW 20 tons are the largest in number (Japan Automobile Dealers Association 2019) according to the 2019 number of sales of general freight cars by gross vehicle weight (GVW), a vehicle corresponding to T11 in the fuel classification number was determined as the representative vehicle for heavy duty vehicles.

Concerning light duty vehicles, a compact car was selected because the number of sales is relatively large in Japan, the vehicle has disclosable fuel map data for

**Table 1.** Simulation condition

| Item | Content |
| --- | --- |
| Road type | Highway 9,473.8 km<br>National highway 55,684.9 km<br>Prefectural road 129,003 km |
| Pavement type | LRRAP(5)<br>Porous(13) |
| Vehicle type | Heavy duty vehicle<br>Light duty vehicle |

Evaluation of the $CO_2$ Reduction Effect of Low Rolling Resistance Asphalt Pavement    591

performing simulation, and the fuel economy is relatively high. The specifications for heavy and light duty vehicles are shown in Table 2.

**Table 2.** Specification sheet

| Item | | Heavy duty vehicle | Light duty vehicle |
|---|---|---|---|
| Vehicle type | | T11 | – |
| Vehicle weight (kg) | | 9,193 | 988 |
| Maximum loading capacity (kg) | | 14,844 | – |
| Gross vehicle weight (kg) | | 24,147 | 1,245 |
| Vehicle weight at test (kg) | | 17,412 | 1,098 |
| Capacity (people) | | 2 | 5 |
| Transmission | 1st | 10.127 | Continuously Variable Transmission (CVT) 2.386~0.426 |
| | 2nd | 8.054 | |
| | 3rd | 6.414 | |
| | 4th | 5.101 | |
| | 5th | 4.038 | |
| | 6th | 3.211 | |
| | 7th | 2.507 | |
| | 8th | 1.994 | |
| | 9th | 1.588 | |
| | 10th | 1.263 | |
| | 11th | 1.000 | |
| | 12th | 0.795 | |
| Final ratio | | 2.928 | 5.833 |
| dynamic rolling radius (m) | | 0.507 | 0.282 |
| Overall Length (m) | | 12.0 | 3.945 |
| Overall Height (m) | | 3.8 | 1.5 |
| Overall width (m) | | 2.49 | 1.695 |
| Coefficient of drag $N/(km/h)^2$ | | 0.2691 | 0.0441 |
| Engine | Type | DI,TI | NA |
| | Total displacement(L) | 12.9 | 1.0 |
| | Maximum power (kW/rpm) | 294/1800 | 51/6,000 |
| | Maximum torque (Nm/rpm) | 2,270/1,100 | 92/4,200 |

(2) Rolling resistance coefficient

The skid resistance measurement vehicle for measuring the rolling resistance coefficient is shown in Fig. 1. The existing data (Shirai et al. 2019) measured at the driving velocity of 20 km/h, 40 km/h, 60 km/h, and 80 km/h was used.

**Fig. 1.** Tire/road rolling resistance measuring car

Measurements were taken on the test road surface constructed at the National Institute for Land and Infrastructure Management. The types of test road surface, International Roughness Index (*IRI*), Mean Profile Depth (*MPD*), and the rolling resistance coefficient (*RRC*), and the *RRC* reduction rate are shown in Table 3. The test road surface was LRRAP (5), the newly constructed New Porous (13), and the old Porous (13) as the road surface in service. The *RRC* reduction rate shows the reduction rate based on Old Porous (13).

**Table 3.** Pavement type

| Pavement type | | New LRRAP(5) | New Porous(13) | Old Porous(13) |
|---|---|---|---|---|
| *IRI* (mm/m) | | 0.92 | 1.26 | 1.39 |
| *MPD* (mm) | | 0.70 | 1.63 | 1.84 |
| *RRC* | 20 km/h | 0.0140 | 0.0150 | 0.0163 |
| | 40 km/h | 0.0156 | 0.0169 | 0.0180 |
| | 60 km/h | 0.0172 | 0.0186 | 0.0206 |
| | 80 km/h | 0.0196 | 0.0213 | 0.0229 |
| *RRC* reduction rate (%) | 20 km/h | 14.1 | 8.0 | – |
| | 40 km/h | 13.3 | 6.1 | – |
| | 60 km/h | 16.5 | 9.7 | – |
| | 80 km/h | 14.4 | 7.0 | – |

The rolling resistance coefficient was calculated by removing the effect of lateral force caused by slight steering movements from the traction force $R_c$ when the fifth wheel was rolling freely. And $R_0$ was obtained using Eq. (1). The coefficient of rolling resistance *RRC* was calculated by dividing $R_0$ by the wheel load $F_z$. (Shirai et al. 2016)

$$R_0 \approx \left(R_C - \frac{F_C^2}{k_y}\right) \Big/ \left\{1 + \left(\frac{F_C}{F_Z}\right)^2\right\}^{\frac{1}{2}} \tag{1}$$

$$RRC = \frac{R_0}{F_Z} \tag{2}$$

where *RRC*: rolling resistance coefficient, $R_0$: traction force (N) when the lateral force is zero, $R_c$: traction resistance force (N), $F_z$: cornering force (N), $F_z$: wheel load (N), $k_y$: cornering power (N) $k_y = CC \times F_z$, and *CC*: cornering coefficient (0.23).

(3) Fuel consumption

a) Fuel consumption simulation method for heavy duty vehicles

For the fuel consumption simulation for heavy duty vehicles, fuel consumption was provided based on TRIAS08-003(1)-02 "Fuel consumption rate test (heavy vehicle [corresponding to the FY 2025 fuel consumption standard])" (JH25 Test Method). (National Traffic Safety and Environment Laboratory 2021a, b). In this simulation, the engine speed and engine torque during mode driving were calculated based on the vehicle and engine specifications, and instantaneous fuel consumption was calculated by referring to the normal fuel economy map separately obtained on an engine bench. The air resistance $\mu_a A$ (N/(km/h)$^2$) was calculated using Eq. (3) used in TRIAS 08-003-02 "Fuel consumption rate test (for heavy vehicles)" (hereinafter referred to as the "JH15 Test Method"). (National Traffic Safety and Environment Laboratory 2021a, b).

$$\mu_a A = (0.00299 \times B \times H - 0.000832) \times 9.8 \tag{3}$$

where *B*: overall width (m) and *H*: overall height (m).

The influence of the rolling resistance coefficient was examined using Eq. (4) used in the JH15 Test Method.

$$\mu_r = 0.00513 + \frac{17.6}{W} \tag{4}$$

Where $\mu_r$: tire rolling resistance coefficient, *W*: vehicle mass during the test (kg).

Fuel consumption was then calculated based on the constant of 0.00513 corresponding to the tire rolling resistance coefficient in Eq. (4), with a 30% decrease (0.00359), 15% decrease (0.00436), 15% increase (0.00590), and 30% increase (0.00667) change. Fuel consumption simulation was conducted for the fuel economy classification T11 category (GVW 25 t vehicle) using the fuel economy map for the category's engine owned by the Japan Automobile Research Institute (JARI). The fuel economy map is shown in Fig. 2.

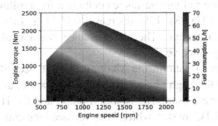

**Fig. 2.** Engine map (heave duty vehicle)

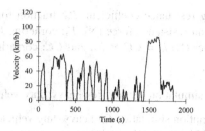

**Fig. 3.** Urban driving mode (heavy duty vehicle)

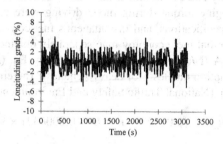

**Fig. 4.** Intercity driving mode (heavy duty vehicle

For the driving pattern, the urban driving modes (average speed 27.3 km/h and speed 0 to 87.6 km/h) and the intercity modes (constant speed 80 km/h), which are integrated into the fuel economy simulation program, were used as is. Each driving mode is shown in Fig. 3 and Fig. 4.

b) Fuel consumption simulation for light duty vehicles.

The fuel consumption simulation for light duty vehicles was conducted in the Worldwide harmonized Light vehicles Test Cycles modes (WLTC modes) for passenger cars and others. $CO_2$ emission was calculated from time-series data of the driving force assumed from the driving speed pattern and vehicle specifications using the disclosable $CO_2$ emission map for speed and driving force owned by JARI. The fuel economy map is shown in Fig. 5.

The fuel economy simulation was conducted based on the constant of 0.0140 corresponding to the tire rolling resistance coefficient for a rolling resistance coefficient, with a 30% decrease (0.0098), 15% decrease (0.00119), 15% increase (0.0161), and 30% increase (0.0182) change. The air resistance was calculated using Eq. (3), the same as that for heavy duty vehicles.

For the driving speed pattern, the Medium phase (speed 0 to 76.6 km/h, average speed 39.5 km/h) and the High phase (speed 0 to 97.4 km/h, average speed 56.7 km/h) in the WLTC modes were used. The Medium phase and the High phase are shown in Figs. 6 and 7. The WLTC modes, which have replaced the conventional driving modes JC08 since October 2018, are worldwide unified test modes close to the actual fuel economy used in the official fuel economy calculation.

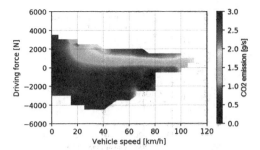

**Fig. 5.** Engine map (light duty vehicle)

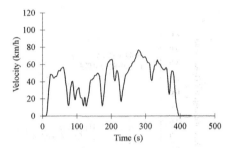

**Fig. 6.** Medium phase (light duty vehicle)

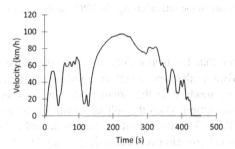

**Fig. 7.** High phase (light duty vehicle)

## 3 The Relationship Between RRC and $CO_2$ Reduction Rate

Figure 8 shows the relationship between the *RRC* change rate and driving fuel consumption in heavy duty vehicles, and Fig. 9 shows the relationship between the *RRC* change rate and $CO_2$ emission reduction rate in light duty vehicles.

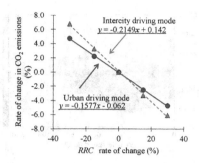

**Fig. 8.** Relationship between the rate of change in *RRC* and the rate of change in $CO_2$ emissions (Heavy duty vehicle)

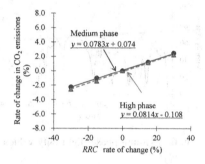

**Fig. 9.** Relationship between the rate of change in *RRC* and the rate of change in $CO_2$ emissions (Light duty vehicle)

Figure 8 indicates that for heavy duty vehicles, the slope of intercity modes is larger than that of urban modes, and the effect of rolling resistance on fuel consumption is greater the intercity mode than the urban mode. In addition, from the simulation results, the fuel consumption when the rolling resistance coefficient is set to the reference value (0.00513) is 3.17 km/L in the urban mode and 4.28 km/L in the intercity mode. Using the gas oil emission original unit 2.58 kg-$CO_2$/L (Ministry of the Environment materials 2021), the $CO_2$ emission coefficient is 0.814 $_{kg}$-$CO_2$/km for urban modes and 0.603 $_{kg}$-$CO_2$/km for intercity modes.

On the other hand, Fig. 9 shows that it is almost equal for the Medium phase and the High phase in light duty vehicles. This indicates that the speed does not significantly affect the relationship between the *RRC* reduction rate and $CO_2$ emission. The $CO_2$ emissions based on the rolling resistance coefficient reference value (0.0140) are 0.093$_{kg}$-$CO_2$/km in the Medium phase and 0.072 $_{kg}$-$CO_2$/km in the High phase.

In addition, in studying the $CO_2$ reduction rate by pavement type, the $CO_2$ reduction rate was calculated when the pavement type is based on Old Porous (13), and New Porous (13) and LRRAP (5) are applied. The calculation result is shown in Table 4. Table 4 indicates that since a rolling resistance coefficient is reduced by refreshing even the same pavement type, the $CO_2$ emission reduction rate is reduced by 1.3 to 1.5% for heavy duty vehicles and by 0.5 to 0.6% for light duty vehicles. It is also confirmed that the $CO_2$ emission reduction rate is reduced by 2.2 to 3.1% for heavy duty vehicles and by approximately 1.0 to 1.2% for light duty vehicles by applying LRRAP (5).

**Table 4.** Pavement type and $CO_2$ emission reduction rate

| Item | | Heavy duty vehicle | | Light duty vehicle | |
|---|---|---|---|---|---|
| Driving mode | | Urban | Intercity | Medium | High |
| Average velocity (km/h) | | 27.3 | 80.0 | 39.5 | 56.7 |
| New Porous(13) | *RRC* reduction rate (%) | 8.0 | 7.0 | 6.1 | 7.0 |
| | $CO_2$ reduction rate (%) | 1.3 | 1.5 | 0.5 | 0.6 |
| New LRRAP(5) | *RRC* reduction rate (%) | 14.1 | 14.4 | 13.3 | 14.4 |
| | $CO_2$ reduction rate (%) | 2.2 | 3.1 | 1.0 | 1.2 |

## 4 Effect Evaluation of Application to Major Roads Across Japan

The effect of application to major roads across Japan was evaluated. Old Porous (13) was used as the reference pavement type assuming in-service routes. In the evaluation, the $CO_2$ emission from an automobile was then calculated for the cases of Porous (13) and LRRAP (5) being applied. The evaluation conditions are shown in Table 5. The various conditions shown in Table 5 are derived from FY 2015 traffic census data. (Ministry of Land, Infrastructure, Transport and Tourism 2015). Also, since the driving speed varies with the route type, the driving modes close to the traffic census data's traveling speed were applied.

**Table 5.** Calculation conditions (Trunk roads all over Japan)

| Road type | | Highway | National highway | Prefectural road |
|---|---|---|---|---|
| Travel speed (km /h) | | 78.4 | 35.4 | 28.5 |
| Total length (km) | | 9,474 | 55,685 | 129,003 |
| 24-h average traffic (Unit/day) | Light duty vehicle | 19,359 | 9,366 | 4,284 |
| | Heavy duty vehicle | 9,803 | 1,952 | 615 |
| Driving mode | Light duty vehicle | High | Medium | Medium |
| | Heavy duty vehicle | Intercity | Urban | Urban |

598    Y. Shirai et al.

In the evaluation, first, the $CO_2$ emission for Old Porous (13), a reference, was calculated. Subsequently, the $CO_2$ reduction was obtained from the $CO_2$ reduction rate. Finally, the annual $CO_2$ reduction across Japan was obtained from the total extension length and number of days for each road type. The rate of the reduction to the annual $CO_2$ emission of the entire transportation sector in Japan was then obtained. The annual $CO_2$ emission of the entire transportation sector in Japan was regarded as 206,000,000 t. (Ministry of Land, Infrastructure, Transport and Tourism 2021). The evaluation result is shown in Table 6.

Table 6 indicates that applying New Porous (13) and LRRAP (5) can annually reduce approximately 1,230,000 t and approximately 2,270,000 t of $CO_2$ emission, respectively. Also, the rates of the reduction to the annual $CO_2$ emission of the entire transportation sector in Japan for Porous (13) and LRRAP (5) are 0.60% and 1.1%, respectively.

**Table 6.** Calculation result

| Road type | | | Highway | | National highway | | Prefectural road | | Total |
|---|---|---|---|---|---|---|---|---|---|
| Total extension (km): $A$ | | | 9,474 | | 55,685 | | 129,003 | | 194,162 |
| Vehicle type | | | Light duty vehicle | Heavy duty vehicle | Light duty vehicle | Heavy duty vehicle | Light duty vehicle | Heavy duty vehicle | |
| 24 hour average traffic (Unit/day): $B$ | | | 19,359 | 9,803 | 9,366 | 1,952 | 4,152 | 596 | |
| $CO_2$ emission factor (kg-$CO_2$/km·Unite): $C$ | | | 0.072 | 0.603 | 0.093 | 0.814 | 0.093 | 0.814 | |
| $CO_2$ emissions amount (kg-$CO_2$/km·day): $D=B \times C$ | | | 1,396 | 5,911 | 871 | 1,589 | 386 | 485 | |
| Porous(13) | $CO_2$ emission reduction rate (%) : $E$ | | 0.60% | 1.50% | 0.50% | 1.30% | 0.50% | 1.30% | |
| | $CO_2$ emission reduction amounts (kg-$CO_2$/km·day) : $F=D \times E$ | | 8.4 | 88.7 | 4.4 | 20.7 | 1.9 | 6.3 | |
| | $CO_2$ emission reduction amounts (t-$CO_2$/km·year) : $G=F \times 365$ | | 3.07 | 32.38 | 1.61 | 7.56 | 0.69 | 2.30 | |
| | $CO_2$ emission reduction amounts (t-$CO_2$·year) : $H=G \times A$ | | 29,047 | 306,719 | 89,430 | 420,727 | 89,464 | 296,642 | 1,232,029 |
| | $CO_2$ reduction rate for the entire transportation sector (206,000,000 ton) (%) | | 0.163 | | 0.248 | | 0.187 | | 0.60 |
| LRRAP(5) | $CO_2$ emission reduction rate (%) : $E$ | | 1.20% | 3.10% | 1.00% | 2.20% | 1.00% | 2.20% | |
| | $CO_2$ emission reduction amounts (kg-$CO_2$/km·day) : $F=D \times E$ | | 16.8 | 183.2 | 8.7 | 35 | 3.9 | 10.7 | |
| | $CO_2$ emission reduction amounts (t-$CO_2$/km·year) : $G=F \times 365$ | | 6.13 | 66.87 | 3.18 | 12.78 | 1.42 | 3.91 | |
| | $CO_2$ emission reduction amounts (t-$CO_2$·year) : $H=G \times A$ | | 58,093 | 633,494 | 176,827 | 711,375 | 183,636 | 503,821 | 2,267,246 |
| | $CO_2$ reduction rate for the entire transportation sector (206,000,000 ton) (%) | | 0.336 | | 0.431 | | 0.334 | | 1.10 |

# 5    $CO_2$ Reduction Effect of the Difference in Traffic Volume

Subsequently, the $CO_2$ reduction when LRRAP (5) is constructed in a section with a relatively high traffic volume in expressways was calculated, and the effect of application was verified. The target section was the Tomei Expressway between Ebina junction (JCT) and Atsugi interchange (IC). The 24-h traffic volume (FY 2010) of weekdays and holidays by automobile type was used as the traffic volume data. (Ministry of Land, Infrastructure, Transport and Tourism 2010). Other evaluation conditions were the same as those for Highway in Table 5. The evaluation result is shown in Table 7. Table 7 indicates that the annual reduction is 559 t-CO$_2$/km-year.

The result comparing the entire expressway and Ebina JCT and Atsugi IC in CO$_2$ reduction is shown in Fig. 10. Figure 10 indicates that the CO$_2$ reduction effect in applying LRRAP (5) to the heavy traffic section of an expressway is approximately 7.7 times that of applying it to the entire expressway. This quantitatively clarifies that applying LRRAP (5) to sections with a high traffic volume improves the CO$_2$ emission reduction.

Table 7. Calculation conditions (Difference traffic volume)

| Vehicle type | Light duty vehicle | Heavy duty vehicle |
|---|---|---|
| 24 h average traffic (Unit/day): $B$ | 196,048 | 72,844 |
| CO$_2$ emission factor (kg-CO$_2$/km·Unite): $C$ | 0.072 | 0.603 |
| CO$_2$ emissions amount (kg-CO$_2$/km·day): $D = B \times C$ | 14,135 | 43,925 |
| CO$_2$ emission reduction rate (%): $E$ | 1.20% | 3.10% |
| CO$_2$ emiss reduction amounts (kg-CO$_2$/km·day): $F = D \times E$ | 169.6 | 1361.7 |
| CO$_2$ emiss reduction amounts (t-CO$_2$/km·year): $G = F \times 365$ | 61.90 | 497.02 |

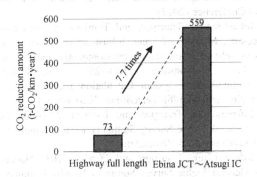

Fig. 10. Calculation result (Difference traffic volume)

## 6 Conclusions

In this study, evaluation was conducted by applying the method used in the fuel consumption simulation in the automobile field to verify the LRRAP's CO$_2$ reduction effect. The acquired knowledge is as follows:

a) For heavy duty vehicles, the impact of *RRC* on CO$_2$ emissions was greater in high-speed intercity mode than in low-speed urban mode. On the other hand, for light duty vehicles, the influence of the difference in speed is minor.

b) It was clarified that rolling resistance can be reduced and $CO_2$ emissions can be reduced simply by renewing the existing pavement of the same type, and it was clarified that the application of LRRAP (5) has a reduction effect of about twice that of New Porous (13).

c) Concerning the reduction effect in application to major roads across Japan, a reduction of 0.60% and 1.1% can be obtained for the $CO_2$ emission of the entire transportation sector in Japan by applying Porous (13) and LRRAP (5), respectively.

d) The reduction efficiency is improved by applying LRRAP (5) to the expressways with a high traffic volume.

From the above examination, it was shown that the application of LRRAP is effective in reducing $CO_2$ by utilizing the actual *RRC* of LRRAP and the accurate fuel simulation method. In the future, the $CO_2$ emission reduction effect will now be studied in terms of cost-effectiveness.

# References

Ishigaki, T., Kawakami, A., Kubo, K.: Development of asphalt pavement technique for low tire/road rolling resistance. In: The 30th Japan Road Conference (2013)

Japan Automobile Dealers Association: New car sales status (monthly report) January-December issue 2019.2 ~ 2020.1 (2019)

Kawakami, A., Terada, M., Yabu, M., Shirai, Y., Ishigaki, T.: Relationship between performance required for low fuel consumption pavement and performance index of road surface. In: The 32nd Japan Road Conference (2017)

Ministry of Land, Infrastructure: Transport and Tourism: "Carbon dioxide emissions in the transportation sec-tor", Ministry of Land, Infrastructure, Transport and Tourism web site (2020). https://www.mlit.go.jp/sogoseisaku/environment/soseienvironment_tk_000007.html

Ministry of the Environment materials: Greenhouse gas emission calculation/reporting/publication system Calculation method/emission factor list (2021). https://ghg-santeikohyo.env.go.jp/files/calc/itiran_2020_rev.pdf

Ministry of Land, Infrastructure, Transport and Tourism: 2010 National Road/Street Traffic Situation Survey General Traffic Survey Summary Results Summary Table (2010). https://www.mlit.go.jp/road/census/h22-1/index.html

Ministry of Land, Infrastructure, Transport and Tourism: 2015 National Road/Street Traffic Situation Survey General Traffic Survey Summary Results Summary Table (2015). https://www.mlit.go.jp/road/census/h27/index.html

Ministry of Land, Infrastructure, Transport and Tourism website: Carbon dioxide emissions in the transportation sector (2021). https://www.mlit.go.jp/sogoseisaku/environment/sosei_environment_tk_000007.html

National Traffic Safety and Environment Laboratory: Fuel consumption rate test ((Heavy vehicle) Compliant with 2025 fuel efficiency standards) TRIAS 08-003(1)-02, National Agency of Vehicle Inspection website (2021a). https://www.naltec.go.jp/publication/regulation/fkoifn0000000ljxatt/fkoifn0000006er2.pdf

National Traffic Safety and Environment Laboratory: Fuel consumption rate test (Heavy vehicle) TRIAS 08-003-02, National Agency of Vehicle Inspection web site (2021b). https://www.naltec.go.jp/publication/regulation/fkoifn0000000ljx-att/fkoifn00000060qj.pdf

Prime Minister's Office of Japan: 203rd Prime Minister Suga's statement of belief at the Diet, Prime Minister's Office of Japan website (2020). https://www.kantei.go.jp/jp/99_suga/statement/2020/1026shoshinhyomei.html

Shirai, Y., Ishigaki, T., Omoto, S., Kawakami, A., Terada, M., Yabu, M.: Relationships between tire-pavement rolling resistance and fuel consumption on asphalt pavement. J. Jpn. Soc. Civil Eng. E1 72(3) (2016)

Shirai, Y., Kawakami, A., Terada, M., Yabu, M., Himeno, K.: A study of methods for indirectly evaluating the rolling resistance coefficient using a road surface profile. J. Jpn. Soc. Civil Eng. E1 75(2) (2019)

# Green Energy Sources Based on Thermo-Electrochemical Cells for Electricity Generating from Transport, Engineering Buildings and Environment Waste Heat

Igor N. Burmistrov[1,2]([✉]), Nikolay V. Kiselev[1,2],
Elena A. Boychenko[1,2], Nikolay V. Gorshkov[3],
Evgeny A. Kolesnikov[2], and Stanislav L. Mamulat[4,5,6]

[1] Plekhanov Russian University of Economics, Moscow, Russia
[2] National University of Science and Technology MISiS, Moscow, Russia
burmistrov.in@misis.ru
[3] Yuri Gagarin State Technical University of Saratov, Saratov, Russia
[4] The Siberian State Automobile and Highway University, Omsk, Russia
[5] Research and Innovation Center Under EC of CTM CIS, Moscow, Russia
[6] Belt and Road International Transport Alliance (BRITA), Beijing, China

**Abstract.** Harvesting of low-grade waste heat and conversion it into electricity is a promising strategy for increasing the efficiency of various engineering systems in road construction and of renewable energy development direction. Thermo-electrochemical cells can directly convert low-grade thermal energy (Temperature of heat sources lower than 150 °C) into electricity without contamination of the environment and consuming any materials. The new type of electrode material based on nickel hollow microspheres has been presented in this work. It was demonstrated, that nickel hollow microspheres-based thermo-electrochemical cells can provide the highest Seebeck coefficient for aqueous electrolytes-based thermo-cells. The results of the investigation of the composition and structure of nickel hollow microspheres-based electrodes are presented in the work. Because of the low cost of nickel microspheres, developed thermocells may become commercially viable for harvesting low-grade thermal energy. It is supposed that the effect obtained with Ni electrodes characterized with well-developed surface area allows producing the commercial thermo-electrochemical cells for waste heat harvesting.

**Keywords:** Thermo-electrochemical cell · Renewable energy · Seebeck coefficients · Waste heat harvesting · Alternative energy sources

## 1 Introduction

A new and fast-developing trend in improving the efficiency of exothermic technical process and alternative energy in general is the harvesting and conversion of low-grade waste heat into electricity through thermoelectrochemical cells (Hu et al. 2010; Gunawan et al. 2014; Artyukhov et al. 2021). A wide variety of technical solutions for

© The Author(s), under exclusive license to Springer Nature Switzerland AG 2022
A. Akhnoukh et al. (Eds.): IRF 2021, SUCI, pp. 602–607, 2022.
https://doi.org/10.1007/978-3-030-79801-7_43

Green Energy Sources Based on Thermo-Electrochemical Cells 603

harvesting and converting low-grade waste heat into electricity have been presented in the scientific literature such as systems with salt bridge, flat cells with membrane or a thermal separator, sells in coin Cell cases, tube-in-tube cells, double positive-negative cell matrix and others (Hu et al. 2010; Wu et al. 2017; Buckingham et al. 2020). Several different approaches in the field of thermo-electrochemical conversion were distinguished: based on potassium ferri/ferrocyanide and others with the $Fe^{2+} \leftrightarrow Fe^{3+}$ transition (Artyukhov et al. 2019), phase transition (Zhou and Liu 2018), cobalt-organic (Al-Masri et al. 2018) etc.

A promising direction in the research and development of efficient thermoelectrochemical systems are cells with $Ni(OH)_2$ based electrodes due to their high exchange currents and anomalous high Seebeck coefficient with nonlinear temperature dependence (Burmistrov et al. 2020). Nickel oxide microstructures such as pollen-like, flakes and roselike 3D-structures, with a high specific surface area are used to make composite electrodes for charge storage and energy conversion (Wang et al. 2010; Wang et al. 2017; Yu et al. 2020).

A promising direction in the research and development of efficient thermoelectrochemical systems are cells with nickel-oxide electrodes due to their high exchange currents and anomalous high Seebeck coefficient with nonlinear temperature dependence (Burmistrov et al. 2020).

Thermo-electrochemical cell with metal nickel hollow microspheres-based electrodes was described in our previous work (Burmistrov et al. 2020) and there was shown the highest hypothetical Seebeck coefficient of 4.5 mV/K and accordingly voltage values of up to 0.2 V for based on aqueous electrolyte thermocells. The mechanism of the electrode reaction described in this work includes the main stage of the transformation of $Ni(OH)_2$ to $NiOOH$ (Burmistrov et al. 2020):

$$\beta - Ni(OH)_2 + OH^- \rightarrow \beta - NiOOH + H_2O + e^-$$

The reaction takes place due to the presence of nickel hydroxide, which forms on the surface of nickel microspheres in an alkaline electrolyte. In this regard, it can be assumed that an increase in the proportion of the hydroxide layer on the surface of metallic microspheres can increase the efficiency of the thermocell.

## 2 Materials and Methods

### 2.1 Materials

Nickel nitrate $Ni(NO_3)_2 \cdot 6H_2O$, (AR grade, Reachem, Russia) was used as a precursor for microspheres synthesis. Ethylene vinyl acetate LG28005 trademark (LG Chemical) was applied as composite electrode binder. Potassium hydroxide (KOH, 99.5% pure "Vekton Ltd", Russia) was used for electrolyte obtaining.

## 2.2 Methods for Obtaining Microspheres and Electrodes

The hollow nickel microspheres have been produced by ultrasonic spray pyrolysis followed by reduction in a hydrogen atmosphere. The technique is described in detail in the work (Burmistrov et al. 2020). Parameters of the process of NiO microspheres obtaining: temperature – 1000 °C, the concentration of the $Ni(NO_3)_2$ precursor – 15% mass. Parameters of the reduction of NiO to metal Ni microspheres process: temperature – 375 °C, time about 4 h.

To obtain a composite electrode, microspheres powder were introduced into a solution of ethylene vinyl acetate in xylene and stirred on a magnetic stirrer for 15 min at a temperature of 80 °C. The concentration of the solution of ethylene vinyl acetate in the solution was 5% of the mass. The resulting solution was applied to a nickel foil and dried at 90°.

## 2.3 Materials Characterization and Thermo-Electrochemical Measurements

The microspheres and composite electrode surface topography were examined by scanning electron microscopy (SEM) Tescan Vega 3, (TESCAN, Czech Republic). X-ray diffractometer (XRD) namely "Difray-401" was used for the microspheres analysis, chromium X-ray radiation ($\lambda = 2.2909$ Å) was applied for the $2\theta$ scanning between 10° to 130° at a scanning speed of 1°/min.

A typical salt bridge electrochemical circuit with jacketed electrochemical cell was used to measure the thermoelectrochemical effect. Aqueous electrolytes were prepared via KOH dissolving in distilled water under 2M concentrations. The temperature in the cold and hot cells was controlled with two thermostats (FT-21-25, Russia). The results were obtained using the equipment of the Center for Collective Use of Scientific Equipment of TSU named after G.R. Derzhavin.

The cell characteristics in terms of current-voltage were measured under potentiodynamic mode under 1 mV/s scan rate using potentiostat P-45X ("Elins", Russia).

# 3 Results and Discussion

SEM and XRD study of hollow nikel microspheres is presenter in Fig. 1. It was shown, the ultrasonic spray pyrolysis synthesized powder contains microparticles of average diameter of 3–7 μm. The XRD analyzis confirmed that the synthesized microspheres after the reduction process composed primarily of cubic Ni phase.

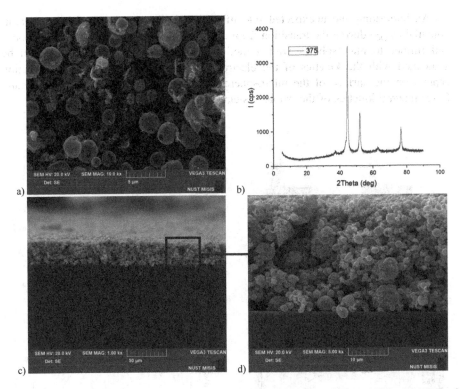

**Fig. 1.** Characterization of hollow microspheres and electrodes: a), b) – SEM and XRD of hollow nickel microspheres respectively; c), d) – SEM electrodes based on hollow nickel microspheres in EVA.

Current-voltage characteristics of three types of electrode materials are shown in Fig. 2. Nickel foil not covered with a composite electrode (microspheres/EVA) is shown in Fig. 2a, nickel foil covered with a composite electrode in Fig. 2b, nickel foil coated with a composite electrode after electrochemical treatment in Fig. 2c. Electrochemical treatment was performed to create a nickel hydroxide on the microspheres surface, which provides the main electrode reaction (Burmistrov et al. 2020).

The short-circuit current density ($J_{SC}$) and open-circuit potential ($V_{OC}$) are increased with increasing temperature difference, resulting in a parabolic power output $P_{MAX}$ curves.

Maximum output power for not covered nickel foil is 55 $\mu W/m^2$, for nickel foil covered with a composite electrode – about 97 $\mu W/m^2$ and for nickel foil coated with a composite electrode after electrochemical treatment is about 12 000 $\mu W/m^2$ (12 $mW/m^2$).

Low overall output power values are associated with the cell design. As it was shown in our previous work (Artyukhov et al. 2019), when changing cell design from "Salt Bridge" to "Coin Cell", the exchange current increases on average by more than 30 times.

An interesting and unexpected scientific result is the increase in the open-circuit potential ($V_{OC}$) during the transition from nickel foil to microspheres based electrodes and further to microspheres with a modified surface. The resulting effect can be associated with the kinetics of the electrode process. A larger number of reaction centers on the surface of the microspheres provide faster charge accumulation and faster integral kinetics of the whole system, correspondingly.

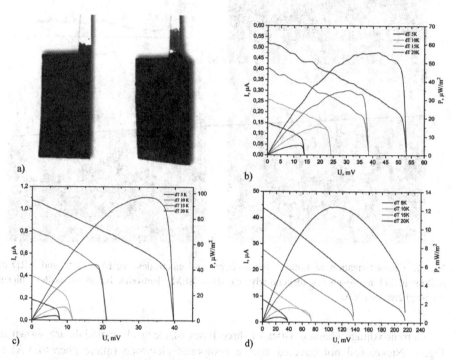

**Fig. 2.** Morphology of electrodes covered with microspheres before (left) and after (right) electrochemical treatment – a) and short-circuit current density ($J_{SC}$) and maximum output power ($P_{MAX}$) dependents on open-circuit potential $V_{OC}$ for tree types of electrodes: b) – not covered nickel foil; c) – nickel foil covered with a composite electrode; d) – nickel foil coated with a composite electrode after electrochemical treatment.

## 4 Conclusions

The new type of composite electrode material based on nickel hollow microspheres and EVA binder has been investigated. The possibility of operation of the nickel hollow microspheres-based thermo-electrochemical system at very low temperature gradients of 5–20 K was shown.

Extremely high hypothetical Seebeck coefficient for aqueous electrolytes-based thermo-cells was demonstrated for electrochemical modified nickel hollow microspheres-based electrodes. Because of the low cost of nickel microspheres based electrodes and alkali electrolyte, versus semiconductor thermoelectric converters or lithium-ion systems, developed thermocells may become commercially viable for harvesting low-grade thermal energy.

The most important results of the study is the creation of thermo-electrochemical cells, operating at a temperature difference of up to 20°, which will make it possible to use them for various engineering systems including transport, road construction, engineering buildings and environment waste heat.

# References

Al-Masri, D., et al.: The electrochemistry and performance of cobalt-based redox couples for thermoelectrochemical cells. Electrochim. Acta **269**, 714–723 (2018)

Artyukhov, D., Kiselev, N., Gorshkov, N., Burmistrov, I.: Research of the influence of electrolyte concentration on thermo-electrochemical cells efficiency. Procedia Environ. Sci. Eng. Manag. **6**, 319–327 (2019)

Artyukhov, D., et al.: Harvesting waste thermal energy using a surface-modified carbon fiber-based thermo-electrochemical cell. Sustainability **13**(3), 1377 (2021)

Buckingham, M.A., et al.: Using iron sulphate to form both n-type and p-type pseudo-thermoelectrics: non-hazardous and 'second life'thermogalvanic cells. Green Chem. **22**(18), 6062–6074 (2020)

Burmistrov, I., Gorshkov, N., et al.: High seebeck coefficient thermo-electrochemical cell using nickel hollow microspheres electrodes. Renewable Energy **157**, 1–8 (2020)

Gunawan, A., et al.: The amplifying effect of natural convection on power generation of thermogalvanic cells. Int. J. Heat Mass Transf. **78**, 423–434 (2014)

Hu, R., et al.: Harvesting waste thermal energy using a carbon-nanotube-based thermo-electrochemical cell. Nano Lett. **10**, 838–846 (2010)

Wang, L., et al.: Preparation of 3D rose-like NiO complex structure and its electrochemical property. J. Alloy. Compd. **495**(1), 82–87 (2010)

Wang, S., et al.: Pollen-inspired synthesis of porous and hollow NiO elliptical microstructures assembled from nanosheets for high-performance electrochemical energy storage. Chem. Eng. J. **321**, 546–553 (2017)

Wu, J., Black, J.J., Aldous, L.: Thermoelectrochemistry using conventional and novel gelled electrolytes in heat-to-current thermocells. Electrochim. Acta **225**, 482–492 (2017)

Yu, J.-H., et al.: The effect of ammonia concentration on the microstructure and electrochemical properties of NiO nanoflakes array prepared by chemical bath deposition. Appl. Surface Sci. **532**, 147441 (2020)

Zhou, H., Liu, P.: High seebeck coefficient electrochemical thermocells for efficient waste heat recovery. ACS Appl. Energy Mater. **1**, 1424–1428 (2018)